Geometry of Derivation with Applications

Geometry of Derivation with Applications is the fifth work in a longstanding series of books on combinatorial geometry (*Subplane Covered Nets, Foundations of Translation Planes, Handbook of Finite Translation Planes,* and *Combinatorics of Spreads and Parallelisms*). Like its predecessors, this book will primarily deal with connections to the theory of derivable nets and translation planes in both the finite and infinite cases. Translation planes over non-commutative skewfields have not traditionally had a significant representation in incidence geometry, and derivable nets over skewfields have only been marginally understood. Both are deeply examined in this volume, while ideas of non-commutative algebra are also described in detail, with all the necessary background given a geometric treatment.

The book builds upon over twenty years of work concerning combinatorial geometry, charted across four previous books and is suitable as a reference text for graduate students and researchers. It contains a variety of new ideas and generalizations of established work in finite affine geometry and is replete with examples and applications.

Norman L. Johnson is an Emeritus Professor (2011) at the University of Iowa where he has had ten PhD students. He received his BA from Portland State University, MA from Washington State University and PhD also at Washington State University as a student of T.G. Ostrom. He has written 580 research items including articles, books, and chapters available on Researchgate. net. Additionally, he has worked with approximately 40 coauthors and is a previous Editor for *International Journal of Pure and Applied Mathematics* and *Note di Matematica*. Dr. Johnson plays ragtime piano and enjoys studying languages and 8-ball pool.

Geometry of Derivation with Applications

Norman L. Johnson

CRC Press
Taylor & Francis Group
Boca Raton London New York

CRC Press is an imprint of the
Taylor & Francis Group, an **informa** business

A CHAPMAN & HALL BOOK

First edition published 2023
by CRC Press
6000 Broken Sound Parkway NW, Suite 300, Boca Raton, FL 33487-2742

and by CRC Press
4 Park Square, Milton Park, Abingdon, Oxon, OX14 4RN

CRC Press is an imprint of Taylor & Francis Group, LLC

Library of Congress Cataloging-in-Publication Data

Names: Johnson, Norman Lloyd, author.
Title: Geometry of derivation with applications / Norman L. Johnson.
Description: Boca Raton : CRC Press, 2023. | Series: Chapman and Hall/CRC
    financial mathematics series | Includes bibliographical references and
    index. | Summary: "This book centers on combinatorial geometry. It
    focuses on derivation over skewfields. By virtue of the combinatorial
    embedding theory is a classification of derivable nets may be given that
    relates the net to a "classical pseudo-regulus net" both of which are
    considered to live in the same ambient affine geometry"-- Provided by
    publisher.
Identifiers: LCCN 2022055812 (print) | LCCN 2022055813 (ebook) | ISBN
    9781032349169 (hardback) | ISBN 9781032349183 (paperback) | ISBN
    9781003324454 (ebook)
Subjects: LCSH: Combinatorial geometry.
Classification: LCC QA167 .J64 2023 (print) | LCC QA167 (ebook) | DDC
    516/.13--dc23/eng20230415
LC record available at https://lccn.loc.gov/2022055812
LC ebook record available at https://lccn.loc.gov/2022055813

ISBN: 978-1-032-34916-9 (hbk)
ISBN: 978-1-032-34918-3 (pbk)
ISBN: 978-1-003-32445-4 (ebk)

DOI: 10.1201/9781003324454

Typeset in CMR10
by KnowledgeWorks Global Ltd.

Publisher's note: This book has been prepared from camera-ready copy provided by the authors.

FOR SHEL
My big brother,
with much love!

Contents

CONTENTS   ix

Acknowledgements

The author thanks and acknowledges Bonnie Lynn Hemenover, for sixty wonderful years together, always my number one supporter, and especially for this unlikely text.

Also, I am thankful for my incredible children Catherine Elizabeth Johnson and Garret Norman Johnson, for tremendous inspiration and encouragement for continuing my work.

Finally, The author is grateful to the Mathematics Department of the University of Iowa for supporting this text, with special thanks to Brian Bacher of the Computer Support group, and to Weimin Han, Chair of the Mathematics Dept.

Preface

This book is the fifth text in a series on combinatorial/incidence geometry that deals with connections with the theory of derivable nets and translation planes in both the finite and infinite cases. Translation planes over non-commutative skew-fields have not traditionally had a significant representation in incidence geometry but will be considered strongly in this book. Derivable nets over skewfields have only been marginally understood. Ideas of non-commutative algebra will be described and all of the necessary background and essentially each of these ideas will be given a geometric framework. The most important theme of these five books is that of the derivable net and the various planes that contain it. Very often these are dual translation planes, although in applications, usually these are translation planes. All of the previous four books are relevant in this work.

These four books are in the same Taylor/Francis series and are Subplane Covered Nets (2000) [**86**], Foundations of Translation Planes (2001) [**16**], Handbook of Finite Translation Planes (2007), [**94**] (the last two coauthored with Mauro Biliotti and Vikram Jha), and Combinatorics of Spreads and Parallelisms (2010) [**88**]. The text Subplane Covered Nets was devoted to understanding the nature of derivable nets, as well as subplane covered nets. This was accomplished by embedding any such net into a projective space written over a skewfield and then a contraction method was developed that reconfigured the net back into an affine space, wherein the net could be seen or represented as a pseudo-regulus net. A pseudo-regulus net is the non-commutative skewfield version of a regulus net. Due to the structure of these nets, the terms are also extended to include the subplane covered net cases. So, a pseudo-regulus net in the more general setting is not necessarily a derivable net. In the subplane covered net case, the projective space into which the net is embedded need not be finite-dimensional. The embedding will play a vital part of this book and some new variations of the theory are also included here.

The non-commutative skewfields of quaternion division rings $(a, b)_K$ (algebras) of dimension 4 over their centers K make their appearance as derivable nets that occupy the same affine space arising from a 3-dimensional projective space over a field extension of dimension 4 of K. Derivable nets in a wide variety of Desarguesian planes shall be developed, many from cyclic division rings. But, cyclic division rings of degree n and center K can be realized as coordinatizing subplane covered nets in $PG(2n - 1, K(\theta))$, for $K(\theta)$ a cyclic Galois field extension of dimension n.

The ideas encountered in the book may be seen as natural extensions and/or generalizations of finite geometries. T.G. Ostrom [**118**] discovered derivable nets in the finite case, and proved that finite dual translation planes of dimension 2 are always derivable. Then the work of Andrè showed that finite Desarguesian planes of dimension 2 can be multiply derived. There are many quaternion division rings that admit central quadratic extensions, which admit involutory automorphism, which is the key ingredient that is necessary for multiple derivation, as indeed there are other division rings that admit such extensions. Using central Galois quadratic extensions of quaternion division rings and cyclic division rings of degree $2k$, as the most notable examples, we completely extend Andrè to show that any central quadratic Galois extension of any skewfield is not only derivable by a pseudo-regulus net, but also multiply derivable. Ostrom then considered a more general construction using what are called 'hyper regulus-nets', the generalization of which in the non-commutative case is called a 'hyper pseudo-regulus net'. The central extensions of cyclic division rings of finite dimension $n > 2$, provide the generalizations, with the ambient space now connected with a classical subplane covered net as the classical hyper pseudo-regulus reference net.

In the second text of the series, foundations of quasifibrations are discussed, often in the context of the algebraic construction technique of 'lifting'. When these are of dimension 2, they also may be lifted, which create α-flock quasifibrations, also of dimension 2, where α has order 2. In the spread case, the lifted translation planes are derivable, and upon derivation create translation planes with blended kernel. Here we generalize the lifting concept to any skewfield spread of dimension 2. Thus from any central quadratic Galois extension of a division ring, a new class of semifields are constructed. So, lifting in this manner constructs classes of semifields with spreads in $PG(3, L(\theta))$, where $L(\theta)$ is a non-commutative division ring, the first such classes of semifields known.

The Handbook was written to provide all known geometric connections to finite translation planes. These connections contain flocks of many types. All flocks are connected to translation planes using derivable nets contained in translation planes admitting certain central collineation groups, usually 'regulus-inducing groups'. But, a regulus net is but one of the various derivable nets that we shall study. In particular, a classification of all of the types of derivable net over skewfields is given and is the focal point of the current book. All of the applications that we consider may be viewed as applications of this classification. Much of the theory can be given over an arbitrary subplane covered net and it is seen that in the finite-dimensional case, where the degree $n > 2$, cyclic division rings often take the place of quaternion division rings in the derivable net case. The central extension concepts are also valid more generally and Ostrom's hyper regulus nets become hyper pseudo- regulus nets and multiple replacement is then valid over higher dimensional central cyclic extensions of division ring translation planes.

The second core feature of the book is the generalization of the ideas of flocks of quadratic cones and of hyperbolic flocks. One main type of derivable net is the twisted regulus net, twisted by a non-trivial automorphism α; so flocks over α-cones and twisted hyperbolic flocks are shown to be equivalent to certain translation planes admitting twisted-regulus inducing central collineation groups. The ideas of hyperbolic fibrations with constant back half and equivalently translation planes admitting appropriate central collineation groups are used to construct two new classes of finite twisted hyperbolic flocks from j-planes. This new connection seems unlikely, since j-planes are connected to monomial flocks of quadratic cones. The correspondence between finite flocks of quadratic cones and translation planes of order q^2 admitting regulus-inducing homology groups of order $q + 1$, called 'cyclic

$q + 1$-planes', is further developed with a blueprint for constructions from one geometry to the other. In addition, some of the ideas of t-nests of derivable nets are connected to show that $q + 1$-nest replaceable translation planes and q-nest replaceable translation planes are equivalent to the Fisher flocks of quadratic cones, and therefore equivalent to each other.

This book contains a variety of new ideas and generalizations of established work in finite affine geometry that has far-reaching applications to geometries over arbitrary skewfields and therefore would be of interest both to the student and teacher, but also to the research mathematician as well. Much of the material presented in this text is completely new and original work written in the last three years.

In particular, the material presented will cover the new methods of the theory of translation planes: spreads, quasifields, Klein quadratic and new generalizations, the complete geometry of the theory of derivable nets, as well as the main connections with flocks of cones, generalized quadrangles, hyperbolic fibrations and hyperbolic flocks developed in sufficient depth to begin research in translation planes or any of the connected areas. In addition, the deficiency-one theory of flocks involving Baer groups of translation planes will be given in complete generalization, but now will push the translation planes applicable to the Baer groups to dimension > 2. A new class of quasifibrations is constructed that are not spreads but are s-inversions of a class of bilinear flocks of quadratic cones. This class is the first such class of flock derivations that do not correspond to a flock. The theories of Andrè and Ostrom on multiple sets of finite regulus nets/and or hyper-regulus nets are generalized to arbitrary skewfield planes admitting an automorphism. This general construction theorem is called the Multiple Replacement Theorem. This theorem shows that given any Desarguesian plane coordinatized by a skewfield that admits a non-trivial automorphism, there is a set of mutually disjoint nets that individually admit net replacement. The use of Galois theory of division rings is developed using towers of length $k + 1$ of cyclic field extensions of dimension n producing an Abelian Galois group as a product $\Pi_{k+1} Z_n$ of $k + 1$ cyclic groups of order n. This is complimented by a corresponding Abelian Galois group $\Pi_k Z_n$ of a cyclic division ring of degree n. There are new constructions of semifield planes with spreads in $PG(3, K)$, where K is a non-commutative skewfield. These semifield planes might be considered generalizations of Hughes-Kleinfeld semifield planes, when K is a finite or an infinite field. Extensions of division rings are considered in several constructions using algebraic and transcendental field extensions.

Norman L. Johnson
Emeritus Professor
University of Iowa
2023

Part 1

Classical theory of derivation

In this first part, three main methods of presentations of translation planes will be given; these are by spreads, quasifields (and pre-quasifields) and by use of the Klein quadric. Also, a variety of finite and infinite derivable affine planes shall be given. An important question is whether the extension of a derivable net to an affine plane preserves the derivation of containing nets. Is it possible that the derived structure is not an affine plane? This question shall be considered and answered. This concept shall commence our more general study of infinite derivable nets and affine planes that contain them.

The concept of 'derivation of a finite affine plane' was discovered by T.G. Ostrom, in the early 1960s and there have been an enormous number of developments and applications in the last sixty years. In the beginning, most of the known translation planes were of dimension 2, which means that they may be represented by their 'spreads'; as a set of mutually disjoint 2-dimensional $GF(q)$ subspaces of a 4-dimensional vector $V_4/GF(q)$ that covers V_4. So, there then would be $q^2 + 1$ spread 'components'. The original concept of derivation actually may be considered as arising from that of a 'derivable net', which is defined affinely using 'points', 'lines', and 'Baer subplanes', using incidence and parallelism relations. It turns out that the affine restriction of the dual of the projective extension of a finite affine translation plane of dimension 2 is derivable. In the finite case, every affine plane that contains a derivable net is derivable; derivation and the net containment of a derivable net are equivalent. In would appear that this is a universal result, but, in fact, this is not the case. There exist non-derivable affine planes that contain derivable nets!

So, this part begins with finite affine planes, finite derivable nets, dual translation planes and their affine plane derivates; called 'semi-translation planes.'

Notation for locations: (part, chapter, section), as, for example $(1, 1, 4)$, refers to (part 1, chapter 1, section 4).

Coordinate methods

In this chapter, we develop a method to assign coordinates to basically any affine incidence structure of 'points', 'lines', 'parallel classes' and, in particular, for nets, subplane covered nets, derivable nets, and net extensions. This method is important as the first derivable affine planes were not translation planes but dual translation planes and the Hughes planes. It is important to notice that derivation is an affine process, although the Hughes planes were constructed as projective planes (Hughes [**63**], Johnson [**86**] chapter 6, 57-63). Also, the term 'dual translation plane' implicitly refers to the dual of a projective translation plane.

DEFINITION 1. *A 'projective plane'* $(\mathcal{P}, \mathcal{L}, \mathcal{I})$ *is a set of points* $\mathcal{P}$, *a set lines* $\mathcal{L}$, *and incidence set* $\mathcal{I}$ *such that*

(1) given any pair of distinct points P *and* Q, *there is a unique line* l *which is incident with both* P *and* Q. *The notation is then* $l = PQ$, *and*

(2) given any pair of distinct lines l *and* m, *there is a unique point* P *which is incident with both lines* l *and* m. *The notation is* $P = l \cap m$.

(3) There are at least four points, no three of which are collinear (incident with the same line).

DEFINITION 2. *A 'net'* $\mathcal{N} = (\mathcal{P}, \mathcal{L}, \mathcal{C}, \mathcal{I})$, *is an incidence structure with a set* $\mathcal{P}$ *of 'points,' a set* $\mathcal{L}$ *of 'lines,' a set* $\mathcal{C}$ *of 'parallel classes,' and a set* $\mathcal{I}$, *which is called the 'incidence set' such that the following properties hold:*

(i) Every point is incident with exactly one line from each parallel class, each parallel class is a cover of the points, and each line of $\mathcal{L}$ *is incident with exactly one of the classes of* $\mathcal{C}$ *(parallelism is an equivalence relation and the equivalence classes are called 'parallel classes').*

(ii) Furthermore, lines from two different parallel classes have exactly one common incident point.

(iii) Two distinct points are incident with exactly one line of $\mathcal{L}$ *or are not incident.*

DEFINITION 3. *An 'affine plane' is a net where two distinct points are incident with exactly one line.*

An affine plane may be constructed from a projective plane by the deletion of any given line, and all points incident with it, with the appropriate definitions of parallel classes and incidence.

Similarly, an affine plane may be extended to a projective plane by the adjunction of a 'line at infinity' and adding an 'infinite point' to each line of a parallel class, for each parallel class, with appropriate incidence extensions.

DEFINITION 4.

A 'subplane covered net' $\mathcal{S} = (\mathcal{P}, \mathcal{L}, \mathcal{B}, \mathcal{C}, \mathcal{I})$

is a net $(\mathcal{P}, \mathcal{L}, \mathcal{C}, \mathcal{I})$ together with a set $\mathcal{B}$ of affine subplanes, such that given any two distinct points a and b, there is a subplane $\pi_{a,b}$ of $\mathcal{B}$ containing a and b, whose parallel class set is exactly $\mathcal{C}$.

A 'Baer subplane' π_0 of a net is an affine subplane with the property that every point b of the net is incident with a line of the subplane π_0 and every line ℓ of the net is incident with a point of π_0. These two terms, in order, determine the subplane as a 'point-Baer subplane', and 'line-Baer subplane', respectively. Thus, a Baer subplane is both a point-Baer and a line-Baer subplane.

A 'derivable net' is a subplane covered net whose subplanes in $\mathcal{B}$ are all Baer subplanes.

REMARK 1. *Since this text will consider the 'derivation' of derivable nets, and of affine planes, as well as of partial affine planes, some remarks are given here to clarify the situation. The term 'derivation' shall mean simply the process of interchanging the sets of lines and Baer subplanes of a derivable net, with appropriate definitions of parallelism, as the interchange or replacement of a net needs to also be a net. The definition of what a Baer subplane means changes and is relative to any extension net or affine plane that contains the derivable net. For example, a subplane of a derivable net, is by definition, a Baer subplane of the net, but might not satisfy both Baer conditions of a net properly containing the subplane in question.*

REMARK 2. *The conditions of 'point-Baer' and 'line-Baer' are equivalent for finite incidence structures that contain these such as affine planes. It might be confusing to jump back and forth between affine planes and their projective extension planes, as the definition of a Baer subplane could be made in the projective plane, whereas the derivation process is usually considered only in an affine restriction of a projective plane.*

DEFINITION 5. *A 'translation' of a net is a collineation that fixes each parallel class, fixes some parallel class linewise, and acts fixed-point-free on the affine points. The 'translation group' is the collineation group generated by the set of translations. A 'translation net' is usually considered in the affine form, so the definition is simply that it is a net that admits a transitive group on points, which is generated by translations'. A 'translation plane' is an affine plane which is a translation net.*

DEFINITION 6. *Given a projective plane, the 'dual' of the projective plane is the point-line geometry obtained by interchanging what are called points and lines, while maintaining the incidence relation.*

DEFINITION 7. *The term 'dual translation plane' is the structure obtained as follows: consider the associated projective translation plane, dualize the projective plane. This is the dual translation plane. From here, choose any line to be called the new line at infinity and form an affine plane. However, these affine planes are not necessarily isomorphic and not necessarily translation planes. Then consider a point of the projective plane (∞) which is the line at infinity l_∞ of the translation plane. Each line incident with (∞) may be chosen as the line of infinity of the dual plane to form a dual translation plane.*

PROBLEM 1. *If a projective plane has an affine restriction that contains a derivable net, does every affine restriction of the projective plane contain a derivable net?*

The previous exercise assumes that the reader would immediately recognize that a net is derivable, when viewed in some manner. So, what does a derivable net look like? This question was considered in Johnson [86], which shall be discussed later on in this part. Initially, all known derivable affine planes were finite and could be coordinatized so that the affine plane would have the following representation: Let $(Q, +, \cdot)$ be a ternary coordinate system, which represents coordinates for an affine plane that admits a derivable net. This coordinate structure can be adapted to consider arbitrary nets. This is covered in depth in Chapter 11 of Johnson [86], so here we just point out what a derivable affine plane should look like, in general.

THEOREM 1. (11.17)([86]) *Let D be a derivable net and let A be any net containing D in the sense that the points of A are the points of D and all parallel classes of D are parallel classes of A.*

Then there exists a coordinate system $(Q, +, \cdot)$ for A such that lines of D may be represented in the form

$$x = c, \ y = x \cdot \alpha + b; \ b, c \in Q$$
$$\alpha \in L, \text{ for } Q \text{ a right 2-dimensional vector space over } L,$$
$$\text{where } L \text{ is a sub-skewfield of } Q.$$

Since we are discussing initially finite derivable affine planes, L is a field in this setting and A is a finite affine plane. This was precisely what was assumed by T.G. Ostrom, [120]. So, the lines but not the putative Baer subplanes of D are given. When L is a field, then Q is a 2-dimensional vector space over L, and $x \cdot \alpha = x \begin{bmatrix} \alpha & 0 \\ 0 & \alpha \end{bmatrix}; \forall \alpha \in L$, so the set of lines incident with $(0,0)$ are given by

$$\left\{ x = 0, x \begin{bmatrix} \alpha & 0 \\ 0 & \alpha \end{bmatrix}; \forall \alpha \in L \right\}.$$

Then the set of Baer subplanes $B(D)(a,b)$ for all $(a,b) \neq (0,0)$ in $L \times L$ incident with the zero vector (Baer subspaces) consists of the following:

$$B(D)(a,b) = \left\{ \left((a,b) \begin{bmatrix} \alpha & 0 \\ 0 & \alpha \end{bmatrix}, (a,b) \begin{bmatrix} \beta & 0 \\ 0 & \beta \end{bmatrix} \right) \right\}; \forall \alpha, \beta \in L.$$

PROBLEM 2. *In the finite case, where L is $GF(q)$, determine that each subplane $B(D)(a,b)$ has q^2 points. Also show that there are a total of q^4 (affine) points, $q+1$ parallel classes, q^2 affine points on each line, $q^2(q+1)$ lines and $q^2(q+1)$ Baer subplanes. Each line incident with the zero vector $(0,0)$ is a 2-dimensional vector space over $GF(q)$ and show that this line is incident with exactly $q+1$ Baer subplanes of the form $B(D)(a,b)$.*

PROBLEM 3. *Now show that the interchange of the set of Baer subplanes with the set of lines of D is also a derivable net with the new lines as the old Baer subplanes, and the new Baer subplanes as the old lines. This net is the 'derived net'. Furthermore, the process in the finite case constructs a new affine plane; the plane derived from A.*

We don't yet really know that every derivable net looks like this or that an affine plane containing a derivable net is always derivable, as there is something interesting that can only occur in the infinite case, that we will discuss shortly.

REMARK 3. *We shall be working with vector spaces over skewfields F. In this setting, there is a right vector space over F. The 'opp-space over F' is defined as follows: if x is a vector and the scalar is written on the right as $x\delta$ for $\delta \in K$ then the 'opp' space is defined as $\delta \circ x = x\delta$, where in the skewfield $(F, +, \cdot)$, then $a \cdot b = b \circ a$ makes $(F, +, \circ)$ a skewfield and the vector space is then a left F-vector space.*

We shall be discussing translation planes with coordinate structures called 'quasifields' more specifically 'right quasifields'. A dual translation plane interchanges points with lines and may be coordinatization by a 'left quasifield', with the definition exactly as the 'opp' definition for skewfields. When the translation plane is a Desarguesian plane then the quasifield becomes (can become) a skewfield considered as a right skewfield and the dual translation plane is then coordinatized by the left skewfield. Since translation planes work over underlying vector spaces, we shall here adopt the equivalent 'left/right' terminology instead of the opp-space terminology.

All of the terms used above shall be defined in more detail in the preliminary sections.

1. Translation planes and quasifibrations

In this section, we review some basics of translation planes, quasifibrations, dual translation planes and semi-translation planes. In chapter 1 of this part 1, we discussed a translation group for a derivable net or any net and the definition of a translation plane. While the fundamentals of translation planes will be covered in this section, the reader is directed to Biliotti, Jha, Johnson [**16**] for complete details and additional reading on the foundations of translation planes.

DEFINITION 8. *Let π_S be an affine plane that admits Baer subplanes. π_S is a 'semi-translation plane' if and only if there exists a translation subgroup each of whose orbits is a Baer subplane.*

DEFINITION 9. *For a translation plane π whose ambient vector space $V = W \oplus W$, for W a left vector space over a skewfield K, points shall be $(x, y); x, y \in W$. Since π admits a transitive translation group, the lines are translates of a set S of mutually disjoint left W-subspaces. Then $\cup S$ is a cover of the points/vectors of V and S is said to be a 'spread'.*

Then the group of 'left mappings' $\langle (x, y) \to \delta(x, y) = (\delta x, \delta y) : \forall \delta \in K^* \rangle$ leaves each of the spread elements (components) invariant. Consider the skewfield L containing K whose corresponding set of left mappings leaves each of the spread elements invariant, then L is called the 'kernel group', or the 'kern', which is isomorphic to L^*.

If the dimension n of W/L is finite, the translation plane is said to be of 'dimension n over the kernel L'.

Most of the translation planes, but not all that shall be encountered in this book, will have finite dimension over the kernel.

DEFINITION 10. *A 'Desarguesian translation plane' π is a translation plane whose coordinate system is given by a skewfield K. Let $V = K \oplus K$ and define points as (x, y), for all $x, y \in K$ and define components as follows:*

(1) If V is a left 2-dimensional K vector space and components are given by

$$\{x = 0, y = xm; \forall m \in K\},$$

then the components are left 1-dimensional vector spaces. Further, π is a 'left Desarguesian translation plane of dimension 1 over the kernel K'. We shall denote this translation plane by Σ^L.

(2) If K is a field, the translation plane shall be called a 'Pappian translation plane'. If K is finite, the custom is to use the term Desarguesian plane, when the plane is actually Pappian. In general, the term Desarguesian is used for translation planes coordinatized by non-commutative skewfields.

(3) There is also a representation of a Desarguesian translation plane Σ^R, with components

$$\{x = 0, y = mx; \forall m \in K\}.$$

In this setting, the components are right 1-dimensional K-subspaces and we still speak of the kernel as K, with the kernel group now acting on the right.

(4) Let σ be an automorphism of K. There are two additional Desarguesian spreads that occur: These are Σ_σ^L and Σ_σ^R, where

$$\Sigma_\sigma^L : \{x = 0, y = x^\sigma m; \forall m \in K\},$$

and

$$\Sigma_\sigma^R, : \{x = 0, y = mx^\sigma; \forall m \in K\}.$$

The Desarguesian translation planes of this previous definition will be used in the **multiple replacement theorem** of part 10.

DEFINITION 11. *Let π be a translation plane of dimension 2 over a skewfield K.*

(1) Then the spread for π consists of a set of mutually disjoint 2-dimensional left K-subspaces given as follows:

$$\left\{x = 0, y = x \begin{bmatrix} f(t, u) & g(t, u) \\ t & u \end{bmatrix}; \forall t, u \in K\right\};$$

f and g are functions: $K \oplus K \to K$,

where the union of these disjoint spaces cover $K \oplus K$. In this setting, the spread is said to be a 'matrix spread of dimension 2'. In an analogous setting for n by n matrices, the spread is said to be a 'matrix spread of dimension n'

(2) (a) If there exists a set of 2 by 2 matrices over a skewfield, where the differences of distinct pairs of matrices (including the zero matrix) are non-singular and (b) if the union of the subspaces defined by the set of matrices covers the vector space, a translation plane is obtained.

If (a) is valid but (b) is not (there is not a cover), we use the term 'quasifibration' of dimension 2, instead of translation plane.

To determine non-singularity for a matrix over a skewfield, a modification of the field determinant is required, which shall be discussed later.

There is a certain abuse of terminology when we consider a translation plane of dimension n. For example, a Desarguesian translation plane can appear in the matrix form of a dimension 2 translation plane, when it actually has dimension 1 over a different skewfield. This variation should be clear by the context.

For example, we shall be considering quaternion division ring translation planes that appear as translation planes of dimension 2 over quadratic extension fields $F(\theta)/F$, where F is the center, so 4-dimensional over their center. The dimension is 1 over the skewfield itself.

When considering cyclic division algebras S of dimension n over their centers F and containing a cyclic Galois field extension $F(\theta)$, the corresponding translation planes of finite dimension n uses matrix spreads of n by n matrices over $F(\theta)$. The skewfield is then of dimension n^2 over the center F.

We shall be discussing quasifibrations of various dimensions. The following provides a definition of quasifibrations of dimension n. We use the term 'partial spread' in the obvious manner.

DEFINITION 12. *Assume that a partial spread over a skewfield F is a matrix partial spread set of n by n matrices over F,*

$$\{x = 0, y = 0, y = xM, M \in \lambda \subset GL(n, F)\}.$$

Let W be a left n-dimensional F-vector space. Assume that in the representation of the partial spread, some row of the set of matrices M, runs over all elements of F^n, (the set of all n-tuples of F), where the associated vector space $V = W \oplus W$ is $2n$-dimensional over F. Then the partial spread is said to be an 'F-quasifibration of dimension n'. Consequently, quasifibrations are 'maximal partial spreads' or spreads (cannot be extended to a spread or is a spread). If a quasifibration is not a spread, it is said to be a 'proper quasifibration'.

It will be shown how to construct quasifibrations from quaternion division ring planes, as well as from cyclic division ring planes.

DEFINITION 13. *Let S be a spread of $V = W \oplus W$ over a skewfield K. Then every non-zero vector of S is contained in a unique spread component of S, where each component is K-isomorphic to W. A 'dual spread D' is a set of mutually disjoint K-subspaces each K-isomorphic to W and where each hyperplane of V contains exactly one element of D.*

DEFINITION 14. *Note the terminology here is confusing, as dual spreads and the dual of an affine translation plane are completely different geometric objects. For example, does a dual translation plane refer to a (i) dual spread or to (ii) the affine restriction of the dual of the projective extension? In this text, we shall use the term 'hyperplane dual spread' to mean a dual spread (i) and 'affine dual translation plane' to indicate the second option (ii), affine restriction ...*

Every finite hyperplane dual spread is a spread but for every infinite skewfield K there are hyperplane dual spreads that are not spreads (See, chapter 12, pp 131-136 [86]). What would be of interest for this book is the following corollary to a general result on the questions of when spreads are hyperplane dual spreads and when the hyperplane dual spread is a spread.

COROLLARY 1. *(12.6 (2) [86]). Let S be any spread of $PG(3, K)$, for a skewfield K. Let π_S denote the corresponding translation plane. Then S is a hyperplane dual spread if and only if any 2-dimensional K-vector space that is not a line of π_S is a point-Baer subplane and hence a Baer subplane of π_S.*

Considering the previous corollary, form the affine restriction of the dual of the projective extension of π_S. Assume then we have a point-Baer subplane which is not a line-Baer subplane and hence is not a Baer subplane of the hyperplane dual translation plane. Note point-Baer in π_S becomes line-Baer in the affine dual translation plane. So, even though we have a natural point-Baer situation (See e.g. 12.7 [**86**]), what all of this means for the question of derivation of affine dual translation planes is the following:

THEOREM 2. *There are affine dual translation planes that contain derivable nets, but the plane itself is not derivable.*

On the other hand, we shall be considering flocks of quadratic cones and hyperbolic flocks. There is an associated algebraic representation as well as a Klein quadric representation of each of the associated translation planes. The flock spreads are all hyperplane dual spreads (See problem end of (4,5,3)).

2. Quasifields

In this section, we give a brief reminder of the coordinate structures for translation planes. The reader is directed to [**16**], particularly 12.2.

DEFINITION 15. *A binary system $(B, \circ)$ is a 'quasigroup' if and only if, to each $a, b \in B$, there corresponds unique $x, y \in B$, such that $x \circ a = b$ and $a \circ y = b$. A quasigroup is a 'loop' if some $e \in B$:*

$$\forall x \in B : x \circ e = e \circ x = x.$$

DEFINITION 16. *A system $Q = (V, +, \circ)$, is a 'prequasifield' if and only if the following five properties are satisfied:*

(1) $(V, +)$ is an Abelian group with additive identity 0.
(2) $(V^*, \circ)$ is a quasigroup.
(3) $\forall x \in V : x \circ 0 = 0$.
(4) $\forall x, y, z \in V : (x + y) \circ z = x \circ z + y \circ z$.
(5) If $a, b, c \in V, a \neq b \implies \exists! x \ni x \circ a - x \circ b = c$.

DEFINITION 17. *A prequasifield with a multiplicative identity is a 'quasifield'*
A quasifield which has a multiplicative group is a 'nearfield'.
A quasifield with both distributive laws is a 'semifield'.
A semifield field that is a nearfield is a 'skewfield'.
A skewfield that has a commutative multiplicative group is a 'field'.

DEFINITION 18. *If any of these algebraic structures satisfies all of the requirements but do not have a multiplicative identity 1, the term 'pre-' as in 'prefield', 'preskewfield' is used.*

DEFINITION 19. *A 'Bol' translation plane is a translation plane that admits a coordinatizing quasifield $Q = (V, +, \circ)$ that has the 'Bol axiom': $\forall a, b, c \in Q$: $((c \circ a) \circ b) \circ a = c \circ ((a \circ b) \circ a)$.*

3. Left quasifields

Let π be a translation plane coordinatized by a spread set of n by n matrices over a field K. Choose $x = 0$, $y = 0$, $y = xM$, where $M \in \lambda \subset GL(n, K)$. We shall also eventually deal with the situation where K is a skewfield, but for the moment, K will be a field. For a spread we may choose the set of matrices λ so that the last row of the matrix set is $[t_1, t_2, ..., t_n] \ \forall t_i \in K$. Now define a coordinate system $(W = \oplus^n K, +, \circ)$, where $+$ is vector addition over the associated vector space $W \oplus W$ of dimension $2n$, and where $a \circ b = aM$, where the last row of M is b for all $a, b \in W$, or $a \circ b = 0$ when $b = 0$ as a vector.

Then $(W = \oplus^n K, +, \circ)$ will be a quasifield provided λ contains the I_n (identity matrix) or a pre-quasifield if λ does not contain I_n.

We discussed dual translation planes and here we shall mean that the line at infinity l_∞ and the infinite point (∞) are interchanged under the change to the affine restriction of the dual of the projective extension of an affine translation plane.

DEFINITION 20. *A 'dual/left quasifield' of a (right) quasifield $(Q, +, \circ)$ is $(Q, +, *)$, where $a * b = b \circ a$. It follows that a dual translation plane obtained by mapping $l_\infty \leftrightarrow (\infty)$ may be coordinatized by a dual/left quasifield.*

(1) Noting that a quasifield admits the right distributive law, a dual quasifield will admit the left distributive law and will be also a quasifield if and only if the original quasifield is a semifield with elation axis $x = 0$.

(2) Important note: If any coordinatization other than $x = 0$ is taken for the axis of a semifield plane, the associated quasifield will not be a semifield (unless the quasifield is a skewfield).

Now let π be a translation plane of dimension 2 over a field K. This means that there is a right quasifield $(Q, +, \circ)$, where the elements of Q are 2-vectors over K. So, we have quasifield multiplication as $x \circ (t, u)$, where (t, u) is the second row of the associated matrix spread set. The dual/left quasifield for the associated dual translation plane has multiplication $x * (t, u) = (t, u) \circ x$. Now let $t = 0$ and write $x * (0, u)$ as $xu = (0, u) \circ x$. Then we have the following subset of elements in the dual translation plane: $x = 0, y = xu$, for all $u \in K$. Note that this is a K-regulus net. So, we have shown:

PROPOSITION 1. *The dual translation plane of any translation plane of dimension 2 over a field K contains a classical K-regulus net.*

The same sort of argument also proves the same for a translation plane that is of dimension 2 (left or right) over a skewfield K.

PROPOSITION 2. *The dual translation plane of any translation plane of dimension 2 over a skewfield (left or right) contains a derivable net of the form $x = 0, y = \delta x$ or of the form $x = 0, y = x\delta$ for all $\delta \in K$, called, respectively, a right classical pseudo-regulus net or a left classical pseudo-regulus net.*

We mention again that saying that a dual translation plane contains a derivable net, is not intended to mean that the dual translation plane is derivable. We have discussed hyperplane dual spreads, which are not related to dual quasifields, but are simply sets of S subspaces such that each hyperplane is contained in a unique component.

- The derivable net of a dual translation plane makes the dual translation plane derivable if and only if the associated spread is a hyperplane dual spread.

4. *T*-extension

In this section, we consider a sort of copying of a translation plane using exactly one component T. Just as a note of terminology, we shall later consider another type of copying, by copying an entire spread using the rational function field of the associated field K in question.

Let $\mathcal{D}$ be a derivable net. We know that this net may be given a coordinatizing structure of a classical pseudo-regulus net, which will be shown in the next few chapters. If this net is a regulus net and if the net is contained in a translation plane π of dimension 2 with spread in $PG(3, K)$, there is a construction technique called 'T-extension' given as follows:

THEOREM 3. *Let S be a spread for a translation plane that admits a classical regulus net $\mathcal{D}$. Then the spread may be written as follows:*

$$x = 0, y = x \left[\begin{array}{cc} f(t, u) & g(t, u) \\ t & u \end{array} \right] : \forall t, u \in K,$$

where K is a field and f, g are functions on $K \times K \to K$ such that $f(0, u) = u$ and $g(0, u) = 0$ for all $u \in K$, which defines the regulus net. Let T be any component not in the derivable net $\mathcal{D}$ (given by a fixed element (t, u) for $t \neq 0$).

Then there is a semifield plane of the following form called the 'T-extension', or 'T-extension of a regulus net':

$$
\begin{aligned}
x &= 0, y = x \circ u + xT \circ \alpha \\
&= x \left[\begin{array}{cc} u & 0 \\ 0 & u \end{array} \right] + x \left[\begin{array}{cc} f(t, v) & g(t, v) \\ t & v \end{array} \right] \left[\begin{array}{cc} \alpha & 0 \\ 0 & \alpha \end{array} \right]
\end{aligned}
$$

$\forall u, \alpha \in K$ *and for $t \neq 0$ and v fixed elements in K, representing T.*

Actually, it is not necessary that there is a translation plane containing a regulus net to use the construction, merely that the union of $\mathcal{D}$ and T define a net containing $\mathcal{D}$. After the embedding and contraction theory to follow in chapter 2, showing the structure of any derivable net, we shall note that Knarr [107] was able to show that given any affine plane π containing a derivable net $\mathcal{D}$, and choosing any line l of π then exactly the same technique as T-extension produced a dual translation plane. Again, even in a more general manner, it is not really necessary that $\mathcal{D}$ is contained in an affine plane, merely that there is a set of points l, with certain fundamental properties, making l a 'transversal' to $\mathcal{D}$. We will take up this theory just after the explanation of how to realize a derivable net as a classical regulus net or classical pseudo-regulus net.

CHAPTER 2

Embedding theory of derivable nets

Here we give a summary of how structure is developed for a derivable net. It is first shown that any derivable net can be embedded into a 3-dimensional projective space $PG(3, K)$, where K is a skewfield. We shall use the notation $PG(V, K)_S$ to denote the subspaces corresponding to an associated vector space V over K. We shall eventually use $PG(V, K)_C$ for a combinatorial embedding.

NOTATION 1. *Let V be any vector space over a skewfield K, then the associated projective space of vector subspaces is denoted by $PG(V, K)_S$. If the dimension of V/K is finite of dimension n, then the notation is $PG(n - 1, K)_S$.*

THEOREM 4. *(See 9.1 p. 85 [86]) Let $PG(V, K)_S$ and $PG(V', K')_S$ be any two projective spaces where both dimensions are ≥ 3.*

(1) If σ is an isomorphism from $PG(V, K)_S$ to $PG(V', K')_S$ then K is isomorphic to K' and σ is induced from a semi-linear mapping from V onto V'.

(2) In particular, if (1) is assumed, then the associated vector spaces have the same dimension.

(3) The collineation group (automorphism group) of $PG(V, K)_S$, called the 'projective semi-linear group' $P\Gamma L(V, K)$ and $\Gamma L(V, K)$ is the 'semi-linear group' of $AGL(V, K)_S$, (the associated affine space -see below). The group of linear automorphisms is called the 'projective general linear group', where $GL(V, K)$ is the 'general linear group'.

DEFINITION 21. *Given a projective space $PG(V, K)$ and let Γ denote a projective subspace. Choose a basis B for the vector subspace of V corresponding to Γ within a basis $B \oplus C$ of V/K. The dimension c of C is said to be the 'co-dimension' of Γ or that Γ has co-dimension c. For example, a 'hyperplane' of $PG(V, K)$ is a projective subspace of co-dimension 1.*

DEFINITION 22. *Given a projective space $PG(V, K)$, an 'Affine Space $AG(V, K)$' is a geometry constructed by the deletion of a 'hyperplane at infinity' (co-dimension 1 projective space), with appropriate definitions of incidence. For example, for projective planes, the deletion of a line l_∞ and all points p_∞ incident with l_∞, results in calling the lines incident with a given point p_∞ 'parallel' in the corresponding affine plane.*

The following theorem is called the **'Fundamental theorem of projective Geometry'**.

THEOREM 5. *(See 9.6 p. 89 [86]) Let $(\mathcal{P}, \mathcal{L}, \mathcal{I})$ be a triple of sets of 'points' $\mathcal{P}$, 'lines' $\mathcal{L}$ and 'incidence' $\mathcal{I}$ with the following properties:*

(1) Any line is incident with at least three points.

(2) Two distinct points are incident with a unique line.

(3) Let A, B, C be distinct points such that C is not incident with the line AB. Call the points A, B, C 'vertex' points of the triangle ABC formed by the lines AB, AC and BC.

  If a line l intersects the lines AB and BC in non-vertex points then l intersects AC non-trivially.

(4) There exist at least four points, no three of which are incident with a line (non-collinear).

  Then $(\mathcal{P}, \mathcal{L})$ is the set of points and lines of a projective plane or a projective geometry $PG(V, K)_S$.

There is also a related theorem for affine spaces, which we shall call the **'Fundamental theorem of affine geometry'**.

THEOREM 6. *(9.10 p. 92 [86]) Let $\pi = (\mathcal{P}, \mathcal{L}, \mathcal{I}, Aff)$ be a set of points, lines, incidence and a set of subspaces Aff, which are affine planes and assume that $card\, Aff > 1$. Let l_1 and l_2 be lines which are possibly equal. We shall say that l_1 is 'parallel' to l_2, $l_1 \parallel l_2$, if and only if these lines are parallel lines in some affine plane in Aff.*

Assume that π has the following properties:

(1) There are at least two points incident with every line.

(2) Two distinct points are incident with a unique line.

(3) There are at least four non-coplanar points (points not all in any plane of Aff).

(4) Given three mutually distinct non-collinear points, there is a unique affine plane of Aff incident with the points.

(5) Parallelism is an equivalence relation.

Then π is an affine space $AG(V, K)$.

REMARK 4. *When parallelism is an equivalence relation in the above theorem, the equivalence classes of lines partition the points so that there is a unique line of each class through each point. When α_o is a subplane of Aff, let (A, l) be a non-incident point-line pair of α_o. Then there is a unique line l_A parallel to l and incident with A. The points of l_A are included in the points-set α_o in this situation.*

We shall be mostly interested initially in the embedding/characterization theorem for derivable and subplane covered nets.

This theorem is called the **'Embedding theorem for derivable nets'**.

THEOREM 7. *(See 9.6 p. 89 [86]) Let $D = (\mathcal{P}, \mathcal{L}, \mathcal{B}, \mathcal{C}, \mathcal{I})$ be a derivable net, where the indicated sets denote points $\mathcal{P}$, lines $\mathcal{L}$, Baer subplanes $\mathcal{B}$, parallel classes, $\mathcal{C}$, and incidence, $\mathcal{I}$.*

  (I) Then there exists a skewfield K and a projective space $PG(3, K)_C = \Sigma$, together with a fixed projective line N, such that

   (1) the points $\mathcal{P}$ of D are the lines of Σ, which are skew to N,

   (2) the lines $\mathcal{L}$ of D are the points of $\Sigma - N$,

   (3) the parallel classes of D are the planes of Σ that contain N, and

   (4) the Baer subplanes $\mathcal{B}$ are the planes of Σ that do not contain N.

  (II) Conversely, given any 3-dimensional projective space Σ over a skewfield K, choose any line N and define a set $\mathcal{S} = (\mathcal{P}, \mathcal{L}, \mathcal{B}, \mathcal{C}, \mathcal{I})$ where the indicated sets

denote points $\mathcal{P}$, lines $\mathcal{L}$, Baer subplanes $\mathcal{B}$, parallel classes, $\mathcal{C}$, and incidence $\mathcal{I}$, are defined exactly as above. That is, the point set $\mathcal{P}$ is the set of lines of Σ that are skew to N, the line set $\mathcal{L}$ is the set of points of $\Sigma - N$, the set of Baer subplanes is the set of planes of Σ that do not contain N, the set of parallel classes $\mathcal{C}$ is the set of planes of Σ that contain N. The incidence set $\mathcal{I}$ is the inherited incidence of Σ.

Then S is a derivable net.

The proof uses both of the theorems (See 9.6 p. 89 [**86**]) *above and* (9.10 p. 92 [**86**]).

1. Co-dimension 2 construction

The embedding theory of subplane covered nets proceeds in very similar ways, but has a much more complex and structured proof. In [**86**], there is a study of what sorts of nets can be embedded in either affine or projective spaces.

DEFINITION 23. *Let D be any net and choose any fixed parallel class C of the set of parallel classes. Then a '1-parallel class retraction by C' of D is the net obtained by deleting C from the set of parallel classes.*

Cofman [**28**] realized a 1-parallel class contraction of a derivable net could be embedded in an affine geometry $AG(3, K)$, for a skewfield K. The embedding theory as given in [**86**] uses the ideas of Cofman in the following manner: There is really a set of affine spaces, indexed by the parallel class $\mathcal{C}$, the extension to the projective space $PG(3, K)_C$ basically weaves these together geometrically.

A larger question regarding embedding is to determine the set of all nets that can be embedded into an affine or projective geometry.

DEFINITION 24. *Let $\mathcal{Z} = (\mathcal{P}, \mathcal{L}, \mathcal{C}, \mathcal{I})$ be a net. We shall say that the net is 'embedded into an affine or projective geometry' if and only if the points of the net become the lines of the geometry and the lines of the net become the points of the geometry, with incidence in the geometry inherited from $\mathcal{I}$.*

It is then shown in [**86**] that the nets embeddable in a projective geometry are exactly the subplane covered nets. Note that the associated projective geometries could be of infinite dimension; the associated vector spaces are of infinite dimension over a skewfield K. Furthermore, any skewfield then appears as the coordinatizing structure of a great variety of subplane covered nets. The nets embed into projective geometry relative to a co-dimension 2 projective subspace N. We shall give the construction in the following. We will be considering points, lines, etc. of projective geometries and formulating the structure of an affine net. So, it will require some diligence to distinguish between a 'point' of a projective space and the point of the net.

The following is called the '**Embedding theorem for subplane covered nets**' or the '**Co-dimension 2 construction**'.

THEOREM 8. *(See 14.4 [**86**]) Let Σ be any projective geometry $PG(V, K)$, where K is a skewfield. Choose any co-dimension 2 projective subspace N.*

(1) Define an incidence structure $\mathcal{S} = (\mathcal{P}, \mathcal{L}, \mathcal{B}, \mathcal{C}, \mathcal{I})$, of points $\mathcal{P}$, lines $\mathcal{L}$, subplanes $\mathcal{B}$, parallel classes $\mathcal{C}$ and incidence $\mathcal{I}$, and defined as follows:

$\mathcal{P}$ is the set of lines skew to N.

$\mathcal{L}$ is the set of points of $\Sigma - N$.

$\mathcal{B}$ is the set of planes of Σ that intersect N in a point.

$\mathcal{C}$ is the set of hyperplanes of Σ that contain N.

$\mathcal{I}$ is the incidence set induced from the incidence of Σ.

Then S is a subplane covered net.

(2) Now consider the affine geometry $AG(V, K)$ obtained by the deletion of any hyperplane H that contains N. Let S^- denote the structure obtained from deleting H from $\mathcal{C}$ as well as the required point and line deletions. Consequently, we obtain $S^- = (\mathcal{P}^-, \mathcal{L}^-, \mathcal{C}^-, \mathcal{I}^-)$, where $\mathcal{P}^-$ is the set of lines of Σ which are not in H and which are not skew to N. $\mathcal{L}^-$ is the set of points of $\Sigma - H$, $\mathcal{C}^- = \mathcal{C} - H$ and $\mathcal{I}^-$ is the inherited incidence.

Then S^- is a net which may be obtained from S by the deletion of a parallel class. We note that the points of S and of S^- are identical. This net is therefore a '1-class retraction' of S.

The main embedding theory for subplane covered nets shows that the previous theorem has a direct converse: Every subplane covered net may be embedded in a projective space $PG(V, K)$ with designated co-dimension 2 subspace that relates points, lines, subplanes, parallel classes and incidence in exactly the same way as in the co-dimension 2 construction from the projective space. Furthermore, any net that can be embedded into an affine space $AG(V, K)$ is a 1-class retraction of a subplane covered net. The reader interested in further investigating these embeddings is directed to chapter 14 of [86].

The embedding theorem characterizes derivable nets and subplane covered nets and it is important to point out that derivable nets or subplane covered nets exist and may be 'coordinatized' by any skewfield K. However, we still do not actually understand or recognize a derivable net. The structure theorem, which we call the 'fundamental contraction theorem', gives an identity to derivable nets. And, when just as the structure is given, questions arise as what different derivable nets may be embedded into isomorphic 3-dimensional projective spaces.

2. Structure theory and contraction of embedded nets

In this section, we give somewhat of an understanding of what a standard representation of a derivable net or subplane covered net can be. In this text, we consider mostly derivable nets, but a more complete analysis of this work can reveal an understanding of certain of the subplane covered nets, as well.

The way that structure can be imposed upon derivable nets arises from the collineation group of the embedded derivable net, for we now understand the following.

THEOREM 9. Let D be a derivable net. Then there is an embedding of D into a 3-dimensional projective space $PG(3, K)_C$, where K is a skewfield. The collineation group of D is isomorphic to $P\Gamma(4, K)_N$, where N is a fixed line of the projective space.

Originally, the derivable net had no imposed structure or coordinatizing set. We may choose certain subgroups of $P\Gamma L(4, K)$, so the derivable net may be represented so as to lie in an ambient 4-dimensional vector space V over K, which can be chosen to be either a left or right space. This means that we need an Abelian subgroup of $P\Gamma(4, K)_N$ that acts regularly on the points of the derivable net. When this occurs, the net will be represented as a subnet of a translation net. When this occurs, we have a translation net so a representation depends only on how the lines of the derivable net that contain the zero vector appear. Initially, in [86], a group was chosen that would show that the representation of the derivable is that of a 'classical regulus net', when K is a field and what is called a 'classical pseudo-regulus net' when K is a non-commutative skewfield. It is part of this text to try to find pseudo-regulus derivable nets that are not regulus nets, as such derivable nets are previously unknown. However, various groups could have been chosen that would have represented the derivable nets in a variety of different ways. In our focus for this book, we have found a way to describe essentially all representations of derivable nets. Realizing that any derivable net that could contract from the embedded net must have an associated skewfield isomorphic to K, then will guide the study of derivable nets into the types of nets in the coming parts and chapters. Once the statement of the classification of subplane covered nets is given, some ideas of how different groups providing different representations of the derivable nets could be given and more fully investigated.

3. Embedding of subplane covered nets

THEOREM 10. *(See 15.37 and 15.38 p. 195-6 [86])*

Let $\mathcal{R} = (\mathcal{P}, \mathcal{L}, \mathcal{B}, \mathcal{C}, \mathcal{I})$ be a subplane covered net. Define the point-line geometry $\Sigma_\mathcal{R}$ as follows:

Call the lines of a given parallel class 'class lines' and call the lines of a given parallel class of a derivable subnet 'class subplanes.' Note that there are equivalence classes of both the set of class lines and of the set of class subplanes. Call equivalence classes of the class lines 'infinite points' and call the equivalence classes of the class subplanes 'infinite lines.'

The infinite points and infinite lines form a projective space $\mathcal{N}$.

The 'points' of $\Sigma_\mathcal{R}$ are of two types:

(i) the 'lines' $\mathcal{L}$ of the net $\mathcal{R}$, and

(ii) the infinite points as defined above.

The lines of $\Sigma_\mathcal{R}$ are of three types:

(i) the set of lines incident with an affine point (identified with the set $\mathcal{C}$),

(ii) the class lines extended by the infinite point containing the class line, and

(iii) the lines of the projective subspace $\mathcal{N}$.

The 'planes' of $\Sigma_\mathcal{R}$ are of three types:

(i) subplanes of $\mathcal{B}$ extended by the infinite point on the equivalence class lines of each particular subplane, where the points and lines of the subplane are now considered as above (another interpretation is this is the affine dual of the subplane extended),

(ii) the affine planes whose 'points' are the lines of a new parallel class and 'lines' the class lines of a derivable subnet of the new parallel class extended by the infinite points and infinite line, and

(iii) the projective planes of the projective space $\mathcal{N} = \mathcal{N}^\alpha$, for all $\alpha \in \mathcal{C}$.

The hyperplanes of $\Sigma_{\mathcal{R}}$ that contain $\mathcal{N}$ are the parallel classes $\mathcal{C}$, extended by the infinite points and infinite lines of $\mathcal{N}$. Then $\mathcal{N}$ becomes a co-dimension two projective subspace of $\Sigma_{\mathcal{R}}$.

Then $\Sigma_{\mathcal{R}}$ is a projective space.

THEOREM 11. *(See 14.14 [86]) Let W be a left vector space over a skewfield K. Let $V = W \oplus W$. Let*

$$(x = 0) \equiv \{(0, y); y \in W\},$$

and

$$(y = \delta x) \equiv \{(x, \delta x); x \in W, \text{ for a fixed } \delta \in K\}.$$

Define a point-line geometry $\mathcal{R} = (\mathcal{P}, \mathcal{L}, \mathcal{B}, \mathcal{C}, \mathcal{I})$ as follows: The set of 'points' $\mathcal{P}$ is V, the set of lines $\mathcal{L}$ is the set of translates of $x = 0$ and $y = \delta x$, for all $\delta \in K$ and these lines incident with $(0,0)$ are representatives of the set of parallel classes $\mathcal{C}$.

The set of subplanes $\mathcal{B}$ is the set of translates of the following set of subplanes $\pi_w = \{(\alpha w, \beta w); \alpha, \beta \in K\}$ for fixed $w \in W - \{(0,0)\}$ and incidence $\mathcal{I}$ is the naturally induced incidence set.

Then $\mathcal{R}$ is a subplane covered net, which is called a right or left 'classical pseudo-regulus net' (just called a 'pseudo-regulus net' in the Subplane Covered Nets book). When the associated skewfield K defining the pseudo-regulus net is a field, then we call the subplane covered net a K-regulus net.

The following is the **Main theorem on subplane covered nets.** The result may also be called the **Contraction theorem of subplane covered nets.** So, by the embedding and contraction of derivable nets, their basic structure is then completely determined.

THEOREM 12. *(15.40 and 15.41 [86])*

(i) A subplane covered net is a pseudo-regulus net.
(ii) A finite subplane covered net is a regulus net.

Finally, we note that in the derivable net situation, we have been referring to the derivable net arising affinely from the combinatorial embedding-contraction theory as the 'right classical pseudo-regulus net', and our classification of derivable nets over skewfields is relative to this type of derivable net. However, the choice of groups of $P\Gamma L(V, K)$ is responsible for the representation. This particular representation was chosen since it was not known at the time that every derivable net/subplane covered net is a pseudo-regulus net or a regulus-net. If another choice of appropriate group would be chosen, the representation of the net would change accordingly.

Before we consider a classification of derivable nets relative to a given classical pseudo-regulus, we provide a few details on how the representation or contraction method from the embedded derivable net is actually accomplished. The representation of subplane covered nets allows more variation but the theory is the same, modulo a more complex representation of points of the projective geometry, especially true when the dimension is infinite. When the dimension n is finite, the projective geometry actually is $PG(n + 1, K)_C$, for a skewfield K. It must be noted that a corresponding normally associated vector space is of dimension $n + 2$, but the contracted target vector space will be of dimension $2n$ over K. When the subplane covered net is a derivable net, $n = 2$, and it would appear that the normally associated vector space producing a projective space of subspaces and the contracted target vector space are the same. However, these two projective spaces are completely different. We now formalize the notation that we have mentioned previously.

NOTATION 2. *For a subplane covered net, the embedded projective space is denoted by $PG(V, K)_C$. For the contraction method, where the subplane covered net has a realized vector space structure, there is then an associated projective space of vector subspaces, which is denoted by $PG(W, K)_S$. When the subplane covered net is a derivable net, then V and W have the same dimension 4 over K, either as a left space or as a right space. In all other cases, the dimensions of V and W are not equal over K.*

So, for a derivable net, we are working in $PG(3, K)_C$, with a designated line $\mathcal{N}$ and have an embedded derivable net $\mathcal{R} = (\mathcal{P}, \mathcal{L}, \mathcal{B}, \mathcal{C}, \mathcal{I})$, where the corresponding points, lines, Baer subplanes, parallel classes of the net are the lines, points, planes that do not contain $\mathcal{N}$ and planes that do contain $\mathcal{N}$, respectively. The intrinsic incidence in the projective space Σ and the incidence of the derivable net are induced.

We first find an appropriate translation group.

DEFINITION 25. *(See 11.2 [86]) Let V be a vector space over a skewfield K.*

A 'transvection' τ is an element of $GL(V, K)$ such that there is a vector d and $\tau(v) - v$ is in $\langle d \rangle$, for all $v \in V$. In this case, τ is said to be 'in the direction of d'

We shall also use the same term for the element in $PGL(V, K)$, induced from τ.

We note that in the situation for derivable nets, transvections are in $GL(4, K)$ and relative to $PG(V, K)_C$ fix a plane pointwise. We shall consider those transvections which fix pointwise planes containing $\mathcal{N}$ and in the direction of d as a point of $\mathcal{N}$. Also, although we are currently speaking of a left space or a right space, the default shall always be a left space or left point.

NOTATION 3. *The notation for a plane in $PG(3, K)_C$ that corresponds to a parallel class $\alpha \in \mathcal{C}$ shall be a_α^+.*

LEMMA 1. *If τ is a transvection of $PG(3, K)_C$ and fixes pointwise a plane $a_\alpha^+ \supset \mathcal{N}$, and has direction $d \in \mathcal{N}$ then τ is a translation of the derivable net.*

PROOF. The proof is a matter of getting used to connecting the two different incidence geometries. Another wrinkle of this theory is that we are working back and forth with a preimage vector space V whose subspace structure is in $PG(V, K)_C$, but there is or will be another vector space, say W, which is being constructed using the collineation group of the net, that is a subgroup of $P\Gamma L(4, K)_N$. Let v be any vector in V, which is not in $\mathcal{N}$, and hence becomes a line l of the derivable net. Let τ be a transvection in the direction $d \in \mathcal{N}$. Then τ will fix all planes containing $\mathcal{N}$ and fix one of these pointwise. This means that there is a collineation of the derivable net that fixes all parallel classes and leaves one of these parallel classes fixed linewise. That is, the transvection group generated by such transvections becomes a natural translation group for the derivable net.   □

DEFINITION 26. *Consider the group*

$$T = \langle \tau : \tau \text{ is a transvection with axis } a_\alpha^+ \text{ and some direction } d \in \mathcal{N} \rangle.$$

Then T is an elementary Abelian translation group acting regularly on the points of the derivable net.

T is called the 'translation group' of $\mathcal{R} = (\mathcal{P}, \mathcal{L}, \mathcal{B}, \mathcal{C}, \mathcal{I})$.

PROPOSITION 3. *(11.6 p. 117 [86]) T is normal in the full collineation group G of $\mathcal{R}$. Furthermore, $G = TG_0$, where 0 is a point of $\mathcal{R}$, called the zero vector.*

The proof of the preceding proposition would appear to be immediate but it requires a careful proof. So, the reader wishing more information is directed to the indicated theorem and page numbers in [86].

In view of the above proposition, if we choose an appropriate scalar multiplication, we would retrieve sufficient information to form an ambient vector space of dimension 4 over K. In section 11 of [86], an appropriate group of the projective space $PG(3, K)_C$ which is or represents a skewfield is given. However, there is a variety of such representations of the derivable net that depend directly of the choice of group. To illustrate the complexity of this idea, we need the framework given by a particular choice of basis elements.

Since 0 is a point of $\mathcal{R}$, let M_0 denote the associated line in Σ. Choose a basis in the preimage vector space V/K so that M_0 is represented as $\langle (0,0,1,0), (0,0,0,1) \rangle$ and $\mathcal{N} = \langle (1,0,0,0), (0,1,0,0) \rangle$, and it is agreed that the scalar action is on the left of the vectors. Let τ be a transvection that fixes the plane $x_3 = 0$ pointwise, when considering vectors (x_1, x_2, x_3, x_4) and has a direction $d = (c_1, c_2, 0, 0)$, then, with some preliminary setup, we have

$$\tau_{c_1, c_2} = \left[\begin{array}{cc} I_2 & 0_2 \\ \begin{bmatrix} c_1 & c_2 \\ 0 & 0 \end{bmatrix} & I_2 \end{array} \right],$$

and acts on

$$v = (x_1, x_2, x_3, x_4) \to^{\tau_{c_1, c_2}} (x_1 + x_3 c_1, x_2 + x_3 c_2, x_3, x_4).$$

So, $\tau(v) - v = (x_3 c_1, x_3 c_2, 0, 0)$, so $\langle d \rangle = \langle (c_1, c_2, 0, 0) \rangle$. Define

$$\left\langle \tau_{c_1, c_2} = \left[\begin{array}{cc} I_2 & 0_2 \\ \begin{bmatrix} c_1 & c_2 \\ 0 & 0 \end{bmatrix} & I_2 \end{array} \right] : \forall c_1, c_2 \in K \right\rangle = T_3.$$

Similarly,

$$\left\langle \rho_{c_3, c_4} = \left[\begin{array}{cc} I_2 & 0_2 \\ \begin{bmatrix} 0 & 0 \\ c_3 & c_4 \end{bmatrix} & I_2 \end{array} \right] : \forall c_3, c_4 \in K \right\rangle = T_4.$$

T_4 fixes $x_4 = 0$ pointwise, as a plane of $\rho_{c_3, c_4}(v) - v$, for $v = (x_1, x_2, x_3, x_4)$, is $(x_4 c_3, x_4 c_4, 0, 0$, so the direction is $\langle (c_3, c_4, 0, 0) \rangle$. It is easy to verify that the group is Abelian and equal to and generated by $T_3 T_4$.

The points of $\mathcal{R}$ are the lines of Σ that are skew to $\mathcal{N}$.

LEMMA 2. *(See 11.8 [86])*

(1) Any line skew to $\mathcal{N}$ has a basis of the following form

$$\{(d_1, d_2, 1, 0), (d_3, d_4, 0, 1)\}.$$

(2) *The element* $\tau_{c_1,c_2,c_3,c_4} = \begin{bmatrix} \begin{bmatrix} c_1 & c_2 \\ c_3 & c_4 \end{bmatrix} & \begin{matrix} 0_2 \\ I_2 \end{matrix} \end{bmatrix}$ *maps the line with basis*

$$\{(d_1, d_2, 1, 0), (d_3, d_4, 0, 1)\}$$

onto the line with the basis

$$\{(d_1 + c_1, d_2 + c_2, 1, 0), (d_3 + c_3, d_4 + c_4, 0, 1)\}.$$

PROBLEM 4. *This proof should be tried as homework.*

NOTATION 4. *To distinguish between points of $\mathcal{R}$ and lines skew to $\mathcal{N}$, we shall use $\mathfrak{P}(d_1, d_2, d_3, d_4)$ to indicate we are considering points of $\mathcal{R}$.*

At this point, we need to consider a group of $\Gamma L(4, K)$, corresponding to the non-zero elements of a skewfield. Hence, the group elements need to be additive. We consider possible groups as subgroups of the following group:

$$\langle g_{\alpha,\beta}^R : (x_1, x_2, x_3, x_4) \to (\alpha x_1, \alpha x_2, \beta x_3, \beta x_4) : \forall \alpha, \beta \in K^* \rangle$$

This group is called the group of 'quasi-skew perspectivities' fixing M_0 and $\mathcal{N}$ pointwise (two skew lines). In [86], the group generated by $g_{\alpha,1}^R$ was used as follows:

Define $\alpha \circ \mathfrak{P}(d_1, d_2, d_3, d_4) = \mathfrak{P}(d_1, d_2, d_3, d_4) g_{\alpha,1}^R$. What occurs is the following:

The points $\mathfrak{P}(d_1, d_2, d_3, d_4)$ form a left vector space over $K^{opp} = K_{right}$, which is also then a right vector space over K_{right}.

Then the representation of the derivable net may be given by providing the lines of the net that are incident with the affine point 0. Given such a line l of the net incident with 0, the set of net points of this line is a set of lines skew to $\mathcal{N}$ that intersect M_0 in a point. Using the representation of a line skew to $\mathcal{N}$ in the form

$$\{(d_1, d_2, 1, 0), (d_3, d_4, 0, 1)\},$$

means there must exist $\alpha, \beta \in K$ such that

$$\alpha((d_1, d_2, 1, 0) + \beta(d_3, d_4, 0, 1) = (0, 0, \alpha, \beta).$$

Consequently,

$$\alpha d_1 + \beta d_3 = \alpha d_2 + \beta d_4 = 0.$$

Consider first the case that $\beta \neq 0$. Then $d_3 = \beta^{-1}\alpha d_1$ and $d_4 = \beta^{-1}\alpha d_2$. So, there exists an element $\delta = \beta^{-1}\alpha \in K$, such that

$$y = \delta x, \text{ as } \{(d_1, d_2, \delta d_1, \delta d_2); \forall (d_1, d_2) \in K \oplus K$$

in the derivable net, representing points first by $\mathfrak{P}(d_1, d_2, d_3, d_4)$, and then dropping the fraktur type and writing points of the net as (x_1, x_2, x_3, x_4) (note the confusion possibility here with points/vectors of the embedded net). The other case when $\beta = 0$, will force $d_1 = d_2 = 0$ and the net line appears as $x = 0$.

Note that when K is a field, this is the form

$$\{x = 0, y = \delta x; \delta \in K\},$$

providing the components (lines incident with the zero vector) of a classical regulus net. When K is a non-commutative skewfield, this defines the components of a right pseudo-regulus net. However, we have not completed the representation of what a Baer subspace (incident with 0) must look like. The Baer subplanes of the net that share the zero vector 0 are the planes of Σ that contain the line M_0 and does not contain the line $\mathcal{N}$. Let $M_0 \oplus \langle (d_1, d_2, 0, 0) \rangle$ denote the preimage of such a plane in the associated vector space V. Then the set of lines that are skew to $\mathcal{N}$ have bases of the form

$$\{(\alpha d_1, \alpha d_2, 1, 0), (\beta d_1, \beta d_2, 0, 1) : \forall \alpha, \beta \in K\}.$$

Consequently, the point set in the derivable net for a Baer subspace has the following form:

$$\pi_{d_1, d_2} = \{\alpha d_1, \alpha d_2, \beta d_1, \beta d_2\); \forall \alpha, \beta \in K$$

THEOREM 13. *Consequently, the derivable net has the following form: Let W be a left 2-dimensional vector subspace over the skewfield K. Let $V = W \oplus W$. Let*

$$(x = 0) \equiv \{(0, y); y \in W\},$$

and

$$\{y = \delta x \equiv (x, \delta x); \forall \delta \in K\},$$

represent the set of components, as right 2-dimensional vector spaces over K.

The set of Baer subspaces (incident with the zero vector), as left 2-dimensional vector spaces over K, has the following form:

$$\left\{ \begin{array}{c} \pi_{d_1, d_2} = (\alpha d_1, \alpha d_1, \beta d_1, \beta d_2); \forall \alpha, \beta \in K, \\ \text{for each pair } (d_1, d_2) \neq (0, 0) \text{ of } K \oplus K \end{array} \right\}.$$

We are now most interested in what happens to the representation of the derivable net, if we vary the definition of scalar multiplication by changing the corresponding group. But, before we do this, note that the original choice of group was important to realize that all derivable nets are classical in the sense that they are all pseudo-regulus nets when K is a skewfield and regulus-nets when K is a field.

Now, we may see that there are other reasonable representations. For example, consider the following group

$$\left\langle \begin{array}{c} g_{\alpha, 1}^{R, \sigma} : (x_1, x_2, x_3, x_4) \to (x_1 \alpha^\sigma, x_2 \alpha, x_3, x_4) : \forall \alpha, \in K^*, \\ \text{where } \sigma \text{ is a non-identity automorphism of } K \end{array} \right\rangle.$$

Everything proceeds in an analogous manner:

$$\alpha \circ \mathfrak{P}(d_1, d_2, d_3, d_4) = \mathfrak{P}(d_1, d_2, d_3, d_4) g_{\alpha, 1}^{R, \sigma} = \mathfrak{P}(d_1 \alpha^\sigma, d_2 \alpha, d_3 \alpha^\sigma, d_4 \alpha).$$

Now, by an similar calculation, the representation of the derivable net is as follows, and the elements will have the following form:

$$(d_3, d_4) = (\delta^\sigma d_1, \delta d_2).$$

$$x = 0, y = \delta x = (\delta^\sigma x_1, \delta x_2) : \forall \delta \in K.$$

So, this can be understood in a field context. Let K be a field, then we would have the components of the derivable net have the form

$$x = 0, y = x \begin{bmatrix} \delta^\sigma & 0 \\ 0 & \delta \end{bmatrix} : \forall \delta \in K$$

and the set of Baer subplanes has the following form:

$$\left\{ \begin{array}{c} \pi_{d_1, d_2} = ((d_1, d_2) \begin{bmatrix} \delta^\sigma & 0 \\ 0 & \delta \end{bmatrix}, (d_1, d_2) \begin{bmatrix} \gamma^\sigma & 0 \\ 0 & \gamma \end{bmatrix}) : \\ \forall \delta, \gamma \in K \text{ and } \forall (d_1, d_2) \neq (0, 0) \text{ of } K \oplus K. \end{array} \right\}$$

So, this is what is called a 'twisted regulus net'. However, we know that this is actually a regulus net. The distinction between a regulus net and a twisted regulus is simply the representation, as they are isomorphic. For each representation, there is a field (or skewfield in the general case) so that the above representation is a regulus (pseudo-regulus) over that field (skewfield). We shall observe various other representations in part 2 on classifying derivable nets.

The interested reader looking for more details is referred to section 11 of [**86**].

4. Transversals to derivable nets

Some of the results of this chapter appear in Johnson [**88**] and connect with the work of Knarr [**107**]. These ideas are reinterpreted here in view of the skewfield classification of derivable nets, and in view of a mistake in Jha and Johnson [**70**], asserting that type 1 derivable nets over fields (See (3,5,11)) require characteristic 2, which we shall see is incorrect. There are several questions that might be open and seem appropriate to mention in this section.

PROBLEM 5. *Given any skewfield K, form $PG(3, K)$. Is there always a spread in $PG(3, K)$?*

PROBLEM 6. *Given any derivable net $\mathcal{D}$, is there always an affine plane containing $\mathcal{D}$? And, is there always a derivable affine plane containing $\mathcal{D}$?*

PROBLEM 7. *Given any quaternion division ring (See (2,3,2)) and an associated derivable net, is there always an extension quaternion division ring representing a spread? Related to this, given any field K, is there always a field extension of K?*

The last part of the last problem has the obvious answer that the rational function field of K, $K(z)$, where z is an independent variable, is a field extension of K. This idea will be used in part 12 to prove the same result for quaternion division rings. However, the answers to the first two problems are most likely no to both. Suppose that given any skewfield K, there is a spread S in $PG(3, K)$. In this section, we shall show that choosing a line N of S and forming the associated spread from the contraction principle of part 1, we obtain a derivable net $\mathcal{D}$ and a so-called 'transversal' to $\mathcal{D}$ and this is enough to construct a dual translation plane containing $\mathcal{D}$. In part 1, it was also shown that there are dual translation planes containing derivable nets but which are not themselves derivable. So, the first two problems are quite related. But, what skewfields or fields K are there so that there is no spread in $PG(3, K)$? If the first thought is let K be the field of complex numbers, is there a spread in $PG(3, K)$? Yes, the Hamiltonians over the field of real numbers R, $(-1, -1)_R$ coordinatize a Desarguesian spread in $PG(3, K(\sqrt{-1}))$.

We begin by connecting spreads S in $\Sigma = PG(3, K)$, for K any skewfield, with derivable nets. A 'projective spread' S, in this setting, is a set of mutually disjoint lines of $PG(3, K)$ such that every point of $PG(3, K)$ is incident with a line of S. Let N be any line of S, and consider the derivable net $\mathcal{D}$ whose points and lines are, respectively, the lines and points of $\Sigma - N$. Also the Baer subplanes of $\mathcal{D}$ are the planes of Σ that intersect N in a point and the parallel classes are the planes of Σ that share N. Consequently, the lines of $S - N$ are points of the derivable net $\mathcal{D}$ that every line of $\mathcal{D}$ is incident with a unique point of $S - N$, as a set of points of $\mathcal{D}$. We know that D is either a left or right classical pseudo-regulus net and without loss of generality, it may be assumed that the zero vector 0 is a line m_0 of S. Choose any parallel class σ and note that the line of σ must intersect $S - N$ in a unique point and hence these intersections cover $S - N$. Noting that $S - N$ in the affine space $AG(3, K)$, is a 2-dimensional (right or left) subspace over K, it follows that the union and $\mathcal{D} \cup \{(S - N) = T\}$ is a net containing a derivable subnet. Such a line T in this setting is said to be a 'transversal to the derivable net $\mathcal{D}$'. Assume that the derivable net is a right classical pseudo-regulus net given by:

$$\{x = 0, y = \delta x; \forall \delta \in K; components\ are\ right\ spaces\}$$

and the set $B(R^K)_{left}$ of Baer subplanes incident with the zero vector $B(R^K)_{left}(a, b)$, where $(a, b) \in K \times K - \{(0, 0)\}$.

$$B(R^K)_{left} = \left\{ \begin{array}{c} B(R^K)_{left}(a, b) \\ = \{(ca, cb, da, db); \forall c, d \in K\} \end{array} \right\};$$

Baer subspaces are left spaces.

So, T is a set of points (x, y). If points (x_1, y_1) and (x_2, y_1) are in T. Then the line $y = y_1$ uniquely intersects the line $x = 0$ in the points $(x_1, 0)$ and $(x_2, 0)$ so $x_1 = x_2$, and therefore, T may be represented as the set of points $(x, f(x))$ for all $x \in W$, where W is a 2-dimensional left K-space and $W \oplus W = V_4$ represents the points, and the associated function f maps W to W.

The points $(c, f(c))$ and $(d, f(d))$ are points of T, and the previous argument also shows that f is injective. Take any line $y = b$ for any $b \in W$, then there is a unique intersection with T, which shows that T is surjective. Now consider $(c - d, f(c) - f(d))$. Assume that $f(c) - f(d) = \delta(c - d)$ for $c \neq d$, for $\delta \in K$; that the vectors are linearly dependent. Then there exists a component $y = \delta x$ which contains $(c, f(c))$ and $(d, f(d))$, a contradiction, since the components of the derivable net intersect T only at the zero vector. Now consider a vector $\alpha c + d$, where $\alpha \in L$. We know that $y = \alpha x + d$ is a line of the parallel class defined by $y = \alpha x$. Since this line intersects T in a unique point $(e, f(e)) = (e, \alpha e + d)$. Consequently, $f(e) = \alpha e + b$. This is exactly the definition of a vector transversal function.

DEFINITION 27. $\mathcal{D}$ *be a derivable net. A 'transversal to $\mathcal{D}$' is a set of points T such that every line of the net intersects T in a unique point of T and, given any parallel class ρ of $\mathcal{D}$, the intersections $\{l \cap T : \forall l \in \rho\} = T$.*

When $\mathcal{D}$ is written as a right classical pseudo-regulus net over K, a 'vector space transversal' T is a transversal that can be written using a bijective function on W such that $f(c) - f(d)$ and $c - d$ are K-left linearly independent for all c, d of W and $c \neq d$ and such that, for each $b \in W$ and $\alpha \in K$, there exists a unique $c \in W$ such that $f(c) = \alpha c + b$.

When T.G. Ostrom was developing the theory of finite derivable nets, transversals to derivable nets were only considered when a derivable net was written as a classical regulus net and the transversal was taken to be a vector space transversal. When this occurred, Ostrom was able to extend the derivable net to a derivable dual translation plane. The generalization to the infinite field or skewfield case requires some modification. We have noted previously that there are dual translation planes containing derivable nets, but the planes themselves are not always derivable. That is, the associated translation planes may not be dual spreads (1,2,1).

Consequently, we have proved (1) of the following theorem.

THEOREM 14.

(1) *Spreads S in $PG(3,K)$ for a skewfield K, correspond to vector space transversals T of a derivable net, each of which, in turn, corresponds to a dual translation plane, and which then is the dual translation plane S^D of the translation plane corresponding to S and where S^D contains a derivable net.*

(2) *The spread S is a hyperplane dual spread if and only if the dual translation plane of (1) is derivable.*

REMARK 5.

(1) *If a spread S is written as a matrix spread over a field K then the hyperplane dual spread is written as the geometry determined by the transposed matrices in the matrix spread set (K need not be the kernel of the translation plane). When K is a non-commutative skewfield, there is no such corresponding representation.*

(2) *Given a semifield spread, meaning that there is an elation group E with axis L, such that E is transitive on the remaining components of the spread, it is customary that L is chosen to be represented in the form $x = 0$. Any other representation of L will not provide a semifield coordinate system. So, in the contraction theory, we let M_0, the line of the associated 3-dimensional projective space, be coordinatized as $x = 0$ and the special line N as $y = 0$. If S is a semifield spread of $PG(3, K)$, we would need to choose N as $x = 0$. So, one would need to rework the contraction theory by choosing M_0 as $y = 0$ and N as $x = 0$, which is easily accomplished.*

(3) *We have seen in (1,1,4), a method for extending a regulus net D by a transversal T which made the union of the regulus net $D \cup T$ a net, thus generating a semifield plane. This theory of T-extension is more generally valid for any derivable net, as the general form of D is similar over any skewfield, except that there are 'left' and 'right' classical pseudo-regulus nets and therefore left and right transversals. This means as sets of points, the vectors are designed as either a set of left points or a set of right points.*

PROBLEM 8. *Rework the contraction method with the special line N as $x = 0$ and M_0 as $y = 0$. Notice that if a spread of $PG(3, K)$ is a semifield spread then this change is necessary to obtain a coordinate semifield containing the associated derivable net.*

Part 2

Classifying derivable nets over skewfields

In this part, the basic structure theory of derivable nets available from the embedding-contraction theory of part 1 shall be used to understand how derivable nets might be represented and how to recognize a derivable net. We have seen that one way to do this is to find suitable representational groups in $P\Gamma L(V, K)$. However, the method we present is the alternative that uses the classical representation which turns out to have the most advantages when a derivable net is considered relative to the classical derivable nets. This material first appeared in Innovations in Incidence Geometry (Classifying Derivable Nets 19.2 (2022), 59-94), and the author is grateful to the journal for permission to provide a varied version in this text. Since we shall develop a method for classifying a derivable net relative to any given classical pseudo-regulus net over a skewfield, we require familiarity with aspects of non-commutative geometry. We begin with the fundamentals required for our study of the general derivable nets.

CHAPTER 3

Fundamentals & background

To recap and synthesize part 1, it is shown that any derivable net D may be embedded within a 3-dimensional projective space $\Sigma = PG(3, K)$, where K is a skewfield. Where the net points of D are the lines of $\Sigma - N$, where N is a fixed line, the lines of D are the points of $\Sigma - N$. The Baer subplanes of D are the planes of Σ that do not contain N and the parallel classes of D are the planes of Σ that contain N. Incidence is inherited.

With the derivable net occupying the projective space Σ, it then follows that the full collineation group of D is $P\Gamma L(4, K)_N$ and this connection allows a contraction to an affine form for D, within an associated 4-dimensional vector space (left space) over K that shows the derivable net R_{right}^K may be coordinatized as follows:

$$\{x = 0, y = \delta x; \forall \delta \in K; components \ are \ right \ spaces\}$$

and the set $B(R^K)_{left}$ of Baer subplanes incident with the zero vector $B(R^K)_{left}$ (a, b), where $(a, b) \in K \times K - \{(0, 0)\}$.

$$B(R^K)_{left} = \left\{ \begin{array}{c} B(R^K)_{left}(a, b) \\ = \{(ca, cb, da, db); \forall c, d \in K\} \end{array} \right\};$$

Baer subspaces are left spaces.

The main question for other derivable nets, which can also be so embedded in perhaps different 3-dimensional projective spaces $PG(3, L)$, for L a skewfield is as follows: **Which of these may be coordinatized so as to sit within the same 4-dimensional vector space over K?** This derivable net, in this form, is called the 'right classical pseudo-regulus net'. Note that this net is a regulus net if and only if K is a field. The derived net of the embedded-contracted net is

$$R_{left}^K = \{x = 0, y = x\delta; \forall \delta \in K\};$$

components are left spaces;

$$B(R)_{right}^K = \left\{ \begin{array}{c} B(R)_{right}^K(a, b) \\ = \{(ac, bc, ad, bd); \forall c, d \in K\} \end{array} \right\};$$

Baer subspaces are right spaces.

This particular understanding of the derived net is explicated in (2,4,2) when discussing the notation. That is, the indicated form of the derived net needs justification, but will be postponed until after some fundamentals are covered.

1. Uniform representation for quaternion division rings

Our approach for the study of quaternion division rings uses matrix representations, so we will continue this study for this book. Although, the material on quaternion algebras is voluminous and far reaching (See, e.g. [**140**]), this material might have some geometric interest due to the strong connection to derivable nets. For the reader interested in the quaternion algebras, the lectures of K. Conrad ([**32**]), are recommended, wherein there is an informative set of homework problems that suggests a uniform representation for quaternion division rings for any characteristic.

All of the examples for any characteristic $\neq 2$ use non-square quadratic extensions in their constructions. Consequently, there is a representation patterned after the Hamiltonians and Dickson generalizations. On the other hand, for characteristic 2, the non-square quadratic extensions do not provide the necessary involution for the constructions. Hence, the characteristic 2 representations necessarily require a quadratic extension based on $f(x) = x^2 - x\alpha - a$, where $\alpha \neq 0$ (again, see [**140**], sections on characteristic 2). It is possible to find a uniform representation for all skewfields of any characteristic and this is the point of this section.

Thus, we begin with a quaternion division ring construction theory without regard to characteristic and based on the extension of a field F by an irreducible polynomial $f(x)$. In a standard presentation of quaternion division rings, the algebras would be defined and then there would be perhaps problems showing that a matrix representation is available. The approach adopted here is just the opposite, as we begin with the matrix connection. This viewpoint was originated by trying to find derivable nets of various types within a given 3-dimensional projective space $PG(3, K)$, for K a field extension $F(\theta)$ of a field F such that $\theta^2 = \theta\alpha + a$. Here is what may be determined to represent non-commutative division rings with partial spread in $PG(3, K)$, where K is a field: $K = F(\theta)$, a quadratic Galois field extension of a field F and components of the net represented as follows:

$$\left\{ x = 0, y = x \begin{bmatrix} u^\sigma & bt^\sigma \\ t & u \end{bmatrix}; u, t \in F(\theta) \right\},$$

where σ is the automorphism induced by the field extension and maps θ to $-\theta + \alpha$, the second root of $f(x)$. In characteristic 2, we would require $\alpha \neq 0$, but we consider an arbitrary situation and not specify any conditions on α. If $b^\sigma = b$, and the matrices are non-singular, this is a non-commutative skewfield, as the entries in the matrices are additive, there is a zero and a 1. The multiplicativity may be checked by the reader, as well as why the set is non-commutative. We note that the condition for non-singularity is exactly b in $F - \{w^{\sigma+1}; w \in F(\theta)\}$. It might seem that this condition is fairly incidental, but finding special fields that admit this condition is exactly the problem. We verify in the following section that we may get to this representation, just by assuming that we have a non-commutative skewfield as a matrix skewfield within $GL(2, K)$, for K a field.

Here, we simply wish to develop a more uniform representation of quaternion division rings for arbitrary characteristic.

We note that the center $Z(S)$ of this skewfield S clearly is F, which makes this non-commutative skewfield 4-dimensional over $Z(S)$, and it is thus a quaternion division ring (again, check e.g. [**140**]).

We now find a basis for the quaternion division ring as follows: In the following let $u = \theta u_1 + u_2$, $t = \theta t_1 + t_2$, for $u_i, t_i \in F$, for $i = 1, 2$ and consider:

$$\left\{ I_2, \begin{bmatrix} \theta^\sigma & 0 \\ 0 & \theta \end{bmatrix}, \begin{bmatrix} 0 & b\theta^\sigma \\ \theta & 0 \end{bmatrix}, \begin{bmatrix} 0 & b \\ 1 & 0 \end{bmatrix} \right\}.$$

To see that this is a basis, note that

$$\begin{bmatrix} u^\sigma & bt^\sigma \\ t & u \end{bmatrix} = (u_2, u_2)I_2 + (u_1, u_1)\begin{bmatrix} \theta^\sigma & 0 \\ 0 & \theta \end{bmatrix} +$$

$$+ (t_2, t_2)\begin{bmatrix} 0 & b \\ 1 & 0 \end{bmatrix} + (t_1, t_1)\begin{bmatrix} 0 & b\theta^\sigma \\ \theta & 0 \end{bmatrix}.$$

We will now map

$$(u_2, u_2)I_2 + (u_1, u_1)\begin{bmatrix} \theta^\sigma & 0 \\ 0 & \theta \end{bmatrix} + (t_2, t_2)\begin{bmatrix} 0 & b \\ 1 & 0 \end{bmatrix} + (t_1, t_1)\begin{bmatrix} 0 & b\theta^\sigma \\ \theta & 0 \end{bmatrix}$$

to

$$u_2 + u_1 U + t_2 V + t_1 W.$$

Now, we form the squares of the basis elements and work out the inter-relationships between them. Recall that

$$\theta^2 = \theta\alpha + a \text{ and } \theta^\sigma = -\theta + \alpha.$$

We shall use these notations loosely and call

$$U = \begin{bmatrix} \theta^\sigma & 0 \\ 0 & \theta \end{bmatrix}, V = \begin{bmatrix} 0 & b \\ 1 & 0 \end{bmatrix} \text{ and } W = \begin{bmatrix} 0 & b\theta^\sigma \\ \theta & 0 \end{bmatrix}.$$

$$U^2 = \begin{bmatrix} \theta^\sigma & 0 \\ 0 & \theta \end{bmatrix}^2 = \begin{bmatrix} (\theta^\sigma)^2 & 0 \\ 0 & \theta^2 \end{bmatrix} = (\alpha, \alpha)U + (a, a)I_2 = \alpha U + a.$$

$$V^2 = \begin{bmatrix} 0 & b \\ 1 & 0 \end{bmatrix}^2 = (b, b)I_2 = b.$$

$$W^2 = \begin{bmatrix} 0 & b\theta^\sigma \\ \theta & 0 \end{bmatrix}^2 = b\theta^{\sigma+1},$$

$$b\theta^{\sigma+1} = b((-\theta + \alpha)\theta) = b((-(\theta^2) + \alpha)\theta) = b(-(\theta\alpha + a) + \alpha\theta) = -ba.$$

So far, then we have:

$$\begin{aligned} U^2 &= \alpha U + a, \\ V^2 &= b, \\ W^2 &= -ab. \end{aligned}$$

Then inter-relationships:

$$UV = \begin{bmatrix} \theta^\sigma & 0 \\ 0 & \theta \end{bmatrix}\begin{bmatrix} 0 & b \\ 1 & 0 \end{bmatrix} = W$$

$$VU = \begin{bmatrix} 0 & b \\ 1 & 0 \end{bmatrix}\begin{bmatrix} \theta^\sigma & 0 \\ 0 & \theta \end{bmatrix} =$$

$$\begin{bmatrix} 0 & b\theta \\ \theta^\sigma & 0 \end{bmatrix} = (-1, -1)\begin{bmatrix} 0 & b\theta^\sigma \\ \theta & 0 \end{bmatrix} + (\alpha, \alpha)\begin{bmatrix} 0 & b \\ 1 & 0 \end{bmatrix}$$

$$-W + \alpha V = -UV + \alpha W.$$

To see this, just note that $\sigma^2 = 1$, and $\theta^\sigma = -\theta + \alpha$. Hence,

$$VU = -W + \alpha V = -UV + \alpha V.$$

Thus, we have the following uniform definition of a quaternion division algebra/ring (which arises from a set of non-singular matrices).

THEOREM 15. *Every quaternion division algebra Q may be defined independent of characteristic as follows: Let F be a field and let $F(\theta)$ be a Galois quadratic extension of F, with non-identity automorphism σ of order 2 such that $\theta^\sigma = -\theta + \alpha$, with defining irreducible quadratic $x^2 - x\alpha - a$. Then there is a basis:*

$$\{1, U, V, W\}$$

and an element b in F such that

$$U^2 = \alpha U + a,$$
$$V^2 = b,$$
$$W^2 = -ba,$$

and

$$UV = W,$$
$$VW = -W + \alpha V = -UV + \alpha V.$$

NOTATION 5. *Furthermore, Q is a quaternion division ring if and only if*

$$b \in F - (\{w^{\sigma+1}; \forall w \in F(\theta)\} = \{-ax^2 + y^2 + \alpha xy; \forall x, y \in F\}).$$

We shall use the notation $(a, b)_F^\alpha$, for this formulation of quaternion algebras. When $\alpha = 0$, we shall use the standard notation $(a, b)_F$. Later in the text, we shall also use the notation $(a, b)_F^{F(\theta)}$, to denote the quadratic field extension $F(\theta)$ used in the construction of the quaternion division ring.

COROLLARY 2.

(1) When the characteristic is 2, taking $\alpha = 1$, we obtain the normal definition of a quaternion algebra for characteristic 2.

(2) When the characteristic is not 2 and $\alpha = 0$, then we obtain the normal definition of a quaternion algebra for characteristic $\neq 2$.

(3) Also, when the characteristic is not 2, it is possible to find isomorphic representations of the cases when $\alpha \neq 0$ for an irreducible polynomial $x^2 - \alpha x - \beta$, to an irreducible polynomial $x^2 - (\alpha^2 + 4\beta)$.

In part 1 and the beginning of part 2, we discussed the form obtained for any derivable net by the embedding-contraction theory. This will show that any quaternion division ring will admit various derivable nets, a so-called 'twisted regulus net' or type 2 derivable net, which we have noted in part 1 is isomorphic to a regulus-net and an 'irreducible or type 0' derivable net. So, the following will be made clear in parts 2 and 3, but shall be listed here, as it is important to note the variety of derivable nets that sit within any quaternion division ring plane. These particular derivable nets share the components $x = 0, y = 0$. In part 10, we shall see that there are many other derivable nets that are disjoint from $x = 0$ and $y = 0$ and there are sets of mutually disjoint derivable nets that cover the remaining components of the spread.

COROLLARY 3. *Every quaternion division ring* $(a, b)^\alpha_F$ *produces a skewfield translation plane that admits three derivable nets, each of which is isomorphic to a regulus net. Let the matrix representation be given by*

$$\left\{ x = 0, y = x \begin{bmatrix} u^\sigma & bt^\sigma \\ t & u \end{bmatrix} ; u, t \in F(\theta) \right\},$$

where $F(\theta) = K$ *is a Galois quadratic extension of a field* F *and* σ *is the induced involutory automorphism.*

(1) Then the subnet of components

$$\left\{ x = 0, y = x \begin{bmatrix} u^\sigma & 0 \\ 0 & u \end{bmatrix} : \forall u \in K \right\},$$

and the set of Baer subplanes of the following form:

$$\left\{ \begin{array}{c} \pi_{d_1,d_2} = ((d_1, d_2) \begin{bmatrix} u^\sigma & 0 \\ 0 & u \end{bmatrix}, (d_1, d_2) \begin{bmatrix} v^\sigma & 0 \\ 0 & v \end{bmatrix}) : \\ \forall u, v \in K \text{ and } \forall (d_1, d_2) \neq (0, 0) \text{ of } K \oplus K \end{array} \right\}$$

is a derivable net which is not a regulus net in $PG(3, F(\theta))$, *but is a twisted regulus net, and is isomorphic to a regulus net in another 3-dimensional projective space.*

(2) The subnet

$$\left\{ x = 0, y = x \begin{bmatrix} 0 & bt^\sigma \\ t & 0 \end{bmatrix} ; t \in F(\theta) \right\},$$

is also a derivable net isomorphic to a regulus net, which is a twisted regulus net, which we shall designate as a type 2 derivable net relative to the classification results of parts 2 and 3.

(3) The subnet

$$\left\{ x = 0, y = x \begin{bmatrix} u & bt \\ t & u \end{bmatrix} ; u, t \in F \right\},$$

is a derivable net, called an 'open form derivable net'. The set of Baer subplanes have the following form:

$$\left\{ \begin{array}{c} \pi_{d_1,d_2} = ((d_1, d_2) \begin{bmatrix} u & bt \\ t & u \end{bmatrix}, (d_1, d_2) \begin{bmatrix} v & bs \\ s & v \end{bmatrix}) : \\ \forall u, v, t, s \in F \text{ and } \forall (d_1, d_2) \neq (0, 0) \text{ of } F(\theta) \oplus F(\theta) \end{array} \right\}.$$

There are two distinct types:

(a) $F(\sqrt{b})$ *is isomorphic to* $F(\theta)$ *if and only if the derivable net is a type 2 derivable net. For example, this occurs for certain quaternion division rings where* $b = a$ *and* $\theta = \sqrt{a}$.

(b) $F(\sqrt{b})$ *is not isomorphic to* $F(\theta)$ *if and only if the derivable net is of type 0 and* $F(\sqrt{b}, \theta)$ *is a field spread within* $PG(3, F(\theta))$. *This always occurs in the characteristic 2 cases. Here the associated matrix set is a field isomorphic to* $F(\sqrt{b})$ *and* V_4 *is a bimodule over* $F(\sqrt{b})$ *and* $F(\theta)$. *The elements* (d_1, d_2) *are written using a basis over* $F(\sqrt{b})$. *In this setting,* $F(\sqrt{b}, \theta)$ *is not* $(a, b)^\alpha_F$.

The matrix spread corresponding to $F(\sqrt{b}, \theta)$ *is* $\left\{ \begin{bmatrix} u & bt \\ t & u \end{bmatrix} ; u, t \in F(\theta) \right\}$.

(4) The translation planes obtained by derivation of the twisted regulus nets have what is called a 'blended kernel'. The kernel then becomes F, so the derived planes are of dimension 4.

In the type 0 cases of (3) (b), the derived plane is also of dimension 4 but for a different reason; the two fields are non-isomorphic, and the derivable net admits a kernel homology group isomorphic to $F(\sqrt{b})$, fixing each component and would be a regulus net in $PG(3, F(\sqrt{b}))$, and but not a regulus net in $PG(3, F(\theta))$, whereas the set of Baer subspaces will be subspaces over $F(\sqrt{b}) \cap F(\theta) = F$.

The proof of the first assertion is in the previous part 1. By a basis change, the subnet in (2) is isomorphic to the subnet in (1). The subnet of (3) is not obviously a derivable at this point, but this will be clear when the types of derivable nets are developed in parts 2 and 3.

2. Quaternion division ring planes

Now we show where the ideas in the previous section originated. Consider $PG(3, K)$, where K is a skewfield. By the embedding and contraction theory, there is an associated right or left classical pseudo-regulus net in a 4-dimension vector space over K (also there is a right or left vector space, upon choice). In the following result, we set up a situation where we are trying to construct a derivable net coordinatized by a skewfield L with three components common to a classical K-regulus net and both L and K are in $GL(2, K)$. Hence, we have the following situation and assumptions for a derivable net, S^{λ}_{left} :

$$\{x = 0, y = 0, y = xM; \forall M \in \lambda^*\},$$

$$\lambda^* \leqslant GL(2, K)_{left}; components\ are\ left\ spaces.$$

The set of Baer subspaces S^{λ}_{right}

$$S^{\lambda}_{right} = \left\{ S^{\lambda}_{right}(a, b) = \{((a, b)M, (a, b)J); a, b \in K, (a, b) \neq (0, 0)\}, \right.$$

$$\forall M, J \ \in \ \lambda \cup \{0\};$$

Baer subspaces are right spaces over λ with (a, b)

written in the $\lambda \cup \{0\}$ basis.

We are ultimately trying to find derivable nets over non-commutative skewfields, whose associated matrix group is a skewfield L within $GL(2, K)$. The theory in the present part 2, shows that a requirement is that K and L be the same size, dimensionally, over some common subfield. What that means is that we will show that K will need to be an extension field of a Galois quadratic extension $F(\theta)$, which is related to L. In this setting, K and L will be of dimension 4 over F.

We will start with a translation plane with spread in $PG(3, K)$, for K a field, and using only matrix arguments on the matrix spread, and assuming that the plane is a non-commutative skewfield, it will be shown that a quaternion division ring plane appears. The point is not necessarily the conclusion but simply that there is an elementary proof that the quaternion division ring planes are the only non-commutative skewfield planes of dimension 2. The first result establishes this, by use of the previous theorem, realizing the matrix form of a quaternion division ring.

THEOREM 16. *Let π be a non-commutative skewfield plane of dimension 2, with spread in $PG(3, K)$, where K is a field. Then K is a Galois quadratic extension $F(\theta)$ and σ the involutory automorphism.*

Then the spread may be represented in the following form:

$$\left\{ x = 0, \ y = x \begin{bmatrix} u^\sigma & bt^\sigma \\ t & u \end{bmatrix}; \forall t, u \in F(\theta) \right\}$$

$\theta^2 = \theta\alpha + \beta$; *where $\alpha \neq 0$, if the characteristic of K is 2.*

If $\alpha \neq 0$, we may, without loss of generality, let $\alpha = 1$ and let $\beta = a$.

$$\theta^\sigma = -\theta + \alpha \ \text{where}$$

$$b \ \in F - \left\{ w^{\sigma+1}; w \in F(\theta)^* \right\}.$$

Note in any characteristic, b is a non-square. For characteristic 2, the field F is an imperfect field.

Now we establish that an algebraic extension of F of dimension 4 is required to realize that a quaternion division ring can coordinatize a derivable net. We shall see this corollary again when the types of derivable nets are introduced. However, a note that there are two possible derivable nets that are connected is important and is noted in the result.

COROLLARY 4. *Let $K(\tau) = F(\theta, \tau)$ be an extension of F of dimension 4 admitting σ as an automorphism (not necessarily a Galois extension over K).*

Then representing the classical $F(\theta, \tau)$-regulus in $PG(3, F(\theta, \tau))$ as

$$x = 0, y = 0, y = x \begin{bmatrix} w & 0 \\ 0 & w \end{bmatrix}; w \in F(\theta, \tau)$$

we have

$$x = 0, y = x \begin{bmatrix} u^\sigma & bt^\sigma \\ t & u \end{bmatrix}; \forall t, u \in F(\theta)$$

is a derivable net within $PG(3, F(\theta, \tau))$. Let

$$\left\{ \begin{bmatrix} u^\sigma & bt^\sigma \\ t & u \end{bmatrix}; \forall t, u \in F(\theta) \right\} = L,$$

with components as left L-subspaces with Baer subspaces defined as follows:

$B(S^L)_{right}$

$$= \left\{ \begin{array}{l} B(S^L)_{right}(a, b) = \left((a, b) \begin{bmatrix} u^\sigma & bt^\sigma \\ t & u \end{bmatrix}, (a, b) \begin{bmatrix} v^\sigma & bs^\sigma \\ s & v \end{bmatrix} \right) \\ \forall a, b \in F(\theta), \ (a, b) \neq (0, 0); \forall t, u, s, u \in F(\theta). \end{array} \right\}.$$

PROOF. The proof of the theorem: It is assumed that the spread is a non-commutative skewfield spread, configured so that $x = 0, y = 0$ and $y = x$ are components. Then the matrices form a non-commutative skewfield L. Let $Z(L)$ denote the center of L. Then the matrices may be represented as

$$\begin{bmatrix} f(t, u) & g(t, u) \\ t & u \end{bmatrix}; \forall t, u \in K, f \text{ and } g \text{ functions on } K \oplus K,$$

and are additive and multiplicative. Hence, L is a skewfield that has a 2-dimensional matrix representation over a field, and so has dimension 4 over the center $Z(L)$. It follows that then $K = F(\theta)$, a quadratic extension of a field F.

Since the matrix set must be additive, let

$$F(t) = f(t, 0), G(t) = g(t, 0), H(u) = f(0, u), \text{ and } R(u) = g(0, u).$$

Thus, we have a set of matrices

$$\left[\begin{array}{cc} H(u) & R(u) \\ 0 & u \end{array} \right].$$

Now, H and R are additive functions and $H(1) = 1, R(1) = 0$. Note that

$$\left[\begin{array}{cc} H(v) & R(v) \\ 0 & v \end{array} \right] \left[\begin{array}{cc} H(u) & R(u) \\ 0 & u \end{array} \right],$$

is commutative since the 2nd row is $[0, vu] = [0, uv]$. The $(1, 1)$-element is

$$H(v)H(u) = H(vu) = H(uv).$$

So, H is a function that is bijective on K (due to the non-singularity of the elements), additive and multiplicative where $H(0) = 0$ and $H(1) = 1$. Thus, H is an automorphism σ so that $H(u) = u^\sigma$. Hence, we have

$$(*) : v^\sigma R(u) + R(v)u = u^\sigma R(v) + R(u)v; \forall u, v \in K.$$

First assume that $\sigma \neq 1$. We claim that

$$R(u) = (u^\sigma - u)r_1,$$

for r_1 a constant. To see this, there exists an element $v_1 \neq v_1^\sigma$. Letting $R(v_1) = k_1$, we see that

$$(*)' : (v_1^\sigma - v_1)R(u) = (u^\sigma - u)k_1,$$

so that $R(u) = (u^\sigma - u)k_1 c$, for $0 \neq v_1^\sigma - v_1 = c^{-1}$ and hence, we may let $r_1 - k_1 c$. Now change bases by conjugation of

$$\left[\begin{array}{cc} 1 & r_1 \\ 0 & 1 \end{array} \right].$$

$$(**) : \left[\begin{array}{cc} F(t) & G(t) \\ t & 0 \end{array} \right] + \left[\begin{array}{cc} u^\sigma & R(u) \\ 0 & u \end{array} \right] = \left[\begin{array}{cc} F(t) + u^\sigma & G(t) + R(u) \\ t & u \end{array} \right].$$

Thus, L is isomorphic to

$$\left[\begin{array}{cc} 1 & r_1 \\ 0 & 1 \end{array} \right] \left[\begin{array}{cc} F(t) + u^\sigma & G(t) + (u^\sigma - u)r_1 \\ t & u \end{array} \right] \left[\begin{array}{cc} 1 & -r_1 \\ 0 & 1 \end{array} \right],$$

Hence, we now have

$$\left[\begin{array}{cc} 1 & r_1 \\ 0 & 1 \end{array} \right] \left[\begin{array}{cc} u^\sigma & (u^\sigma - u)r_1 \\ 0 & u \end{array} \right] \left[\begin{array}{cc} 1 & -r_1 \\ 0 & 1 \end{array} \right] = \left[\begin{array}{cc} u^\sigma & 0 \\ 0 & u \end{array} \right].$$

And,

$$\left[\begin{array}{cc} 1 & r_1 \\ 0 & 1 \end{array} \right] \left[\begin{array}{cc} F(t) & G(t) \\ t & 0 \end{array} \right] \left[\begin{array}{cc} 1 & -r_1 \\ 0 & 1 \end{array} \right]$$

$$= \left[\begin{array}{cc} F(t) + r_1 t & (F(t) + r_1 t)(-r_1) + G(t) \\ t & -t r_1 \end{array} \right].$$

Now add

$$\begin{bmatrix} (tr_1)^\sigma & 0 \\ 0 & tr_1 \end{bmatrix},$$

to obtain

$$\begin{bmatrix} F(t) + r_1 t + (r_1 t)^\sigma & (F(t) + r_1 t)(-r_1) + G(t) \\ t & 0 \end{bmatrix}.$$

Now let

$$M(t) = F(t) + r_1 t + (r_1 t)^\sigma \ and \ N(t) = (F(t) + r_1 t)(-r_1) + G(t),$$

so we now have a matrix skewfield isomorphic to L as

$$\begin{bmatrix} M(t) + u^\sigma & N(t) \\ t & u \end{bmatrix} ; \forall u, t \in K.$$

Hence,

$$\begin{bmatrix} M(t) & N(t) \\ t & 0 \end{bmatrix} \begin{bmatrix} u^\sigma & 0 \\ 0 & u \end{bmatrix} = \begin{bmatrix} M(t)u^\sigma & N(t)u \\ tu^\sigma & 0 \end{bmatrix} ; \forall u, t \in K.$$

Hence, we must have

$$M(tu^\sigma) = M(t)u^\sigma \text{ and } N(t)u = N(tu^\sigma).$$

Letting $t = 1$, we see that

$$M(u^\sigma) = M(1)u^\sigma, \text{ and } N(1)u = N(u^\sigma).$$

Letting $M(1) = m_1$ and $N(1) = b$, we have the following matrix representation:

$$\begin{bmatrix} m_1 t^\sigma + u^\sigma & bt^{\sigma^{-1}} \\ t & u \end{bmatrix} ; \forall u, t \in K.$$

But, look again at $M(tu^\sigma) = M(t)u^\sigma$, which says $m_1(tu^\sigma)^\sigma = m_1 t^\sigma u^\sigma$. If $m_1 \neq 0$, then $u^{\sigma^2} = u^\sigma$, which shows that $v^\sigma = v$, replacing u^σ by v. This implies that $m_1 = 0$. Now we multiply on the left, as follows:

$$\begin{bmatrix} u^\sigma & 0 \\ 0 & u \end{bmatrix} \begin{bmatrix} 0 & bt^{\sigma^{-1}} \\ t & 0 \end{bmatrix} = \begin{bmatrix} 0 & bt^{\sigma^{-1}} u^\sigma \\ ut & 0 \end{bmatrix} ; \forall u, t \in K,$$

which, in turn, implies that $b(ut)^{\sigma^{-1}} = bt^{\sigma^{-1}} u^\sigma$, and since $b \neq 0$, we have $\sigma = \sigma^{-1}$, so that $\sigma^2 = 1$ and $\sigma \neq 1$. Thus, we would have $\begin{bmatrix} u^\sigma & bt^\sigma \\ t & u \end{bmatrix} ; \forall u, t \in K$. Now, Let F denote the fixed field of σ, since $\sigma^2 = 1$, F is a quadratic subfield of K and $F(\theta) = K$. Let $\theta^2 = \theta\alpha + \beta$. Then $\theta^\sigma = -\theta + \alpha$. If $\alpha = 0$, the assumption would exclude characteristic 2 in this case. In characteristic 2, when $\alpha \neq 0$, let $\alpha = 1$ and let $\beta = a$. So, the final representation of L is $\begin{bmatrix} u^\sigma & bt^\sigma \\ t & u \end{bmatrix} ; \forall u, t \in F(\theta); \theta^2 = \theta + \beta$, and $\theta^\sigma = \theta + 1$, and $b \notin \{w^{\sigma+1}; w \in F(\theta)\}$. It remains to show that $b \in Z(L) = F$. To see this, just consider

$$\left(\begin{bmatrix} 1 & b \\ 1 & 1 \end{bmatrix} \begin{bmatrix} 0 & b \\ 1 & 0 \end{bmatrix} = \begin{bmatrix} b & b \\ 1 & b \end{bmatrix} \right) + \begin{bmatrix} 0 & b \\ 1 & 0 \end{bmatrix} = \begin{bmatrix} b & 0 \\ 0 & b \end{bmatrix},$$

which shows that $b^\sigma = b$. If $\alpha = 0$, then the characteristic is not 2 and everything is still analogous. This is the representation in the statement of the theorem. It remains to consider when $\sigma = 1$. Furthermore, then

$$(***)_{left} \quad : \quad \begin{bmatrix} F(t) & G(t) \\ t & 0 \end{bmatrix} \begin{bmatrix} u & R(u) \\ 0 & u \end{bmatrix}$$

$$= \begin{bmatrix} F(t)u & F(t)R(u) + G(t)u \\ tu & tR(u) \end{bmatrix}.$$

Then

$$(***)_{right} \quad : \quad \begin{bmatrix} u & R(u) \\ 0 & u \end{bmatrix} \begin{bmatrix} F(t) & G(t) \\ t & 0 \end{bmatrix}$$

$$= \begin{bmatrix} F(t)u + R(u)t & uG(t) \\ ut & 0 \end{bmatrix}.$$

Now subtract $(***)_{left}$ from $(***)_{right}$ to obtain: $\begin{bmatrix} R(u)t & -F(t)R(u) \\ 0 & -R(t)u \end{bmatrix}$. This says that $R(u)t = -R(t)u$. Let $t = 1$ to obtain $R(u) = -R(1)u$, let $R(1) = c$. Then, $-cut = -(-ctu)$ so either $c = 0$ or the characteristic is 2. Assume that characteristic is 2 and

$$R(u) = cu; \forall u \in K.$$

But then

$$R(u^2) = uR(u) + R(u)u = 2uR(u) = 0 = cu^2.$$

Hence, $c = 0$.

$$F(ut) = F(t)u \text{ and } G(ut) = uG(t); \forall u, t \in K.$$

So, $F(u) = F(1)u = f_1 u$, and $G(u) = ug_1$, for constants f_1, and $g_1 \neq 0$. Thus, we have the following representation for an isomorphism of

$$L : \begin{bmatrix} u + f_1 t & g_1 t \\ t & u \end{bmatrix}; \forall u, t \in K, \text{ for } K \text{ a field.}$$

However, this is a field, which cannot be isomorphic to L. This completes the proof of the theorem.

The proof of the corollary: To see that a derivable net is obtained as represented shall be given in the more general material on classifying derivable nets over skewfields. The main points are to show that the set of Baer subspaces and the set of components cover each other; are replacements for each other and to note that any two Baer distinct subspaces restricted to $x = 0$, a 2-dimensional $F(\theta, \tau)$-space, are 1-dimensional $F(\theta, \tau)$-spaces and thus generate the vector space $x = 0$ and therefore, are 'Baer subspaces'. This completes the proof of the corollary. □

So, the question becomes, what happens to the quaternion algebra definition if $\alpha \neq 0$, since historically, all of the known quaternion algebras for characteristic not 2, do not require this restriction.

We now discuss quaternion division rings $(a, b)_F^{\alpha \neq 0}$. It is shown that for characteristic $\neq 2$, the normal quaternion algebras $(a, b)_F^{\alpha = 1}$ arising from quadratic extensions of F using non-square elements a, so $F(\sqrt{a})$ can be obtained from quadratic extensions arising from irreducible quadratics of the form $x^2 - x - \delta$, such that $\delta = (a - 1)/4$. Since we now may use the quadratic equation, we see that a root

$\theta = (1 \pm \sqrt{1 + 4\delta})/2 = (1 \pm \sqrt{a})/2$, then shows that the polynomial is irreducible and we may obtain a quadratic extension $F((1 + \sqrt{a})/2$, with automorphism group σ mapping θ to $-\theta + 1 = (1 - \sqrt{a})/2$. Note that $1 + 4\delta$ is nonsquare so a is nonsquare.

Now we look at the matrix version again, to see how to connect $(a, b)_F$, using $F(\sqrt{a})$ and the quaternion division rings $((a - 1)/4, b)_F^{\alpha=1}$, using $F(\theta)$, with polynomial $x^2 - x - \delta$. These two quaternion division rings are isomorphic. Here is the proof: Consider $\begin{bmatrix} u^\sigma & bt^\sigma \\ t & u \end{bmatrix}$, which defines a quaternion algebra over F, $(a, b)_F$, where $t, u \in F(\sqrt{a})$. Write $t = \sqrt{a}t_1 + t_2$ over $F(\sqrt{a})$, and then rewrite the same element t as $(1 + \sqrt{a})/2 \; (2t_1) + t_2 - t_1 = \theta t_1^* + t_2^*$ Similarly, write $u = \sqrt{a}u_1 + u_2$ as $\theta u_1^* + u_2^*$. Note that the automorphism σ mapping $\sqrt{a}$ to $-\sqrt{a}$ will now map θ to $-\theta + 1$, call this σ_θ. Define $s^* = \theta s_1^* + s_2^*$. Thus, what this means, is that by this mapping, we may transform $\begin{bmatrix} u^\sigma & bt^\sigma \\ t & u \end{bmatrix}$ into $\begin{bmatrix} u^{*\sigma_\theta} & bt^{*\sigma_\theta} \\ t^* & u^* \end{bmatrix}$. Hence, for characteristic $\neq 2$, one may use either field $F(\theta)$ or $F(\sqrt{a})$, where $1 + 4\delta = a$ to define the quaternion algebra and quaternion division rings, as this sort of analysis is reversible.

COROLLARY 5. *For quaternion algebras/quaternion division rings of characteristic $\neq 2$, we may use either use the uniform definition or the non-square definition as either method will produce isomorphic algebras.*

In most treatments of the construction of a quaternion division ring, the question of the 'b' is central. In the matrix viewpoint, there is an involutory automorphism σ of a quadratic Galois extension of a field F, $F(\theta)$, where the corresponding irreducible quadratic is $x^2 - \alpha x - \beta$. A quaternion division ring is obtained from a quaternion algebra if and only if $b \in F - \{w^{\sigma+1}; w \in F(\theta)\}$. There are two cases: $\alpha = 0$ and $\alpha \neq 0$. And these lead to the following two sets

$$\{w^{\sigma+1}; w \in F(\theta)\} = \{-\beta r^2 + n^2; \; r, n \in F\}, \text{ when } \alpha = 0 \text{ and}$$
$$= \{-\beta r^2 + n^2 + \alpha rn; \; r, n \in F\}, \text{ when } \alpha \neq 0.$$

Since there are no Galois extensions for non-square quadratic extensions when the characteristic is 2, then there is only the second choice, in this setting. However, what about when the characteristic is not 2? Then note that

$$\{-\beta r^2 + n^2 + \alpha rn; \; r, n \in F\}$$
$$= \left\{-(\alpha^2 + 4\beta)(\frac{r}{2})^2 + (n + \alpha\frac{r}{2})^2; \; r, n \in F\right\},$$

that is, the 'β—' in the first expression is not the 'β' is the second expression! So, this is another way to notice that the choice of b works or does not, for every quadratic extension of F, when the characteristic is not 2, regardless of which type of irreducible quadratic is used. Since we shall be interested in the matrix forms of spreads, we have chosen to show that the spreads in $PG(3, K)$ for K a field, that may be represented as non-commutative skewfield spreads/Desarguesian spreads are quaternion division ring spreads. The fact that non-commutative skewfields of dimension 2 are quaternion division rings is basic theory in the general theory of quaternion division rings (eg. [**140**]).

3. Matrices and determinants over skewfields

In terms of background information regarding non-commutative algebra, the reader will need to be aware that one can treat matrices over skewfields F essentially like matrices over fields with the exception that multiplication should be strictly in order. We shall be considering 2 by 2 matrices over F and the action is xM, for x a 2-vector over F. Let a putative spread be $\{x = 0, y = 0, y = x, y = xM; M \in \lambda\}$, elements of $GL(n, K)_{left}$, for F a left skewfield. To be a spread, given any vector (x, y) where x and y are non-zero, there must be a unique $M \in GL(n, F)$ such that $y = xM$. In this setting, we would also be asking if M has an inverse that acts on the right of M; a 'right inverse'. For 2 by 2 matrices, the question of solving for vectors $(x = (x_1, x_2), y = (y_1, y_2))$ is simple enough so that this can be resolved into a 'determinant question' and that a skewfield determinant may be determined in the non-commutative case, simply by using the set of simultaneous equations that forces a determinant definition by requiring a right inverse. Since we are assuming the action of $(x_1, x_2)M$ is on the left of the matrix elements, we would consider the determinant a 'skewfield determinant' finding conditions for a right inverse'.

DEFINITION 28. *In this setting, the determinant of M is as follows: Let* $\begin{bmatrix} e & f \\ g & h \end{bmatrix}$ *be any 2 by 2 matrix over a left skewfield L. It is noted that we assume that matrix multiplication proceeds left to right. This also assumes that the multiplications of matrix entries are also taken left to right. Define the skewfield determinant δ_L of* $\begin{bmatrix} e & f \\ g & h \end{bmatrix}$ *by $(eg^{-1}h - f)g = eg^{-1}hg - fg$, when g is nonzero and eh, when $g = 0$. When L is a field, this becomes $eh - fg$, just the field determinant. We call this the 'skewfield determinant', as $y = xM$ will become a left subspace over the left scalar group F^*_{left}. When the matrix has a non-zero skewfield determinant, there is a 'right inverse' M_R^{-1} of M, so that an equation such as $(*) : y = xM$ may be solved as $yM_R^{-1} = x$. The reason for choosing this particular determinant is that is possible to solve to $x = (x_1, x_2)M = y = (y_1, y_2)$ uniquely for x in terms of y.*

We note that the same determinant choice possibly would not work for equations $y^t = Nx^t$, producing 'left inverses', as the analogous determinant would be something like $(hg^{-1}e - f)g \neq 0$ to solve for x^t in terms of y^t. So, there is no unique choice for a skewfield determinant to obtain a right inverse or a left inverse of a matrix. Moreover, a matrix M, where the skewfield determinant is non-zero, could produce a transpose M^t, where the skewfield determinant is 0. For example, if $ut^{-1} \neq t^{-1}u$, consider $M = \begin{bmatrix} 1 & ut^{-1} \\ t & u \end{bmatrix}$. Then M is non-singular and has a non-zero skewfield determinant but M^t is singular, in the sense that it has a zero skewfield determinant. So, we choose to work with this form as a determinant, with this caution. Most of the inverses of matrices that we will encounter will have obvious inverses, but the formal inverse (right inverse) is as follows: If δ_L denotes the skewfield determinant, and is non-zero, then the matrix has a right inverse and a *unique solution to the above equation* (∗) *exists. Then the right inverse is as follows:*

$$\begin{bmatrix} e & f \\ g & h \end{bmatrix}^{-1} = \begin{bmatrix} g^{-1}hg\delta_L^{-1} & g^{-1}(1 - hg\delta_L^{-1}eg^{-1}) \\ -g\delta_L^{-1} & g\delta_L^{-1}eg^{-1} \end{bmatrix}$$

$$= \begin{bmatrix} g^{-1}h(eg^{-1}h - f)^{-1} & g^{-1}(1 - h(eg^{-1}h - f)^{-1}eg^{-1}) \\ -(eg^{-1}h - f)^{-1} & (eg^{-1}h - f)^{-1}eg^{-1} \end{bmatrix},$$

for g not zero,

$$= \quad for \ a \ field =$$

$$\begin{bmatrix} h\delta_L^{-1} & g^{-1}(1 - hg\delta_L^{-1}eg^{-1}) = -f\delta_L \\ -g\delta_L^{-1} & \delta_L^{-1}e = e\delta_L^{-1} \end{bmatrix},$$

where now $\delta_L = \delta = eh - fg$

$$\begin{bmatrix} e & f \\ 0 & h \end{bmatrix}^{-1} = \begin{bmatrix} h\delta_L^{-1} & -e^{-1}f\delta_L^{-1}e \\ 0 & \delta_L^{-1}e \end{bmatrix}$$

$$= \quad and, \ if \ L \ is \ a \ field =$$

$$= \begin{bmatrix} e^{-1} & -e^{-1}fh^{-1} \\ 0 & h^{-1} \end{bmatrix} = \begin{bmatrix} h\delta^{-1} & -f\delta^{-1} \\ 0 & \delta^{-1}e = e\delta^{-1} \end{bmatrix},$$

for $\delta = eh$.

REMARK 6. *For n by n matrices over skewfields, the general determinants can be determined as in Dieudonnè* [**44**], *which explicitly provides a 'skewfield determinant' in the 2 by 2 case. For the work in the text, we are mostly interested in the 2-dimensional case, as this is what drives the ideas of derivable nets. On the other hand, we shall also be interested in subplane covered nets which work with cyclic division rings, and this opens up the n-dimensional cases over skewfields. Therefore, we might wish to consider more general left and right determinants and corresponding right and left inverses. Note that with derivable nets we work with left vector spaces over skewfields, but the analysis could be equally considered over right vector spaces over skewfields, and therefore right determinants (note, could use the same determinant form) and left inverses would then need to be considered.*

All of this may be considered more generally. In particular, in Draxl [**48**] *chapters 19, 20 are devoted to the Bruhat normal form and the Dieudonnè determinant. For a set of n by n matrices over a skewfield F, to form a spread, under standard conditions, we require the matrices and their differences to be non-singular. We have seen the left and right determinants are obtained depending if one is writing the spread as a set of left subspaces or a set of right subspaces. Furthermore, the sets of rows/columns of matrix spread sets must be left or right linearly independent. With that in mind, we list the following theorem:*

Theorem 3 p. 131 [**48**]. *Let* $M_n(D)$ *be the set of n by n matrices over a skewfield D. Define* $GL_n(D)$ *as the set of invertible matrices of* $M_n(D)$. *Then the following are equivalent:*

(a) $A \in GL_n(D)$;

(b) $AB = 1$ *for some* $B \in GL_n(D)$; *A has a right inverse;*

(c) $CA = 1$ *for some* $C \in GL_n(D)$; *A has a left inverse;*

(d) the rows of A are left linearly independent over D;

(e) the columns of A are right linearly independent over D.

PROBLEM 9. *Given* $\begin{bmatrix} e & f \\ 1 & h \end{bmatrix}$, *find the skewfield determinant required to solve* $y = x \begin{bmatrix} e & f \\ 1 & h \end{bmatrix}$ *by using simultaneous equations, and find the right inverse, without using the above definition.*

The determinants and inverses that shall be used in this text are either 2 by 2 matrices over skewfields, as in the Lifting skewfields of part 8 or actually are field determinants and inverses, as in the field matrices describing quaternion division rings and cyclic division rings. Due to lack of space, we cannot present a general theory of spreads of dimension n over skewfields, but the reader is directed to Draxl [48] for more information. However, there are ideas in that direction in part 12.

4. Classifying derivable nets

In the author's text on derivable nets, [86], it is shown that every derivable net may be embedded into a 3-dimensional projective space over a skewfield $PG(3, K)_C$, where K is the corresponding skewfield. The notation indicates this is a combinatorial embedding.

The points and lines of the derivable net are, respectively, the lines and points of $PG(3, K)_C$ not incident with, or not equal to, a fixed line N. The Baer subplanes and parallel classes are, respectively, the planes not containing, and containing N. Using this characterization, the full collineation group of a derivable net is $P\Gamma L(4, K)_N$, the stabilizer subgroup leaving N invariant. There is a contraction/retraction method to a 4-dimensional vector space over K, where the derivable net may be presented in the form of what we are calling a classical pseudo-regulus net.

Actually, derivable nets may be constructed from any skewfield by this contraction method. The method works just as well for the construction of subplane covered nets. Choose any skewfield K, choose any projective space $PG(|V| - 1, K)$ of dimension ≥ 3 over K, where the notation just indicates the space of subspaces of an associated vector space V. Let N be any co-dimension 2 projective subspace and form the following incidence structure: Consider the set of points which are the points of $PG(|V| - 1, K) - N$, the set of lines are those not contained in N, and the co-dimension 1 subspaces are of two types, those that do not contain N and those that do contain N. Then, the contraction, constructs a subplane covered net within an associated vector space. In the finite-dimensional case, the projective space is $PG(n + 1, K)$; this construction produces an associated vector space of dimension $2n$ over F, and the subplane covered net has the pseudo-regulus form (extending the definitions of regulus, and pseudo-regulus accordingly). This points out that the construction technique of retraction is completely distinct from simply returning to the normal associated affine space of dimension $n + 2$.

When F is finite and the projective space is isomorphic to $PG(3, GF(q))$, the classification of subplane covered nets was given by De Clerck and Johnson [39], when it was realized that finite dual nets satisfying the axiom of Pasch are or correspond to subplane covered nets. Thus, the work of De Clerck and Thas [41] and of Thas and De Clerck [138], combined with the author's embedding of derivable nets into 3-dimensional projective spaces ([**75**], [**80**]) becomes the framework for the more general work over arbitrary subplane covered nets.

Finite derivable nets and affine or projective planes that contain them as well as a variety of interconnected geometries to such planes are of considerable interest in finite geometry. It will be shown that the more general theory of subplane covered nets also has considerable application.

The possible extensions to analogous infinite geometries are also of interest and these extensions that are connected to derivable nets are invariably those associated with an infinite field.

In this text, we give a classification of the types of derivable nets that are possible over arbitrary skewfields. The classification will provide certain techniques that will enable the theory of derivation to extend into infinite geometries based on non-commutative skewfields.

A major application of this work will be devoted to the study of quaternion division ring planes. Although, it is known that many quaternion division rings have what is called a 'central quadratic extension', we shall show that there is a natural geometric way, using derivable nets, to obtain this result. This concept will extend to cyclic division algebras, where the geometry becomes a subplane covered net.

We also note that realizing quaternion division rings in their matrix form has shown that all such skewfields are derivable and recognized as corresponding to flocks of σ-cones, where σ is an automorphism of order 2. Furthermore, with a slight change in the spread set, quaternion division rings also produce twisted hyperbolic flocks. Additionally, skewfields that admit central quadratic extensions are extraordinary in at least two ways. First, these skewfield planes may be multiply derived in that there is a set of mutually disjoint pseudo-regulus nets that together with two components form the spread. Also, the lifting procedure for 2-dimensional spreads defined over fields K has been shown to apply to every skewfield F that has a central Galois quadratic extension. Such skewfields may be 'lifted' to a derivable semifield field spread in $PG(3, F)$ by Johnson and Jha [**93**]. The lifting process originated in Hiramine, Matsumoto, and Oyama [**62**], for finite translation planes with spreads in $PG(3, GF(q))$.

In order to classify the general derivable nets over skewfields, we study how the collineation subgroups of $P\Gamma L(4, F)_N$ work in practice, when considering derivable translation planes over non-commutative 3-dimensional projective spaces.

Although the work is essentially fundamental, it provides the tools to establish a complete extension of the seminal work of Andrè on mutually disjoint sets of finite derivable nets in finite Desarguesian planes over $GF(q^2)$, to any skewfield that admits a central Galois quadratic extension.

The work of Andrè, in turn, inspired Ostrom to consider analogous work for finite Desarguesian planes over $GF(q^n)$, where a derivable net is replaced by what Ostrom called a 'hyper regulus net', which bears certain similarities to a derivable net, but without the uniqueness of replacement nets. Since all such finite extensions are cyclic Galois extensions, the question is whether there are cyclic central Galois extensions over skewfields, as the work on multiple derivation on central quadratic extensions of skewfields extends naturally to cyclic central extensions. Of course, there are such extensions of fields.

To distinguish between a derivable net embedded into a 3-dimensional projective space over a field, a regulus net, and the case when the projective space is over a non-commutative skewfield, we shall use the term 'pseudo-regulus net' in the latter case. There is a left/right method of working with pseudo-regulus nets and the planes that contain them, which is given by our study of derivable nets, in general.

Here we give the foundation for these works by establishing a classification of the types of derivable nets that are possible.

We shall provide a classification of subplane covered nets of finite dimension n in part 11 of the book.

CHAPTER 4

Classification theory over skewfields

In this section, we restate the main embedding-retraction results from part 1, in a lengthy form so as to begin to set the notation, which is crucial for an accurate reading of this theory.

THEOREM 17. *Let D be any derivable net. Let D be embedded into the 3-dimensional projective space $PG(3,F)_C$, for a skewfield F. Using the embedding, the full collineation group of D is $P\Gamma L(4,F)_N$, where N is a fixed line of the projective space. Then there is a contraction to a 4-dimensional left (or right) vector space V over F, such that the derivable net D has the following form, which shall be called the right 'classical pseudo-regulus' net over F. In this form, D shall be denoted by R^F_{right} and the set of Baer subspaces will be denoted by $B(R)^F_{left}$. This notation is intended to be a reminder that the components are right 2-dimensional F-subspaces, where the right scalar mappings are denoted by F^*_{right}. Similarly, the Baer subspaces are left 2-dimensional F-subspaces, where the left scalar mappings are denoted by F^*_{left}.*

Since we are considering the derivable nets within a 4-dimensional F-vector space, we then denote points/vectors by (x, y), where x and y are 2-vectors for most spread representations. It is emphasized that all of the derivable nets share three common components $x = 0$, $y = 0$, and $y = x$, all of which are 2-dimensional F-bimodules. Thus, each of the partial spreads will contain

$$\{x = 0, \ y = 0, \ y = x\}.$$

Our standard notation for a right classical pseudo-regulus net shall be:

$$R^F_{right} \quad : \quad \{x = 0, \ y = \delta x; \forall \delta \in F\}; \ components \ are \ right \ spaces;$$

$$B(R)^F_{left} \quad = \quad \left\{ \begin{array}{c} B(R)^F_{left}(a,b) = \{(ca, cb, da, db); \forall c, d \in F\} : \\ \forall (a,b) \neq (0,0) \ in \ F \oplus F \end{array} \right\};$$

Baer subspaces are left spaces.

Notice that since 0 and 1 are elements of F, the set of common components is always contained in the partial spread, as represented. Note that the Baer subspaces in

$$\{B(R)^F_{right}(1,0), \ B(R)^F_{right}(0,1), \ B(R)^F_{right}(1,1)\}$$

are also 2-dimensional F-modules.

Similarly, our standard notation for a left pseudo-regulus net shall be:

$$R^F_{left} \quad : \quad \{x = 0, \ y = x\delta; \forall \delta \in F\}; \ components \ are \ left \ spaces;$$

$$B(R)^F_{right} \quad = \quad \left\{ \begin{array}{c} B(R)^F_{right}(a,b) = \{(ac, bc, ad, bd); \forall c, d \in F\} : \\ \forall (a,b) \neq (0,0) \ in \ F \oplus F \end{array} \right\};$$

Baer subspaces are right spaces.

If the vector space is considered as a right F-vector space, the right and left classical pseudo-reguli would naturally be reversed. So, the classical pseudo-regulus is left or right exactly when the components are left or right subspaces, respectively. Note again that the Baer subspaces in

$$\left\{ B(R)^F_{right}(1,0), B(R)^F_{right}(0,1), B(R)^F_{right}(1,1) \right\}$$

are right and left 2-dimensional F-subspaces as are $x = 0$, $y = 0$, $y = x$. Now we wish to formally consider the derivation process. Recall that the derived net of the right classical pseudo-regulus net simply interchanges what are called the components and Baer subspaces of the derivable net. Since we are dealing with non-commutative coordinate systems, what actually occurs is that a right classical pseudo-regulus net when derived becomes a left classical pseudo-regulus net.

So, we shall carefully show how a right classical pseudo-regulus net 'derives' to a left classical pseudo-regulus. We shall use 'new' and 'old', for simplicity. When the net is derived, the 'new components' are the old Baer subspaces and the 'new Baer subspaces' are the old components. Note that since we are in a vector space version of a derivable net, the other lines and Baer subplanes are translates of the components and Baer subspaces, respectively. So, it suffices to consider the interchange of components and Baer subspaces. At this point, the coordinate system used for the new components is not a skewfield, so we need to choose coordinates so that we have three new components that can be represented in the forms of $x = 0$, $y = 0$, and $y = x$, with their intended understanding, as above. Since we have three old Baer subspaces that are F-bimodules and three old components that are F-bimodules, we agree to continue this representation. Thus, to see that the derived net has the left classical pseudo-regulus form upon derivation, a basis change must be made that maps the ordered sets as follows:

$$\left\{ B(R)^F_{left}(1,0), B(R)^F_{left}(0,1), B(R)^F_{lft}(1,1) \right\} \to \{x = 0, y = 0, y = x\}$$

and

$$\{x = 0, y = 0, y = x\} \to \left\{ B(R)^F_{right}(1,0), B(R)^F_{right}(0,1), B(R)^F_{right}(1,1) \right\},$$

where the first set indicated by $\{x = 0, y = 0, y = x\}$ refers to the common subspaces of the left classical pseudo-regulus net and the second set indicated by $\{x = 0, y = 0, y = x\}$ refers to the common subspaces of the right classical pseudo-regulus net. Note, these subspaces represented in these two sets are not equal! This note will become important when we give our definitions below. In the vector space, if a vector $(x_1, x_2, x_3, x_4) \to^\rho (x_1, x_3, x_2, x_4)$, then each of these mappings above are realized. It is then required to verify that the remaining new components have the left form (they need to be represented as $y = xe$ for all $e \in F$. Consider then any new component (old left Baer subspace) $B^F_{Left}(a,b) = \{(ca, cb, da, db); \forall c, d \in F\}$ where now $a \neq 0$. Using the mapping ρ, the old left Baer subspace has the form

$$\{(ca, da, cb, db); \forall c, d \in F\} = \{(ca, da, ca(a^{-1}b), da(a^{-1}b)); \forall c, d \in F\}$$

Note that the abstract form of this new component is $y = x(a^{-1}b)$. Letting $a^{-1}b = e$, it can now be seen that the new components have the correct form. A similar argument shows that the new Baer subspaces (now right spaces) also have the correct form. Hence, we see that a right classical pseudo-regulus net derives to a left classical pseudo-regulus nets and conversely.

In the following, we shall compare left derivable nets with right classical pseudo-regulus nets.

DEFINITION 29. *Assume that R_{right}^F is a right classical pseudo-regulus net over a skewfield F. Then a non-singular matrix written on the right is a left-2-dimensional vector space over F, so lies in $GL(2, F)_{left}$. We note the components of the right classical regulus over F are right 2-dimensional subspaces over F. This tension between right and left F-subspaces is intentional. The Baer subspaces of a derivable net are the Baer subplanes incident with the zero vector. Now assume that we have a left derivable net S_{left}^K; the components are left subspaces:*

$$\{x = 0, \ y = 0, \ y = xM, \}; \ for \ M \ in \ a \ subset \ \lambda \ of \ GF(2, F)_{left}.$$

Then λ union $\{0\}$ is a skewfield K. And, for a derivable net, it must be that case that V_4 is a right/left vector space over both F and K. The idea is that we test the right (respectively, left) vector space over F against the left (respectively, right) vector space over K. The set of Baer subspaces $B(S)_{right}^K$ of the left derivable net is

$$B(S)_{right}^K = \left\{ \begin{array}{c} B(S)_{right}^K(a, b) = \{((a, b)M, (a, b)N) : \\ \forall (a, b) \neq (0, 0) \ in \ F \oplus F \end{array} \right\}; \forall M, N \in K\},$$

where (a, b) is written in the basis over K.

Note that we are assuming that the two derivable nets share three common components. This is never the case for two derivable nets that might lie in a translation plane. Since the contraction method will always show the derivable net sharing the three common components, we begin by including this condition, in order to determine the possible types of nets. The more common situation will be discussed briefly in the following sections, and when discussing carrier derivable nets (3,5,12).

The following definition is fairly complicated. There is an important connection between left and right scalar mapping groups that is essential to understanding what we are trying to accomplish in this section. So, before we give the formal definition of the classification scheme, we provide some discussion on notation. We first give a detailed list of the notation that we shall use consistently for the book.

1. Notation

NOTATION 6. *Notation for the 'classical derivable net' and the associated 'suspect derivable net'.*

(1) We use R_{right}^F (pseudo-regulus) to indicate both a right classical pseudo-regulus net and the components as right 2-dimensional F-subspaces. We use S_{left}^K (suspect derivable net) to indicate both a left derivable net and the components which are left 2-dimensional F-subspaces are within the same 4-dimensional vector space V_4 over F.

Notation for the set of Baer subspaces of the classical net and the associated suspect derivable net:

(2) We use $B(R)_{lefr}^F$ to denote the set of all Baer subspaces $B(R)_{left}^F(a, b)$ for all $(a, b) \neq (0, 0)$, all of which are left 2-dimensional F-subspaces within the right classical pseudo-regulus net R_{right}^F. And, we shall use $B(S)_{right}^K$ to denote the set

of right 2-dimensional K-subspaces within the left suspect derivable net S^K_{left}, where (a, b) is written over the K-basis.

(3) We define the 'left F-scalar mappings' as
$$F^*_{left} = \{(x, y) \rightarrow (ex, ey); e \in F^*\},$$
and the 'right F-scalar mappings' are defined by
$$F^*_{right} = \{(x, y) \rightarrow (xf, yf); e \in F^*\},$$
where x and y are 2-vectors over F. To indicate that a subspace is a left or right F-subspace simply means that the subspace is invariant under F^*_{left} or F^*_{right}, respectively. The left scalar group leaves invariant each component of the form $y = xM$, as $(x, xM) \rightarrow (ex, e(xM) = (ex)M)$. The right scalar group leaves invariant each component of the right classical pseudo-regulus net $y = dx$, as $(x, dx) \rightarrow (xf, (dx)f = d(xf))$ (a complete proof of the linearity of these mappings is given in (3,5,1) on type 0 derivable nets (See below for the definition of types).

(4) Then, the left scalar group F^*_{left} leaves invariant each Baer subspace of the right classical pseudo-regulus net. To see this formally, note that
$$B(R)^F_{left}(a, b) = \{(ca, cb, da, db); \forall c, d \in F\},$$
and
$$(ca, cb, da, db) \rightarrow (e(ca, cb), e(da, db)).$$
Noting that $ex = e(x_1, x_2)$, for $x_1, x_2 \in F$, we obtain $e(x_1, x_2) = (ex_1, ex_2)$, and
$$(e(ca, cb), e(da, db)) = ((ec)(a, b), (ed)(a, b)) = (eca, ecb, eda, edb),$$
so $B(R)^F_{left}(a, b)$ is invariant under F^*_{left} and hence is a 2-dimensional left F-subspace.

(5) For the left derivable nets S^K_{left}, we have the same/analogous left and right groups K^*_{left} and K^*_{right}, acting on the components and Baer subspaces of $B(S)^K_{right}(a, b)$. However, although we theoretically understand that there needs to be a known action on the matrix components in $GL(2, F)_{left}$ by K^*_{left}, this cannot be given a priori. This is only known specifically, if the suspect derivable net does actually lie in the associated 4-dimensional vector space over F. However, we do know the action of K^*_{right} on the set of Baer right subspaces $B(S)^K_{right}$ as (a, b) is written over the K-basis.

(6) **Important Note:** Since F^*_{left} leaves each of the components of S^K_{left} invariant and since there is a unique set of Baer subspaces that covers the component set, it follows that F^*_{left} must permute the set of right subspaces $B(S)^K_{right}$; we have no idea as to the permutation group action. This point on Baer subplanes is exactly what we wish to consider as a classification theme. Subplane covered nets in $PG(n + 1, F)$ have a similar form and the permutation of the little subplanes (subspaces) by F^*_{left} occurs as a natural consequence of V_{2n} acting as an (F, K)-bimodule.

In the formal proof, we shall show/note that both of the left groups F^*_{left} and K^*_{left} commute with both of the right groups F^*_{right} and K^*_{right}. Furthermore, all of these groups, by definition, leave invariant each of the common subspaces in

$\{x = 0, y = 0, y = x\}$. We shall use the notation *Com* to denote any of these subspaces. Since we know the action of K^*_{right} on the set of Baer subspaces $B(S)^K_{right}$, as well as the knowledge that F^*_{left} does permute the same set of Baer subspaces $B(S)^K_{right}$, with the understanding that K^*_{right} also permutes the set $B(R)^F_{left}$ of Baer subspaces of R^F_{right}, we see that there a symmetry in play as we have a pair of commuting scalar groups, left and right, that act faithfully on the sets of intersection subspaces on any 2-dimensional left/right F-subspace *Com*.

The reader will need to become very familiar with the left/right notation, as this is necessary for an understanding of the material and the proofs. Armed with the extensive notational preparation, we may define our classification scheme.

When F is a field and K is a field, we may dispense with the right/left notation, and use R^F and $B(R)^F$, for the components and Baer subspaces of the classical regulus net and S^K and $B(S)^K$ for the components and Baer subspaces, respectively, of the suspect derivable net.

DEFINITION 30. *With regard to the set of Baer subspaces $B(S)^K_{right}$, we define the net S^K_{left} to be of 'type i', if and only if the two derivable nets R^F_{right} and S^K_{left} share exactly i Baer subspaces, $i = 0, 1, 2$. If the two derivable nets share at least three Baer subspaces, we shall use the term 'type 3',*

Another wrinkle of this definition is that a left derived classical pseudo-regulus net $\overline{R^F_{right}} = R^F_{left}$ of a right classical pseudo-regulus net R^F_{right} becomes a left pseudo-regulus net upon a basis change. We have mentioned this idea above. So, the derived net of the right classical pseudo-regulus net is the left classical pseudo-regulus net and shares at least three Baer subspaces with the original net. Even though this may seem to be unusual, we may then classify the derived net with respect to our classification scheme to be of type 3. Let $Z(F)$ denote the center of F. In this setting, the cardinality of common Baer subspaces is the $\text{card}Z(F) + 1$, which could be exactly 3, if $\text{card}Z(F) = 2$, where $Z(F) = GF(2)$.

DEFINITION 31. *If two left derivable nets $S^K_{1,left}$ and $S^K_{2,left}$ (suspects #1 and #2) are of the same type $i = 1, 2$ or 3 with respect to a right classical pseudo-regulus net R^F_{right} then the two nets each share i Baer subspaces with R^F_{right}. However, these two sets of shared Baer subspaces may not be the same. Using the collineation group $\Gamma L(4, F)_N$ of R^F_{right}, we shall show that there is a collineation in $\Gamma L(4, F)_N$ that fixes $x = 0, y = 0$ and $y = x$ and maps the shared subspaces of one left derivable net $S^K_{1,left}$ of type i to the shared subspaces of the second left derivable net $S^K_{2,left}$ of type i. When the shared subspaces of one of these left derivable nets are contained in*

$$\left\{ B(S)^K_{right}(0, 1), \ B(S)^K_{right}(1, 0), \ B(S)^K_{right}(1, 1) \right\},$$

and therefore would be also be contained in

$$\left\{ B(R)^F_{left}(0, 1), \ B(R)^F_{left}(1, 0), \ B(R)^F_{left}(1, 1) \right\},$$

*with **order** observed, in both sets, then that derivable net is said to be 'generic' of type i.*

To be clear, a generic type 1 derivable net shares

$$B(S)^K_{right}(0, 1) = B(R)^F_{left}(0, 1)$$

and a generic type 2 derivable net shares

$$\left\{ B(S)_{right}^{K}(0,1),\ B(S)_{right}^{K}(1,0) \right\} = \left\{ B(R)_{left}^{F}(0,1),\ B(R)_{left}^{F}(1,0) \right\}.$$

Generic derivable nets of type i are convenient since if such a derivable net of type i is contained in a translation plane with spread in $PG(3,F)_S$, (the projective space of vector subspaces), there is an isomorphic translation plane containing a generic derivable net of type i.

It is apparent that the type 0 derivable nets are very different from the other types. The reason for this is that if the spread is written as a left spread, then all component matrices are in $GL(2,F)_{left}$, which is another way of saying that the translation plane is in $PG(3,F)_S$. Then the mapping function leaves the right classical pseudo-regulus net invariant and hence leaves invariant $PG(3,F)_S$; therefore, the derivable net can be assumed to be generic in translation planes.

2. Extension of skewfields theorem/Skewfield bimodules

For convenience of reading, the classification will be stated in two parts; the first part shall discuss all of the types of derivable nets and isolate on the types $1, 2, 3$. The second part will deal exclusively with derivable nets of type 0 and the **extension of skewfields theorem.**

It is important to point out that all Baer subspaces need to be written relative to the basis for the particular skewfield indicated by the components. Here are some important instances of this. Suppose that we have a field $K(\theta)$, a Galois extension of a field K. We write the classical regulus net $PG(3,K(\theta))$ obtained from the embedding/contraction theory. Then realizing that $K(\theta)$ has a so-called open form representation as

$$\left\{ \begin{bmatrix} u + \alpha t & t\beta \\ t & u \end{bmatrix} : t, u \in K \right\} \text{ and where } x^2 - \alpha x - \beta \text{ is irreducible over } K,$$

then in $PG(3, K(\theta))$ we have a derivable net where the set components is

$$Compontents : \left\{ x = 0, y = x \begin{bmatrix} u + \alpha t & t\beta \\ t & u \end{bmatrix} : u, t \in K \right\}$$

and the set of Baer subspaces, for each (e, f) written over the associated matrix in the required basis, is

$$Baer\ subspaces$$
$$= \left\{ B(e,f) = ((e,f) \begin{bmatrix} u + \alpha t & t\beta \\ t & u \end{bmatrix}, (e,f) \begin{bmatrix} v + \alpha s & s\beta \\ s & v \end{bmatrix}) \right\},$$
$$\forall u, t, v, s \ \in \ K.$$

LEMMA 3. *We claim that a basis for the open form representation is*
$$\{(0,1,0,0),(0,\theta,0,0),(0,0,0,1),(0,0,0,\theta)\}.$$

PROOF. It suffices to show that $\{(0,1),(0,\theta)\}$ is a basis for $y = 0$ (or $x = 0$). So consider the set

$$(0,1)\begin{bmatrix} u+\alpha t & t\beta \\ t & u \end{bmatrix} + (0,\theta)\begin{bmatrix} v+\alpha s & s\beta \\ s & v \end{bmatrix})$$

$$= (t,u) + (\theta s, \theta v) = (t+\theta s, u+\theta v)$$

It is an easy exercise to see that we have bijectivity and linear independence. $\qquad\square$

COROLLARY 6. *Any derivable net containing* $\{x = 0, y = 0, y = x\}$ *as a subset of components, with open form*

$$\left\{ x=0, y=x\begin{bmatrix} f(t,u) & g(t,u) \\ t & u \end{bmatrix}; \forall t, u \in K \right\} \quad :$$

$$f, g \text{ functions on } K \oplus K \;\to\; K$$

in $PG(3, K(\theta))$ *has a basis as*

$$\{(0,1,0,0),(0,\theta,0,0),(0,0,0,1),(0,0,0,\theta)\}.$$

COROLLARY 7. *We have discussed quaternion division rings* $(a,b)_K$, *where their matrix forms are in* $PG(3, K(\theta, \tau))$, *where* $K(\theta)$ *is a Galois quadratic extension of* K *with induced automorphism* σ *and* $K(\tau)$ *is a quadratic extension (necessarily Galois if the characteristic is* $\neq 2$). *These matrix nets can be shown to be derivable nets as open matrix forms*

$$x = 0, y = x\begin{bmatrix} u^\sigma & bt^\sigma \\ t & u \end{bmatrix} : \text{for all } t, u \in K(\theta).$$

A basis over the skewfield is

$$\{(0,1,0,0),(0,\tau,0,0),(0,0,0,1),(0,0,0,\tau)\}.$$

COROLLARY 8. *In* $PG(3, K(\theta, \tau))$, *there could be fields* $K(\theta, \rho)$ *that are non-isomorphic to* $K(\theta, \tau)$ *but still the open form matrices will generate a derivable net that is relative to the classical* $K(\theta, \tau)$-*regulus net and also relative to the classical* $K(\theta, \rho)$-*regulus net with appropriate bases; one is of type* 0 *and one is of type* 2 .

3. Preliminary types $1, 2, 3$

THEOREM 18. **Part I; The types of derivable nets.** *A left derivable net* S_{left}^K *within the affine space of a right classical pseudo-regulus net* R_{right}^F *is one of four types* i, *for* $i = 0,1,2,3$. *Where the type* i *indicates, for* $i = 0,1,2$, *the number of shared Baer subspaces and, for* $i = 3$, *when there are at least 3 shared Baer subspaces. The ambient 4-dimensional vector space* V_4 *over* F *is also a 4-dimensional vector space over* K.

F_{left}^* *and* K_{right}^* *act on any Com 2-dimensional-*(F, K)-*bimodule.*

Furthermore, the right 1-dimensional K_{right}-*subspaces are the Com-intersections with the set* $B(S)_{right}^K(a,b)$ *of right Baer subspaces and the left 1-dimensional* F_{left}-*subspaces are the Com-intersections with the set of* $B(R)_{left}^F(a,b)$ *Baer subspaces.*

The skewfield mappings F_{left}^* *and* K_{right}^* *are linear transformation groups on the opposite 2-dimensional vector space.*

Both groups are reducible if and only if one group is reducible if and only if the type is $1, 2,$ *or* 3.

Both groups are irreducible if and only if one group is irreducible if and only if the type is 0.

(i) *When $i = 1$, the left derivable net is either a derivable net with a derivative function (a derivation) or is a derivable net with a twisted derivative function (a τ-derivation for the automorphism τ).*

(ii) *When $i = 2$, the left derivable net is a twisted pseudo-regulus net. In particular, the open form of any right classical pseudo-regulus net coordinatized by a central Galois quadratic extension is always of type 2.*

(iii) *When $i = 3$, the left derivable net is a semi-classical pseudo-regulus net. When the right classical pseudo-regulus net is a field net, the left and right nets are each classical regulus nets of type 3.*

We shall postpone the full statement of Part II on derivable nets of type 0 until we state and discuss the standard framework result.

4. Standard framework

THEOREM 19. **Standard framework theorem.** *Let S_{left}^K be a left derivable net within the 4-dimensional vector space V_4 over F that realizes the right classical pseudo-regulus net R_{right}^F. Then the scalar groups F_{left}^* and K_{right}^* act on the corresponding sets of Baer subspaces $B(S)_{right}^K$ and $B(R)_{left}^F$, respectively. Each of these scalar groups fix any Com 2-dimensional (F_{left}, K_{right})–bimodule and the sets of 1-dimensional K_{right}-subspaces of $B(S)_{right}^K(a,b) \cap Com$ and 1-dimensional F_{left}-subspaces of $B(R)_{left}^F(a,b) \cap Com$ as permuted by F_{left}^* and K_{right}^* are permutation isomorphic to the permutation groups acting on the corresponding sets of Baer subspaces.*

The skewfield mappings F_{left}^* and K_{right}^* are linear transformation groups on the opposite 2-dimensional vector space. Both groups are reducible if and only one group is reducible if and only if the type is 1, 2, or 3. Both groups are irreducible if and only if one group is irreducible if and only if the type is 0.

PROOF. Most of the foundational theorems have been proven with the exception of permutation aspects of the two group actions on type 0 derivable nets, which is covered more completely in (3,5,1). The groups are faithful on *Com* and we know that there is a unique covering set of Baer subspaces for any derivable net. All of the other parts of this theorem are consequences of the material proved in the following extension of skewfields theorem.   □

5. Generalized quaternions over skewfields

As a preliminary explication of the extension theorem, we shall provide some elements of non-commutative algebra and a discussion of what we are calling, in this text, 'generalized quaternion division rings'.

We have previously mentioned that there are matrix methods available to use for quaternion division rings of arbitrary characteristic. The following generalization is straightforward and basically defines what might be called a 'generalized quaternion division ring $\mathcal{G}(a,b)_F$', where F is an arbitrary non-commutative skewfield.

The following notation is used to denote the quaternion division rings $(a,b)_F$, (or $(a,b)_F^\alpha$, when an irreducible quadratic is an issue, as in characteristic 2) where

F is a field. In this section, we extend these constructions to the case when F is an arbitrary skewfield. Our treatment shall require that the elements a and b are in $Z(F)$. The division rings over F are 4-dimensional vector spaces over F and the centers of the generalized structures are all equal to $Z(F)$.

First we review the basics on quaternion division rings but using a skewfield instead of a field.

Let $V = G(a,b)_F = F + Fu + Fv + Fw$, where F is a skewfield, u^2, v^2 are nonzero elements in F, such that $u^2 = a, v^2 = b, w = uv = -vu$, and assume that the characteristic $\neq 2$. The quaternion algebra V is a 4-dimensional vector space over F, with basis $B = \{1, u, v, w\}$, representing an element v in the form (x_1, x_2, x_3, x_4) over the basis B, as $x_1 + x_2 u + x_3 v + x_4 w$.

In Conrad (4.26) ([**32**]), when F is a field, the conditions (slightly modified) are as follows: (1) the equation b cannot be represented in the form $x^2 - ay^2$ for elements x, y in F and (2) the set of elements of the form $x^2 - ay^2$ can never be zero unless $(x, y) = (0, 0)$ for all elements $x, y \in F$, which is equivalent to a is a non-square in F.

THEOREM 20. *((3.8) Conrad [**32**]) Let Q denote the field of rationals. Let a be an integer and p an odd prime such that a is not congruent to a square modulo p. Then $(a, p)_Q$ is a quaternion division ring.*

Using this result, Conrad proves that $(2,3)_Q$, $(2,5)_Q$ are quaternion division rings and also $(-1, p)_Q$, for p not congruent to 3 mod 4, is a quaternion division ring. But, also it turns out that $(3, 11)_Q$ is a quaternion division ring, as $(11, 3)_Q$ is isomorphic to $(3, 11)_Q$.

In part 8, we discuss 'lifting skewfields', where we take a central extension $F(\theta)$ of a skewfield F and construct a class of semifields in $PG(3, F)$. The matrix representation of these semifields looks similar to a quaternion division ring matrix representation except that the matrix entries in the $(1, 2)$-entries are $\theta\delta + \gamma$ where δ and γ are in $Z(F)$, and $\delta \neq 0$. But suppose there were a situation where δ could be 0. Then we would have an interesting situation.

6. Matrix skewfields are generalized quaternion

Assume that F is a non-commutative skewfield of characteristic $\neq 2$ and, in the following, all matrices are in $GL(2, F)_{left}$. Let $F(\sqrt{a})$ be a central Galois quadratic extension of F (in particular, $x^2 - a$, for $a \in Z(F)$ is irreducible over F). Central extensions of skewfields are defined in part 8 and this section uses ideas from part 8 as well as ideas on extensions of skewfields in part 12. So these few sections are a preview of the later parts of this book.

We begin with the matrix version of the central Galois quadratic extension results.

PROPOSITION 4. *Let F be a central Galois quadric extension of a non-commutative skewfield F by a central non-square and assume that the characteristic is $\neq 2$. Consider the set of matrices:*

$$\left\{ \begin{bmatrix} v & as \\ s & v \end{bmatrix} ; \forall s, v \in F, a \in Z(F) \right\},$$

and a non-square in F. The automorphism σ is induced from the quadratic extension and maps $\sqrt{a}s + v$ onto $-\sqrt{a}s + v$, by taking $\sqrt{a}$ to commute with F, we have an isomorphism from this form $F(\sqrt{a})$ (the closed form) to the set of matrices (the open form).

PROOF. To see this, for s nonzero, the determinant of the above matrix is $(vs^{-1}v - as)s = 0$ if and only if $(vs^{-1}vs^{-1} - a)s^2 = 0$, if and only if for $vs^{-1} = x$, and $y = s$, we have the condition that $(x^2 - a)y^2 = 0$, if and only if a in $Z(F)$ is also a square in F, for some elements x and $y \in F(\sqrt{a})$, contrary to our assumptions. If $s = 0$, the determinant is v^2, which is 0 if and only if $(s, v) = (0, 0)$. In part 8, it is noted that any quasifibration that also defines a skewfield is a spread. Therefore, we have a skewfield spread

$$\left\{ x = 0, y = x \begin{bmatrix} v & as \\ s & v \end{bmatrix} ; \forall s, v \in F, a \in Z(F) \right\}.$$

We have an automorphism σ, where $\sigma^2 = 1$, and σ is induced from the quadratic extension and maps $\sqrt{a}s + v$ onto $-\sqrt{a}s + v$, by noting that $Z(F(\sqrt{a})) = Z(F)(\sqrt{a})$, so that $\sqrt{a}$ commutes with F. It is then easy to determine that we have an isomorphism from this form $F(\sqrt{a})$ (the closed form) to the set of matrices (the open form). This completes the proof.   □

THEOREM 21. *Let F be a skewfield and $F(\sqrt{a})$ a central Galois quadratic extension of F, where $x^2 - a$ is irreducible over $Z(F)$. Then*

$$\left\{ \begin{bmatrix} u^\sigma & bt^\sigma \\ t & u \end{bmatrix} ; t, u \in F(\sqrt{a}),$$

such that $b \in Z(F)$ is not of the form $x^2 - ay^2$; $xy = yx$, for x and $y \in F$, and $a, b \in Z(F)\}, a, b$ non-squares in F, is a generalized quaternion division ring 4-dimensional over F, whose center is $Z(F)$, which shall be denoted by $\mathcal{G}(a, b)_F$.

PROOF. $\begin{bmatrix} u^\sigma & bt^\sigma \\ t & u \end{bmatrix}$ is non-singular if and only if the skewfield determinant is not zero. Clearly, we may assume that $t \neq 0$. The determinant is

$$(4.1) \qquad\qquad (u^\sigma t^{-1}u - bt^\sigma)t = 0,$$

if and only if $u^\sigma t^{-1}u - bt^\sigma = 0$. We note that $(t^{-1})^\sigma = (t^\sigma)^{-1}$. Since $\sigma^2 = 1$, the expression

$$\begin{aligned} u^\sigma t^{-1}u(t^\sigma)^{-1} &= (u(t^{-1})^\sigma)^\sigma (u(t^\sigma)^{-1} = (u(t^{-1})^\sigma)^\sigma u(t^{-1})^\sigma = v^\sigma v, \\ \text{for } u(t^{-1})^\sigma &= v \in F(\sqrt{a}). \end{aligned}$$

So, we need to show that $b \neq v^\sigma v$, for all $v \in L$. Letting $u(t^{-1})^\sigma = v = \sqrt{a}s + w$, for $s, w \in F$, then if $b = v^\sigma v$ then

$$b = w^2 - as^2 + \sqrt{a}(sw - ws),$$

and since $b \in Z(F(\sqrt{a}))$, it follows that $sw = ws$, and $b = w^2 - s^2$, so we have a contradiction. When $t = 0$, the determinant is $u^\sigma u = 0$, so that $u = 0$. To be certain of this, let $u = \sqrt{a}u_1 + u_2$ for $u_1, u_2 \in F$, so that

$$0 = u_1^2 - au_2^2 + \sqrt{a}(u_1u_2 - u_2u_1),$$

which shows that $u_1u_2 = u_2u_1$ and $u_1^2 = au_2^2$, so that for non-zero u, and $a \in Z(F)$, so that $u_2^{-1}u_1 = u_1u_2^{-1}$, or $u_1u_2 = 0$. Clearly, $u_1 = 0$ if and only if $u_2 = 0$. Therefore, it follows that $u_1^2u_2^{-2} = (u_1u_1u_2^{-1}u_2^{-1} = (u_1u_2^{-1})(u_1u_2^{-1}) = a$, a contradiction.

We claim that a basis is

$$\left\{ \begin{bmatrix} 1 & 0 \\ 0 & 1 \end{bmatrix}, \begin{bmatrix} -\sqrt{a} & 0 \\ 0 & \sqrt{a} \end{bmatrix}, \begin{bmatrix} 0 & b \\ 1 & 0 \end{bmatrix}, \begin{bmatrix} 0 & -\sqrt{ab} \\ \sqrt{a} & 0 \end{bmatrix} \right\}.$$

To see this, just note that for $u = \sqrt{a}u_1 + u_2$, and $t = \sqrt{a}t_1 + t_2$, for $u_i, t_i \in F$, $i = 1, 2$ and

$$\begin{bmatrix} -\sqrt{a}u_1 + u_2 & -\sqrt{ab}t_1 + bt_2 \\ \sqrt{a}t_1 + t_2 & \sqrt{a}u_1 + u_2 \end{bmatrix} \text{ is}$$

$$\begin{bmatrix} 1 & 0 \\ 0 & 1 \end{bmatrix} u_2 + \begin{bmatrix} -\sqrt{a} & 0 \\ 0 & \sqrt{a} \end{bmatrix} u_1 + \begin{bmatrix} 0 & b \\ 1 & 0 \end{bmatrix} t_2 + \begin{bmatrix} 0 & -\sqrt{ab} \\ \sqrt{a} & 0 \end{bmatrix} t_1.$$

This set is clearly linearly independent so we have a basis of dimension 4 over F. We now establish that we have a generalized quaternion division ring 4-dimensional over the skewfield F.

Denote

$$\begin{bmatrix} 1 & 0 \\ 0 & 1 \end{bmatrix} \text{ by } 1,$$

$$\begin{bmatrix} -\sqrt{a} & 0 \\ 0 & \sqrt{a} \end{bmatrix} \text{ by } U,$$

$$\begin{bmatrix} 0 & b \\ 1 & 0 \end{bmatrix} \text{ by } V,$$

and

$$\begin{bmatrix} 0 & -\sqrt{ab} \\ \sqrt{a} & 0 \end{bmatrix} \text{ by } W.$$

Then

$$U^2 = aI = a, \ W^2 = \begin{bmatrix} 0 & b \\ 1 & 0 \end{bmatrix} \begin{bmatrix} 0 & b \\ 1 & 0 \end{bmatrix} = \begin{bmatrix} b & 0 \\ 0 & b \end{bmatrix} = bI = b, W^2 = -ab,$$

$$UV = \begin{bmatrix} -\sqrt{a} & 0 \\ 0 & \sqrt{a} \end{bmatrix} \begin{bmatrix} 0 & b \\ 1 & 0 \end{bmatrix} = -\begin{bmatrix} 0 & -\sqrt{ab} \\ \sqrt{a} & 0 \end{bmatrix} = -VW = W,$$

which is equal to

$$-\begin{bmatrix} 0 & b \\ 1 & 0 \end{bmatrix} \begin{bmatrix} -\sqrt{a} & 0 \\ 0 & \sqrt{a} \end{bmatrix} = \begin{bmatrix} 0 & -\sqrt{ab} \\ \sqrt{a} & 0 \end{bmatrix}.$$

Hence, with our conditions, we have completed the proof and we have shown there is a generalized quaternion division ring over a skewfield F. $\qquad\square$

7. Generalized $\mathcal{G}(a,b)_F$ contains $(a,b)_{Z(F)}$

In this small section, we note that in any generalized quaternion division ring over a skewfield F, $\mathcal{G}(a,b)_F$, there are always quaternion division rings over $Z(F)$, $(a,b)_{Z(F)}$. To see this we recall that we have a generalized quaternion division ring of the form

$$\left\{ \begin{bmatrix} u^\sigma & bt^\sigma \\ t & u \end{bmatrix} ; u, t \in F(\sqrt{a}) \right\}, \sigma^2 = 1, \sigma \neq 1,$$

where F is a skewfield, $b \in Z(F)$, and σ is induced from the quadratic extension and maps $\sqrt{a}s + v$ onto $-\sqrt{a}s + v$, by noting that $\sqrt{a}$ commutes with F. The elements

a and b are non-squares in F and as elements of $Z(K)$, there are non-squares in $Z(K)$. We note that a, b are in $Z(F)$, and b is not in $F(\sqrt{a})^\sigma F(\sqrt{a}) \cap Z(F)$. Hence, b is not in $(Z(F)(\sqrt{a}))^{\sigma+1} \cap Z(F)$. Accordingly, we obtain a quaternion division ring $(a,b)_{Z(F)}$ contained in the generalized quaternion division ring $\mathcal{G}(a,b)_F$.

We mentioned previously the idea of lifting skewfields, using central Galois quadratic extensions of skewfields which are considered in part 8. The constructed semifields have the following form:

$$\left\{ \begin{bmatrix} u^\sigma & (\sqrt{a}\delta + \gamma)t^\sigma \\ t & u \end{bmatrix} \right\}; u,t \in F(\sqrt{a})\}, \sigma^2 = 1, \sigma \neq 1,$$

$$\text{for any } \delta \neq 0, \gamma \in Z(F).$$

The arguments do not seem to show that $\delta = 0$ is a possibility, but if so, we would obtain a generalized quaternion division ring. We leave this topic with an open question.

PROBLEM 10. *Do generalized quaternion division rings exist? Note, the question then is $\delta = 0$, in an associated semifield plane, possible as a replacement set?*

PROBLEM 11. *Also, the result given is not the most general version of a 'generalized quaternion division ring'. What would another version look like?*

PROBLEM 12. *In (12,14,4), on extending finite-dimensional division rings D over their center F by forming*

$$D \otimes F(z_1, z_2) = D_{F(z_1,z_2)},$$

where z_1, z_2 are commuting independent variables, it is shown that $D_{F(z_1,z_2)}$ is a division ring with center $F(z_1, z_2)$. Then considering $\sigma : z_1 \longleftrightarrow z_2$ and $\sigma = 1_D$ (fixing D pointwise) it follows that σ is an automorphism of order 2 and $F(z_1, z_2)/Fix\sigma$ is a Galois quadratic extension.

(1) *Show that $D_{F(z_1,z_2)}$ is a central Galois quadratic extension of $D_{F(z_1 z_2)} = D \otimes F(z_1 z_2)$.*

(2) *Could such a central Galois quadratic extension lead to a generalized quaternion division ring?*

8. Artin-Wedderburn theorem & Brauer groups

Although the following material will be used only slightly in this text, we provide this information as background.

DEFINITION 32. *Let K be a field and let C be an associative K-algebra with unit element such that C is simple as a ring (no proper (two-sided) ideals) and let the center of C be K. Then C is said to be a 'central simple algebra'.*

We list some results of Draxl [48], as necessary background, that basically lead to/are the Artin-Wedderburn theorem.

THEOREM 22. *(Theorem 1, p. 59 Draxl [48]) The notation $M(n, D)$ denotes the ring of n by n matrices with elements in D, for D a skewfield. The following are equivalent:*

(a) C is a central simple K-algebra;
(b) C is a finite-dimensional K-algebra without proper two-sided ideals such that $K = Z(C)$;
(c) C is a simple ring which is finite dimensional over $K = Z(C)$;
(d) $C \otimes_K L$ is a central simple L-algebra (L/K a field extension, not necessarily finite);
(e) Let $\overline{K}$ denote any algebraic closure of K, then

$$C \otimes_K \overline{K} \simeq M(n, \overline{K}),$$

and then

$$[C : K] = n^2;$$

(f) $C \simeq M(r, D) \simeq D \otimes_K M(r, K)$, where D is a division ring, with unique r, and unique D, up to isomorphism. D is called the 'skewfield component' of C and n (above) is said to be the 'reduced degree'(index) of C, and $r^2[D : K] = [C : K] = n^2$.

Therefore, C is a skewfield if and only if $r = 1$ and index$^2 = n^2$.

DEFINITION 33. *The central simple algebras (CSA) are characterized by the Artin-Wedderburn theorem, which states that*

(1) *every finite-dimensional central simple algebra C over a field K is isomorphic to a matrix algebra $M(r, D)$, for some skewfield (division algebra) D.*
(2) *$M(n, D)$ is isomorphic to $M(m, E)$, if and only if $n = m$ and D and E are isomorphic.*
(3) *Taking K as the center of D, the dimension of $M(r, D)$, as a K-vector space, is a finite square r^2, and the center of $M(r, D)$ is isomorphic to the center of D.*

DEFINITION 34. *Let C and R be CSAs over the same field K and isomorphic to $M(n, D)$ and $M(m, E)$ respectively. Define an equivalence relation on all central simple algebras over K by identifying all matrix algebras $M(n, D)$ and $M(m, E)$, such that D and E are isomorphic. This is called the 'Brauer equivalence relation' on central simple algebras finite-dimensional over the field K.*

Now to make a group out of the Brauer equivalence classes, let R and S be central simple K-algebras and form the tensor product $R \otimes S$, which is also a central simple K-algebra. The inverse of a central simple K-algebra C is the opposite algebra C^{opp}. And, in the notation in this text, we use C_{left} for C and C_{right} for C^{opp}, noting that $C_{left} \otimes C_{right} = M(n^2, K)$, which is equivalent to the identity (1-dimensional K algebra), where the (reduced) degree of C_{left} is n.

This group is said to be the 'Brauer group of K'. Define the 'period or exponent of a central simple algebra C' to be the order of C as an element of the Brauer group. We see by Theorem 2.7.1 of Jacobson [64] that the period divides the degree, so that the period is finite. Accordingly, the Brauer group is an 'Abelian torsion group' (all elements have finite order).

Therefore, for a central simple algebra C over a field K, there is a minimal positive integer e so that $C^{\otimes e} = C \otimes C ... \otimes C$ (e times), the 'exponent of C', which is isomorphic to a matrix algebra over K, that is, equivalent to 1 in the Brauer group of K.

If K is a 'number field', that is, an algebraic extension field of the field of rational numbers $\mathbb{Q}$, then the exponent e is equal to the (reduced) degree/index n.

REMARK 7. *If D is a central division algebra of degree $n > 2$ then $D_{right} \not\simeq D_{left}$, and $D_{right} \simeq D_{left}$ when $n = 1$ or 2.*

The reader is directed Draxl [48] for a presentation of the proof of the Artin-Wedderburn theorem.

9. Extending skewfields

This section, will act as, if not a background section, then perhaps just a way of explaining why a matrix method could produce an extension skewfield. The derivable net method, coupled with Schur's lemma, does construct the quaternion division ring extensions. The following shows how the proof would work using non-commutative algebra, and, in particular, we also show how the proof would proceed using aspects of Brauer groups. Actually, just using matrix methods of derivable nets, it is possible to demonstrate using the matrix equivalency of quaternion division rings that this can be deduced directly from the material on derivable nets.

We shall give a slightly more general theorem in part 3, in our discussion of type 0 derivable nets.

The formal definition of a generalized quaternion division ring is as follows:

DEFINITION 35. *Let F be an arbitrary skewfield. Consider the structure for a quaternion division ring has relative to a field H (for example, $(c,d)_H^\beta$). If there is a division ring of the same structural form $(a,b)_F^\alpha$, we shall call this a 'generalized quaternion division ring'. We have seen previously that such division rings, should they exist, require a central extension of a skewfield F.*

The ideas of 'central extensions of skewfields' were originated by Cohn and Dicks [30], part of which we used in the classification theorems.

DEFINITION 36. *A 'central extension of a skewfield' D is an extension skewfield E of D, which is generated as a skewfield by $\langle D, Z(E) \rangle$, where $Z(E)$ is the center of E.*

We take a less general definition here, which is similar to the extensions referred to in Proposition 3 of [30], in the section on 'regular central extensions of skewfields'.

DEFINITION 37. *A 'central Galois extension of dimension n' of a skewfield F is a finite Galois extension of a skewfield based on an irreducible n-dimensional polynomial over the center $Z(F)$. We have proved that certain quaternion division rings are central Galois quadratic extensions of quaternion division rings. The extension of skewfields theorem provides a way to view these extensions using derivable nets.*

In the proof given on extending skewfields theorem that is developed using derivable nets, we have a skewfield F of dimension 4 over $Z(F)$. The derivable net as a left derivable net sits in a 4-dimensional vector space over a field L, representing a right pseudo-regulus net (regulus net, in this case). Since F is not isomorphic to L, the group action on the derivable nets is Baer subplane fixed free. Since we may realize the matrix action of $*F_{right}$ as within $GL(2,L)_{left}$, then F_{right}^* is irreducible on any 2-dimensional bimodule over F and L. Any of the common components provides the particular 2-dimensional space V_2.

Then F^*_{right} in $GL(2, L)_{left}$ is irreducible on V_2/L_{left}, and L^*_{left} is in $GL(2, F)_{right}$ on V_2/F_{right}. Schur's Lemma shows that

$$(1) \ C_{GL(2,L)_{left}}(F^*_{right}) \ \& \ (2) \ C_{GL(2,F)_{right}}(L^*_{left})$$

are skewfields. Note that L^*_{left} and F^*_{right} commute as linear transformation groups.

Brauer group theory allows that in this setting (exponent 2), F is a quaternion division ring $(a, b)_W$. The matrix method does the same. In order that we have both of these skewfields in the same 4-dimensional left vector space over L, we need $(a, b)_W$ and L to have the same dimension over a field; L needs to be 4-dimensional over the common field W. In the characteristic $\neq 2$ case, note that a is a non-square in F, which forces $L = W(\sqrt{a}, \sqrt{c})$, where both $W(\sqrt{a})$ and $W(\sqrt{c})$ are quadratic extensions of W. However, we realize that $W(\sqrt{a}, \sqrt{c}) \neq W(\sqrt{a}, \sqrt{b})$, as it is possible that $a = b$ and, in case would not be dimension-wise large enough. The composition L of these two quadratic extensions of W is Galois over W, as L is a splitting field of a product of separable polynomials.

Noting that we now have a skewfield within $GL(2, W(\sqrt{a}, \sqrt{c}))$, in case (2), above, it follows, again using Brauer group theory, that we must have a quadratic quaternion extension of $(a, b)_W$, namely $(a, b)_{W(\sqrt{c})}$.

Furthermore, since the quadratic extension contains L and F_{left} (noting that F_{left} commutes with F_{right}), and sits in $GL(2, L)$, since L is a field, we see that the quadratic extension is generated by the set product $((a, b)_W)_{left} \cdot L$. In other words, we see that

$$(2) \ C_{GL(2,F)_{right}}(L^*_{left}) \ \text{is isomorphic to} \ ((a, b)_W)_{left} \otimes_W L.$$

Noting that F^{opp} is the inverse of F in the Brauer group, we see that

$$(1) C_{GL(2,L)_{left}}(F^*_{right}) \ \text{is isomorphic to} \ L \otimes_W ((a, b)_W)_{right}.$$

Since the degree is 2, the order of a quaternion division ring is 2 in the torsion Brauer group, so that (1) and (2) are isomorphic, in this case, to $(a, b)_{W(\sqrt{c})}$, a quadratic extension of $(a, b)_W$ and $W(\sqrt{a}, \sqrt{c})$. (Note the quaternion division ring $(a, b)_{W(\sqrt{c})}$ may be considered a quadratic extension of $W(\sqrt{a}, \sqrt{c})$, in the second sense).

When the characteristic is 2, one starts with a Galois quadratic extension $W(\theta)$ and the use of a quadratic extension $W(\tau)$, which need not be Galois then forms $W(\theta, \tau)$, where similar remarks apply. $W(\theta, \tau)$ inherits the involutory automorphism, and one would obtain $(a, b)^\alpha_{W(\tau)}$ as a quadratic extension of both $(a, b)^\alpha_W$ and $W(\theta, \tau)$. Hence, we have essentially proved the following theorem, but see the more general form $(3, 5, 1)$ that discusses the faithful action of groups.

THEOREM 23. *Extension of skewfields theorem (Restricted form)* Let R^L be a left classical regulus net in $PG(3, L = W(\theta, \tau))$, where L is a 4-dimensional field extension of W admitting an automorphism σ arising from the Galois quadratic extension $W(\theta)$, and let $W(\tau)$ denote a quadratic extension of W. Let V_4 denote the corresponding 4-dimensional L space containing R^L. Let S^F_{right} be a right suspect derivable net, where $Z(F) = W$, that lives in V_4 so that V_4 is a (L, F)-bimodule.

(1) Then L^*_{right} acts as a group of linear transformations within

$$GL(2, F)_{Left}$$

and F^*_{left} acts as a group of linear transformations within $GL(2, L)_{right}$, such that, as linear transformation groups, the groups centralize each other.

(2) S^F_{right} is a derivable net of of type 0, and is coordinatized by a quaternion division ring $(a, b)^\alpha_W$.

(3) Furthermore, the groups F^*_{right} and $L^*_{left} = L^*$ act irreducibly on intersections with the sets of their respective Baer subspaces, as in the General Framework Result, on any of the common components

$$Com \in \{x = 0, y = 0, y = x\}.$$

There are centralizer skewfields

(2) $C_{GL(2,F)_{right}}(L^*_{left})$ is isomorphic to $((a, b)_w)_{left} \otimes_W L_{right}$.

Noting that F^{opp} is the inverse of F in the Brauer group, we see that

(1)$C_{GL(2,L)_{left}}(F^*_{right})$ is isomorphic to $L_{left} \otimes_W ((a, b)_W)_{right}$.

(4) Both centralizer subgroups are isomorphic to the central quadratic skewfield extension $(a, b)_{W(\tau)}$ of both $(a, b)_W$ and $W(\theta, \tau)$.

PROPOSITION 5. **Commutative linear transformation groups.** *In the more general result in $(3, 5, 1)$, we need to show the commutativity between two arbitrary skewfields F and K, coordinating a right suspect derivable net S^K and a left classical derivable net R^F, both in the same 4-dimensional vector space V_4 acting as a (F, K)-bimodule.*

Then the sets of left mappings F^*_L and K^*_L commute with the sets of right mappings F^*_R and K^*_R.

PROOF.

PROBLEM 13.

(1) For $\gamma \in F^*$, define $f_\gamma \in F^*_{left}$ by

$$f_\gamma : f_\gamma(x_1, x_2, x_3, x_4) = (\gamma x_1, \gamma x_2, \gamma x_3, \gamma x_4).$$

Similarly, for $M \in K^*$ define $k_M \in K^*_{right}$ by

$$k_M : k_M(x, y, z, w) = ((x, y)M, (z, w)M),$$

for all $M \in K^*$, $x, y, z, w \in F$. Note that the Baer subspaces of R^F_{right} are

$$B(R)^F_{left}(a, b) = \{((\gamma a, \gamma b), (\rho a, \rho b)); \forall \gamma, \rho \in L\}.$$

Now restrict to $y = 0$, noting that both K and F leave each of the common components invariant. Apply, in order, left to right f_γ then k_M,

$$k_M f_\gamma : (x, y) = (\gamma x, \gamma y)M = \gamma(x, y)M,$$

and then just note that

$$f_\gamma k_M : (x, y) = \gamma((x, \gamma)M) = \gamma(x, y)M,$$

as M is a matrix with elements of F. This shows that these two mappings centralize each other. Since the γ's pull out to the left and the M's pull *out to the right, we claim that this implies that F^* is in $GF(2, K)_{Left}$ and K^*_{left} is in $GL(2, F)_{right}$. The arguments are symmetric: We will show that K^*_{right} is in $GL(2, F)_{Left}$, acting on $y = 0$ as a left 2-dimensional vector space. Suppose we consider a linear set $(\gamma(x, y) + \rho(z, w))$ for γ and ρ in F, and apply the mapping k_M. Since M is a linear transformation within $GL(2, F)_{Left}$, by assumption, we see that k_M is additive and*

$$
\begin{aligned}
k_M(\gamma(x, y) + \rho(z, w)) &= (\gamma x + \rho z, \gamma y + \rho w)M \\
&= (\gamma x, \gamma y)M + (\rho z, \rho w)M \\
&= \gamma(x, y)M + \rho(z, w)M \\
&= \gamma k_M(x, y) + \rho k_M(z, w).
\end{aligned}
$$

The analogous argument is as follows: First note that $(x, y)M + (z, w)N$, for any M, $N \in K$, is just an element (r, k), a 2-vector of V over F, for M, N in K. Using this simple fact will then show that on any common component Com, K^*_{right}, a group of right mappings acts as a group of linear transformations of $GL(2, F)_{Left}$ and F^*_{Left}, a group of left mappings, acts as a group of linear transformations of $GL(2, K)_{right}$ on Com. This completes the proof of the proposition.

$\square$

Continuing with general remarks, now since the group K^*_{right} acting on a common component Com is an irreducible group of linear transformations within $GL(2, F)_{Left}$ and Com is a 2-dimensional left vector space over F, by Schur's lemma, the centralizer $C_{GL(2,F)_{Left}}(K^*_{right})$ is a skewfield. Since K^*_{right} is centralized by the left mappings of F, it is not always the case that F^*_{left} is a subgroup of the centralizer, since, in general, only the right mappings F^*_{right} are in $GL(2, F)_{Left}$. However, $Z(F)$ is in $GL(2, F)_{Left}$, as are the following mappings: $f_{\gamma,\rho}$:

$$
f_{\gamma,\rho}(x, y) = (x \begin{bmatrix} \gamma & 0 \\ 0 & \gamma \end{bmatrix}, y \begin{bmatrix} \rho & 0 \\ 0 & \rho \end{bmatrix}); \forall \gamma, \rho \in Z(F)^*,
$$

each of which leaves both $x = 0$, $y = 0$ invariant and maps $y = \delta x$ to $y = \delta x$ and so also fixes $y = x$. Furthermore, this is a right mapping and so is in $GL(2, F)_{Left}(K^*_{right})$. In addition, $B(R)^F_{left}(a, b)$ maps to $B(R)^F_{left}(\gamma a, \rho b)$, so the Baer left F^*_{left}-subspaces are permuted. So, $C_{GL(2,F)_{Left}}(K^*_{right})$ contains

$$
< f_{\gamma,\rho}, Z(K)^* ; \gamma, \rho \in Z(F)^* > .
$$

Part 3

Types i of derivable nets

In this part, we give an in-depth analysis of the four types of derivable nets.

CHAPTER 5

The types

1. Type 0

In this section, we give a more complete version of the extension of skewfields theorem, where all possibilities are listed.

THEOREM 24. **Part II; Derivable nets of type 0, and Extension of skewfields theorem.**

(iv) When $i = 0$, the groups K^*_{left} and F^*_{right} act faithfully and irreducibly on Com as linear groups acting on the intersections $Com \cap B(R)^F_{right}$ and $Com \cap B(S)^K_{left}$, respectively. Applying Schur's Lemma, there are two centralizer skewfields and for these, there are the following three possibilities:

(1) $C_{GL(2,F)_{Left}}(K^*_{right})$ and $C_{GL(2,K)_{right}}(F^*_{left})$, the centralizer skewfields, are simply K and F, respectively.

(2) If $F^*_{right} \subset C_{GL(2,F)_{Left}}(K^*_{right})$, then there is a quadratic skewfield extension of F within $GL(2,F)_{left}$.

(3) If $K^*_{right} \subset C_{GL(2,K)_{right}}(F^*_{left})$, then there is a quadratic skewfield extension of K within $GL(2,K)_{right}$.

There are then three alternatives:

(a) K and F are both fields and we obtain a quadratic field extension of both K and F, which is a spread in $PG(3,F)$.

(b) K and F are both non-commutative skewfields.

Examples would be two mutually non-isomorphic central quadratic extensions of a non-commutative skewfield L, where $K = L(\theta)$ and $F = L(\tau)$, where $Z(L(\theta)) = Z(L)(\theta)$ not isomorphic to $Z(L(\tau)) = Z(L)(\tau)$. Known specific examples are two mutually non-isomorphic central extensions of $L = (a, b)^\alpha_H$, a quaternion division ring. In general, the centralizer skewfields are isomorphic to $L(\tau, \theta)$.

(c) One of K and F is a field and the other is a non-commutative skewfield. In this setting, the non-commutative skewfield is a quaternion division ring $(a, b)^\alpha_H$. F is a 4-dimension vector space over H and K is a 4-dimensional vector space over H. There must be one Galois quadratic extension $H(\theta)$, from which F uses the involutory automorphism for the matrix representation. Since K is of dimension 4, there must be another quadratic extension, $H(\tau)$, which makes $K = H(\theta, \tau)$ Galois over K if the characteristic is not 2. However, K need not be Galois over H, when the characteristic is 2, but requires that there are two quadratic subfields and inherits the automorphism from $H(\theta)$. In either case, there is an extension $(a, b)^\alpha_{H(\tau)}$ of $(a, b)^\alpha_H$ and $(a, b)^\alpha_{H(\tau)}$ is a spread within $PG(3,K)$, if $F \simeq (a, b)^\alpha_H$. The extension is always a central quadratic extension *and is a central Galois quadric extension if and only if $H(\theta, \tau)$ is Galois over H. Therefore, $(a, b)^\alpha_{H(\tau)}$ is a spread in both $PG(3, H(\theta, \tau))$ and $PG(3, (a, b)^\alpha_H)$, even as these projective spaces are non-isomorphic.*

Although there are resultant centralizer skewfields of (b) and (c) that could produce isomorphic skewfields, the methods of construction are essentially different.

$C_{GL(2,F)_{left}}(K^*_{right}) \simeq C_{GL(2,K)_{right}}(F^*_{left}) \simeq (a, b)^\alpha_{H(\tau)}$. *This is the extension quaternion division ring of $(a, b)^\alpha_H = F$, and is the tensor product of $(a, b)^\alpha_H$ with $H(\tau)$ over H. Specifically, we have the isomorphism:*

$$C_{GL(2,K)_{right}}(F^*_{left}) \simeq ((a, b)^\alpha_H)_{left} \bigotimes_H H(\tau).$$

Noting that $K^{opp} = K_{right}$,

$$C_{GL(2,F)_{Left}}(K^*_{right}) \simeq H(\tau) \bigotimes_H ((a, b)^\alpha_H)_{right},$$

and these division rings are both isomorphic to the quaternion division ring $(a, b)^\alpha_{H(\tau)}$.

- When K has characteristic 2, as noted, there are two possible extensions of $(a, b)^\alpha_H$, when the skewfield represents a derivable net in $PG(3, H(\theta, \tau))$.

  The first possibility occurs when $H(\theta, \tau)$ is separable and so the extension to $(a, b)^\alpha_{H(\tau)}$ is a central Galois quadratic extension.

  The second possibility is when $H(\theta, \tau)$ is inseparable and admits non-squares (as the endomorphism $x \to x^2$ is not surjective). Then if θ satisfies the irreducible polynomial $x^2 + x + \beta$, and $\tau = \sqrt{c}$, for c a non-square, $(a, b)^{\alpha=1}_H$ then extends to $(a, b)^{\alpha=1}_{H(\sqrt{c})}$ and is a central extension but not a central Galois quadratic extension of $(a, b)^{\alpha=1}_H$.

PROOF. Note that if either K or F is a field, then the assumptions of (1) or (2) are valid. And if (1) or (2) is valid then it follows that one or more of $C_{GL(2,F)_{Left}}(K^*_{right})$ and $C_{GL(2,K)_{right}}(F^*_{left})$ are skewfields that are quadratic extensions over both K and F. It follows that the centralizer skewfields are thus spreads that may be considered within $PG(3, F)$ and/or $PG(3, K)$. Note that this is merely convention to consider subspaces of the spreads to be lines of the corresponding projective space corresponding to the 4-dimensional vector space wherein lives the partial spreads.

For example, let F be a field. Then $GL(2, K)_{right} = GL(2, K)_{Left}$, and we know that F^*_{left} and F^*_{right} centralize K. Hence, the centralizer of F^*_{right} in $GL(2, K)$ now also contains K and F_{left} since F_{right} is in $GL(2, K)_{Left} = GL(2, K)$.

We claim that F and K are not equal. If so then since R^F_{right} now has the matrix form $\begin{bmatrix} u & 0 \\ 0 & u \end{bmatrix}$; $u \in F$, and K is a field of matrices whose nonzero elements are in $GL(2, F)_{Left}$, it follows that the derivable net in question is the derived net of the right classical pseudo-regulus net, the left classical pseudo-regulus net. However, these two derivable nets share at least three Baer subplanes, $B(R)^F(0, 1)$, $B(R)^F(1, 0)$, and $B(R)^F(1, 1)$; type 3.

Accordingly, F and K_{right} are not equal and centralize each other by the proposition following the restricted extension of skewfields theorem. And, the nonzero elements of both F and K_{right} are both in $GL(2, F)_{left}$.

It is straightforward to see that the centralizer acts identically on each of the three common components, due to the structure of the Baer subspaces. Therefore, F^*_{left} is irreducible if and only if K^* is irreducible and we shall prove the action is faithful on each Com, due to the symmetry of the derivable net structures. So, since both K^* and F^* fix all common components, the centralizers subgroups fix all common components. The dimension of the common component Com is 2 over K and 2 over F, and Com is a vector space over the centralizer, which implies that Com is of dimension 1 over $C_{GL(2,F)_{Left}}(K^*)$, unless possibly there is an element $g \neq 1$ of the centralizer that fixes a vector of Com. However, this cannot occur as if there is an element, assume that the vector space over the centralizer $C_{GL(2,L)_{Left}}(K^*)$ has dimension > 2 and acts on any common component Com. The centralizer is transitive on the K-subspaces on each common component, so g fixes V_4 vectorwise, due to the symmetry of the derivable net structures, which implies that $g = 1$, a contradiction. Hence, we have the proofs to (1) and (2) and the case when at least one of K or F is a field.

Therefore, the centralizer is a quadratic extension of both K and F. If both are fields, we clearly obtain the proof of (3).

Assume that K is a non-commutative skewfield and F is a field. Then K is a quaternion division $(a, b)_H$ which is defined over $H(\theta)$, a Galois quadratic extension of H (for example, see (2,3,1&2)). This forces $F = H(\theta, \tau)$ a 4-dimensional extension of H that admits the automorphism σ induced by $H(\theta)$. Since the non-commutative centralizer skewfields are 4-dimensional over their centers, they are quaternion division rings by noting that they are spreads in $PG(3, F)$ again by (2,3,1&2) or (also see, for example, Voight [140]). The extension skewfield S must contain $F(\theta, \tau)$ and be a quaternion division ring, since it sits in $PG(3, F(\theta, \tau))$, this then implies that the center is $F(\tau)$, then S can only be $(a, b)_{H(\tau)}$, which is a central quadratic extension of $(a, b)_F$.

Note that if $F(\theta, \tau)$ is a extension field of dimension 4 over F and contains two fields $F(\theta)$, $F(\tau)$ only one of these needs to be Galois, and this can occur only when the characteristic is 2. So, when the characteristic is not 2, all such extensions are central Galois quadratic extensions of skewfields.

So, we have completed parts (a) and (c), so we have only to consider when K and F are non-isomorphic non-commutative skewfields. We then illustrate the situation when K and F are both central quadratic extensions of a non-commutative skewfield L.

We mentioned the tensor product characterization as well the fact that constructed skewfields are quaternion division rings previously, where we gave a version of the proof with certain details added. Now choose one $L(\tau)$ to write in open form (the full matrix version):

$$L(\tau) = \left\{ \begin{bmatrix} u + \alpha t & \beta t \\ t & u \end{bmatrix}; u, t \in L \right\}; \alpha, \beta \in Z(L).$$

Then the other $L(\tau)(\theta)$ will have an analogous open form

$$L(\tau)(\theta) = \left\{ \begin{bmatrix} u + \delta t & \gamma t \\ t & u \end{bmatrix} ; u, t \in L(\tau) \right\} ; \delta, \gamma \in Z(L).$$

Accordingly, writing $L(\theta)$ in closed form, since all individual matrix entries are in L, we see that $L(\theta) = K$ and $L(\tau) = F$ are not isomorphic; K produces a left type 0 derivable net relative to the right classical pseudo-regulus net corresponding to F, and F produces a left type 0 derivable net relative to the right classical net corresponding to K as $L(\theta, \tau) = L(\theta)(\tau) = L(\tau)(\theta)$. This completes the proofs of all parts of the theorem.   □

REMARK 8. *The central Galois extension tower of a non-commutative skewfield mentioned in part (b) previously did not use the centralizer argument to establish the results, for it seems that either K or F needs to be a field for that analysis to work. In the next section, we will see the same sort of argument with two fields.*

2. Double regulus type 0 derivable nets

THEOREM 25. *For the double field option, we have:*
If both K and F are fields, $K \simeq H(\theta)$ and $F \simeq H(\tau)$, and $H(\theta)$ and $H(\tau)$ are not isomorphic. Then K is the 'open form' Pappian spread of $H(\theta)$ defined by the matrix field within the classical regulus net over $H(\tau)$, where

$$K = \left\{ \begin{bmatrix} u + \alpha t & \beta t \\ t & u \end{bmatrix} ; u, t \in H \right\}$$

K^* is irreducible and faithful on any common component and the centralizer fields are both isomorphic to $H(\theta, \tau)$, which is defines a Pappian spread in $PG(3, H(\theta))$ and also in $PG(3, H(\tau))$.
These 3-dimensional projective spaces are mutually non-isomorphic.

REMARK 9. *Note that when K and F are not both fields, and one is a quaternion division ring $(a, b)_H^\alpha$, then the division ring is 4-dimensional over H. Then, if the other skewfield is a field, the field must also be of dimension 4 over H, so $K = H(\theta, \tau)$. When K and F are both fields, we would need only two non-isomorphic field extensions $H(\theta)$ and $H(\tau)$ to construct type 0 derivable nets, using open forms of say $H(\theta)$, relative to the classical regulus net relative to $H(\tau)$. We shall come back to this idea later in the book, when we consider type 2 derivable nets defined by open and closed forms of the same pseudo-regulus net in (3,5,6).*

One interesting point is that 'derivation' of the open form derivable net of $H(\theta)$, which cannot be a regulus net within $PG(3, H(\tau))$, and which is not within the spread of $H(\theta, \tau)$, produces a 'open form derived translation plane' with a 'blended kernel' .

To see that examples actually exist, we need to ensure that we have an appropriate environment to have this type of derivable net. That is to say, since quaternion division rings $K \simeq (a, b)_H^\alpha$ are of dimension 4 over their centers H, one way to ensure that such a skewfield type 0 left derivable net actually occurs with the associated matrix representation in $GL(2, F)$, we need to have $F = H(\theta, \tau)$, a 4-dimensional H-field extension admitting an involutory automorphism σ_θ, where F does not actually need to be a Galois extension. Although this has been done in the quaternion division/field cases, we actually need to consider a more general/different situation that allows that F and K could both be skewfields.

3. The ambient space

In this section, we concentrate on central Galois quadratic extensions of skewfields K. These quadratic extensions may be field or non-commutative division ring extensions of a skewfield K. This shows that it is possible to find an ambient 4-dimensional vector space V_4 over a skewfield F that admits a variety of skewfields K for which V_4 is also a K-space; we are verifying that such (F, K)-bimodules exist. For example, this idea works for fields as well as non-commutative skewfields.

If $(a, b)_H^\alpha$ is a quaternion division ring viewed as a left derivable net S_{left}^K, then the right classical pseudo-regulus net R_{right}^F could be relative to $F = H(\theta, \tau)$, or $H(\theta, \rho)$, provided there exist such extensions of H degree 4, admitting the automorphism group $\langle \sigma \rangle$ of order 2. In this case, we clearly have a vector space V that when considered as a 4-dimensional $F = H(\theta, \tau)-$space could also be a 4-dimensional $K = (a, b)_H^\alpha-$space, since the dimensions of K and F are 4-dimensional over H. This is the main idea of the extension of skewfields theorem.

Now consider the cases when $K = H(\rho)$ and $F = H(\tau)$, are both central extensions of a skewfield H. When both are fields, we have noted previously that if non-isomorphic then the centralizer skewfields are isomorphic to $H(\rho, \tau)$.

If $H(\tau)$ not isomorphic to $H(\rho)$ then there are two centralizer skewfields, one isomorphic to $H(\tau) = (a, b)_{H(\tau)}^\alpha$ and one isomorphic to $H(\rho) = (a, b)_{H(\rho)}^\alpha$, that are not isomorphic. The point is that the centralizer skewfields, thinking of $(a, b)_K$ as a derivable net in $PG(3, K(\theta, \tau))$, will create $(a, b)_{K(\tau)}$ and thinking of $(a, b)_K$ as a derivable net in $PG(3, K(\theta, \rho))$ will create $(a, b)_{K(\rho)}$. But if we consider $(a, b)_{K(\tau)}$ as a derivable net in $PG(3, K(\theta, \tau, \rho))$, then using the extension of semifields theory will then produce $(a, b)_{K(\tau, \rho)}$. Similarly, if $(a, b)_{K(\rho)}$ is realized as a derivable net in $PG(3, K(\theta, \tau, \rho))$ will also produce $(a, b)_{K(\rho, \tau)}$. This means that $(a, b)_K \otimes_K K(\tau, \rho)$ is a quaternion division ring. That is, when $K(\theta, \tau, \rho)$ is a Galois extension of degree 2^3, then $(a, b)_K \otimes_K K(\tau, \rho)$ is a Galois skewfield extension of $(a, b)_K$ of degree 2^2. In this way, we may then consider derivable nets relative to L or relative to M both non-isomorphic non-commutative skewfields in an oblique manner, as in $(a, b)_{K(\rho)}$ and $(a, b)_{K(\tau)}$.

We could then consider the case when have there are two mutually non-isomorphic central extensions of a non-commutative skewfield. Letting K_1 be $(a, b)_{H(\tau)}^\alpha$ and F_1 be $(a, b)_{H(\rho)}^\alpha$, and we have two central extensions of the same skewfield. They are both of the following form.

$$\begin{bmatrix} f(t, u) = u + \delta t & g(t, u) = \varepsilon t \\ t & u \end{bmatrix}; t, u \in (a, b)_H^\alpha, \rho, \gamma \in Z((a, b)_H^\alpha) = H,$$

where f, g are suitable functions on $(a, b)_H^\alpha \oplus (a, b)_H^\alpha$.

We are representing central extensions and hence, $f(t, u) = u + \delta t$, and $g(t, u) = \varepsilon t$ for $\delta, \varepsilon \in H$. Then the centralizer skewfields act in a non-obvious manner to produce $(a, b)_{H(\tau, \rho)}^\alpha$ even when reversing the roles of K_1 and F_1. So, this would be a situation where the vector space V is 4-dimensional over $L(\rho)$ and 4-dimensional over $L(\tau)$, both of the same dimension over L.

Accordingly, for all of the cases we considered, there are skewfields K and L that have the same dimension over a common skewfield and also could act on any 2-dimensional K or L common subspace Com as a (L, K)-bimodule.

By results of Conrad [32],[34], Pickert [125] and Yaqub [143], there are infinitely many mutually non-isomorphic quaternion division rings and these exist in all characteristics. And, there exist infinitely many quaternion division rings that admit central Galois quadratic extensions. Hence, there are infinitely many mutually non-isomorphic examples of type 0 derivable nets over non-commutative skewfields. Of course, the same can be said of the field type examples of type 0 derivable nets; there are infinitely many types that are mutually non-isomorphic.

The previous result then creates central Galois quadratic extensions from skewfields that appear as an open form matrix skewfield relative to a right classical pseudo-regulus net given by a skewfield written over a Galois quadratic extension. The quaternion division rings satisfy these conditions, but it might be possible to have other non-commutative skewfields that admit central Galois quadratic extensions, as we shall see in the later parts of the book. The following corollary to the extension of skewfields theorem considers this possibility. The use of the extension of skewfields theorem requires an intervening field that does not appear in the following corollary.

COROLLARY 9. *Let $(a, b)_H^\alpha$ be a quaternion division ring that admits central Galois quadratic extensions $(a, b)_{H(\tau)}^\alpha$ and $(a, b)_{H(\theta)}^\alpha$, whose centers $H(\tau)$ and $H(\theta)$ are not isomorphic. Then there is a quaternion division ring $(a, b)_{H(\tau,\theta)}^\alpha$, with center $H(\tau, \theta)$.*

In $(11,13,3)$ and $(12,14,4)$, we develop additional central Galois quadratic extensions related to cyclic division rings and division ring extensions over rational function fields of many variables.

4. Derivable nets of type 3

From the standard framework result $(2,4,6)$, we now consider a right classical pseudo-regulus R_{right}^F with the set of Baer subspaces $B(R)_{left}^F(a, b)$. Furthermore, we allow that there is a suspect left derivable net S_{left}^K with the set of Baer subspaces $B(S)_{right}^K(a, b)$, where there are three (at least) common Baer subspaces. This will imply that K and F are isomorphic skewfields. Assume these assumptions and conditions in the following theorems in this section, where the same basis for all of the Baer subspaces may be used throughout. First a note regarding left and right inverses of C, where $D \in GL(2, F)_{left}$. A right inverse D_R^{-1} acts on the right of D so that the $DD_R^{-1} = I$. A left inverse D_L^{-1} of an element $D \in GL(2, F)_{right}$ then acts on the left of D so that $D_L^{-1}D = I$. Therefore, $D = (D_R^{-1})_L^{-1} = (D_L^{-1})_R^{-1}$.

THEOREM 26. *The right classical pseudo-regulus net R_{right}^F has a group isomorphic to $\Gamma L(4, F)_N$, where N is a line of the combinatorially embedded net $PG(3, F)_C$. The subgroup*

$$\left\{ \begin{bmatrix} D & 0 \\ 0 & D \end{bmatrix} ; D \text{ in } GL(2, F)_{Left} \right\}$$

preserves the form of the right classical pseudo-regulus net and is transitive on the set of distinct triples of Baer subspaces. Also, note that we may determine that D is non-singular, by using a skewfield adaptation of the determinant. We then may, first of all, consider when

$$B(S)^K_{right}(a,b) = \{(\alpha a, \alpha b, \beta a, \beta b)\}; \forall \alpha, \beta \in F\} = B(R)^F_{left}(a,b).$$

Apply the mapping indicated, to obtain

$$\{((\alpha a, \alpha b)D, (\beta a, \beta b)D); \forall \alpha, \beta \in F\} = (\alpha(a,b)D, \beta(a,b)D).$$

Since $(a,b)D$, is just another (c,d), we see that the Baer subplanes $B(S)^K_{right}(a,b)$ are mapped to $B(S)^K_{right}(a,b)D$.

Since the Baer subspaces are generated as 2-dimensional right or left subspaces, we see that there are the following possibilities for (a,b), a short notation for $B(R)^F_{left}(a,b)$. These are the elements of $\{(0,1),(1,b); \forall b \in F\}$. For triples, there are two distinct types, those that share $(0,1)$, and those that are all in the form $(1,c)$, for three distinct values of c.

Then, for each triple T, either $(i) = \{(0,1),(1,b),(1,c); b \neq c\}$ or $(ii) = \{(1,d),(1,b),(1,c); d, b, c \text{ distinct}\}$ are sets of distinct Baer subplanes, and there is a non-singular matrix $D = \begin{bmatrix} e & f \\ g & h \end{bmatrix}$ such that the

$$\left\{ \begin{bmatrix} D & 0 \\ 0 & D \end{bmatrix}; D \text{ in } GL(2,F)_{Left} \right\},$$

maps T onto $\{(0,1),(1,0),(1,1)\}$ in order.

The components of $B^K(S) : x = 0, y = 0, y = x$ are fixed and

$$\{y = xM; \text{ for } M \in K^*\}$$

is mapped onto

$$\{y = xD_R^{-1}MD; M \in K^*\} = \{y = x \begin{bmatrix} u & 0 \\ 0 & u \end{bmatrix}; u \in F\}.$$

Furthermore,

$$\{y = xM; M \in K^*\} = \{y = x(D_R^{-1})_L^{-1} \begin{bmatrix} u & 0 \\ 0 & u \end{bmatrix} D_R^{-1}; u \in F\},$$

which is

$$\{y = xM; M \in K^*\} = \{y = xD \begin{bmatrix} u & 0 \\ 0 & u \end{bmatrix} D_R^{-1}; u \in F\},$$

Case (i) $D = \begin{bmatrix} c-b, & -b \\ 0 & 1 \end{bmatrix}$ and $D_R^{-1} = \begin{bmatrix} (c-b)^{-1} & (c-b)^{-1}b \\ 0 & 1 \end{bmatrix}$.

Case (ii)

$$D = \begin{bmatrix} -d(c-d)^{-1}(c-b) & -b \\ (c-1)^{-1}(c-b) & 1 \end{bmatrix}.$$

PROOF. **Case (i)**. Mapping $(0,1)$ to $(0,1)$ requires $g = 0$, and $h = 1$, as we are mapping 1-spaces to 1-spaces. Mapping $(1,b)$ to $(1,0)$ directly one obtains (e,f), $e^{-1}(f+b) = 0$ so that $f = -b$. Then mapping $(1,c)$ to $(1,1)$ directly, one obtains $(e,(f+c))$ and then $e^{-1}(f+c) = 1$. Accordingly, $c - b = e$. Hence, we may take $e = 1$ so that $D = \begin{bmatrix} c - b & -b \\ 0 & 1 \end{bmatrix}$. A direct calculation shows that $D_R^{-1} = \begin{bmatrix} (c-b)^{-1} & (c-b)^{-1}b \\ 0 & 1 \end{bmatrix}$.

Case (ii). Applying a similar argument

$$D = \begin{bmatrix} -d(c-d)^{-1}(c-b) & -b \\ (c-d)^{-1}(c-b) & 1 \end{bmatrix}.$$

It will be left as a homework problem to show that D for (ii) has a non-zero skewfield determinant. But, as a note, we see that $(1, d) \to (0, d - b)$ which is equivalent to $(0, 1)$ as $d - b \neq 0$. And, $(1, b) \to ((b - d)(c - d)^{-1}(c - b), 0) \equiv (1, 0)$, as b, d, c are all distinct. Then $(1, c) \to ((c - d)(c - d)^{-1}(c - b), c - b) \equiv (1, 1)$, since $c \neq b$.

It remains to show that the form of the right classical pseudo-regulus is left invariant. We have seen that the set of Baer subspaces is left invariant as a set. Furthermore, the first two components of $x = 0, y = 0, y = \delta x$, are mapped to $x = 0, y = 0$, and $(x, \delta x)$ is mapped to $(xD, (\delta x)D = \delta(xD))$, so that $y = \delta x$ is also fixed by the mapping. Now we have $\{y = xD_R^{-1}MD; M \in K^*\}$ and $\{y = \delta x; \delta \in F\}$ are the left K-components and right F-components, respectively, and K and F are isomorphic. Furthermore, these two derivable nets share three Baer subspaces that are denoted by $(0, 1)$, $(1, 0)$ and $(1, 1)$. This implies that the components of $\{y = xD_R^{-1}MD; M \in K^*\} = \left\{ y = x \begin{bmatrix} u & 0 \\ 0 & u \end{bmatrix} ; u \in F \right\}$. This completes the proof.

□

COROLLARY 10. *When $D = I$, we obtain the left classical pseudo-regulus net and the derived net of the right classical F-pseudo-regulus net; generic semi-classical left derivable nets are left classical pseudo-regulus nets.*

PROBLEM 14. *Show in the types (i) and (ii) situations in the previous theorem that the matrix D is non-singular.*

PROBLEM 15. *Determine D_R^{-1} for D of the previous exercise of type (ii).*

PROBLEM 16. *In both cases (i) and (ii) completely determine*

$$K = D \begin{bmatrix} u & 0 \\ 0 & u \end{bmatrix} D_R^{-1}.$$

This determines the matrix representation of type 3 derivable nets.

DEFINITION 38. *The set of left derivable nets S_{left}^K of type 3, which includes R_{left}^F; the derived net of R_{right}^F, is said to be the set of 'semi-classical' left derivable nets or classical when equal to R_{left}^F. When F is a field, all type 3 left derivable nets are classical.*

5. Order in type 3 derivable nets

In the previous results, we classified type 3 derivable nets relative to the three distinct Baer subspaces of the right classical pseudo-regulus that are shared with the derivable net. We did this by mapping the triple in question onto $\{(0, 1), (1, 0), (1, 1)\}$, using the short notation for Baer subplanes. But, what we did not mention is that the mapping needs to be on 'ordered' triples, which was observed in the proofs. As is the case in Theorem 25, triple (ii), $\{(1,d),(1,b),(1,c)\}$, the order matters, which means that by ordering the three $\{d,b,c\}$, we actually obtain 3! different derivable nets, all sharing the same triple of Baer subspaces of the right classical net. However, there is a catch to that statement as well, which is only true provided all of the mutually distinct differences $(d-b),(d-c),(b-c)$, are all not in $Z(F)$, the center of F. If all elements of the set are elements of $Z(L)$, then the above matrices collapse to $\begin{bmatrix} u & 0 \\ 0 & u \end{bmatrix}$; $u \in L$. So, for non-commutative skewfields, there are infinitely many mutually distinct type 3 derivable nets, which although all isomorphic, need to be considered different within a classification scheme.

DEFINITION 39. *Let Baer subplanes of the right classical pseudo-regulus be denoted more simply by their associated 2-vectors (a_i, b_i), and consider a distinct triple for $i = 1, 2, 3$, and a distinct order, notated by $(a_i, b_i)\eta; \eta \in S_3$ (the group of permutations on 3 elements of order 3!). This type of derivable net shares $\{(a_i, b_i)$; in ordered form (this means that the triple $(a_i, b_i)\sigma$ maps to $\{(0,1),(1,0),(1,1)\}$, for σ in S_3, acting on $\{1,2,3\}$. We shall use the terminology*

$$Type\ 3(a_i, b_i)\eta,$$

to denote the subtypes within type 3 derivable nets. To use the proper form for K, one may simply relabel the matrices D relative to η.

6. Derivable nets of type 2

In this section, we consider type 2 derivable nets. We begin by formulating the basic generic form, where there are two common Baer subspaces $B(S)^K_{right}(0,1)$ and $B(S)^K_{right}(1,0)$.

THEOREM 27. *A type 2 left derivable net within $PG(3,F)_S$ and related to the right classical pseudo-regulus net such that the two shared Baer subspaces are $B(S)^K_{right}(0,1)$ and $B(S)^K_{right}(1,0)$, has the following form:*

$$Components \quad : \quad x = 0, y = x \begin{bmatrix} u^\sigma & 0 \\ 0 & u \end{bmatrix} \in K; \ u \in F.$$

$$Baer\ subspaces \quad : \quad B(S)^K_{right}(a,b) =$$

$$\left\{ \left((a,b) \begin{bmatrix} u^\sigma & 0 \\ 0 & u \end{bmatrix}, (a,b) \begin{bmatrix} v^\sigma & 0 \\ 0 & u \end{bmatrix} \right) \right\};$$

$$u, v \ \in \ F, \sigma\ a\ nonidentity\ automorphism\ of\ F.$$

Furthermore, for each nonzero element c of L, cuc^{-1} is not u^σ;

σ is not an inner automorphism of F. The basis is standard.

PROOF. By assumption, we know that the matrices involved form a skewfield K, which has the following form: $\begin{bmatrix} f(u) & 0 \\ 0 & u \end{bmatrix}$; $u, f(u)$ in F, f a function on F. Clearly then, K and F are isomorphic. Since we have a skewfield, f is an additive and multiplicative function such that $f(uv) = f(u)f(v)$. Since the derivable net shares $y = x$, and $y = 0$, $f(1) = 1$ and $f(0) = 0$. Hence, f is an endomorphism σ of L. Since the intersections of the Baer subplanes on Com are 1-dimensional K-subspaces, it follows that f is surjective and hence is an automorphism σ. Furthermore, as there are exactly two shared Baer subspaces, σ is not 1. The last line of the assertion is considered below. But this completes the proof of the theorem.   □

We now consider the more general cases.

Using the short notation, the pairs T_2 of distinct Baer subspaces are

$$(i) = \{(0,1),(1,b)\}, \text{ where we may assume that } b \neq 0,$$

by the previous theorem and

$$(ii) = \{(1,d),(1,b); d \neq b\}$$

$\{(0,1),(1,0)\}$ is the image of T_2 by matrices $diag D$, for D non-singular, by the results of (3,5,3) on derivable nets of type 3. The following case (i) $D = \begin{bmatrix} c-b, & -b \\ 0 & 1 \end{bmatrix}$ and $D_R^{-1} = \begin{bmatrix} (c-b)^{-1} & (c-b)^{-1}b \\ 0 & 1 \end{bmatrix}$, where $T = \{(0,1),(1,b),(1,c)$. In this setting, one may choose c in any manner so that the matrix is non-singular; $c \neq b$. Therefore, $c - b = 1$ will suffice, or $c = 0$.

Case (ii)

$$D = \begin{bmatrix} -d(c-d)^{-1}(c-b) & -b \\ (c-d)^{-1}(c-b) & 1 \end{bmatrix},$$

where $T = \{(1,d),(1,b),(1,c)\}$, where d, b, c are distinct elements of F. Therefore the matrices will still work for T_2, indicating the first two elements of each set T. Here, $\{d,b,c\}$ must be a set of three distinct elements. So, given d and b, $d \neq b$, c can be chosen to be any other element.

To summarize:

THEOREM 28. *Let S_{left}^K be a type 2 derivable net which*

(i) *shares the F-Baer subplanes $\{(0,1),(1,b)\}$ has the following form with $D = \begin{bmatrix} b & -b \\ 0 & 1 \end{bmatrix}$ or*

(ii) *shares the F-Baer subplanes $\{(1,d),(1,b)\}$ has the following form with $D = \begin{bmatrix} -d(c-d)^{-1}(c-b) & -b \\ (c-d)^{-1}(c-b) & 1 \end{bmatrix}$.*

Components are

$$x = 0, y = xD \begin{bmatrix} u^\sigma & 0 \\ 0 & u \end{bmatrix} D_R^{-1}; u \in F;$$

$$B(S)_{right}^K(a,b)D$$

$$= \{((a,b)D(D_R^{-1} \begin{bmatrix} u^\sigma & 0 \\ 0 & u \end{bmatrix} C, (a,b)D(D_R^{-1} \begin{bmatrix} v^\sigma & 0 \\ 0 & v \end{bmatrix} D))\};$$

$u, v \in F, \sigma$ *a nonidentity automorphism,*

where $D = \begin{bmatrix} -b & b \\ 0 & 1 \end{bmatrix}^{-1}$.

(ii) *And, when S_{left}^K and R_{right}^F share the F-Baer subplanes $\{(1,b),(1,c); b$ not equal to $c\}$ then the form for the derivable net is*

Components are

$$\left\{ x = 0, \; y = xD \begin{bmatrix} u^\sigma & 0 \\ 0 & u \end{bmatrix} D_R^{-1}; u \in F \right\};$$

$$\left\{ \begin{array}{c} Baer \; subplanes : B(S)_{right}^K(a,b)D = \\ \{((a,b)D(D_R^{-1} \begin{bmatrix} u^\sigma & 0 \\ 0 & u \end{bmatrix} D_R^{-1}, (a,b)D(D_R^{-1} \begin{bmatrix} v^\sigma & 0 \\ 0 & v \end{bmatrix} D)) \end{array} \right\};$$

$u, v \in F, \sigma$ *a nonidentity automorphism.*

7. Fake type 2 derivable nets

While it might seem obvious that the derivable nets of this previous form are of type 2, when σ is not 1, recall that we are dealing with a skewfield and not necessarily a field. For example, consider the following construction: $\{c^{-1}uc \; ; b$ is a fixed nonzero element in $F, \forall u \in F\}$. Since the elements are additive and multiplicative, and when $u = 1$, we have an identity and a zero, this set defines an inner automorphism of F, say σ. If this is the case, consider the type (i) $\begin{bmatrix} c - b, & -b \\ 0 & 1 \end{bmatrix}$. Take $b = 0$ and so that $c \neq 0$.

$$\begin{bmatrix} c & 0 \\ 0 & 1 \end{bmatrix} \begin{bmatrix} u^\sigma = c^{-1}uc & 0 \\ 0 & u \end{bmatrix} \begin{bmatrix} c^{-1} & 0 \\ 0 & 1 \end{bmatrix} = \begin{bmatrix} u & 0 \\ 0 & u \end{bmatrix},$$

which may be observed with a short computation.

- What was thought to be an obvious type 2 derivable net is a fake type 2 net, which really is a type 3 derivable net.

  Here is the proof of that fact directly. If the derivable net $x = 0, y = x \begin{bmatrix} u^\sigma & 0 \\ 0 & u \end{bmatrix}; \forall u \in F$, which shares $B(S)_{right}^K(0,1)$ and $B(S)_{right}^K(1,0)$ also shares another Baer subspace of the right classical pseudo-regulus net, then it must have the form $B(S)_{right}^K(1,c)$, for some non-zero element c. This would say that since $(1,c)$ maps to (u^σ, cu), the entire left 2-dimensional subspace maps to the corresponding left 2-dimensional subspace so that $(\gamma, \gamma c) = (u^\sigma, cu)$, in the sense that for each element γ of F, there exists an element u in F such that the previous relation is valid. It now follows that for $\gamma = u^\sigma$, $\gamma c = cu$, or rather $cuc^{-1} = u^\sigma$. So, this is a fake type 2 derivable net that becomes a type 3 net.

COROLLARY 11. *Fake type 2 appearing nets sharing*

$$B(S)_{right}^K(0,1), B(S)_{right}^K(1,0)$$

$$\left\{ x = 0, y = x \begin{bmatrix} u^\sigma & 0 \\ 0 & u \end{bmatrix}; \forall u \in F \right\},$$

are type 3 nets and share $B(S)_{right}^K(1,c)$ *if and only if* $cuc^{-1} = u^\sigma$.

- Recall that type 3 derivable nets have at least three shared Baer subspaces, and we note here that if e is in $Z(F)$, then $(1,be)$ is also a shared Baer subspace, provided $(1,b)$ is shared. We have noted in the statement of the classification of derivable nets over skewfields that type 3 derivable nets share exactly three Baer subspaces if and only if $Z(F) = GF(2)$.

This is true in the above particular case, but it is true in general. The way to see this is that when a mapping takes a generator to an image, the entire space maps along with the generator, the right-left mappings will force $bub^{-1} = u$, for any shared Baer subspace $B(S)_{right}^{K}(1,b)$, showing that b is in the center of $Z(F)$ of F. We see that $y = x \begin{bmatrix} u & 0 \\ 0 & u \end{bmatrix}$ and $y = ux$, will share, in short notation, $(1,0), (0,1)$ and $(1,1)$ and will share $(1,c)$ for $c \neq 1$ or 0 if and only if $(1,c)$ in the first set of Baer subspaces of $y = ux$, is $(1,c)$ in the second set of Baer subspaces of $y = x \begin{bmatrix} u & 0 \\ 0 & u \end{bmatrix}$.

Therefore, for each $\gamma \in F$, $(\gamma, \gamma c) = (u, uc)$ for some $u \in F$. So $uc = cu$, for all $u \in F$, implying that $c \in Z(K)$. If we have a set of four common Baer subspaces, map the first three to $(0,1), (1,0)$, and $(1,1)$ then the fourth one will map to an element in the center $Z(F)$. Therefore, the set of common Baer subspaces is in a set of $1 + card\ Z(F)$ Baer subspaces.

8. Open form derivable nets of type 2

Before we leave the type 2 situation, it is important to mention a special case; when F is a central Galois quadratic extension $L(\theta)$ of a skewfield L.

THEOREM 29. *Let K be the open form of a central Galois quadratic extension $F(\theta) = \{\theta t + u : t, u \in F\}$ of a skewfield F. Then K is of type 2 with respect to the right classical pseudo-regulus net coordinatized by $F(\theta)$. The common Baer subspaces are $(1,\theta)$ and $(1,\theta^\sigma)$, using the short notation, and σ is the generated automorphism.*

PROOF. Consider the open form of $F(\theta) = K$ as follows: We have S_{left}^{K} as

$$Components \quad : \quad \left\{ x = 0, y = x \begin{bmatrix} u + \alpha t & \beta t \\ t & u \end{bmatrix} = M_{u,t} : t, u \in F \right\};$$

$$x^2 \pm \alpha x - \beta, \text{ irreducible over } K, \ \alpha, \beta \in Z(F).$$

Baer subspaces :

$$\left\{ \begin{array}{c} B(S)_{right}^{K}(a,b) = \\ \{((a,b)M_{t,u}, (a,b)M_{s,v})\}; \forall u, t, s, v \in F \end{array} \right\}$$

$$\forall (a,b) \ \in \ F \oplus F - \{(0,0)\}.$$

And, the closed form R_{right}^{F} is

$$Components : \{x = 0, y = \delta x : \delta \in L(\theta)\},$$

Baer subspaces :

$$\left\{ B(R)_{left}^{F}(a,b) = \{(c(a,b), d(a,b)) : \forall c, d \in F\} \right. : \quad (a,b) \in F \oplus F - \{(0,0)\}.$$

For a common Baer subplane $B(R)_{left}^{F}(a,b)$, it is almost immediate that $ab \neq 0$, so we may take $a = 1$. We now have $u, t \in F$, $\alpha, \beta \in Z(F)$, and given any $c \in K = F(\theta)$, there must be a unique pair (t, u) such that the equations

$$(1,b) \begin{bmatrix} u + \alpha t & \beta t \\ t & u \end{bmatrix} = (c, cb)$$

have a solution. Consider what would be the requirement for b if $c \in Z(F(\theta)) - Z(F)$. Then, this changes the equation slightly to

$$(1, b) \begin{bmatrix} u + \alpha t & \beta t \\ t & u \end{bmatrix} = (c, bc) = (1, b) \begin{bmatrix} c & 0 \\ 0 & c \end{bmatrix}.$$

Let $c = \theta c_1 + c_2$, for $c_1 \neq 0$ for $c_1, c_2 \in Z(F)$. Although this does not say anything about whether u and t might then be in $Z(L(\theta))$, what it does allow is that

$$(1, b) \begin{bmatrix} u + \alpha t - c & \beta t \\ t & u - c \end{bmatrix} = (0, 0).$$

We don't have a viable way to use the determinant, since the skewfield could be non-commutative. But, we do know that

$$(*) \quad \begin{aligned} u + \alpha t - (\theta c_1 + c_2) + bt &= 0 \\ \beta t + b(u - (\theta c_1 + c_2)) &= 0. \end{aligned}$$

Hence, for $c_1 \neq 0$, there is a solution (t, u) if there is a common Baer subspace.

$$b(u - (\theta c_1 + c_2)) = -b(b + \alpha)t = -\beta t,$$

using both equations of $(*)$, which implies that

$$(b^2 + \alpha b - \beta)t = 0.$$

If $t = 0$ then $u = \theta c_1 + c_2$, a contradiction, as $u \in F$ and $\theta c_1 + c_2 \in Z(F(\theta)) - Z(F)$. Hence, $t \neq 0$ which implies that b satisfies $x^2 - \alpha x + \beta$ so that $b = \theta$ or θ^σ. Since this element is in $Z(F(\theta))$, the analysis is now valid for all $c \in F(\theta)$ (we now know that $(c, bc) = (1, b) \begin{bmatrix} c & 0 \\ 0 & c \end{bmatrix}$), so c_1, c_2 are now just in F, not necessarily in $Z(F)$). This gives a necessary condition for a common Baer subspace. Choose $b = \theta$ and given any c, we need to find a solution (t, u). Since $b = \theta$, and $u, t \in F$ with $\alpha, \beta \in Z(F)$, it follows that $t = -c_1$ and $u + \alpha t = c_2$, and $u = c_2 + \alpha c_1$ by the first equation of $(*)$. The converse also follows and shows the connection between c and (t, u) is established. The case when $b = \theta^\sigma$ is analogous and therefore, we have the proof of the theorem.

COROLLARY 12. *When $K = F(\theta)$ is a field and defines a regulus then K is the field open form of F as a type 2 twisted regulus net.*

$\square$

9. Order in type 2 derivable nets

Order is also important in type 2 derivable nets, as the mapping taking $(0, 1)$ to $(1, b)$ and $(1, 0)$ to $(1, c)$ is potentially different than when b and c are interchanged. So, order matters.

DEFINITION 40. *There are then subtypes with the type 2 derivable nets, we shall use for shared nets (a_i, b_i), we have types $2 - (\sigma, (a_i, b_i))$ and type $2 - (\sigma, (b_i, a_i))$, where σ denotes the associated automorphism of L;*

type $2 - (\sigma, (a_i, b_i)\eta)$, where η is 1 or $\begin{bmatrix} 0 & 1 \\ 1 & 0 \end{bmatrix}$, for $i = 1, 2$.

REMARK 10. *We note that σ is not restricted to finite order, so if a field K has an infinite automorphism, than we still have a type 2 derivable net.*

As an example of an automorphism group which has elements of both finite and infinite order, let $K = F(z)$, the rational function field over the field of real numbers. The automorphism group has defining elements $x \to \frac{ax+b}{cx+d}$, where $ad - bc \neq 0$. Then the automorphism $x \to 2x$ has infinite order, whereas the automorphism $x \to -x$ has finite order 2.

10. Derivable nets of type 1

Much of the work on type 1 derivable nets is from Jha and Johnson [**70**], (also see (12,15,10) on corrections to that work). There is very similar material, developed more generally from rings in Cohn [**29**], chapter 2, p. 49-52, but without the derivable net connection. The specific connections are mentioned more completely in the section on examples.

THEOREM 30. *Let S_{left}^F be a derivable net of type 1 and assume that the shared derivable net is $(0,1)$, again using the short notation. Then the net has the following form:*

$$Components: \left\{ x = 0, y = x \begin{bmatrix} u^\sigma & A(u) \\ 0 & u \end{bmatrix} ; u \in F \right\},$$

Baer Subspaces :

$$: \left\{ \begin{array}{c} B(S)_{right}^K(a,b) = \{((a,b) \begin{bmatrix} u^\sigma & A(u) \\ 0 & u \end{bmatrix}), (\begin{bmatrix} v^\sigma & A(v) \\ 0 & v \end{bmatrix})) \\ : u,v \in F \} : (a,b) \in F \times F - \{(0,0)\} \end{array} \right\};$$

$$A(0) = A(1) = 0,$$

σ an automorphism, possibly 1,

A a non-identically zero function on F, such that

$$u^\sigma A(v) + A(u)v = A(uv);$$

A *is defined*

as a twisted derivative function

or $\sigma - $ derivation.

When F is a field then $\sigma = 1$. In general, when $\alpha = 1$,

A is called a derivative function

or a derivation.

PROOF. We have a derivable net with matrix representation in the form

$$\begin{bmatrix} f(u) & A(u) \\ 0 & u \end{bmatrix} ; u \in F.$$

The functions f and A are additive functions with $f(uv) = f(u)f(v)$, $f(1) = 1$, $A(1) = A(0) = 0$. Hence, f is an automorphism of F, σ possibly 1 (the endomorphism is surjective, mentioned previously, due to the derivable net assumptions).

$$\begin{bmatrix} u^\sigma & A(u) \\ 0 & u \end{bmatrix}\begin{bmatrix} v^\sigma & A(v) \\ 0 & v \end{bmatrix} = \begin{bmatrix} (uv)^\sigma & u^\sigma A(v) + A(u)v = A(uv) \\ 0 & uv \end{bmatrix}.$$

We need to determine, in spite of the appearance of the matrices, that there cannot be an additional shared Baer subspace. So, assume that there is, say $B(S)^K_{right}(1, b)$ is a common Baer subspace. Since A is not identically zero, b cannot be zero. Then we have $(1, b)$ maps to $(u^\sigma, A(u) + bh)$, which implies that $(\gamma, \gamma b) = (u^\sigma, A(u) + bu)$, in the sense that given $\gamma \in F$, there exists an element $u \in F$, such that the previous relationship holds. Hence, we have

$$u^\sigma b = A(u) + bu \ ; \forall u \in F, \text{ so that } A(u) = u^\sigma b - bu.$$

We note that

$$\begin{bmatrix} 1 & -b \\ 0 & 1 \end{bmatrix}\begin{bmatrix} u^\sigma & 0 \\ 0 & u \end{bmatrix}\begin{bmatrix} 1 & b \\ 0 & 1 \end{bmatrix} = \begin{bmatrix} u^\sigma & u^\sigma b - bu \\ 0 & u \end{bmatrix}.$$

Since this is a type 2 derivable net by results of (3,5,6), this completes the first part of the proof, which is very similar to the following. It remains to show that when F is a field, we may assume that $\sigma = 1$. So, we now have

$$u^\sigma A(v) + vA(u) = A(uv) = v^\sigma A(u) + uA(v).$$

Fix v_0 so that v_0^σ is not v_0. Let $A(v_0) = k$. Then we have:

$$A(u)(v_0^\sigma - v) = (u^\sigma - u)k, \ \forall u \in F.$$

If $k = 0$ then A is identically zero, a contradiction to type 1. Hence, $A(u) = (u^\sigma - u)c$, for c a non-zero constant. In this setting, change bases by $\mathrm{diag}\begin{bmatrix} 1 & c \\ 0 & 1, \end{bmatrix}$, so that

$$y = x \begin{bmatrix} u^\sigma & (u^\sigma - u)c \\ 0 & u \end{bmatrix}$$

maps onto the equation

$$y = x \begin{bmatrix} 1 & -c \\ 0 & 1, \end{bmatrix}\begin{bmatrix} u^\sigma & (u^\sigma - u)c \\ 0 & u \end{bmatrix}\begin{bmatrix} 1 & c \\ 0 & 1, \end{bmatrix},$$

noting that the inverse of $\begin{bmatrix} 1 & c \\ 0 & 1, \end{bmatrix}$ is $\begin{bmatrix} 1 & -c \\ 0 & 1, \end{bmatrix}$. But, this is

$$y = x \begin{bmatrix} u^\sigma & 0 \\ 0 & u \end{bmatrix},$$

a type 2 derivable net. Since $(0,1)\begin{bmatrix} 1 & -c \\ 0 & 1, \end{bmatrix} = (-c, 1) = (1, -c^{-1})$, as a 1-dimensional subspace (an intersection, representing the Baer subspace $B(S)^K_{right}$ $(1, -c^{-1})$), we see that the original net is also of type 2 with $\{(0, 1), (1, -c^{-1})\}$, as the two Baer subspaces. Hence, since we are in a type 1 situation, it follows that $\sigma = 1$, when F is a field. Hence, in this situation A is a derivative function. $\square$

THEOREM 31. *A derivable net S_{left}^K of type 1 with fixed Baer subspace $B(S)_{right}^K$ $(1, b)$, has the following form:*

$$x = 0, y = x \begin{bmatrix} 1 & 0 \\ -b^{-1} & b^{-1} \end{bmatrix} \begin{bmatrix} u^\sigma & A(u) \\ 0 & u \end{bmatrix} \begin{bmatrix} 1 & 0 \\ 1 & b \end{bmatrix} ; u \in L,$$

where σ is an automorphism, possibly 1, A is a non-identically zero function on F, such that

$$u^\sigma A(v) + A(u)v = A(uv);$$

A is defined to be a twisted derivation function. When F is a field then $\sigma = 1$ and A is a derivative function. The form then is as follows:

$$x = 0, y = x \begin{bmatrix} u^\sigma + A(u) & A(u)b \\ b^{-1}(u - u^\sigma - A(u)) & b^{-1}(u - A(b))b \end{bmatrix} ;$$

$$\forall u \in F.$$

PROOF. The general map that takes $(1, b)$ to $(0, 1)$ is

$$D = \begin{bmatrix} -bg & 1 - bh \\ g & h \end{bmatrix}.$$

Take $g = -b^{-1}$ and $h = b^{-1}$ to obtain the mapping used in the theorem. The determinant is $(-bg)g^{-1}h - (1 - bh))g = -1g$. This proves the theorem. □

11. Examples of type 1 derivable nets

First, we show a powerful construction involving fields of characteristic 2.

THEOREM 32. *Let F be a field of characteristic 2 that admits a non-square γ. So, we may assume that F is an inseparable field with prime field $GF(2)$. Then there is an open form construction of a quadratic extension field $F(\sqrt{\gamma})$ as $\begin{bmatrix} v & \gamma s \\ s & v \end{bmatrix}$, for all $v, s \in F$, as the determinant is $v^2 - \gamma s^2$ and cannot be zero for $(s, v) \neq (0, 0)$. Let $K = F(\sqrt{\gamma})$ and consider the (right) classical regulus net in $PG(3, K)$. Then the open form derivable net*

$$x = 0, y = x \begin{bmatrix} u & A(u) \\ 0 & u \end{bmatrix}, \forall u \in K,$$

and

$$A(u) = \delta/\delta\sqrt{\gamma}(u) = s), \text{ with } u = \sqrt{\gamma}s + v : s, v \in L$$

is a type 1 derivable net, where the notation represents formal partial differentiation with respect to $\sqrt{\gamma}$

PROOF. Just check that

$$(**) : uA(w) + A(u)w = A(uw),$$

for all $u, w \in F$. First writing $u = \sqrt{\gamma}s + v$ and $w = \sqrt{\gamma}k + m$. Then we obtain $uk + ws$ must equal $\delta/\delta\sqrt{\gamma}(uv)$. Work out uv, to obtain $\gamma s + vm + \sqrt{\gamma}(vk + sm)$, whose (partial) derivative with respect to $\sqrt{\gamma}$ is just $(vk + sm)$, as the right side of $(**)$. The left side is $(\sqrt{\gamma}s + v)k + (\sqrt{\gamma}k + m)s$, and note that in characteristic 2, the $\sqrt{\gamma}$-terms drop out to obtain $(vk + ms)$, thus proving equation $(**)$. Just to determine that this example, is, in fact, type 1. We look at the Baer subspaces of R^F,

$$B(R)^F(a, b) = \{((ca, cb), (da, db)) : c, d \in F\}.$$

We apply the right mappings of $\begin{bmatrix} u & A(u) \\ 0 & u \end{bmatrix}$ to $B(R)^F(a, b)$, just restricted to $y = 0$. There is a invariant Baer subspace if and only if

$$(a, b) \begin{bmatrix} u - \gamma & A(u) \\ 0 & u - \gamma \end{bmatrix} = (0, 0),$$

since we have a field situation. This is true if and only if $u = \gamma$, for all u, we have $(aA(u), 0) = (0, 0)$, so that $a = 0$, which implies that $B(R)^F(0, 1)$, is the unique shared Baer subspace; the derivable net is of type 1. This completes the proof. $\square$

EXAMPLE 1. *Now let F be any field and let $K = F(x_1, x_2, ...x_n)$ be the rational function field on n commuting indeterminates x_i. Let the (right) classical regulus net $x = 0$, $y = zx$; $\forall z \in F$ let*

$$x = 0, y = x \begin{bmatrix} u & A(x) \\ 0 & u \end{bmatrix}; \forall u \in K = F(x_1, x_2, ...x_n),$$

such that

$$A(u) = (\delta/\delta x_i)u \text{ for } u \in K,$$

where the notation is meant to indicate the formal partial derivative with respect to x_i. Then, this provides an example of a type 1 derivable net with derivation.

The following example shows another characteristic 2 example.

EXAMPLE 2. *For another example, we require a quadratic extension of $F(x_1, x_2, ...x_n)$, of characteristic 2. Note that any x_i is a non-square, so we may form the quadratic extension*

$$F(x_1, x_2, ...x_n)(\sqrt{x_i}).$$

Consider the following open form $\begin{bmatrix} u & xt \\ t & u \end{bmatrix}$; for all $u, t \in F(x_1, x_2, ...x_n)$. Then the determinant is $u^2 - xt^2 = 0$ if and only if x is a square, for t nonzero and equal to zero for $t = 0$ if and only if $(t, u) = (0, 0)$. Now consider the closed form (right) classical regulus net $x = 0$, $y = \delta x$, for all $\delta \in F(x_1, x_2, ...x_n)(\sqrt{x_i})$. And the corresponding (left) derivable net, in open form

$$y = x \begin{bmatrix} u & A(x) \\ 0 & u \end{bmatrix}; \forall u \in W = F(x_1, x_2, ...x_n)(\sqrt{x_i})$$

and let

$$A(x) = \delta/\delta\sqrt{x_i}(u) : u \in W.$$

In this situation, let $u = \sqrt{x_i}t + v$, for $t, v \in F(x_1, x_2, ...x_n)$. Then the formal partial derivative is just t. So, we have the form

$$\begin{bmatrix} \sqrt{x_i}t + v & t \\ 0 & \sqrt{x_i}t + v \end{bmatrix}.$$

This is another example of a derivable net of type 1 with a derivative function.

We shall return to descriptions of type 1 derivable nets in (12,14,2-5) and an open problem, after a brief description of skew polynomial rings.

12. Carrier nets

There are many derivable nets that we have not considered within the main classification. There are some ways to deal with such nets, using 'carrier derivable nets'.

DEFINITION 41. *Let R denote a classical pseudo-regulus net in $PG(3, K)$, where K is a skewfield. Hence, in the contraction representation, R has three special components $x = 0, y = 0$, $y = x$. Let S be a derivable net in $PG(3, K)$ that shares i of the three components, in ordered form, left to right, for $i = 0, 1, 2, 3$. Then S is called a 'carrier derivable net of class i'.*

Our classification discussed left derivable nets that shared three common components $x = 0, y = 0, y = x$. These three components are called 'carriers'. If a left derivable net in a given common vector space wherein the right classical pseudo-regulus lives, we required that it had three carriers. But, there are a variety of left derivable nets that share two carriers and one carrier and actually 0 carriers..

Consider a left derivable net sharing two carriers. We also consider that bases have been chosen so that the two carriers are $x = 0$ and $y = 0$. This will force the remaining components to have the form $y = xM$, for all $M \in \lambda$, where λ is a 'pre-skewfield' within $GL(2, L)_{left}$. For example, consider an open form Desarguesian translation plane, coordinatized by L, a quadratic skewfield extension of L.

$$x = 0, y = x \begin{bmatrix} u + \alpha t & \beta t \\ t & u \end{bmatrix}; \alpha, \beta \neq 0 \in Z(L), \text{ for all } t, u \in L. \text{ Now we look the}$$

derivable net

$$x = 0, y = x \begin{bmatrix} \alpha t & \beta t \\ t & 0 \end{bmatrix}; \forall t \in L. \text{ This is a 2-carrier derivable net within}$$

$PG(3, L_{left})_S$. To see that this is, in fact, a derivable net, we may post multi-

ply by $\begin{bmatrix} \alpha & \beta \\ 1 & 0 \end{bmatrix}^{-1} = \begin{bmatrix} 0 & 1 \\ \beta^{-1} & -\alpha\beta^{-1} \end{bmatrix}$, to obtain $\begin{bmatrix} t & 0 \\ 0 & t \end{bmatrix}$ (the right inverse, when

$\alpha, \beta \in Z(L)$, is just the commutative (field) inverse). This is isomorphic to the derived net, the left classical pseudo-regulus net. There is a nice bonus with these sorts of 2-carrier nets. What would be thought to be a type 3, derivable net, in the 3-carrier situation, here is actually derived by an irreducible net. To see this, we ask, if there are any common Baer subplanes, could this be possible, in the net case? We have seen that just because two derivable nets share three carrier components and a Baer subplane or not, this does not mean that they share all components. So, it might be possible that a 2-carrier left derivable net shares a Baer subplane with a 3-carrier net. An easy example of this sort of net would be

$$x = 0, y = 0, y = x \begin{bmatrix} u\rho & 0 \\ 0 & u \end{bmatrix}, \rho \neq 1, \forall u \in L.$$

So, for a 2-carrier net, by post or pre-multiplying by a left inverse of any one of the matrices in $GL(2, L)_{left}$, the net would become a 3-carrier net of type $j = 0, 1, 2$, or 3.

DEFINITION 42. *(1) A 2-carrier net D is defined to be of 'type $2 - j$', for $j = 1, 2, 3$ if and only if post multiplication by $\begin{bmatrix} I & 0 \\ 0 & M^{-1} \end{bmatrix}$ where one of the components* *of D has the form $y = xM$, transforms the net to type j. The notation will be:*

$$(2-j) \begin{bmatrix} I & 0 \\ 0 & M^{-1} \end{bmatrix}.$$

This transformation maps the 2-carrier net to a 3-carrier net. The classification scheme now applies to classify the 3-carrier nets of type j. Note the classification is applied after the transformation.

(2) A 1-carrier net is defined to be of type $1-(2-j)$, if and only if post multiplication by $\begin{bmatrix} I & -M \\ 0 & I \end{bmatrix}$ is of type $2-j$. The notation is

$$(1-(2-j)) \begin{bmatrix} I & -M \\ 0 & I \end{bmatrix}.$$

Assuming that the 1-carrier is $x = 0$, the transformation maps the 1-carrier net to a 2-carrier net of type $2-j$. A subsequent transformation, as in (1) transforms the type $2-j$ net into a 3-carrier net of type j.

(3) A 0-carrier net is defined to be of type $0-(1-(2-j))$ if and only if using premultiplication by $\begin{bmatrix} I & 0 \\ -M & I \end{bmatrix}$ transforms the net to a type 1 net. The notation is

$$\begin{bmatrix} I & 0 \\ -M & I \end{bmatrix} (0-(1-(2-j))).$$

The indicated transformation produces a type $(1-(2-j)$, a second transformation of type as in (2) (with a new component $y = xN$, and corresponding mapping), by using $\begin{bmatrix} I & -N \\ 0 & I \end{bmatrix}$, gives a type $(2-j)$ derivable net and finally a third transformation as in (1) using $y = xJ$ and mapping $\begin{bmatrix} I & 0 \\ 0 & J^{-1} \end{bmatrix}$ produces a 3-carrier net of type j.

(4) When $j = 2$, use the order type $2\text{-}(\sigma, (a_i, b_i)\eta)$, where η is 1 or $\begin{bmatrix} 0 & 1 \\ 1 & 0 \end{bmatrix}$, for $i = 1, 2$.

(5) When $j = 3$, use the order type

$$Type\ 3(a_i, b_i)\eta, i = 1, 2, 3; \eta \in S_3.$$

Reversing the ideas, here are the connections between carrier classes.

THEOREM 33.

(1) Let $\mathcal{D}$ be a derivable net of carrier class 3. Then $\mathcal{D} \begin{bmatrix} I & 0 \\ 0 & M \end{bmatrix}$ for M not in the matrix skewfield is a derivable net of carrier class 2.

(2) Let $\mathcal{D}$ be a derivable net of carrier class 2. Then $\mathcal{D} \begin{bmatrix} I & M \\ 0 & I \end{bmatrix}$, for M not in the matrix set of $\mathcal{D}$, is a derivable net of carrier class 1. Let $\mathcal{D}$ be a derivable net of carrier class 1. Then $\begin{bmatrix} I & 0 \\ M & I \end{bmatrix} \mathcal{D}$, for M not in the matrix set of $\mathcal{D}$, is a derivable net of carrier class 0.

REMARK 11. *There is a method using 'circle geometry' that provides a convenient method for describing derivable nets of Pappian translation planes over full* *fields $K(\sqrt{\gamma})$, where K is a full field. All of these derivable nets are regulus nets, and would appear as carrier nets. See (9,11,11) on flocks of elliptic quadrics. There are sets of carrier nets of class 0 that partition the spread minus $0, 1$ or 2 components.*

13. Derivable nets in translation planes

- Carrier Class 0 derivable nets are found in Desarguesian planes admitting an automorphism of order 2 by the Multiple Replacement Theorem of part 10, or in a translation plane corresponding to a flock of an elliptic quadric, part 9. Also found in any translation plane of order q^2 admitting a cyclic homology group of order $q+1$ and the infinite generalizations both of planes and of quasifibrations (5,7,9). Also, by lifting quasifibrations of dimension 2 or by lifting non-commutative skewfields (7,9,2).

- Carrier Class 1 derivable nets are found in any translation plane corresponding to a flock of a quadratic cone or an α-cone. Also, unusual Pappian spreads of classes 0 obtained using flocks of elliptic quadrics can become of class 1.

- Carrier Class 2 derivable nets are found in any translation plane corresponding to a flock of a hyperbolic quadric or twisted hyperbolic quadric.

- Carrier Class 0 derivable nets are found in any Desarguesian plane admitting an automorphism of order 2.

- In part 10, we provide a general multiple replacement procedure that is valid for any Desarguesian plane coordinatized by a skewfield that admits an automorphism τ. This procedure provides a set of mutually disjoint replacement nets that together with $x = 0$ and $y = 0$ covers the spread. Applying this shows that given any quaternion division ring $(a, b)_K^{\alpha}$, written as a matrix skewfield over $K(\theta)$, there is a class of mutually disjoint derivable nets in the associated Desarguesian translation plane using an inner automorphism, where each derivable net is a type 2 derivable net relative to the classical left regulus net relative to $K(\theta)$. This would be a class 0 example.

- Using the same procedure as above for $(a, b)_{K(\tau)}^{\alpha}$, provides a Desarguesian translation plane admitting a class of mutually disjoint derivable nets each isomorphic to the classical left pseudo-regulus net relative to $(a, b)_K^{\alpha}$. This also would also be a class 0 example.

- Given any finite flock of a quadratic cone, of order q^2, there is an associated translation plane of order q^2 that admits an affine cyclic homology group H_{q+1} of order $q + 1$. The orbits of H_{q+1} are regulus nets, so that multiple derivation is possible. There derivable nets are of carrier class 0. Hence, finite translation planes q^2 admitting a set of $q - 1$ mutually disjoint derivable nets with associated cyclic homology group of order $q+1$ are ubiquitous when regarding flocks of quadratic cones. In (5,7,8) the correspondence between such cyclic group translation planes and quadratic flock planes is explicated and applied in (6,8,4) to find new classes of α-twisted hyperbolic flocks.

Part 4

Flocks of α-cones

In this part, we consider a generalization of flocks of quadratic cones to flocks of what are called α-cones, a variant using an automorphism of the associated field. Just as there are associated translation planes admitting regulus-inducing elation groups, the generalization of α-flocks has corresponding translation planes admitting twisted regulus-inducing elation groups. Much of this part follows Cherowitzo, Johnson, and Vega [26], with some slight variation.

To see how the ideas connect, we make variations on the Klein quadric.

CHAPTER 6

Klein quadric and generalization

Many of the translation planes that shall be encountered in this book can be obtained using the Klein quadric, as they are of so-called dimension 2, with ambient space a 4-dimensional vector space over a field K. The spread then is in $PG(3, K)$. Considered in the projective space, homogeneous coordinates $(x_1, x_2, x_3, x_4) \equiv \delta(x_1, x_2, x_3, x_4)$, for all $\delta \in K^*$ are employed, $x_i \in K, i = 1, 2, 3, 4$.

DEFINITION 43. *The Klein quadric Ω_5 in $PG(5, K)$ given by homogeneous coordinates $(z_1, z_2, z_3, z_4, z_5, z_6)$ is a hyperbolic quadric with defining relations:*

$$z_1 z_6 - z_2 z_5 + z_3 z_4 = 0.$$

The way that quadratic cones, hyperbolic quadrics, and elliptic quadrics are considered in $PG(3, K)$ may be defined as follows:

Embed the **hyperbolic quadric** Ω_3 of $PG(3, K)$ with homogeneous coordinates (x_1, x_2, x_3, x_4) such that

$$x_1 x_2 - x_3 x_4 = 0,$$

into Ω_5 by letting $(x_1, x_2, x_3, x_4) \to (x_1, x_3, 0, 0, x_4, x_2)$.

For a **quadratic cone** in Ω_3, defined by $x_1 x_2 = x_3^2$, in the plane $x_3 = x_4$ with vertex $(0, 0, 0, 1)$ then may also be embedded in Ω_3 as $x_4 = x_3$, so also in Ω_5.

For an **elliptic quadric** $x_3 x_4 - \gamma x_1^2 + x_2^2 = 0$, for γ a non-square, embed as follows:

$$(x_1, x_2, x_3, x_4) \to (x_3, x_1, x_2, x_2, \gamma x_1, x_4).$$

There is a bijection between lines of $PG(3, K)$, and the points of the Klein quadric. Considering the lines as 2-dimensional vector spaces of the associated 4-dimensional K-vector space of the form $y = x \begin{bmatrix} a & b \\ c & d \end{bmatrix}$ with non-zero determinant Δ, the mapping is in this manner: flocks of quadratic cones are in bijective correspondence with translation planes in $PG(3, K)$ admitting regulus-inducing elation groups. But the manner of construction involves not the components but the subspaces defined by the set of Baer subspaces.

Hyperbolic flocks are in bijective correspondence with translation planes in $PG(3, K)$ admitting regulus-inducing homology groups.

Elliptic $i-$flocks (defined as a partition of an elliptic quadric by conics except for at most two points) are in bijective correspondence with Pappian spreads 2-dimensional over K, that admit a cover of mutually disjoint regulus nets except for possible for $i = 0$, 1, or 2 components.

The construction using regulus-inducing elation and homology groups shows that the point orbits under the groups are connected under the Klein quadric. These point orbits form the Baer subplanes of the corresponding regulus nets.

Since there are twisted regulus nets and translation planes admitting twisted regulus net inducing elation and homology groups, there are connections then with flocks of certain varieties that use the same terminology, but are, of course, completely different structures. We consider two possibilities in the following section.

1. α-Klein quadric

This type of construction as above can be formed analogously using a variation on the Klein quadric, called the α-Klein quadric.

Consider a 6-dimensional vector space V_6 over K in the standard manner and over K^α with special scalar multiplication $\cdot$ as follows:

$$(x_1, x_2, x_3, x_4, x_5, x_6) \cdot \delta = (x_1\delta^\alpha, x_2\delta, x_3\delta^\alpha, x_4\delta, x_5\delta^\alpha, x_6\delta),$$

and written over the associated standard basis, we obtain vectors as

$$(x_1^{\alpha^{-1}}, x_2, x_3^{\alpha^{-1}}, x_4, x_5^{\alpha^{-1}}, x_6),$$

then the standard hyperbolic form Ω_5 becomes

$$x_1^\alpha x_6 - x_2 x_5^\alpha + x_3^\alpha x_4 = 0.$$

Now embed $PG(3, K)$ $(x_1^{\alpha^{-1}}, x_2, x_3^{\alpha^{-1}}, x_4) \to (x_1^{\alpha^{-1}}, x_3, 0, 0, x_4^{\alpha^{-1}}, x_2)$. We note that the group E^α now acts on $PG(3, K)$. We consider the associated α-hyperbolic form, which can be given by $x_1^\alpha x_6 = x_2 x_5^\alpha$, without loss of generality. Now consider the set of points within $x_2 = x_5$, to obtain $x_1^\alpha x_6 = x_2^{\alpha+1}$, an α-conic. We realize that E^α leaves the associated homogeneous points on the α-regulus of Baer subspaces pointwise fixed (as 'points'). Identifying the α-quadric projectively as the α-twisted hyperbolic quadric, we see that we may consider the α-conical flock as a set of points within the α-twisted quadric in $PG(3, K)$ as follows: Consider a point of $PG(3, K)$ say $v_0 = (0, 0, 0, 0, 0, 1)$ as the vertex of the cone, then a set of planes of $PG(3, K)$ that partitions the points of the lines from v_0 to the α-conic $x_1^\alpha x_6 = x_2^{\alpha+1}$, correspond to the K-components of a translation plane with spread in $PG(3, K)$ that admit an α-regulus-inducing elation group E^α. The converse that from a translation plane with spread in $PG(3, K)$ with group E^α produces an α-flock of the α-quadratic cone is now immediate.

Although, the analogies are apparent, what we are calling the α-Klein quadric is not the Klein quadric on the field $\left\{ \begin{bmatrix} u^\alpha & 0 \\ 0 & u \end{bmatrix} ; u \in K \right\}$, where K is a field admitting the automorphism α. We recall also that this derivable net, the twisted regulus net in $PG(3, K)$, or rather existing in the associated 4-dimensional K-space, is also a regulus net in $PG(3, \left\{ \begin{bmatrix} u^\alpha & 0 \\ 0 & u \end{bmatrix} ; u \in K \right\})$. So, there would be a Klein quadric relative to this field. Why this consideration is important is that the method of using the Klein quadratic for constructions of flocks of quadratic cones, and hyperbolic quadrics showing that associated translation planes are constructed or rather are equivalent, also works for the α-Klein quadric.

2. Construction of general flocks

The original Klein quadric construction of Thas [137] and Walker [142], used the Klein quadric to associate translation planes with the flocks of a quadric cone and of a hyperbolic quadric in $PG(3, K)$. The analysis that is considered here is from a slightly different point of view. Moreover, these methods are valid for α-flocks, for non-commutative flocks and non-commutative α-flocks. For all of these general flocks, there are associated translation planes with spreads in $PG(3, K)$, where now, K is a skewfield, possibly commutative but also possibly non-commutative, which admit what might be called 'derivable net'-inducing groups.

We shall use the notation developed in the parts involving the classical theory of derivation and of the classifying derivable nets. The use of the Klein quadric generalizations have not been completed for all types of derivable nets, but our definition of the possibility of these generalizations is shown in the following definition.

DEFINITION 44. *Let K be a skewfield and let V_4 be a 4-dimensional vector space over K, which is a K-bimodule. Let D be a derivable net with components in $PG(3, K)_S$. Thinking of the contraction method from the embedding theory, writing the derivable net with components of the form $y = xM$, for $M \in GL(2, K)_{left}$, means that M is a left 2-dimensional K-subspace. In his setting D is a left pseudo-regulus net but not necessarily classical. If there is a translation plane with spread in $PG(3, K)_{S,left}$ admitting a central collineation group one of whose orbits, together with the axis or axes, is D, we shall call this group a 'derivable net-inducing central collineation group'. There are two possible groups, an elation group or a homology group, for each type of derivable net.*

DEFINITION 45. *Assume the conditions of the previous definition. Assume that a translation plane with mixed or blended (left/right) kernel in $PG(3, K)$, that admit a D-derivable net-inducing Baer elation or Baer homology group. By this we mean that one of whose orbits not incident with the subnet containing the Baer axis or axes, together with this axis or axes, is a D-derivable net.*

So, there are conceivably central collineation derivable net-inducing groups and Baer derivable net-inducing groups for any derivable net. We have determined the forms of the possibilities for the types $0, 1, 2, 3$, when the derivable nets share three components $x = 0, y = 0, y = x$, which ensures that the derivable nets are coordinatized by skewfields.

We have also discussed the carrier nets of carrier class $k = 0, 1, 2$, where a derivable net might only contain k of the fundamental components listed. In the part 10 of this text, called Multiple Replacement Theorem, there is a construction technique given that constructs in any Desarguesian plane, coordinatized by any skewfield that admits a non-trivial automorphism, a set of mutually disjoint replaceable nets that cover the components except for $x = 0$ and $y = 0$, so that technically all replaceable nets when they are derivable nets are of carrier class 0.

When the automorphism has order 2, we have a set of mutually disjoint derivable nets. In part 12, it is shown that every quaternion division ring Desarguesian plane may be extended to a quaternion division ring Desarguesian plane that admits an automorphism of order 2 and hence there is a set of mutually isomorphic derivable nets of carrier class 0. These are derivable nets in $PG(3, (a, b)_F = K)$, using the notation of part 2.

The general construction of flocks from group orbits considers the point orbits of any of these derivable net-inducing groups, now called 'points'. If the set of these points can be organized into a set of derivable nets containing the axis or axes of the groups and if these 'points' are meaningful in $PG(3, K)$ (left or right), then an associated flock of $PG(3, K)$ by planes (left or right) can be determined.

The details of this construction technique will be made clear in part 4 on α-quadrics.

3. The field case

We shall illustrate this idea using a field K so that the derivable net is a regulus net in $PG(3, K)$.

We first consider the quadratic cone case, where the associated translation plane admits a regulus-inducing elation group. The form of the group is

$$\left\langle \begin{bmatrix} I_2 & \begin{bmatrix} u & 0 \\ 0 & u \end{bmatrix} \\ 0_2 & I_2 \end{bmatrix} ; u \in K \right\rangle = E$$

The idea then is to show that the point orbits are 2-dimensional K-subspaces that share a 1-dimensional subspace on the axis of the elation group, $x = 0$. In the projective space $PG(3, K)$, the point-orbits are 'points'. Each derivable net is defined by E-orbits of the spread of the translation plane. The sets of point orbits, however, are the Baer subplanes of the derivable nets. Putting this another way, the point-orbits have been organized into the sets of Baer subplanes of the derivable nets, that share a common component. Each such set of Baer subplanes is then described by a plane of $PG(3, K)$, which has a relationship to the derivable net from which the set of Baer subplanes arise. There is a cover of the quadratic cone by planes exactly when the derivable nets in the affine space define a translation plane. Since the translation plane is defined 2 by 2 matrices, we then have a Klein quadric type of construction. Although this was not apparent in the original manner of presentation, we have that flocks of quadratic cones are equivalent to translation planes admitting regulus-inducing elation groups with spreads in $PG(3, K)$.

The hyperbolic flocks work exactly the same way, except that the translation planes use regulus-inducing homology groups and the derivable nets share two components, the axes (axis and coaxis) of the group.

Baer elation groups that are regulus-inducing act essentially the same way, except that what occurs is an equivalent partial flock of a quadratic cone missing one plane; a deficiency-one partial flock of a quadratic cone is obtained. Baer homology groups that are regulus-inducing correspond to a deficiency-one partial hyperbolic flock.

In a way, the affine approach using regulus-inducing groups or derivable-net inducing groups is more general and could help guide the way for research in the general theory of flocks.

Again, more details will given in the sections that deal with α-flocks and non-commutative flocks.

PROBLEM 17. *Using the Klein quadric,*

(1) *Show that a spread in $PG(3, K)$, for K a field, corresponding to a flock of a quadratic cone, is also a hyperplane dual spread.*

(2) *Show that a spread in $PG(3, K)$, for K a field, corresponding to a flock of a hyperbolic quadric, is also a hyperplane dual spread.*

4. Algebraic construction for α-cones

Here we show that the algebraic construction connects flocks of quadratic cones and translation planes with regulus-inducing collineation groups. However, we may do this for α-quadratic cones and translation planes admitting α-regulus-inducing collineation groups, obtaining the flock situation when $\alpha = 1$.

In this section, we are interested in extending the ideas of deficiency-one flocks of quadratic cones in $PG(3, q)$ and the associated Baer group theory of translation planes with spreads in $PG(3, q)$ to their generalizations, α-flokki of cones C_α, where $C_\alpha = \{(x_0, x_1, x_2, x_3) \mid x_0^\alpha x_1 = x_2^{\alpha+1}\}$ is a cone in $PG(3, q)$ with vertex $(0, 0, 0, 1)$ with $\alpha \in Aut(K)$. An α-flokki is a 'flock' of this cone, that is, a set of planes of $PG(3, q)$ which partition the points of C_α except for the vertex. There are corresponding translation planes here, as in the quadratic cone case, which we will call α-flokki planes.

First, we mention the Payne-Thas [123] extension theorem for partial flocks of a quadratic cone.

THEOREM 34. *A partial flock of a quadratic cone of deficiency-one in $PG(3, q)$ has a unique extension to a flock of a quadratic cone.*

The theory of Baer groups of Johnson [77] connects such partial flocks with translation planes admitting Baer groups.

THEOREM 35. *Translation planes with spreads in $PG(3, q)$ that admit Baer groups of order q are equivalent to deficiency-one partial flocks of a quadratic cone.*

In this setting, we have $q - 1$ planes of a partial flock of deficiency-one of a quadratic cone, which may be extended to a flock by the theorem of Payne-Thas. Hence, we have:

THEOREM 36. *Let π be a translation plane of order q^2 with spread in $PG(3, q)$. If π admits a Baer group of order q then the partial spread of degree $q + 1$, whose components are on the Baer axis, is a regulus partial spread. Derivation of this spread constructs a translation plane of order q^2 with spread in $PG(3, q)$, admitting an elation group of order q whose orbits together with the elation axis are reguli; a conical translation plane.*

Hence, translation planes of order q^2 with spread in $PG(3, q)$ admitting Baer groups of order q are equivalent to flocks of quadratic cones in $PG(3, q)$.

The Payne-Thas result in the odd order case involves the idea of 'derivation' of a conical flock (5,7,3), whereas the proof in the even order case used ideas from extensions of k-arcs. We have seen that are several different uses of this word. So, the term 'derivation', in this context, refers to a construction using BLT sets that shall be discussed in part 5. These two constructions have been called 'Ostrom derivation' and 'flock derivation' to help distinguish between the two concepts. However, a proof of this result by Sziklai [133] is independent of order. In this section, it is realized that this proof may be adapted to prove the same theorem for α-flokki. The Baer group theory that applies is then an extension of the work of Johnson [77].

We also consider the cones C_q in $PG(3, q^2)$ and algebraically lifting the spreads in $PG(3, q)$ (See part 7). Such lifted spreads automatically give rise to q-flokki of the cones C_q (also see Kantor and Penttila [102]). A *bilinear* flock is a flock in which each plane passes through at least one of two distinct lines of $PG(3, q)$; these lines (called *supporting lines*) may either meet or be skew. A result of Thas [**137**] (also see (9,11,1 &2) shows that flocks of quadratic cones whose planes share a point must be linear in the even characteristic case and either linear or Knuth-Kantor in odd characteristic. Furthermore, bilinear flocks cannot exist in the finite case, since if so, since there are q planes, there would be a linear subflock of degree at least $(q-1)/2+1$ sharing a line. By Johnson [**81**] the only possible flocks of odd order are linear and the Fisher flocks and q is odd, neither of which are bilinear. However, Biliotti and Johnson [**15**] show that bilinear flocks can exist in $PG(3, K)$, where K is an infinite field. In fact, the situation is much more complex for infinite flocks, for example, there are n-linear flocks for any positive integer n (the planes, as a set ,pass through exactly n lines). Recently, Cherowitzo and Holder [**23**], found an extremely interesting bilinear q-flokki using ideas from blocking sets (a set of points of a finite projective that every line intersects and that contains no line completely). It might be suspected that the Cherowitzo-Holder q-flokki might be algebraically lifted from a translation plane, and indeed, in these sections, we show that this is, in fact, the case (See part 7 for the concept of algebraic lifting).

We also show how work on extensions and Baer groups gives results that allow a reverse procedure that identifies certain translation planes of order q^4 that admit Baer groups of orders q^2 and $q+1$ as lifted and derived spreads.

5. Elation groups and flokki planes

In this section, we will be considering finite translation planes admitting twisted regulus-inducing elation groups, mostly in the finite case. In a more general setting, we could be considering derivable net-inducing elation groups in $PG(3, K)$ for a skewfield K. If we consider the four types of derivable nets that we have considered in parts 2 and 3, and recalling that if we wish to embed the derivable nets into translation planes, we realize that, up to isomorphisms, we could consider generic derivable nets, which reduce the complexity of the form of the derivable nets. We shall take this up in (6,8,3), and consider the α-conical flocks in this section.

Let π denote a translation plane of order q^2 with spread in $PG(3, q)$ that admits an elation group E such that some orbit Γ union the axis is a derivable partial spread. We know from Johnson [**73**], that the derivable partial spread may be represented in the form

$$\left\{ x = 0, \ y = x \begin{bmatrix} u & 0 \\ 0 & u^\alpha \end{bmatrix} ; \ u \in GF(q) \right\},$$

where α is an automorphism of $GF(q)$, and we have chosen the axis of E to be $x = 0$ and Γ to contain $y = 0$ and $y = x$. Here, as usual, $x = (x_1, x_2)$, $y = (y_1, y_2)$, for $x_i, y_i \in GF(q)$, $i = 1, 2$ and vectors in the 4-dimensional vector space over $GF(q)$ are (x_1, x_2, y_1, y_2).

Since Γ is an orbit, this means that E has the form

$$E = \left\{ \begin{bmatrix} 1 & 0 & u & 0 \\ 0 & 1 & 0 & u^\alpha \\ 0 & 0 & 1 & 0 \\ 0 & 0 & 0 & 1 \end{bmatrix} ; \ u \in GF(q) \right\}.$$

Let $y = x \begin{bmatrix} g(t) & f(t) \\ t & 0 \end{bmatrix}$ be a typical component of the spread of π for $t \in GF(q)$ and the $(2,2)$-entry equal to zero, where g and f are functions on $GF(q)$, with $g(0) = f(0) = 0$. This is always possible by a basis change allowing that $y = 0$ and $y = x$ represent components of π. Hence, the action of E on $y = x \begin{bmatrix} g(t) & f(t) \\ t & 0 \end{bmatrix}$ produces components

$$y = x \begin{bmatrix} u + g(t) & f(t) \\ t & u^\alpha \end{bmatrix}; \ t, u \in GF(q).$$

Now assume that π is a translation plane with spread in $PG(3, K)$, where K is an infinite field, and π admits an elation group such that the axis and some orbit Γ is a derivable partial spread. In this case, a derivable partial spread has the form

$$\left\{ x = 0, \ y = x \begin{bmatrix} u & A(u) \\ 0 & u^\alpha \end{bmatrix}; \ u \in K \right\},$$

where α is an automorphism of K and such that $\left\{ \begin{bmatrix} u & A(u) \\ 0 & u^\alpha \end{bmatrix}; \ u \in K \right\}$ is a field. Also, if there are at least two Baer subplanes that are K-subspaces, then $A \equiv 0$ (that is type 1 derivable nets) (See (6,8,3).

DEFINITION 46. *A translation plane π with spread in $PG(3, K)$, where K is a field, is said to be an 'α-flokki plane' if and only if there are functions g and f on K so that $f(0) = g(0) = 0$, and*

$$S = \left\{ x = 0, \ y = x \begin{bmatrix} u + g(t) & f(t) \\ t & u^\alpha \end{bmatrix}; \ t, u \in K \right\}$$

is the spread for π, and α is an automorphism of K. We will also say that S is an α-spread.

LEMMA 4. *Let K be an infinite field. If there is a representation of a partial spread in $PG(3, K)$, then this partial spread is a maximal partial spread/quasifibration.*

PROOF. To see that such a partial spread is maximal in $PG(3, K)$, we assume not, then there is a matrix M so that $y = xM$ is mutually disjoint from the other components, which means, since $x = 0$ and $y = 0$ are components that $\begin{bmatrix} u + g(t) & f(t) \\ t & u^\alpha \end{bmatrix} - M$ is non-singular for all t, u in K. Letting $M = \begin{bmatrix} a & b \\ c & d \end{bmatrix}$, choose $t = c$ and $u^\alpha = d$, forces $a = b = 0$. $\square$

A maximal partial spread as described previously will be said to be *injective but not bijective*. We will also mention α, as in α-partial spreads, when reference to the automorphism α is needed.

The next theorem follows immediately.

THEOREM 37. *A translation plane π with spread in $PG(3, K)$, for K a field, is an α-flokki plane if and only if there is an elation group E one of whose orbits is a derivable partial spread containing at least two Baer subplanes that are K-subspaces.*

REMARK 12. *It makes sense to believe that linear α-flokki should correspond to Desarguesian α-flokki spreads. But this is not necessarily the case. For instance, let $\alpha(x) = x^q$ and consider the α-flokki spread in $PG(3, q^2)$ given by*

$$\left\{ x = 0, \ y = x \begin{bmatrix} u & \gamma s^q \\ s & u^q \end{bmatrix}; \ s, u \in GF(q^2) \right\},$$

q odd and γ a non-square. Then the associated α-flokki is given by

$$\rho_t: \ x_0 t - x_1 \gamma^q t - x_3 = 0$$

for all $t \in K$. This is a linear α-flokki with an associated translation plane that is a semifield plane.

EXAMPLE 3. *Let K be any ordered (infinite) field and let α be an automorphism of K. Consider the α-partial flokki spread in $PG(3, K)$:*

$$\mathcal{S} = \left\{ x = 0, \ y = x \begin{bmatrix} u & -t^{3\alpha^{-1}} \\ t & u^\alpha \end{bmatrix}; \ t, u \in K \right\}.$$

We first determine that $\phi_u : t \to u^{\alpha+1} t^3$, is injective. Now

$$(t - s) + u^{\alpha+1}(t^3 - s^3) = 0, \ for \ t \neq s,$$

if and only if

$$1 + u^{\alpha+1}(t^2 + st + s^2) = 0, \ for \ t \neq s.$$

Consider the quadratic in t, $t^2 + st + s^2 + u^{-(\alpha+1)} = 0$, the discriminant of which is $s^2 - 4(s^2 + u^{-(\alpha+1)}) = -3s^2 - u^{-(\alpha+1)}$. But in any ordered field, $u^{\alpha+1} > 1$. Thus, the discriminant is negative, which is never a square. Hence, $\mathcal{S}$ is an α-partial spread, which because of the previous lemma must be maximal and a proper quasifibration.

Given any element r of K, the question is whether there is a solution to $t + u^{\alpha+1} t^3 = r$. Suppose that K is a subfield of the field of real numbers that does not contain all cube roots of elements of K. Then by Cardano's equations the roots will involve cube roots of elements of K and so the $t + u^{\alpha+1} t^3$ will not be surjective. Hence, there are subfields K of the field of real numbers for which the maximal partial spread is not a spread.

6. Maximal partial spreads and α-flokki

In this section, we connect α-flokki translation planes and maximal α-partial spreads/quasifibrations.

DEFINITION 47. *Let K be any field and let $\alpha \in Aut(K)$. Considering homogeneous coordinates (x_0, x_1, x_2, x_3) of $PG(3, K)$, we define the α-cone $\mathcal{C}_\alpha$ as $x_0^\alpha x_1 = x_2^{\alpha+1}$, with vertex $v_0 = (0, 0, 0, 1)$. A set of planes of $PG(3, K)$ which partition the non-vertex points of $\mathcal{C}_\alpha$ will be called an α-flokki. The intersections are called α-conics.*

Now we show that there are maximal partial spreads in $PG(3, K)$, associated with α-flokki, which are called α-*partial spreads* or, if the context is clear, *flokki partial spreads* or flokki quasifibrations. When K is finite, these partial spreads are the spreads arising from α-flokki. In this text, as mentioned, we use also use the term 'quasifibrations' when the injective partial spreads are not spreads.

THEOREM 38. *Let K be any field, and f and g be functions from K to K such that $f(0) = g(0) = 0$. Then*
(1)(a)

$$\left\{ x = 0, \ y = x \begin{bmatrix} u + g(t) & f(t) \\ t & u^\alpha \end{bmatrix}; \ t, u \in K \right\}$$

is a maximal α-partial spread/quasifibration if and only if,

$$\phi_u : \ t \to t - u^{\alpha+1} f(t)^\alpha + u g(t)^\alpha,$$

is injective for all $u \in K$.
(1)(b)

$$\left\{ x = 0, \ y = x \begin{bmatrix} u + g(t) & f(t) \\ t & u^\alpha \end{bmatrix}; \ t, u \in K \right\}$$

is injective if and only if

$$\left\{ x = 0, \ y = x \begin{bmatrix} u + g(t) & t \\ f(t) & u^\alpha \end{bmatrix}; \ t, u \in K \right\}$$

is injective.
(2)(a) *An injective maximal α-partial spread/quasifibration is equivalent to a partial α-flokki of $\mathcal{C}_\alpha$, having defining equations for the planes as follows:*

$$\rho_t : \ x_0 t - x_1 f(t)^\alpha + x_2 g(t)^\alpha - x_3 = 0$$

for all $t \in K$.
  (2)(b) *The two sets of functions*

$$\mathcal{F} = \{\phi_u : \ \phi_u : \ t \to t - u^{\alpha+1} f(t)^\alpha + u g(t)^\alpha, \ \text{for all } u \in K\}$$
$$\mathcal{F}^\perp = \{\phi_u^\perp : \ \phi_u^\perp : \ t \to f(t) - u^{\alpha+1} t^\alpha + u g(t)^\alpha, \ \text{for all } u \in K\}$$

both consist of injective functions if and only if either set consists of injective functions.
  (3) *An α-flokki of $\mathcal{C}_\alpha$ is obtained if and only if ϕ_u is bijective for all $u \in K$.*
  (4) *When K is finite, the set of α-flokki planes is equivalent to the set of α-flokki of $\mathcal{C}_\alpha$.*
  (5) *If $\alpha^2 = 1$, $g \equiv 0$, and ϕ_u is bijective, then for any field K, this subset of α-flokki planes is equivalent to the corresponding set of α-flokki of $\mathcal{C}_\alpha$.*

PROOF. Consider $\Gamma_u(t) = t u^{\alpha+1} - f(t)^\alpha + u^\alpha g(t)^\alpha$, and note that $u^{\alpha+1} \Gamma_{u^{-1}} = \phi_u$. It is immediate to check that Γ_u is injective if and only if ϕ_u is injective. So, we first show that Γ_u is injective for all u, if and only if

$$\mathcal{S} = \left\{ \begin{bmatrix} u + g(t) & f(t) \\ t & u^\alpha \end{bmatrix}; \ u, t \in K \right\}$$

is a set of non-singular matrices or identically zero, whose distinct differences are also non-singular. This will show that the injectivity of the functions ϕ_u will prove that there are associated injective maximal partial α-flokki. If we let $tu = w^\alpha$, we get

$$\begin{aligned}
\Gamma_u(t) &= t u^{\alpha+1} - f(t)^\alpha + u^\alpha g(t)^\alpha \\
&= t^{-\alpha} (w^{\alpha+1} - t f(t) + w^\alpha g(t))^\alpha \\
&= t^{-\alpha} \left(det \begin{bmatrix} w + g(t) & f(t) \\ t & w^\alpha \end{bmatrix} \right)^\alpha.
\end{aligned}$$

Since u may be varied, and by looking at $\Gamma_u(t) - \Gamma_u(s)$, we see that the matrices in S are non-singular (or zero) and the differences of distinct pairs of matrices are non-singular if and only if Γ_u is injective. This proves $(1)(a)$.

Part $(1)(b)$ is immediate as the determinants of a matrix and its transpose are equal. Also, using an argument similar to that proving $(1)(a)$, but now with $\Gamma_u^{\perp}(t) = f(t)u^{\alpha+1} - t^{\alpha} + u^{\alpha}g(t)^{\alpha}$ and $\phi_u^{\perp} = f(t) - u^{\alpha+1}t^{\alpha} + ug(t)^{\alpha}$ one obtains a 'transpose' analogue to $(1)(a)$. This proves $(2)(b)$.

Now let $P = (x_0, x_1, x_2, x_3)$ (homogeneous coordinates) be a point on $\mathcal{C}_{\alpha} \cap \rho_t$. If $x_0 = 0$, as $x_0^{\alpha}x_1 = x_2^{\alpha+1}$, then so is x_2. So, we get $P = (0, 1, 0, -f(t)^{\alpha})$. If $x_0 \neq 0$ then we get

$$0 = t - x_2^{\alpha+1}f(t)^{\alpha} + x_2 g(t)^{\alpha} - x_3 = \phi_{x_2}(t) - x_3$$

and so $P = (1, x_1, x_2, \phi_{x_2}(t))$. It follows that ϕ_u being injective, for all u, is equivalent to all the intersections $\mathcal{C}_{\alpha} \cap \rho_t$ being disjoint. This proves $(2)(a)$. Now take a point $(u^{\alpha+1}, 1, u^{\alpha}, 0)$ in $\mathcal{C}_{\alpha}$ and form the line through $(0, 0, 0, 1)$. The point where this line intersects ρ_t is $(u^{\alpha+1}, 1, u^{\alpha}, tu^{\alpha+1} - f(t)^{\alpha} + u^{\alpha}g(t)^{\alpha})$. Hence, $tu^{\alpha+1} - f(t)^{\alpha} + u^{\alpha}g(t)^{\alpha}$ is bijective, for each u in K, if and only if we have an α-flokki of $\mathcal{C}_{\alpha}$, thus proving part (3). Since finite injective functions are bijective, we also have the proof of part (4).

Finally, consider $\alpha^2 = 1$ and $g(t) = 0$ for all t. Then, we consider for $x_1 x_2 \neq 0$, (x_1, x_2, y_1, y_2) is on $y = x \begin{bmatrix} u & f(t) \\ t & u^{\alpha} \end{bmatrix}$, if and only if

$$x_1 u + x_2 t = y_1 \text{ and } x_1 f(t) + x_2 u^{\alpha} = y_2.$$

We now multiply the first equation by x_2^{α}, and apply the automorphism α to the second followed by a multiplication times x_1. We get

$$x_1 x_2^{\alpha} u + x_2^{\alpha+1} t = y_1 x_2^{\alpha} \text{ and } x_1^{\alpha+1} f(t)^{\alpha} + x_1 x_2^{\alpha} u = y_2^{\alpha} x_1.$$

Hence, subtracting and multiplying by $x_1^{-(\alpha+1)}$,

$$(x_1^{-1}x_2)^{\alpha+1}t - f(t)^{\alpha} = (y_1 x_2^{\alpha} - y_2^{\alpha} x_1)x_1^{-(\alpha+1)},$$

which may be re-written as $\Gamma_{x_1^{-1}x_2}(t) = (y_1 x_2^{\alpha} - y_2^{\alpha} x_1)x_1^{-(\alpha+1)}$. It now follows that if the functions Γ_u are all bijective, we obtain a spread for an α-flokki plane. The converse is similar and left to the reader. This proves all parts of the theorem. $\square$

REMARK 13. *Because of the previous theorem, when all the functions in $\mathcal{F}$ are injective, we will say that $\mathcal{F}$ is an 'injective α-partial flokki'.*

We have seen that the set of functions $\mathcal{F}$ produces a maximal partial α-flokki, but to ensure that these objects are equivalent we need the concept of a 'hyperplane dual spread'. Given a spread S in $PG(3, K)$, for K a field, applying a polarity $\perp$ to $PG(3, K)$ transforms S to a set of lines $S^{\perp}$, with the property that each plane of $PG(3, K)$ contains exactly one line of $S^{\perp}$, which is the definition of a 'hyperplane dual spread'. In the finite case, hyperplane dual spreads are also spreads, which may be seen by an easy counting argument. However, when K is infinite, there are spreads that are not hyperplane dual spreads and hyperplane dual spreads that are not spreads, which we have mentioned previously.

DEFINITION 48. *Let K be a field and α an automorphism of K. Choose functions f and g on K, with $f(0) = g(0) = 0$, and consider the set of functions*

$$\mathcal{F} = \{\phi_u \mid \phi_u : \ t \to t - u^{\alpha+1}f(t)^\alpha + ug(t)^\alpha, \ for \ all \ u \in K\}.$$

Then we define the hyperplane dual (spread) of $\mathcal{F}$, $\mathcal{F}^\perp$, as follows:

$$\mathcal{F}^\perp = \{\phi_u^\perp \mid \phi_u^\perp : \ t \to f(t) - u^{\alpha+1}t^\alpha + ug(t)^\alpha, \ for \ all \ u \in K\}.$$

Assume that both $\mathcal{F}$ and $\mathcal{F}^\perp$ consist of bijective functions, then there are corresponding α-flokki by Theorem. In this case, we shall say that $\mathcal{F}$ and $\mathcal{F}^\perp$ are α-flocks, and furthermore use the terminology that the 'hyperplane dual of an α-flokki is an α-flokki'.

REMARK 14. *In the proof of the previous theorem, we used that $u^{\alpha+1}\Gamma_{u^{-1}} = \phi_u$ to get that*

$$\{\phi_u \mid \phi_u : \ t \to t - u^{\alpha+1}f(t)^\alpha + ug(t)^\alpha, \ for \ all \ u \in K\}$$

is a set of bijective functions if and only if

$$\{\Gamma_u \mid \Gamma_u : \ t \to tu^{\alpha+1} - f(t)^\alpha + u^\alpha g(t)^\alpha, \ for \ all \ u \in K\}$$

is a set of bijective functions. Similarly, one may prove that

$$\{\phi_u^\perp \mid \phi_u^\perp : \ t \to f(t) - u^{\alpha+1}t^\alpha + ug(t)^\alpha, \ for \ all \ u \in K\}$$

is a set of bijective functions if and only if

$$\{\Gamma_u^\perp \mid \Gamma_u^\perp : \ t \to f(t)u^{\alpha+1} - t^\alpha + u^\alpha g(t)^\alpha, \ for \ all \ u \in K\}$$

is a set of bijective functions.

Applying a polarity to a spread in $PG(3, K)$ will produce a hyperplane dual spread, which in the α-flokki case, may not be a spread. It turns out that the hyperplane dual spread may be coordinatized by the transpose of matrices defining the spread. In other words, if

$$\pi = \left\{ x = 0, \ y = x \begin{bmatrix} u + g(t) & f(t) \\ t & u^\alpha \end{bmatrix}; \ u, t \in K \right\},$$

is an α-flokki spread then the hyperplane dual spread $\pi^\perp$ is (isomorphic to)

$$\pi^\perp = \left\{ x = 0, \ \begin{bmatrix} u + g(t) & t \\ f(t) & u^\alpha \end{bmatrix}; \ u, t \in K \right\}.$$

We call $\pi^\perp$ the *transposed spread* of π to avoid confusion with the dual of a projective plane. The connections are as follows.

THEOREM 39. *Let K be a field, $\mathcal{F}$ be an injective partial α-flokki, and $\pi_\mathcal{F}$ be the associated maximal partial α-flokki plane. Then, $\mathcal{F}$ is bijective and the dual flokki is bijective if and only if $\pi_\mathcal{F}$ is a spread and the transposed spread/hyperplane dual spread $\pi_\mathcal{F}^\perp$ is a spread.*

PROOF. Assume that $\mathcal{F}$ is bijective and the hyperplane dual flokki is bijective. We first show that

$$\mathcal{S} = \left\{ x = 0, \ y = x \begin{bmatrix} u + g(t) & f(t) \\ t & u^\alpha \end{bmatrix}; \ t, u \in K \right\}$$

is a spread. Consider a vector (x_1, x_2, y_1, y_2). If $x_1 = 0 = x_2$ then $x = 0$ is the unique 2-space containing the vector. If x_1 is not zero but $x_2 = 0$, then there are unique t and u so that $x_1 t = y_1$ and $x_2 u^\alpha = y_2$. Hence, we may assume that x_2 is non-zero. Hence, we need to solve the following simultaneous equations uniquely for t and u.

$$x_1(u + g(t)) + x_2 t = y_1,$$
$$x_1 f(t) + x_2 u^\alpha = y_2.$$

Taking the α-automorphism of the first equation, multiplying the resulting equation by x_2, multiplying the second equation by x_1^α and subtracting the two resulting equations we obtain the following:

$$x_2 x_1^\alpha g(t)^\alpha + x_2^{\alpha+1} t^\alpha - x_1^{\alpha+1} f(t) = y_1^\alpha x_2 - x_1^\alpha y_2.$$

Since $x_2 \neq 0$, divide by x_2^{a+1}, to transform this equation into

$$\left(x_1 x_2^{-1}\right)^\alpha g(t)^\alpha + t^\alpha - \left(x_1 x_2^{-1}\right)^{\alpha+1} f(t) = y_1^\alpha x_2 - x_1^\alpha y_2.$$

Notice that

$$\begin{bmatrix} 1 & 0 \\ 0 & -1 \end{bmatrix} \begin{bmatrix} u + g(t) & f(t) \\ t & u^\alpha \end{bmatrix} \begin{bmatrix} 1 & 0 \\ 0 & -1 \end{bmatrix} = \begin{bmatrix} u + g(t) & -f(t) \\ -t & u^\alpha \end{bmatrix}.$$

This means that we may use this transformed version of the maximal partial α-flokki, which, letting $v = x_1 x_2^{-1}$, turns the previous equation into $v^{\alpha+1} f(t) - t^\alpha + v^\alpha g(t)^\alpha = y_1^\alpha x_2 - x_1^\alpha y_2$. Recall that we are assuming the four sets in the previous remark are all sets of bijective functions. Hence, there is a unique t so that the basic equation has a solution. It is now easily verified that returning to our set of simultaneous equations, there is a unique t and a unique u that solve these equations. This proves that $\mathcal{S}$ is an α-flokki spread.

Now if we repeat the argument for the transposed maximal partial α-flokki plane $\pi_{\mathcal{F}}^\perp$, the assumptions on the α-flokki and its dual show that the transposed maximal partial α-flokki plane is also an α-flokki plane.

Now assume that both $\pi_{\mathcal{F}}$ and $\pi_{\mathcal{F}}^\perp$ are both α-flokki planes. Then by rereading the previous argument, it is immediate that the fact that $\pi_{\mathcal{F}}$ is an α-flokki plane implies that the hyperplane dual $\mathcal{F}^\perp$ of the injective partial α-flokki $\mathcal{F}$ is an α-flokki. Since $\mathcal{F}^{\perp\perp} = \mathcal{F}$, and $\pi_{\mathcal{F}}^\perp$ is an α-flokki plane then also $\mathcal{F}$ is an α-flokki. This completes the proof of the theorem. $\qquad \square$

COROLLARY 13. *Every α-flokki plane is isomorphic to an α^{-1}-flokki plane. In particular,*

$$\begin{bmatrix} u + g(t) & f(t) \\ t & u^\alpha \end{bmatrix} \quad and \quad \begin{bmatrix} u + g(t)^\alpha & t \\ f(t) & u^{\alpha^{-1}} \end{bmatrix}$$

give isomorphic flokki planes.

PROOF. Note that

$$\begin{bmatrix} 0 & 1 \\ 1 & 0 \end{bmatrix} \begin{bmatrix} u + g(t) & f(t) \\ t & u^\alpha \end{bmatrix} \begin{bmatrix} 0 & 1 \\ 1 & 0 \end{bmatrix} = \begin{bmatrix} u^\alpha & t \\ f(t) & u + g(t) \end{bmatrix}.$$

Now let $u + g(t) = v^{-1}$, so that $u^\alpha = v - g(t)^\alpha$, and obtain

$$\begin{bmatrix} u^\alpha & t \\ f(t) & u + g(t) \end{bmatrix} = \begin{bmatrix} v - g(t)^\alpha & t \\ f(t) & v^{-1} \end{bmatrix}.$$

A basis change by

$$\begin{bmatrix} -1 & 0 \\ 0 & 1 \end{bmatrix} \begin{bmatrix} v - g(t)^\alpha & t \\ f(t) & v^{-1} \end{bmatrix} \begin{bmatrix} -1 & 0 \\ 0 & 1 \end{bmatrix},$$

followed by replacing $-v$ by w, gives

$$\begin{bmatrix} w + g(t)^\alpha & t \\ f(t) & w^{\alpha^{-1}} \end{bmatrix}$$

which has the desired form.   □

For a given cone, a *partial flock* is a set of planes which do not intersect on this cone. Any finite flock is a partial flock of q planes. A partial flock of $q - 1$ planes is said to be of *deficiency-one*. The Payne-Thas theorem (Theorem [123]) states that for a quadratic cone, any deficiency-one partial flock can be uniquely extended to a flock. We extend this theorem to α-flokki, for when K is finite, by noting that the proof of the result due to Sziklai [133] is valid in this situation as well.

THEOREM 40. *Let $K \cong GF(q)$. A deficiency-one α-flokki may be extended to a unique α-flokki.*

PROOF. Let $\alpha \in Aut(K)$, and $\mathcal{C}_\alpha$ be the α-cone with vertex $v_0 = (0, 0, 0, 1)$ given by $x_0^\alpha x_1 = x_2^{\alpha+1}$. Consider a partial α-flokki of $\mathcal{C}_\alpha$ of deficiency-one consisting of $q - 1$ planes of the following form:

$$\rho_t : \ x_0 t - x_1 f(t)^\alpha + x_2 g(t)^\alpha - x_3 = 0,$$

for t in a subset λ of K of cardinality $q - 1$. So, the function

$$\phi_u : \ t \to t - u^{\alpha+1} f(t)^\alpha + u g(t)^\alpha$$

is injective in K, for each u in K. Hence, we have $q - 1$ of the elements of K as images for each ϕ_u. We note that the point on the generator $\langle v_0, (u^{\alpha+1}, 1, u^\alpha, 0) \rangle$ on ρ_t is $(u^{\alpha+1}, 1, u^\alpha, u^{\alpha+1}t - f(t)^\alpha + u^\alpha g(t)^\alpha)$. For $q > 2$, we see that $-\sum_{t \in \lambda} t$ is the missing element from K in λ. We note that the point

$$\left(u^{\alpha+1}, 1, u^\alpha, -\sum_{t \in \lambda} (u^{\alpha+1}t - f(t)^\alpha + u^\alpha g(t)^\alpha) \right)$$

is the missing point on each generator other than $\langle (0, 0, 0, 1), (1, 0, 0, 0) \rangle$. The points $(1, 0, 0, t)$ on ρ_t are on the generator $\langle (0, 0, 0, 1), (1, 0, 0, 0) \rangle$ and the missing point is $\left(1, 0, 0, -\sum_{t \in \lambda} t \right)$. Now all of these points lie on the plane

$$x_0 \left(\sum_\lambda t \right) + x_1 \left(-\sum_\lambda f(t)^\alpha \right) + x_2 \left(\sum_\lambda g(t)^\alpha \right) + x_3 = 0,$$

so the missing plane is determined.   □

7. The second cone

An α-flokki is a flock of the cone $\mathcal{C}_\alpha$. There is a projectively equivalent cone $\mathcal{C}'_\alpha = \{(x_0, x_1, x_2, x_3) \mid x_0 x_1^\alpha = x_2^{\alpha+1}\}$ with vertex $(0, 0, 0, 1)$ and it is of interest to ask if an α-flokki can simultaneously be a flock of this '2nd-Cone'. Let an α-flokki be given by the planes $\pi_t : \ x_0 t - f(t)^\alpha x_1 + g(t)^\alpha x_2 - x_3 = 0$, hence $\phi_u : t \to u^{\alpha+1}t - f(t)^\alpha + u^\alpha g(t)^\alpha$ is a bijective function. Note that for this set of planes to be a flock of C'_α, we must have that $\rho_u : t \to t - u^{\alpha+1} f(t)^\alpha + u^\alpha g(t)^\alpha$ is a bijective function. By changing all signs in this function we obtain the bijective function $\tau_u : t \to u^{\alpha+1} f(t)^\alpha - (t^{\alpha^{-1}})^\alpha + u^\alpha(-g(t))^\alpha$. This means that the associated α-flokki spread or maximal partial spread relative to the second cone is given by

$$\left\{ x = 0, \ y = x \begin{bmatrix} u - g(t) & t^{\alpha^{-1}} \\ f(t)^\alpha & u^\alpha \end{bmatrix}; \ t, u \in GF(q) \right\}.$$

This argument may be 'reversed' to prove that a given α-flokki plane produces a flock in both cones as long as τ_u is bijective.

Note that when $g = 0$, if $\phi_u : t \to u^{\alpha+1}t - f(t)^\alpha$ is a bijective function for all elements $u \in K$, then for u non-zero, we can factor out $u^{\alpha+1}$ to obtain that $\rho_{u^{-1}=v} : t \to t - v^{\alpha+1} f(t)^\alpha$ is bijective.

DEFINITION 49. *Given an α-flokki plane relative to the functions $(g(t), f(t), t)$, the set $(g(t^\alpha), f(t^\alpha)^\alpha, t)$ is called the '2nd-cone triple'.*

THEOREM 41.

(1) Given an α-flokki plane relative to $(g(t), f(t), t)$, the 2nd-cone triple $(g(t^\alpha), f(t^\alpha)^\alpha, t)$ corresponds to an α-flokki plane if and only if

$$\phi_u : \ t \to u^{\alpha+1}t - f(t)^\alpha + u^\alpha g(t)^\alpha$$

for all u is a bijective function implies

$$\tau_u : \ t \to t - u^{\alpha+1} f(t)^\alpha + u^\alpha g(t)^\alpha$$

is a bijective function, for all $u \in K$.

(2) If $g = 0$ then any α-flokki of C_α is a flock of C'_α.

We leave it as an open problem to show that if g is not zero then an associated α-flokki of C_α is never a flock of C'_α.

8. Baer groups for flokki planes

The Baer group theory for translation planes with spreads in $PG(3, q)$ of Johnson [**77**], shows that Baer groups of order q produce partial flocks of quadratic cones of deficiency-one. Given any conical flock plane, there is an elation group E, whose orbits union the axis form reguli in $PG(3, q)$. Derivation of one of these regulus nets, produces a translation plane with spread in $PG(3, q)$ admitting a Baer group of order q. Now consider any given α-flokki and corresponding α-flokki plane. Again, there is an elation group E whose orbits union the axis form derivable partial spreads. Derivation of one of the derivable nets produces a translation plane admitting a Baer group of order q, but now the spread for this plane π^* is no longer in $PG(3, q)$ (for $\alpha \neq 1$). The components not in the derivable net are still subspaces in $PG(3, q)$, as are the Baer subplanes of the derivable net of π^*. Hence, we would expect that Baer groups of order q in such translation planes might also produce deficiency-one partial α-flokki.

THEOREM 42. *Let π be a translation plane of order q^2 that admits a Baer group B of order q. Assume that the components of π and the Baer axis of B are lines of $PG(3, q)$.*

(1) Then π corresponds to a partial α-flokki of deficiency-one.

(2) Hence, the Baer partial spread defined by the Baer group is derivable and the Baer subplanes incident with the zero vector are also lines in $PG(3, q)$. The derived plane is the unique α-flokki plane associated with the extended α-flokki.

PROOF. Let $q = p^r$, for p a prime. Let (x_1, x_2, y_1, y_2) represent points, where x_i are r-vectors over the prime field. We may also represent x_i as an element of $GF(q)$. Choose the Baer axis to be $(0, x_2, 0, y_2)$, and obtain the group in the form

$$\left\langle \begin{bmatrix} A & 0 \\ 0 & A \end{bmatrix} \right\rangle,$$

where $A = \begin{bmatrix} I & C \\ 0 & I \end{bmatrix}$ and C is in a field K. The Baer group of order q is elementary Abelian and corresponds to a field by the fundamental results on Baer p-elements of Foulser [**52**]. It follows that the Baer axis together with any orbit of length q forms a derivable partial spread. If the Baer axis and the components not on the Baer net are lines of $PG(3, q)$ and we choose the Baer axis to be $x = 0$, then by Johnson [**77**], we have that the group B may be represented in the form

$$\left\{ \begin{bmatrix} 1 & 0 & u & 0 \\ 0 & 0 & 0 & u^\alpha \\ 0 & 0 & 1 & 0 \\ 0 & 0 & 0 & 1 \end{bmatrix} ; \ u \in GF(q) \right\},$$

where α is an automorphism of $GF(q)$. It follows that the components not on the Baer net have the form

$$x = 0, \ y = x \begin{bmatrix} u + a & b \\ c & u^\alpha \end{bmatrix};$$

for $q - 1$ values of c in $GF(q)$. Let $c = t^\alpha$, then a and b are functions of c, say $g(t)$ and $f(t)$, respectively. Then we have a partial spread of the form

$$\left\{ x = 0, \ y = x \begin{bmatrix} u + g(t) & f(t) \\ t & u^\alpha \end{bmatrix}; \ t \in \lambda, \ u \in GF(q) \right\},$$

where $\lambda \subseteq GF(q)$ has cardinality $q - 1$. Therefore, we have a partial α-flokki of deficiency-one with planes

$$x_0 t - x_1 f(t)^\alpha + x_2 g(t)^\alpha - x_3 = 0$$

for all $t \in \lambda^{\alpha^{-1}}$. Since any partial α-flokki of deficiency-one can be extended to a unique α-flokki, we see that the Baer net must be derivable, the Baer group is the elation group of the derived plane and the Baer subplanes of the Baer net are lines of $PG(3, q)$. This completes the proof of the theorem. □

9. q-Flokki and lifting

When α is the automorphism $x \mapsto x^q$ of $GF(q^2)$, an α-flokki will be referred to as a *q-flokki*. We have that a q-flokki in $PG(3, q^2)$ has an associated flokki plane with spread

$$\left\{ x = 0, \ y = x \begin{bmatrix} u + g(t) & f(t) \\ t & u^q \end{bmatrix}; \ t, u \in GF(q^2) \right\}.$$

The following theorem tells us when an α-flokki plane has been lifted from a plane with spread in $PG(3,q)$. The following theorem considers the concept of 'lifting', which we also be more generally considered in parts 7 and 8.

THEOREM 43. *(Cherowitzo and Johnson [25]) A translation plane with spread*

$$\left\{ x = 0, \ y = x \begin{bmatrix} u + G(s) & F(s) \\ s & u^\alpha \end{bmatrix}; s, u \in GF(q^2) \right\},$$

is a lifted spread if and only if there is a coordinate change so that $G(s) = 0$ and $\alpha = q$.

The bilinear q-Flokki of Cherowitzo and Holder [23] yields a flokki plane with spread

$$\left\{ x = 0, \ y = x \begin{bmatrix} u & \gamma(s^q)^{(q^2+1)/2} \\ s & u^q \end{bmatrix}; \ s, u \in GF(q^2) \right\},$$

where q is odd and γ is a non-square in $GF(q^2)$ such that γ^2 is a non-square in $GF(q)$. The flokki planes of Cherowitzo of order q^4 may be lifted from regular nearfield planes of order q^2.

Now we consider a lifted translation plane of order q^4. Because of the previous theorem, we know that it has a spread of the form

$$\left\{ x = 0, \ y = x \begin{bmatrix} u & F(t) \\ t & u^q \end{bmatrix}; \ t, u \in GF(q^2) \right\}.$$

Since

$$\begin{bmatrix} 1 & 0 \\ 0 & e^q \end{bmatrix} \begin{bmatrix} u & 0 \\ 0 & u^q \end{bmatrix} \begin{bmatrix} e & 0 \\ 0 & 1 \end{bmatrix} = \begin{bmatrix} ue & 0 \\ 0 & (ue)^q \end{bmatrix},$$

the Baer group of order $q + 1$,

$$B = \left\langle \begin{bmatrix} 1 & 0 & 0 & 0 \\ 0 & e & 0 & 0 \\ 0 & 0 & e & 0 \\ 0 & 0 & 0 & 1 \end{bmatrix}; e^{q+1} = 1 \right\rangle$$

maps $y = x \begin{bmatrix} u & 0 \\ 0 & u^q \end{bmatrix}$ to $y = x \begin{bmatrix} ue & 0 \\ 0 & (ue)^q \end{bmatrix}$. If we now derive the associated net, the Baer group B is still a Baer group of order $q + 1$, but the elation group of order q^2 is now a Baer group E of order q^2 (See e.g. Johnson, Jha and Biliotti [94], Theorem 35.18 for more details).

THEOREM 44. *Let π be a translation plane of order q^4 that admits a Baer group E of order q^2 and a non-trivial Baer group B such that $[E, B] \neq 1$. If the axis of E and its non-trivial orbits are lines of $PG(3, q^2)$ then π is a derived q-flokki translation plane; a translation plane that has been algebraically lifted and then derived.*

PROOF. The plane is derivable since it corresponds to a partial α-flokki. Hence, the derived plane has the following spread set

$$\left\{ x = 0, \ y = x \begin{bmatrix} u + G(t) & F(t) \\ t & u^\alpha \end{bmatrix}; \ t, u \in K \right\}.$$

The group E is now an elation group E and since B and E do not centralize each other, it follows that the plane is a lifted plane. Hence, $\alpha = q$ and $G(t) = 0$, that is, once we know that there is a non-trivial Baer group, this group is forced to have order $q + 1$. □

REMARK 15. *The theorem on lifting may be extended more generally showing that all lifted Andrè planes provide bilinear α-flocks. This theory is also valid in the infinite case. In particular, this is reintroduced in part 6.*

10. Collineations and isomorphisms of α-flokki planes

We now want to study the full collineation group of a finite α-flokki plane π with spread

$$\left\{ x = 0, \ y = x \begin{bmatrix} u + g(t) & f(t) \\ t & u^\alpha \end{bmatrix}; \ t, u \in GF(q) \right\}.$$

We will start by looking at the elation group

$$E = \left\{ \begin{bmatrix} 1 & 0 & u & 0 \\ 0 & 1 & 0 & u^\alpha \\ 0 & 0 & 1 & 0 \\ 0 & 0 & 0 & 1 \end{bmatrix}; \ u \in GF(q) \right\}.$$

This next section involves detailed group theory on finite translation planes, for which the reader may wish to look more completely at the supporting articles. Space considerations do not allow anything like a complete understanding of these concepts in this text. For those interested, here are some of the details and references: Let E' be any elation group of π, different from E. Assume that E' does not have axis $x = 0$. Then, by the Hering-Ostrom theorem (See [94]), $\langle E, E' \rangle$ is isomorphic to $SL(2, p^t)$, $S_z(2^{2e+1})$ and q is even, or $SL(2, 5)$ and 3 divides q. In the $SL(2, p^t)$ case by Johnson [74], the group is either $SL(2, q)$ or $SL(2, q^2)$. In the former case, the planes are Desarguesian, Ott-Schaeffer, Hering, Dempwolff of order 16 or Walker of order 25. Since 2^{2e+1} is at least q, the $S_z(2^{2e+1})$ case does not occur by Büttner [20]. Assume that $q > 4$, then $SL(2, 5)$ cannot occur. The Dempwolff plane is not in $PG(3, q)$. Hence, this leaves the Hering and Ott-Schaeffer planes. The Hering planes are not derivable and the Ott-Schaeffer planes admit reguli, so the plane is Desarguesian.

Therefore, assume that E' does have axis $x = 0$. We now assume $E' \cap E \neq \langle 1 \rangle$, and get derivable nets sharing $y = 0$, and $y = x \begin{bmatrix} u_0 & 0 \\ 0 & u_0^\alpha \end{bmatrix}$, for some $u_0 \in GF(q)$. Note that a change of basis that allows us to represent this last component by $y = x$ does not change the general form of the derivable net

(6.1) $$\left\{ x = 0, \ y = x \begin{bmatrix} u & 0 \\ 0 & u^\alpha \end{bmatrix}; \ u \in GF(q) \right\}.$$

Hence, by Johnson [73], we have two distinct derivable nets of π: the one in (6.1) and

(6.2) $$\left\{ x = 0, \ y = x \begin{bmatrix} v & 0 \\ 0 & v^\sigma \end{bmatrix}; \ v \in GF(q) \right\},$$

where $\sigma \in Aut(GF(q))$. But, in this case, there are differences of matrices that are non-singular, a contradiction. Hence, $E' \cap E = \langle 1 \rangle$.

First notice that if E is normal in $Aut(\pi)$ then the α-derivable nets are permuted by $Aut(\pi)$.

Now assume that E is not normal in $Aut(\pi)$. Thus, there exists an element $\phi \in Aut(\pi)$ such that the elation group $\overline{E} = \phi E \phi^{-1}$ is different from E. Clearly ϕ must fix $x = 0$, and thus representing a generic element of $\overline{E}$ by

$$\begin{bmatrix} E_{11} & E_{12} \\ 0 & I \end{bmatrix}$$

we obtain $E_{11} = I$. It follows that $\overline{E}$ and E commute with each other and belong to an elation group with axis $x = 0$. So, the group $\langle \overline{E}, E \rangle$ contains the set product $\overline{E}E$, of order q^2, which means that the α-flokki plane is a semifield plane.

Now, having that $\overline{E}E$ has order q^2 allows us to assume that ϕ fixes both $x = 0$ and $y = 0$, which then has the general form

$$(x, y) \rightarrow (x^\rho, y^\rho) \begin{bmatrix} A & 0 \\ 0 & B \end{bmatrix},$$

where $A, B \in GL(2, q)$ and $\rho \in Aut(GF(q))$. Then ϕ maps $y = x$ to $y = xA^{-1}B$. Change basis by $(x, y) \rightarrow (x, yB^{-1}A)$ to find a derivable net with components $x = 0$, $y = 0$, and $y = x$. This net must be of the general form in Equation 6.2 by Johnson [73]. But the image of

$$y = x \begin{bmatrix} u & 0 \\ 0 & u^\alpha \end{bmatrix},$$

for $u \in GF(q)$, under ϕ, and the posterior change of basis

$$(x, y) \rightarrow (x, yB^{-1}A),$$

is

$$y = xA^{-1} \begin{bmatrix} u^\rho & 0 \\ 0 & u^{\rho\alpha} \end{bmatrix} A.$$

So, we get that either $A = \begin{bmatrix} a & 0 \\ 0 & d \end{bmatrix}$, $v = u^\rho$ and $\sigma = \alpha$, or $A = \begin{bmatrix} 0 & b \\ c & 0 \end{bmatrix}$, $v = u^{\rho\alpha}$ and $\sigma = \alpha^{-1}$.

Similarly, since ϕ maps $y = x$ to $y = xA^{-1}B$ we change basis using the transformation $(x, y) \rightarrow (xA^{-1}B, y)$ to find a derivable net with components $x = 0$, $y = 0$, and $y = x$. Then again, this net must be of the general form in Equation (6.2) by Johnson [73]. But the image of

$$y = x \begin{bmatrix} u & 0 \\ 0 & u^\alpha \end{bmatrix},$$

for $u \in GF(q)$, under ϕ, and the posterior change of basis $(x, y) \rightarrow (xA^{-1}B, y)$, is

$$y = xB^{-1} \begin{bmatrix} u^\rho & 0 \\ 0 & u^{\rho\alpha} \end{bmatrix} B.$$

So, we get that either $B = \begin{bmatrix} \overline{a} & 0 \\ 0 & \overline{d} \end{bmatrix}$, $v = u^\rho$ and $\sigma = \alpha$, or $B = \begin{bmatrix} 0 & \overline{b} \\ \overline{c} & 0 \end{bmatrix}$, $v = u^{\rho\alpha}$ and $\sigma = \alpha^{-1}$. It follows that $A^{-1}B$ is either a diagonal or an anti-diagonal matrix.

Since $A^{-1}B$ being diagonal would yield a contradiction with ϕ not normalizing E we get that

$$B^{-1}A = \begin{bmatrix} 0 & m \\ n & 0 \end{bmatrix},$$

for some $m, n \in GF(q)^*$.

We now make the change of basis $(x, y) \to (x, yB^{-1}A)$, to get a derivable net in π of the form

$$\left\{ x = 0, \ y = x \begin{bmatrix} 0 & vm \\ v^\sigma n & 0 \end{bmatrix}; \ v \in GF(q) \right\}.$$

Let $t = v$, and get the orbits of this net under the group E to obtain:

$$\left\{ x = 0, \ y = x \begin{bmatrix} u & tm \\ t^\sigma n & u^\alpha \end{bmatrix}, \ t, u \in GF(q) \right\}.$$

Use $t^\sigma n = s^\sigma$ to transform the spread set into the form

$$\left\{ x = 0, \ y = x \begin{bmatrix} u & fs \\ s^\sigma & u^\alpha \end{bmatrix}, \ s, u \in GF(q) \right\},$$

where f is a constant, and $\sigma = \alpha^{\pm 1}$. If $\sigma = \alpha$ then this is a Hughes-Kleinfeld semifield plane (See [94]; 91.22), and if $\sigma = \alpha^{-1}$ then this is a dual/transposed Hughes-Kleinfeld semifield plane. Since there are α-flokki spreads of this form exactly when the flock is linear, we obtain the following theorem. Note that the case where $x = 0$ is not left invariant leads to the Desarguesian plane and $\alpha = 1$.

THEOREM 45. *Let π be an α-flokki plane of order q^2, where $q > 5$. Then one of the following occurs:*

(1) the full collineation group permutes the q α-derivable nets,
(2) $\alpha = 1$ and the plane is Desarguesian or
(3) $\alpha \neq 1$ and the α-flokki plane is a Hughes-Kleinfeld, or transposed Hughes-Kleinfeld semifield plane that corresponds to a linear flokki.

REMARK 16. *The above theorem is the generalization of the analogous theory of flocks of quadratic cones when $\alpha = 1$. These ideas are also discussed in comparing the theories finite and infinite flocks in (12,15,5).*

Now assume that π_1 and π_2 are two isomorphic α-flokki planes of order q^2, for $q > 5$, neither of which are Hughes-Kleinfeld (transposed or not) or Desarguesian. Let σ be an isomorphism from π_1 to π_2, so we may assume that $x = 0$ is left invariant and that the elation group E of order q is normalized by σ. Initially, assume that π_i has spread

$$\left\{ x = 0, \ y = x \begin{bmatrix} u + g_i(t) & f_i(t) \\ t & u^\alpha \end{bmatrix}; \ t, u \in GF(q) \right\},$$

for $i = 1, 2$. By following with an element of E, if necessary, we may assume that σ maps $y = x$ to $y = x \begin{bmatrix} 1 + g_2(t_0) & f_2(t_0) \\ t_0 & 1 \end{bmatrix}$. Then, by changing the basis using the

map $\begin{bmatrix} 1 & 0 & -g_2(t_0) & -f_2(t_0) \\ 0 & 1 & -t_0 & 0 \\ 0 & 0 & 1 & 0 \\ 0 & 0 & 0 & 1 \end{bmatrix}$, the resulting matrix spread set has the following form:

$$\left\{ x = 0, \ y = x \begin{bmatrix} u + g_2(t) - g_2(t_0) & f_2(t) - f_2(t_0) \\ t - t_0 & u^\alpha \end{bmatrix}; \ t, u \in GF(q) \right\},$$

for $i = 1, 2$. Note that this spread, after a suitable selection of f_2 and g_2, maintains the form of the original matrix spread set. By this recoordinatization we may assume that σ fixes $x = 0$ and $y = x$, leaving invariant

(6.3) $$\left\{ x = 0, \ y = x \begin{bmatrix} u & 0 \\ 0 & u^\alpha \end{bmatrix}; \ u \in GF(q) \right\}.$$

Since E is transitive on each derivable α-net, it follows that σ fixes $x = 0$ and $y = 0$ (though not necessarily fixing $y = x$ anymore), and is of the general form:

$$(x, y) \to (x^\beta, y^\beta) \begin{bmatrix} A & 0 \\ 0 & B \end{bmatrix},$$

where $\beta \in Aut(GF(q))$, and the matrices $A, B \in GL(2, q)$ with $A^{-1}B = \begin{bmatrix} u_0 & 0 \\ 0 & u_0^\alpha \end{bmatrix}$, for some $u_0 \in GF(q)^*$. Then, we must have

$$\begin{bmatrix} v + g_2(s) & f_2(s) \\ s & v^\alpha \end{bmatrix} = A^{-1} \begin{bmatrix} u^\beta + g_1(t)^\beta & f_1(t)^\beta \\ t^\beta & u^{\alpha\beta} \end{bmatrix} B$$

$$= A^{-1} \begin{bmatrix} u^\beta + g_1(t)^\beta & f_1(t)^\beta \\ t^\beta & u^{\alpha\beta} \end{bmatrix} A \begin{bmatrix} u_0 & 0 \\ 0 & u_0^\alpha \end{bmatrix}.$$

We let $s = 0$ and use that (6.3) is fixed by σ to get that $A = \begin{bmatrix} a & 0 \\ 0 & d \end{bmatrix}$ or $\begin{bmatrix} 0 & b \\ c & 0 \end{bmatrix}$.

Thus, $B = A \begin{bmatrix} u_0 & 0 \\ 0 & u_0^\alpha \end{bmatrix} = \begin{bmatrix} au_0 & 0 \\ 0 & du_0^\alpha \end{bmatrix}$ or $\begin{bmatrix} 0 & cu_0^\alpha \\ bu_0 & 0 \end{bmatrix}$, respectively. Moreover, by applying a scalar mapping, we may assume that $a = 1$ or $b = 1$. In the first situation,

$$\begin{bmatrix} v + g_2(s) & f_2(s) \\ s & v^\alpha \end{bmatrix} = A^{-1} \begin{bmatrix} u^\beta + g_1(t)^\beta & f_1(t)^\beta \\ t^\beta & u^{\alpha\beta} \end{bmatrix} A \begin{bmatrix} u_0 & 0 \\ 0 & u_0^\alpha \end{bmatrix}$$

$$= \begin{bmatrix} 1 & 0 \\ 0 & d^{-1} \end{bmatrix} \begin{bmatrix} u^\beta + g_1(t)^\beta & f_1(t)^\beta \\ t^\beta & u^{\alpha\beta} \end{bmatrix} \begin{bmatrix} u_0 & 0 \\ 0 & du_0^\alpha \end{bmatrix}$$

$$= \begin{bmatrix} (u^\beta + g_1(t)^\beta)u_0 & f_1(t)^\beta du_0^\alpha \\ t^\beta d^{-1} & u^{\alpha\beta}u_0^\alpha \end{bmatrix}.$$

Let $u_0 = w_0^\beta$. Then for $v = (uw_0)^\beta$, $s = t^\beta d^{-1}$, we have

$$f_2(s) = f_1(s^{\beta^{-1}}d^{\beta^{-1}})^\beta du_0^\alpha \qquad \text{and} \qquad g_2(s) = g_1(s^{\beta^{-1}}d^{\beta^{-1}})^\beta u_0.$$

In the second situation, we have

$$\begin{bmatrix} v + g_2(s) & f_2(s) \\ s & v^\alpha \end{bmatrix} = A^{-1} \begin{bmatrix} u^\beta + g_1(t)^\beta & f_1(t)^\beta \\ t^\beta & u^{\alpha\beta} \end{bmatrix} A \begin{bmatrix} u_0 & 0 \\ 0 & u_0^\alpha \end{bmatrix}$$

$$= \begin{bmatrix} 0 & c^{-1} \\ 1 & 0 \end{bmatrix} \begin{bmatrix} u^\beta + g_1(t)^\beta & f_1(t)^\beta \\ t^\beta & u^{\alpha\beta} \end{bmatrix} \begin{bmatrix} 0 & cu_0^\alpha \\ u_0 & 0 \end{bmatrix}$$

$$= \begin{bmatrix} c^{-1}u^{\alpha\beta}u_0 & t^\beta u_0^\alpha \\ f_1(t)^\beta u_0 & (u^\beta + g_1(t)^\beta)cu_0^\alpha \end{bmatrix}.$$

We may assume that $g_i(0) = f_i(0) = 0$, for $i = 1, 2$ (this is not affected by all the manipulations done to get that σ fixes $y = 0$). Therefore, taking $s = 0$, we see that $t = 0$, implying that $v = (c^{-1}u^{\alpha\beta}u_0)$, and $v^\alpha = (u^\beta c u_0^\alpha)$, which implies that $c^{-\alpha}u^{\alpha^2\beta} = cu^\beta$, for all u, thus $\alpha^2 = 1$ and $c^{\alpha+1} = 1$. Letting $v = 0$ implies $u = 0$, so that $g_2(s) = 0$ and $f_2(s) = t^\beta u_0^\alpha$, where $s = f_1(t)^\beta u_0$. Therefore, $t = f_1^{-1}((su_0^{-1})^{\beta^{-1}})$, so that $f_2(s) = f_1^{-1}((su_0^{-1})^{\beta^{-1}})^\beta u_0^\alpha$. Moreover, it also follows that $g_1(t) = 0$. Assume that q is a non-square then $\alpha = 1$, a contradiction. Hence, $q = h^2$, and $\alpha = h$. Therefore, we have the spreads

$$\left\{ x = 0, \; y = x \begin{bmatrix} u & f_i(t) \\ t & u^h \end{bmatrix} ; \; t, u \in GF(h^2) \right\},$$

for $i = 1, 2$, where $f_2(t) = f_1^{-1}((tu_0^{-1})^{\beta^{-1}})^\beta u_0^\alpha$.

We now have the following theorem:

THEOREM 46. *Two finite α-flokki planes, π_i, for $\alpha \neq 1$ and not Hughes-Kleinfeld or their transposed spreads with spreads*

$$\left\{ x = 0, \; y = x \begin{bmatrix} u + g_i(t) & f_i(t) \\ t & u^\alpha \end{bmatrix} ; \; t, u \in GF(q) \right\},$$

for $i = 1, 2$, are isomorphic if and only if one of the following occurs:

(1) There is an automorphism β, a constant t_0 and a constant u_0 so that

$$f_2(s + t_0) - f_2(t_0) = f_1(s^{\beta^{-1}}d^{\beta^{-1}})^\beta du_0^\alpha,$$
$$g_2(s + t_0) - g_2(t_0) = g_1(s^{\beta^{-1}}d^{\beta^{-1}})^\beta u_0.$$

(2) There is an automorphism β, a constant t_0 and a constant u_0 so that

$$f_2(s + t_0) - f_2(t_0) = f_1^{-1}((tu_0^{-1})^{\beta^{-1}})^\beta u_0^\alpha,$$
$$g_2(s + t_0) - g_2(t_0) = 0.$$

Moreover, $q = h^2$ and $\alpha = h$.

COROLLARY 14. *If $\alpha \neq 1$ and π is a finite α-flokki plane that is not Hughes-Kleinfeld or the transposed spread, then the transposed α-flokki plane of π is isomorphic to π if and only if there is an automorphism β and constants t_0 and u_0 so that*

(1)

$$f_1^{-1}(s + t_0) - f_1^{-1}(t_0) = f_1(s^{\beta^{-1}}d^{\beta^{-1}})^\beta du_0^\alpha,$$
$$g_1 f_1^{-1}(s + t_0) - g_1 f_1^{-1}(t_0) = g_1(s^{\beta^{-1}}d^{\beta^{-1}})^\beta u_0,$$

or

(2)

$$f_1^{-1}(s + t_0) - f_1^{-1}(t_0) = f_1^{-1}((tu_0^{-1})^{\beta^{-1}})^\beta u_0^\alpha,$$
$$g_1(s + t_0) - g_1(t_0) = 0,$$

where $q = h^2$ and $\alpha = h$.

PROOF. Note that $g_2 = g_1 f_1^{-1}$ and $f_2 = f_1^{-1}$, in the transposed plane. The parts of the functions involving t_0 simply ensure that an isomorphism leaves the derivable net in (6.3) invariant. □

We finish this section by taking up the problem of determining when a finite α-flokki plane π_1 is isomorphic to a δ-flokki plane π_2. Since we know that this is always the case when $\delta = \alpha^{-1}$, for $g_2(t) = g_1(t)^\alpha$, and $f_2(t) = f_1(t)$, assume that δ is not α^{-1}, that neither plane is Hughes-Kleinfeld (transposed or not), and that α and δ are both not equal to 1. The analogous equations in the first situation are

$$
\begin{bmatrix} v + g_2(s) & f_2(s) \\ s & v^\delta \end{bmatrix} = A^{-1} \begin{bmatrix} u^\beta + g_1(t)^\beta & f_1(t)^\beta \\ t^\beta & u^{\alpha\beta} \end{bmatrix} A \begin{bmatrix} u_0 & 0 \\ 0 & u_0^\delta \end{bmatrix}
$$

$$
= \begin{bmatrix} 1 & 0 \\ 0 & d^{-1} \end{bmatrix} \begin{bmatrix} u^\beta + g_1(t)^\beta & f_1(t)^\beta \\ t^\beta & u^{\alpha\beta} \end{bmatrix} \begin{bmatrix} u_0 & 0 \\ 0 & du_0^\delta \end{bmatrix}
$$

$$
= \begin{bmatrix} (u^\beta + g_1(t)^\beta)u_0 & f_1(t)^\beta du_0^\delta \\ t^\beta d^{-1} & u^{\alpha\beta}u_0^\delta \end{bmatrix}.
$$

But, this implies that $v^\delta = u^{\alpha\beta}u_0^\delta$, $v = u^\beta u_0$, so that $v^\delta = u^{\delta\beta}u_0^\delta = u^{\alpha\beta}u_0^\delta$, for all v, so that $\delta = \alpha$. In the second situation, the analogous equations are

$$
\begin{bmatrix} v + g_2(s) & f_2(s) \\ s & v^\delta \end{bmatrix} = A^{-1} \begin{bmatrix} u^\beta + g_1(t)^\beta & f_1(t)^\beta \\ t^\beta & u^{\alpha\beta} \end{bmatrix} A \begin{bmatrix} u_0 & 0 \\ 0 & u_0^\delta \end{bmatrix}
$$

$$
= \begin{bmatrix} 0 & c^{-1} \\ 1 & 0 \end{bmatrix} \begin{bmatrix} u^\beta + g_1(t)^\beta & f_1(t)^\beta \\ t^\beta & u^{\alpha\beta} \end{bmatrix} \begin{bmatrix} 0 & cu_0^\delta \\ u_0 & 0 \end{bmatrix}
$$

$$
= \begin{bmatrix} c^{-1}u^{\alpha\beta}u_0 & t^\beta u_0^\delta \\ f_1(t)^\beta u_0 & (u^\beta + g_1(t)^\beta)cu_0^\delta \end{bmatrix}.
$$

Then, similar to the previous arguments, $v^\delta = u^\beta cu_0^\delta$, and $v = c^{-1}u^{\alpha\beta}u_0$, implying that $v = u^{\delta^{-1}\beta}c^{\delta^{-1}}u_0 = c^{-1}u^{\alpha\beta}u_0$. This implies that $c^{\delta^{-1}} = c^{-1}$ and $\delta = \alpha^{-1}$, contrary to our assumptions. Hence, we have the following theorem.

THEOREM 47. *Let π_1 and π_2 be α-flokki and δ-flokki planes respectively, of order q^2. If π_1 and π_2 are isomorphic then $\delta = \alpha^{\pm 1}$.*

PROOF. If π_1 and π_2 are isomorphic and one is a Hughes-Kleinfeld (or transposed) semifield plane then the other is also a Hughes-Kleinfeld semifield plane and the result follows (for example, from Johnson and Liu [95]), the other case follows from our previous remarks. □

Part 5

Flock geometries

In this part, a number of representations of flocks of quadratic cones are discussed, which we group together as 'flock geometries'. This material, for the most part, represents well known connections, there are also generalizations into what might be called 'flock quasifibrations', where an associated geometry admitting certain sets of derivable nets might be quasifibrations and not fibrations (spreads).

The number of flock geometries of flocks of these types seems to be increasing. Since this text is concerned with derivable nets, flock geometries essentially are then relevant and because of this, we give some short descriptions of some of the great variety of flock geometries that are currently known, as well as how quasifibrations are related.

CHAPTER 7

Related geometries

There are a number of connections with derivable nets living in or related to a variety of incidence geometries. Thas [**135**], [**136**], [**137**] determined that to every finite flock of a quadratic cone, there corresponds a generalized quadrangle of order (q^2, q), that can be given as a coset geometry. It is not our intent to discuss generalized quadrangles in any particular depth but to show connections. We give a coordinate method to connect flocks of quadratic cones and generalized quadrangles with finite translation planes admitting finite regulus-inducing elation groups. This material is based on the articles of H. Gevaert and N.L. Johnson [**56**] and that of Gevaert, Johnson, and Thas [**57**], which are modified for this text.

When connecting translation planes to flocks, there are often different ways to do this. In particular, we shall provide Klein quadric constructions as well as coordinate methods. There are then the natural questions if such different constructions provide isomorphic geometries.

Generalizing flocks of quadratic cones to flocks of α-cones there are also what might be called α-Klein constructions and algebraic constructions of translation planes. For α-flocks, there may not be connections to generalized quadrangles. Similarly, in the infinite case, there have been just a few studies connecting infinite generalized quadrangles to infinite flocks of quadratic cones, although the connections to infinite translation planes are rich and also involve quasifibrations.

DEFINITION 50. *A 'generalized quadrangle' is a set of points and lines, with an incidence relation, such every two distinct points are incident with at most one line, and two distinct lines are incident with at most one line, and one additional axiom: Given a point-line pair (P, l), where P is not incident with l, there is 'a unique' line m and a unique point Q incident with l such that P and Q are incident with m.*

A 'partial generalized quadrangle' is a set of points and lines, with an incidence relation, such every two distinct points are incident with at most one line, and two distinct lines are incident with at most one line and one additional axiom: Given a point-line pair (P, l), where P is not incident with l, there is 'at most one' line m incident with P, where m is concurrent with l.

Thus, a generalized quadrangle may be either infinite or finite. In the finite case, there can be two additional requirements.

DEFINITION 51. *A generalized quadrangle with parameters (s, t), for s and t both positive integers ≥ 1, is a generalized quadrangle such that: There are exactly $s + 1$ points incident with every line and there are exactly $t + 1$ lines incident with every point. Therefore, counting provides that there are $(s + 1)(st + 1)$ points and $(s + 1)(t + 1)$ lines. It is possible to consider infinite generalized quadrangles with* *these parameters, but here we consider that both s and t are finite, and denoted by* $GQ(s,t)$.

A generalized quadrangle of type (q^2, q), where q is a prime power, has $(q^3 + 1)(q^2+1)$ points and $(q^2+1)(q+1)$ lines. A finite flock of a quadratic cone, where the defining conic C in $PG(3,q)$ has $q+1$ points and center O, has $q+1$ lines OC, and each line has $q+1$ points, producing $q(q+1)$ non-vertex points. A finite translation plane with spread in $PG(3,q)$ has q^2+1 spread components. Considering the affine viewpoint, there is a regulus-inducing elation group of order q, where the orbits on the line at infinity union the axis point of the elation group form a set of q regulus nets, sharing a component, and therefore $q(q+1)$ Baer subplanes.

To see an algebraic connection between these three incidence geometries, we use the coset geometry method of Kantor [104] to construct certain $GQ(s,t)$, where we only use the $q^2 = s$ and $q = t$, type, and we follow Gevaert and Johnson [56], to describe the technique. Moreover, we shall be interested in infinite generalized quadrangles, infinite flocks of quadratic cones and infinite translation planes admitting regulus-inducing elation groups. We first make some definitions, required in the following.

DEFINITION 52. *Let K be a field, finite or infinite of characteristic $\neq 2$. K is said to be a 'full field' if and only if for any two elements both of which are non-square, their product is a square. When K is a field of characteristic 2, let $C_1(K)$ denote the set of all elements k such that $x^2 + x + k$ has no solutions in K. Then K is a 'full field of characteristic 2' if and only if the sum of two elements in $C_1(K)$ is never in $C_1(K)$.*

DEFINITION 53. *A generalized (or partial generalized) quadrangle Q is said to be an 'elation generalized quadrangle' (respectively, partial elation generalized quadrangle) if and only if there is an automorphism group of Q that fixes a point p (the elation point), fixes all lines incident with p and which acts regularly on the points opposite p (not collinear with p).*

1. Kantor's coset technique

In this section, we consider a generalization of Kantor's coset technique for finite generalized quadrangles $GQ(q^2, q)$, adapted to $GQ(K \times K, K)$, for K a field (See Kantor [104]).

We are looking for a procedure to construct $(q^3 + 1)(q^2 + 1)$ points and $(q^2 + 1)(q + 1)$ lines. This is accomplished using a group of order q^5, as follows.

This method will be used to show that in the finite case, there are generalized quadrangles associated with every flock of a quadratic cone. Moreover, there is a procedure (BLT-sets) due to Bader-Lunardon-Thas [6], whereby any finite flock of a quadratic cone produces $q + 1$ flocks. However, all of these flocks correspond to a single flock generalized quadrangle. This method may be extended to partial structures using any field K, and the BLT-sets will be generalized in a following section.

The point of the following is to verify that using the method of Kantor, over infinite fields, it is possible to construct a partial generalized quadrangle, which we call a 'quasi-partial generalized quadrangle. The method to be described ultimately depends on certain irreducibility requirements of $GF(q)$, which are also valid in the so-called 'flock quasifibrations', which are quasifibrations with partial spreads in $PG(3, K)$, for K a field, that admit an elation regulus-inducing elation group. This will establish that associated with any flock quasifibration, there is a set of flock quasifibrations corresponding to certain recoordinatizations of the partial generalized quadrangle in question.

The following is from Gevaert and Johnson [56], drawn from the ideas of W.M. Kantor but modified using an arbitrary field K.

We will discuss the difference between finite (q^2, q)-generalized quadrangles of flock type, and when an arbitrary field K is used in place of $GF(q)$. We begin with a quasifibration of flock type:

THEOREM 48. *Let K be any field and let π be a quasifibration of dimension* 2 *that admits a regulus-inducing elation group. Then the partial spread for π may be represented in the following form:*

$$x = 0, y = x \begin{bmatrix} u + g(t) & f(t) \\ t & u \end{bmatrix}; \forall t, u \in K, \text{ and } g, f \text{ functions on } K.$$

What this means is that the differences of distinct pairs of the matrices (including a zero matrix) are non-singular. There are two versions of this result, the first is that

$(*)_1 : u^2 + u(g(t) - g(s)) - (t - s)(f(t) - f(s))$ is irreducible for all $t \neq s$.

There is an equivalent version Cherowitzo-Johnson-Vega [26] that is given by the functions

$$(*)_2 : \phi_u : t \to t - u^2 f(t) + u g(t)$$

are injective for all u.

For our study of partial generalized quadrangles of characteristic 2, arising from quasifibrations, we also consider that g is injective.

This version (when the functions are bijective) can be seen also in De Clerck and Van Maldeghem [40] (2.2) (which lists two requirements), when discussing when partial generalized quadrangles are generalized quadrangles.

Let $G = \{(\alpha, c, \beta); \alpha, \beta \in K \times K \text{ and } c \in K\}$. Let $u \cdot v$ denote the ordinary dot product in $K \times K$. Define an operation in G

$$(\alpha, c, \beta) \cdot (\overline{\alpha}, \overline{c}, \overline{\beta}) = (\alpha + \overline{\alpha}, c + \overline{c} + \beta \cdot \overline{\alpha}, \beta + \overline{\beta}).$$

Then, over a field K, G is a group. Also let

$$A(\infty) = \{(0, 0, \beta) : \beta \in K \times K\}$$

and

$$A_t = \begin{bmatrix} t & g(t) \\ 0 & -f(t) \end{bmatrix}; t \in K,$$

where the functions g and f correspond to the quasifibration listed above. Now define

$$A(t) = \{(\alpha, \alpha A_t \alpha^T, \alpha(A_t + A_t^T)); \alpha \in K \times K\},$$

where α^T denotes the transpose of α.

Let
$$C = \{(0, c, 0) : c \in K\}.$$
Then define
$$A^*(t) = A(t)\,C : t \in K \cup \{\infty\}.$$
Then 'points', 'lines' and 'incidence' are defined as follows:

Points, of three types:
(i) elements of G,
(ii) the right cosets $A^*(t)g$ for $g \in G$, for all $t \in K \cup \{\infty\}$, and
(iii) the symbol $\{\infty\}$.

Lines, of two types:
(i) the right cosets $A(t)g$ for $g \in G$, for all $t \in K \cup \{\infty\}$, and
(ii) the symbols $[A(t)]$ for all $t \in K \cup \{\infty\}$.

Incidence:
A point h of type (i) is incident with a line of type (i) $A(t)h$, for all $t \in K \cup \{\infty\}$.
A point $A^*(t)h$ of type (ii) is incident with a line of type (i) $A(t)g$ if and only if $A(t)g \subset A^*(t)h$.
A point $A^*(t)h$ of type (ii) is incident with the line $[A(t)]$ of type (ii).
Finally, the point $\{\infty\}$ of type (iii) is incident with all lines $[A(t)]$ of type (ii), for all $t \in K \cup \{\infty\}$.

Then using work of Kantor [104] and Payne [121], a generalized quadrangle of type/order $(K \times K, K)$ for $K = GF(q)$ is constructed under certain conditions satisfied by the set of 2 by 2 matrices A_t. Since we are interested in these conditions, in general fields K, we consider when the conditions arise, and wish to show that the conditions are those required by a flock quasifibration. Consider the line $A(t)$ and point $A^*(s)$, when $s \neq t$, for $s, t \in K$. We note that the line $A(t)$ contains the identity element $(0, 0, 0)$ in the group, as a point, and the points $A^*(t)h : h \in G$ for $h = (0, 0, \beta)$, so that $A(t) \subset A^*(t)h$. The lines incident with $A^*(s)$ are the lines $A(s)(0, c, 0)$ for all $c \in K$ and $[A(s)]$. Notice that the two lines $A(t)(0, 0, 0)$ and $A(s)(0, 0, 0)$ share a point $(0, 0, 0)$. For a partial generalized quadrangle, these two lines can share no other point. Since $h = (0, 0, \beta)$, $A(t)C(0, 0, \beta)$ contains $A(t)$, for every $\beta \in K \times K$. Consequently, the question becomes can
$$A^*(t)h = A^*(s)g,$$
for some h and $g \in G$, noting that both h and g may be taken within $A(\infty)$? Since $A(t)Ch = A(s)Cg$, it follows immediately that $h = g$ and $A(t) = A(s)$. Noting that

$$A(t) = \{(\alpha, \alpha \begin{bmatrix} t & g(t) \\ 0 & -f(t) \end{bmatrix} \alpha^T, \alpha \begin{bmatrix} 2t & g(t) \\ g(t) & -2f(t) \end{bmatrix})\} =$$

$$A(s) = \{(\alpha, \alpha \begin{bmatrix} s & g(s) \\ 0 & -f(s) \end{bmatrix} \alpha^T, \alpha \begin{bmatrix} 2s & g(s) \\ g(s) & -2f(s) \end{bmatrix})\}$$

Consequently, we have for each α that

$$(\#)_1 : \alpha \begin{bmatrix} t & g(t) \\ 0 & -f(t) \end{bmatrix} \alpha^T - \alpha \begin{bmatrix} s & g(s) \\ 0 & -f(s) \end{bmatrix} \alpha^T = 0$$

and

$$(\#)_2 : \alpha \left(\begin{bmatrix} 2t & g(t) \\ g(t) & -2f(t) \end{bmatrix} - \begin{bmatrix} 2s & g(s) \\ g(s) & -2f(s) \end{bmatrix} \right) = (0,0).$$

Consequently, we have letting $\alpha = (\alpha_1, \alpha_2)$, we obtain:

$$(\#)'_1 : \alpha_1^2(t-s) + \alpha_1\alpha_2(g(t) - g(s)) - \alpha_2^2(f(t) - f(s)) = 0.$$

And:

$$(\#)'_2 : \alpha \begin{bmatrix} 2(t-s) & g(t) - g(s) \\ g(t) - g(s) & -2(f(t) - f(s)) \end{bmatrix} = (0,0),$$

therefore,

$$(\#)''_2 : -4(t-s)(f(t) - f(s)) - (g(t) - g(s))^2 = 0.$$

Which in characteristic 2 implies that $t = s$, as g is assumed to be injective.

In $(\#)'_1$ if $\alpha_2 = 0$, then $t = s$ or $\alpha_1 = 0$. For $\alpha_2(t-s) \neq 0$, then letting $u = \frac{\alpha_1(t-s)}{\alpha_2}$ transforms the equation into

$$(\#)''_1 : u^2 + u(g(t) - g(s)) - (t-s)(f(t) - f(s)) = 0,$$

which is a contradiction to $(*)_1$, of the quasifibration, so we have a contradiction.

Hence, we have seen one part of the proof when K is $GF(q)$ that a flock spread will construct an order (q^2, q) generalized quadrangle. In general, the rest of the proof that a flock quasifibration will construct a partial $(K \times K, K)$-generalized quadrangle is straightforward. This completes our discussion.

In (9,11,2) on bilinear flocks of quadratic cones, it is shown that there exist infinite bilinear flocks of quadratic cones over ordered fields. Given a flock quasifibration, with some additional assumptions, it is possible to show that there are related quasifibrations called s-inverted quasifibrations. Furthermore, a flock spread can produce an s-inverted quasifibration that is not a spread. When this occurs, the flock spread cannot correspond to a flock generalized quadrangle. That is, we have what we shall call a 'quasi-generalized quadrangle' (or quasi-partial GQ), some of whose derivations are associated with flocks and some are associated with proper flock quasifibrations. We shall also consider quasi-BLT-sets in the following section, all of which produce quasifibrations, some of which can be flock spreads, where others do not.

In the infinite case, flocks of quadratic cones and translation planes admitting regulus-inducing elation groups are equivalent. Moreover, whenever the associated field used in $PG(3, K)$ is a full field, it is shown in De Clerk and Van Mandehelm [40] that coordinate constructions of corresponding generalized quadrangles, constructed by the same coset technique, work analogously, just as for the finite case. Many of the known infinite examples are generalizations of finite flocks. What has not been clearly established is a study of flocks of quadratic cones, where there are no finite analogues. There are a number of ways to go about such an analysis. Here we consider two ideas, which culminate in the same class of examples.

2. Quasi-BLT-sets

In this section, we consider BLT-sets and their infinite generalizations, which we shall call 'quasi-BLT sets'.

This work in the finite case was due to Bader, Lunardon and Thas [**136**], [**8**]. The infinite case was considered by Knarr [**106**], and here will be explicated using quasifibrations. We recall the form:

THEOREM 49. *Let K be any field and let π be a quasifibration of dimension 2 that admits a regulus-inducing elation group. Then π may be represented in the following form:*

$$x = 0, y = x \begin{bmatrix} u + g(t) & f(t) \\ t & u \end{bmatrix}; \forall t, u \in K, \text{ and } g, f \text{ functions on } K.$$

DEFINITION 54. *π above is a proper quasifibration if and only if the partial spread does not cover the vectors $K \oplus K$. If π does have the covering condition, it is a translation plane corresponding to a flock of a quadratic cone. In either case, we shall use the term 'quadratic flock quasifibration'.*

DEFINITION 55. *Let $Q(4, K)$ be a quadric (parabolic quadric) in $PG(4, K)$ where K has characteristic 0 or odd and where the points are given by homogeneous coordinates $(x_0, x_1, x_2, x_3, x_4)$ and the parabolic quadric is given by*

$$x_0^2 + x_1 x_2 + x_2 x_3 + x_3 x_4 = 0.$$

A 'quasi-BLT-set' over a field K is a set of points of $Q(4, K)$ indexed by $K \cup \{\infty\}$ such that no distinct triple of points are collinear. A 'BLT-set is a quasi-BLT-set when K is $GF(q)$, for q odd.

Let $x_0^2 = x_1 x_2$ be a quadratic cone with vertex $v_0 = (0, 0, 0, 1)$. Let π be quadratic flock quasifibration so there is a set of planes of $PG(3, K)$, indexed by K, that correspond to a maximal partial flock.

We shall provide some definitions regarding proper quasifibrations and spreads.

DEFINITION 56. *A partial flock in $PG(3, K)$ of planes indexed by K is called a 'full K-maximal partial flock'. This is a flock if and only if the associated quasifibration is a spread.*

DEFINITION 57. *A partial flock in $PG(3, K)$ of 'deficiency-one' is a partial flock whose planes cover all but one point on each line of the quadratic cone. The deficiency-one partial flocks correspond to a partial spread that may always be extended to a translation plane admitting a Baer elation group. If the Baer axis is contained in a regulus net, then the derived translation plane is a flock translation plane.*

To distinguish between the two maximal partial flocks: A proper quasifibration defines a maximal partial flock; the full K-maximal partial flock, which can never be a deficiency-one partial flock. A deficiency-one partial flock is also maximal if and only if the Baer axis net in the associated translation plane is not a regulus net.

Let β denote the bilinear form associated with the parabolic quadric, where

$$\beta(x, y) = Q(x + y) - Q(x) - Q(y),$$

for points x and y.

LEMMA 5. *Embed the quadratic cone together with the center into the parabolic quadric mapping $(x_0, x_1, x_2, x_3) \to (x_0, x_1, -x_2, x_3, x_2)$, so that $x_0^2 + x_1 x_2 + x_2 x_3 +$*

x_3x_4 becomes $x_0^2 = x_1x_2$ in the associated $PG(3, K)$. For $P_\infty = (0, 0, 0, 1)$ then in $PG(4, K)$,

$$P_\infty^\perp = \{(x_0, x_1, -x_2, x_3, x_2); \forall x_0, x_1, x_2, x_3 \in K\}$$

and $P_\infty^\perp \cap Q(4, K)$ is the quadratic cone $x_0^2 = x_1x_2$ with vertex P_∞.

PROOF. Let $x = (x_0, x_1, x_2, x_3, x_4)$ and $y = (y_0, y_1, y_2, y_3, y_4)$. Then using $\beta(x, y) = Q(x + y) - Q(x) - Q(y)$, we have

$$\beta(x, y) = 2x_0y_0 + (x_1y_2 + y_1x_2) + (x_2y_3 + y_2x_3) + (x_4y_3 + y_4x_3).$$

Letting $P_\infty = (0, 0, 0, 1, 0$, then $\beta(x, P_\infty) = 0$ if and only if $x_2 = -x_4$, which is $P_\infty^\perp$. This completes the proof of the lemma. □

Corresponding to the quasifibration partial spread, there is a set of planes of $PG(3, K)$. We then connect this set of planes with $\{P_t^\perp, P_\infty^\perp; t \in K\}$ whose intersections are in the associated $PG(3, K)$ embedded in $PG(4, K)$.

THEOREM 50. *The isometry group $Q(4, K)$ of $PG(4, K)$ is transitive on the singular points, so consider points $\{P_t, P_\infty; t \in K\}$ for all $t \in K$, whose planes in $PG(3, K)$ corresponding to the quasifibration are $\{P_t^\perp, P_\infty^\perp; t \in K\}$, where $P_s^\perp = \{points\ N; \beta(N, P_s) = 0\}$ for $s = t \in K$ or ∞ and where $P_\infty = (0, 0, 0, 1, 0)$.*

LEMMA 6. *Then $\{P_t, P_\infty; t \in K\}$ is a quasi-BLT-set over K.*
Conversely, any quasi-BLT-set over K may be represented in this form.

PROOF. Let $(x_0, x_1, x_2, x_3, x_4)$ be a singular point N and consider the line NP_∞. We now calculate the lines that lie in $Q(4, K)$. The points of the line may be represented in the form $(x_0, x_1, x_2, x_3 + \beta, x_4)$. It follows immediately that the lines in $Q(4, K)$ joining P_∞ force $x_4 = -x_2$. Consequently, $(x_0, x_1, -x_2, x_3, x_2)$ where $x_0^2 = x_1x_2$ is a quadratic cone with vertex P_∞, all of whose points are in $Q(4, K)$.

Then $P_\infty^\perp \cap (P_t^\perp \cap Q(4, K); \forall t \in K)$ represents the partial flock over K corresponding to the quasifibration.

Conversely, if $\{P_s; s \in K \cup \{\infty\}\}$ is a quasi-BLT-set over K then,

$$P_z^\perp \cap (P_t^\perp \cap Q(4, K); \forall t \in K \cup \{\infty\} - \{z\})$$

is a full K-maximal partial flock of a quadratic cone. Otherwise, if $P_z^\perp \cap (P_t^\perp \cap P_r^\perp \cap Q(4, K))$ for z, t, r mutually distinct, is non-empty, let N be a singular point in the three hyperplanes in question. Note that there are three quadratic cones and N is incident with P_z, P_t, and P_r. Then, P_z, P_t, P_r are singular points that are collinear, a contradiction. Since such maximal partial spreads are indexed by K, and they admit regulus-inducing elation groups, it follows that there is an associated quadratic flock quasifibration obtained for each P_z considered. This completes the proof. □

THEOREM 51. *Given any quasi-BLT-set $\{P_s; s \in K \cup \{\infty\}\}$, a quasi-BLT-set over K, then for each P_s, $P_s^\perp$ is a $PG(3, K)$ such that $P_s^\perp \cap Q(4, K)$ is a quadratic cone with vertex P_s and*

$$\left\{P_s^\perp \cap (P_t^\perp \cap Q(4, K)); t \in K \cup \{\infty\} - \{s\}\right\}$$

is a full K-maximal partial flock. This set of K-maximal partial flocks constructs an equivalent set of quasifibrations in $PG(3, K)$ that admits a regulus-inducing elation group.

3. s-Inversion & s-square

For each finite BLT-set, then K is $GF(q)$ and there are $q+1$ associated flocks of quadratic cones. And, there are associated flock translation planes and flock generalized quadrangles.

For quasi-BLT sets over infinite fields K, there are then infinitely many constructed quadratic flock quasifibrations. At this point, a given quasifibration could be a spread or a proper quasifibration. However, Payne and Rogers [**122**] were able to show that BLT-sets over $GF(q)$ gave rise to exactly one flock generalized quadrangle, but $q+1$ flocks of quadratic cones and the same number of flock spreads. Moreover, a flock generalized quadrangle of type (q^2, q) is an elation generalized quadrangle with a fixed center (∞). There are $q+1$ lines incident with (∞) and a particular line $[A(\infty)]$. A recoordinatization choosing different lines on (∞) produces the $q+1$ flocks of the BLT-sets, when q is odd and also produces a set of $q+1$ flocks when q is even. This means that, for every flock/flock translation plane of order q^2 with spread in $PG(3, q)$, there is a set of $q+1$ such flocks/flock translation planes of order q^2. The associated spreads of the translation planes are given by what are termed 's-inverted planes', when q is odd and 's-square planes' when q is even, and we use the same terminology for quasifibrations of characteristic $\neq 2$ or equal to 2.

DEFINITION 58. *Let π be any quadratic flock quasifibration with spread given as follows:*

$$x = 0, y = x \begin{bmatrix} u + g(t) & f(t) \\ t & u \end{bmatrix};$$

$$\forall t, u \in GF(q), \; g, f \text{ functions on } K.$$

(1) *Assume that the characteristic of K is $\neq 2$. Then there is a set of related quadratic flock quasifibrations called the set of 's-inverted quasifibrations' which have the following form: For any fixed $s \in K$,*

$$x = 0, y = x \begin{bmatrix} v & 0 \\ 0 & v \end{bmatrix},$$

$$y = x \left\{ \begin{bmatrix} -(g(t) - g(s))/2 & f(t) - f(s) \\ t - s & (g(t) - g(s))/2 \end{bmatrix}^{-1} + \begin{bmatrix} u & 0 \\ 0 & u \end{bmatrix} \right\};$$

$$\forall t \neq s, u, v \in K, \; g, f \text{ functions on } K.$$

(2) *If the characteristic is 2 noting that g is injective then there is a set of related quadratic flock quasifibrations called the set of 's-square' quasifibrations' which have the following form: For any fixed $s \in K$,*

$$x = 0, y = x \begin{bmatrix} v & 0 \\ 0 & v \end{bmatrix},$$

$$y = x \left\{ \frac{1}{(g(t) + g(s))^2} \begin{bmatrix} (g(t) + g(s)) & f(t) + f(s) \\ t + s & 0 \end{bmatrix} + \begin{bmatrix} u & 0 \\ 0 & u \end{bmatrix} \right\};$$

$$\forall t \neq s, u, v \in K, \; g, f \text{ functions on } K.$$

We have also noted that Kantor's coset construction, for arbitrary fields K will produce a partial flock generalized quadrangle, when there is a corresponding quadratic flock quasifibration. Moreover De Clerk and Van Maldeghem [**40**], show that every flock of a quadratic cone does have an associated flock generalized quadrangle, when the associated field is a full field. In that setting the s-inversions and the s-squares are quadratic flock planes. In all cases, the s-inversions and the s-squares are then quadratic flock quasifibrations.

The s-inversions that are found that are proper quasifibrations may be lifted (See (7,9,3)) provided there are quadratic field extensions of K.

To summarize the results generalizing the

$$flock/\ spread/\ generalized\ quadrangle/\ BLT-sets\ connections,$$

we note the following general theorem.

DEFINITION 59. *For an arbitrary field K, and associated quasifibration of flock type has a representation as a partial spread as follows:*

$$x\ =\ 0, y = x \begin{bmatrix} u + g(t) & f(t) \\ t & u \end{bmatrix}:$$

$$t, u\ \in\ K, f,\ functions\ on\ K,$$

and the functions $\phi_u\ :\ t \to t - u^2 f(t) + ug(t)\ are\ injective\ for\ all\ u$.

If the characteristic of K is 2, assume that g is injective. If all functions ϕ_u are bijective then the quasifibration is a spread of flock type. The quasifibration is 'proper' if and only if some function ϕ_u is not bijective.

In De Clerck and Van Maldeghem [**40**] (2.2), there are two conditions given that show that a quaternary ring defining a putative generalized quadrangle has the point-line property for non-incident point-line pairs. In Cherowitzo-Johnson-Vega [**26**] show that the first condition provides the second condition.

DEFINITION 60. *Consider the Kantor coset construction method of generalized quadrangles over $GF(q)$, generalized over an arbitrary field K that admits a quasifibration of flock type. We use the term 'partial generalized quadrangle of flock type' (or quasi-generalized quadrangle over K'. Note that the 2 by 2 matrices over K* $\begin{bmatrix} t & g(t) \\ 0 & -f(t) \end{bmatrix}$ *satisfies the conditions given in the previous results if and only if there is an associated quasifibration of flock type.*

DEFINITION 61. *Given a quasifibration of flock type over K, there is an associated partial flock of a quadratic cone. We shall call this a 'quasi-flock'.*

DEFINITION 62. *Given a quasifibration of flock type of K, then the construction of Bader-Lunardon-Thas in the finite case and of Knarr in the infinite case constructs a set of quasi-flocks and a set of quasifibrations indexed by $K \cup \{\infty\}$. These quasifibrations are connected using the construction techniques of s-inversion when the characteristic is $\neq 2$ or s-square when the characteristic is 2. The generalized BLT-sets are called 'quasi-BLT-sets', when the characteristic is odd or 0. More generally, the term 'quasi-flock derivation' shall also be used.*

The term derivation here should not be confused with a derivable net derivation. We note that an s-inverted or s-square quasifibration could be a proper quasifibration even if it arises from a flock spread.

THEOREM 52.

(1) The combinatorial incidence geometries of quasifibrations of flock type, quasi-flock and quasi-generalized quadrangles are equivalent.

(2) The quasi-BLT-sets for characteristic odd or 0 are equivalent to the s-inverted quasifibrations of any associated quasifibration. These sets correspond to the s-inverted quadratic flock quasifibrations.

(3) For arbitrary characteristic, the analogous coordinate changes for the partial generalized quadrangles produce the s-inverted quadratic flock quasifibrations and/or the s-square quadratic flock quasifibrations for characteristic $\neq 2$ and equal to 2, respectively.

4. A census

In this final section of flocks of quadratic and α-flocks, we simply provide a census, of the geometries and partial geometries that we have covered.

NOTATION 7. *For quasifibrations related to α-flocks, where α is an automorphism, possibly the identity, we have a partial spread of the following form:*

$$x = 0, y = x \begin{bmatrix} u^{\alpha} + g(t) & f(t) \\ t & u \end{bmatrix} : t, u \in K, \ g \ and \ f \ functions \ on \ a \ field \ K.$$

(1) 'Quadratic α-flock quasifibrations'. We shall use the notation

$$\begin{bmatrix} g(t) & f(t) \\ t & 0 \end{bmatrix}_{E}^{\alpha}$$

for a representation, where E denotes that there is a regulus-inducing elation group, and the fundamental twisted regulus net, is represented as $\begin{bmatrix} u^{\alpha} & 0 \\ 0 & u \end{bmatrix}$ (instead of perhaps as $\begin{bmatrix} u & 0 \\ 0 & u^{\alpha} \end{bmatrix}$).

(2) 'Baer group quadratic α-flock quasifibrations'. It is possible to have deficiency 1 partial α-flocks, where the twisted regulus-inducing group is a Baer group. The notation shall then be $\begin{bmatrix} g(t) & f(t) \\ t \neq 0 & 0 \end{bmatrix}_{B}^{\alpha}$, where the term B indicates that there is a Baer group and $t \neq 0$ indicates that there is a missing plane of the flock, whereas, there is a translation plane.

(3) 'Partial quadratic α-flocks'

  (a) There are partial flocks of planes

$$\rho_t : x_1 t - x_2 f(t)^{\alpha^{-1}} + x_3 g(t)^{\alpha - 1} - x_4; t \in K.$$

  corresponding to the quasifibrations, we shall use the notation

$$(t, -f(t)^{\alpha^{-1}}, g(t)^{\alpha^{-1}}),$$

  and

(b) for the deficiency 1 partial flocks, we shall use the notation that

$$(t, -f(t)^{\alpha^{-1}}, g(t)^{\alpha^{-1}})_{t \neq 0}$$

to indicate that one plane is missing.

For $\alpha \neq 1$, there are therefore the following interconnected geometries

$$\left[\begin{array}{cc} g(t) & f(t) \\ t & 0 \end{array} \right]_E^{\alpha} \Leftrightarrow (t, -f(t)^{\alpha^{-1}}, g(t)^{\alpha^{-1}}),$$

and

$$\left[\begin{array}{cc} g(t) & f(t) \\ t \neq 0 & 0 \end{array} \right]_B^{\alpha}, (t, -f(t)^{\alpha^{-1}}, g(t)^{\alpha^{-1}})_{t \neq 0}.$$

(4) *For $\alpha = 1$, we also have the 'Partial $(K \times K, K)$-generalized quadrangles', where the*

$$\left[\begin{array}{cc} t & g(t) \\ 0 & -f(t) \end{array} \right],$$

shall also represent the geometry, where $\alpha = 1$ to connect to the other types of geometries.

Using quasi-BLT-sets and recoordinatization, there are families of quasifibrations and partial flocks.

(a) When the characteristic of $K \neq 2$, there are the quasifibrations

$$\mathfrak{Q}_s^E : \left[\begin{array}{cc} -(\frac{g(t)-g(s)}{2}) & f(t) - f(s) \\ t - s & (\frac{g(t)-g(s)}{2}) \end{array} \right]_E^{-1}$$

and

$$\mathfrak{Q}_s^B : \left[\begin{array}{cc} -(\frac{g(t)-g(s)}{2}) & f(t) - f(s) \\ (t - s) \neq 0 & (\frac{g(t)-g(s)}{2}) \end{array} \right]_B^{-1},$$

for all $s \in K$, and the latter expression to indicate the associated Baer group in the deficiency-one cases.

(b) When the characteristic of $K = 2$, there are the quasifibrations

$$\mathfrak{T}_s^E = \left[\begin{array}{cc} \frac{(g(t)+g(s)}{(g(t)+g(s))^2} & \frac{f(t)+f(s)}{(g(t)+g(s))^2} \\ \frac{t+s}{(g(t)+g(s))^2} & 0 \end{array} \right]_E$$

to denote the quasifibration admitting the regulus-inducing elation group E and

$$\mathfrak{T}_s^B = \left[\begin{array}{cc} \frac{(g(t)+g(s)}{(g(t)+g(s))^2} & \frac{f(t)+f(s)}{(g(t)+g(s))^2} \\ \frac{t+s}{(g(t)+g(s))^2} \neq 0 & 0 \end{array} \right]_B$$

to denote the quasifibration admitting the regulus-inducing Baer group.

The indicated geometries are equivalent, in the sense that each $(K \times K, K)$-quasi generalized quadrangle is equivalent to the other geometries, as K has characteristic $\neq 2$ or equal to 2, respectively.

(5) The determinant for characteristic $\neq 2$ of

$$\left[\begin{array}{cc} -(\frac{g(t)-g(s)}{2}) & f(t) - f(s) \\ t - s & (\frac{g(t)-g(s)}{2}) \end{array} \right]_E^{-1}$$

is

$$\frac{-1}{4}((g(t) - g(s))^2 + 4(t - s)(f(t) - f(s))) \neq 0$$

and with a basis change, letting

$$D_{t,s} = ((g(t) - g(s))^2 + 4(t - s)(f(t) - f(s))) \neq 0$$

an isomorphic quasifibration may be represented in the same form for all/any characteristic as

$$\mathfrak{U}_s^E = \begin{bmatrix} \frac{g(t)-g(s)}{D_{t,s}} & \frac{f(t)-f(s)}{D_{t,s}} \\ \frac{t-s}{D_{t,s}} & 0 \end{bmatrix}_E^-$$

and for the Baer type as

$$\mathfrak{U}_s^B = \begin{bmatrix} \frac{g(t)-g(s)}{D_{t,s}} & \frac{f(t)-f(s)}{D_{t,s}} \\ \frac{t-s}{D_{t,s}} \neq 0 & 0 \end{bmatrix}_B^-.$$

   Then the following describes when the sets define a partial generalized quadrangle or a generalized quadrangle of $(K \times K, K)$-*type.*

   (a) Proper flock quasifibration $\Leftrightarrow$ *partial generalized quadrangle* $\Leftrightarrow$

$$\phi_u \; : \; t \to t - u^2 f(t) + u g(t) \; and$$

$$\Gamma_u^s \; : \; \frac{t-s}{D_{t,s}} \to \frac{t-s}{D_{t,s}} - u^2 \frac{f(t) - f(s)}{D_{t,s}} + u \frac{g(t) - g(s)}{D_{t,s}}$$

are all injective, for all $u, s \in K$.

   For the spreads admitting a Baer group, restrict the mappings by one entry.
   (b) Flock spread $\Leftrightarrow$ *generalized quadrangle* $\Leftrightarrow$

$$\phi_u \; and \; \Gamma_u^s \; are \; all \; bijective.$$

   These quasifibrations $\mathfrak{U}_s^E$ *are also called 'flock derivations' when the quasifibrations are actually spreads. We shall use the term 'quasi-flock derivations', to make the distinction.*

5. Quasi-flock derivations

   There are very few considerations of the flock derivations of flocks of quadratic cones in the infinite case. Must the flock derivations even ever exist? This has not been completely clear in the literature. In this section, we shall discuss a class of flocks of quadratic cones for which none of their flock derivations exist as flocks. However, they do exist as classes of proper quasi-flock derivations. Actually, there are great varieties of such flocks, all of these, of course, are in the infinite case. Later, in the book, we shall consider 'bilinear flocks of quadratic cones', where the planes of the flock share two lines L and M. Such flocks are not possible in the finite case, as has been previously mentioned. The class that is described uses certain bilinear flocks, and their inversions.

THEOREM 53. *Let $\mathbb{Q}$ denote the field of rational numbers. Let P_π denote the following set:*

$$x = 0, y = x \begin{bmatrix} u & 0 \\ 0 & u \end{bmatrix} : u \in \mathbb{Q}, \text{ and}$$

$$N_t \quad : \quad y = x \begin{bmatrix} u & a_1 t \\ t & u \end{bmatrix} : a_1 < 0 \text{ and for all } t > 0,$$

$$S_s \quad : \quad y = x \begin{bmatrix} u & b_1 s \\ s & u \end{bmatrix} : b_1 < 0 \text{ and for all } s < 0, \text{ for } a_1 \neq b_1.$$

Then P_π is a bilinear flock spread for a translation plane π.

Furthermore, the two lines are

$$\{(a_1 x_1, x_1, x_2, 0); x_1, x_2 \in \mathbb{Q}\} \text{ and } \{(b_1 x_1, x_1, x_2, 0); x_1, x_2 \in \mathbb{Q}\},$$

which have a common point $(0, 0, 1, 0)$.

PROOF. We see that N_t and S_s are sub-partial spreads of Pappian translation planes. To see that we have a spread, it suffices to show that the differences between the two subsets of subspaces from N_t (the north) and S_s (the south) are non-singular. Consequently we consider $\begin{bmatrix} u & a_1 t \\ t & u \end{bmatrix} - \begin{bmatrix} 0 & b_1 s \\ s & 0 \end{bmatrix} = \begin{bmatrix} u & a_1 t - b_1 s \\ t - s & u \end{bmatrix}$. The determinant is $u^2 - (t - s)(a_1 t - b_1 s)$. Note that $(t - s) > 0$ and $(a_1 t - b_1 s) < 0$. So that the determinant is > 0 as $t \neq s$ and $a_1 t \neq b_1 s$. The reader might verify that it makes no difference in which order the two matrices are subtracted. This completes the proof. Notice that if b_1 is changed to a_1, we obtain a linear flock and if a_1 is changed to b_1, we obtain a second linear flock. So, there has been a net replacement that shows we have a cover of points and therefore a spread which corresponds to a bilinear flock of a quadratic cone. The linear flocks corresponding to the two Pappian planes in question have the following planes:

$$\pi_t : t x_0 - c_1 d t_1 + x_3 = 0; \forall t \in \mathbb{Q},$$

where c_1 is a_1 or b_1. From here, it is easy to deduce that the two lines are as claimed. This completes the proof. $\qquad\square$

Now, since we have a flock spread P_π, there are classes of s-inversions that are quasifibrations or spreads. We shall show that there exist (b_1, a_1) so that some of the inversions are not spreads.

THEOREM 54. *Choose any rational number $p > 0$ and form the p-inversion of P_π. These are:*

$$p > 0 : x = 0, y = x \begin{bmatrix} w & 0 \\ 0 & w \end{bmatrix} : w \in \mathbb{Q};$$

$$y = x \begin{bmatrix} u & \frac{1}{t-p} \\ \frac{1}{a_1(t-p)} & u \end{bmatrix} ; \forall, u, t > 0, t \neq p \in \mathbb{Q}$$

$$y = x \begin{bmatrix} v & \frac{1}{s-p} \\ \frac{1}{b_1 s - a_1 p} & v \end{bmatrix} ; \forall v, s < 0 \in \mathbb{Q}.$$

And

$$p \; < \; 0 : x = 0, y = x \begin{bmatrix} w & 0 \\ 0 & w \end{bmatrix} : w \in \mathbb{Q};$$

$$y \; = \; x \begin{bmatrix} u & \frac{1}{t-p} \\ \frac{1}{a_1 t - b_1 p} & u \end{bmatrix} ; \forall, u, t > 0, t \in \mathbb{Q}$$

$$y \; = \; x \begin{bmatrix} v & \frac{1}{s-p} \\ \frac{1}{b_1(s-p)} & v \end{bmatrix} ; \forall v, s < 0, s \neq p \in \mathbb{Q}.$$

Then each p-inversion, $p > 0$ is a proper quasifibrations (not spreads) provided the following equation does not hold:

$$(p(a_1 - b_1))^2 - 2pg(a_1 - b_1)(b_1 + 1) + (1 - b_1)^2 g^2$$

is a positive square,

$$\forall g \; \in \; (-\infty, \frac{a_1 p}{1 - a_1}).$$

PROOF. For any fixed $p > 0$, consider $(1, -1, c, d)$, for $cd \neq 0$ and $c + d \neq 0$. It is easier to keep the argument clear if the representation for the matrices is in the original form: Let the determinant of the matrix with the u-elements be Δ_t and Γ_s, where

$$\Delta_t = -a_1 (t - p)^2 > 0$$

and

$$(I)_t : \begin{bmatrix} u & \frac{-a_1(t-p)}{\Delta_t} \\ \frac{-(t-p)}{\Delta_t} & u \end{bmatrix} = \begin{bmatrix} u & \frac{1}{t-p} \\ \frac{1}{a_1(t-p)} & u \end{bmatrix}.$$

And, where

$$\Gamma_s = -(b_1 s - a_1 p)(s - p) > 0$$

so

$$(II)_s : \begin{bmatrix} v & \frac{-(b_1 s - a_1 p)}{\Gamma_s} \\ \frac{-(s-p)}{\Gamma_s} & v \end{bmatrix} = \begin{bmatrix} v & \frac{1}{s-p} \\ \frac{1}{b_1 s - a_1 p} & v \end{bmatrix}.$$

For the p-inversion to be a spread, there must be a unique component covering the point $(1, -1, c, d)$, it follows that this point is incident with a type $(I)_t$ exactly when

$$\frac{(t-p)}{\Delta_t} - \frac{a_1(t-p)}{\Delta_t} = c + d.$$

Then

$$(1 - a_1)(t - p) = \Delta_t(c + d) = -a_1(t - p)^2 f,$$

where $f = (c + d)$. Therefore, we have

$$\frac{(1 - a_1)}{-a_1} \frac{1}{f} + p \; = \; t.$$

$$\text{If } t \; < \; p \text{ then}$$

$$0 \; < \; \frac{(1 - a_1)}{-a_1} \frac{1}{f} + p < p$$

This implies that $f < 0$ and

$$-p < \frac{(1 - a_1)}{-a_1} \frac{1}{f} < 0$$

so that

$$f \in (-\infty, \frac{1-a_1}{a_1 p}).$$

If $t > p$ then $f > 0$ and $f \in (0, \infty)$. This means there is the interval $(\frac{1-a_1}{a_1 p}, 0)$ where $(1, -1, c, d)$ where $f = c + d$ must be covered by the type $(II)_s$. Note that if $f \in (\frac{1-a_1}{a_1 p}, 0)$,then $\frac{1}{f} = g \in (-\infty, \frac{a_1 p}{1-a_1})$.

Assuming then that $(1, -1, c, d)$ is covered by the type $(II)_s$ we must have

$$\frac{s-p}{\Gamma_s} - \frac{(b_1 s - a_1 p)}{\Gamma_s} = f$$

for $f \in (\frac{1-a_1}{a_1 p}, 0)$. And hence,

$$g(s - p - (b_1 s - a_1 p) = \Gamma_s = -(b_1 s - a_1 p)(s - p).$$

For a given g, we are then trying to solve a quadratic in s, for s a negative rational number. The quadratic in s, is as follows:

$$b_1 s^2 + (-p(b_1 + a_1) + (1 - b_1)g)s + a_1 p^2 + gp(a_1 - 1) = 0.$$

For a solution in the rational numbers to exist the discriminate must be a rational square, and this must be valid for all $g \in (-\infty, \frac{a_1 p}{1-a_1})$.

So the discriminant is

$$(-p(b_1 + a_1) + (1 - b_1)g)^2 - 4b_1(a_1 p^2 + gp(a_1 - 1)).$$

Recall that $a_1 \neq b_1$. Working out this term, the following is obtained:

$$(*) \quad : \quad p^2(a_1^2 + b_1^2 - 2a_1 b_1) - 2pg(a_1 b_1 + a_1 - b_1 - b_1^2) + ((1 - b_1)g)^2;$$
$$(p(a_1 - b_1))^2 - 2pg(a_1 - b_1)(b_1 + 1) + (1 - b_1)^2 g^2.$$

Consider, for a moment, the case when $a_1 = b_1$. This would correspond to a Pappian spread and notice that $(*)$ now becomes

$$((1 - b_1)g)^2,$$

in which case, the solution would be

$$s = -(-p(2b_1) + (1 - b_1)g) \pm (1 - b_1)g,$$

and the solution involving g is

$$s = 2(b_1 + (1 - b_1)g), \text{ and note this term is negative.}$$

However, $a_1 \neq b_1$, this would mean that there are infinity many rational squares varying as does rational g over $(-\infty, \frac{a_1 p}{1-a_1})$, where the rational squares have the form $k^2 + rg + m^2 g^2$, where k and m are non-zero, for constant rationals k, r and m, which is a contradiction, by our assumptions. This completes the proof of the theorem for $p > 0$. □

PROBLEM 18. *(Seems to be partially open). Prove if $a_1 \neq b_1$, $p > 0$ then the following condition is never valid.*

$$(p(a_1 - b_1))^2 - 2pg(a_1 - b_1)(b_1 + 1) + (1 - b_1)^2 g^2$$

is a positive rational square,

$$\forall g \quad \in \quad (-\infty, \frac{a_1 p}{1 - a_1}).$$

COROLLARY 15. *For example, if $b_1 = -1$, we obtain $(p(1 + a_1))^2 + 4g^2$ must be a positive rational square for all $g \in (-\infty, \frac{a_1 p}{1 - a_1})$. Choose a_1 so that $(1 + a_1)^2 + 4$ is a non-square. Then the condition is never valid.*

PROOF. $\left| \frac{a_1 p}{1 - a_1} \right| < p$ so $g = -p < \frac{a_1 p}{1 - a_1}$, and choose $g = -p$. It follows that $(1 + a_1)^2 + 4$ must be square, a contradiction.   □

PROBLEM 19. *When $p < 0$, work out the condition so that a proper quasifibration is obtained. Once the condition is obtained, find an infinite class of examples of proper quasifibrations.*

So, we have conditions so that there is a flock of a quadratic cone, where all of the associated positive (or negative) flock derivations are not flocks, but are quasi-flocks (also see (9,11,3)). In Biliotti, Johnson [**15**], there is a more general treatment. We shall come back to these associated quasifibrations later in the book (7,9,3), when considering chains of quasifibrations of dimension 2.

6. Herds of hyperovals

For an attempt at some completeness, the following flock geometry is included.

When q is even, every flock of a quadratic cone is equivalent to what are called a 'herd of hyperovals'. A 'hyperoval' is a $q + 2$-arc (no three points incident with a line) in $PG(3, q)$. We use again, $\begin{bmatrix} t & \mathcal{G}(t) \\ 0 & -\mathcal{F}(t) \end{bmatrix}$, the generalized quadrangle notation to denote a flock of a quadratic cone. So, when q is even we have $-\mathcal{F}(t) = \mathcal{F}(t)$.

For the references related to the following theorem, the reader is directed to [**94**] p. 418. The mathematicians originating this material include S.E. Payne, W. Cherowitzo, T. Penttila, I. Pinneri, G.F. Royle, L. Storme, J.A. Thas, see Theorem 54.22, and note the work is attributed to Payne and Penttila.

THEOREM 55. *The set of all flocks of quadratic cones for q even of the form* $\begin{bmatrix} t & \mathcal{G}(t) \\ 0 & \mathcal{F}(t) \end{bmatrix}$ *defines the following set of hyperovals $K(a_1, a_2), \forall a_1, a_2 \in GF(q)$, $(a_1, a_2) \neq (0, 0)$.*

$$K(a_1, a_2) = \left\{ (1, \sqrt{a_1^2 - a_1 a_2 \mathcal{F}(t) + a_2^2 \mathcal{G}(t)} : t \in GF(q) \right\}.$$

A set of $q^2 - 1$ hyperovals of the type listed above is said to be a 'herd of hyperovals'. Herds of hyperovals and flocks of quadratic cones of even order are equivalent.

7. Hyperbolic fibrations

In this chapter, hyperbolic fibrations in $PG(3, K)$, for K a field are considered. Originally, this work, in the finite case, for $K = GF(q)$, is due to Baker, Ebert, and Penttila [**11**]. In this book, we consider the more general case over arbitrary fields, and this follows the material in Johnson [**88**], chapters 17 and 18, although here we provide a brief view of the concepts.

DEFINITION 63. *A 'hyperbolic fibration. over K' is a set λ of mutually disjoint hyperbolic quadrics, indexed by K^*, with two carrying lines L and M in $PG(3, K)$,* *for K a field, such that the union of the hyperbolic quadrics together with L and M is a cover of the points/vectors of the associated 4-dimensional vector space V_4/K.*

DEFINITION 64. *A 'regular hyperbolic fibration over K' is a hyperbolic fibration over K such that, for each of the hyperbolic quadrics, the induced polarity interchanges L and M. In this case, the hyperbolic quadrics have the following form, where points are represented homogeneously as (x_1, x_2, y_1, y_2) : $\forall x_i, y_i \in K$, for $i = 1, 2$.*

$$V_4(x_1^2 a_i + x_1 x_2 b_i + x_2^2 c_i + y_1^2 g_i + y_1 y_2 h_i + y_1^2 k_i) : \forall i \in |K^*| = card K^*.$$

DEFINITION 65. *A 'regular hyperbolic fibration over K with constant back half' is a regular hyperbolic fibration over K such that $g_i = g$, $h_i = h$, $k_i = k$, for g, h, k constants in K for all $i \in card K^*$.*

In the multiple replacement theorem of part 10, when there is a Galois quadratic field extension $K(\theta)$ of K, there is an associated Pappian translation plane with spread in $PG(3, K)$, that has a set λ of mutually disjoint derivable nets together with two carrying lines L and M ($x = 0, y = 0$) such that the union of these K-subspaces is a spread. We see that the derivable nets are K-regulus nets (by the embedding theorem of part 1), and thus are hyperbolic quadrics in $PG(3, K)$. The induced polarity interchanges the components and Baer subspaces of each derivable net and therefore, interchanges $x = 0$ and $y = 0$, Johnson [**78**].

In part 10, we shall see that we have the following structure for the Pappian spread, partitioned into derivable nets/and their derived nets:

$$x = 0, y = 0, \{y = x(bm_i b^{-\sigma}) : \forall b \in K^*\}$$
$$= \{y = (bm_i b^{-\sigma})x^\sigma : \forall b \in K^*\},$$

where σ in the induced automorphism of $K(\theta)$ of order 2,

for $i \in \rho$ an index set.

Since here we have commutativity of multiplication, we have

$$x = 0, y = 0, \{y = xm_i b^{1-\sigma} : \forall b \in K^*\}$$
$$= \{y = x^\sigma m_i b^{1-\sigma} : \forall b \in K^*\},$$

where σ is the induced automorphism of $K(\theta)$ of order 2.

In this setting, we may write $K(\theta)$ in the following form:

$$x = 0, y = 0, y = x \begin{bmatrix} u & t \\ ft & u + gt \end{bmatrix}^* : \forall t, u \in K.$$

So, we see that $(b^{1-\sigma})^{\sigma+1} = 1$, as σ has order 2.

Consequently, $\langle b^{1-\sigma} : \forall b \in K^* \rangle$ is a subgroup of $K(\theta)$. The $m_i = \theta t + u = \begin{bmatrix} u & t \\ \beta t & u - \alpha t \end{bmatrix}$ and let H denote the group $\langle b^{1-\sigma} : b \in K^* \rangle$ in the matrix form. The set $\{m_i\}$ forms a representation set for the quotient group

$$\left\{ \begin{bmatrix} u & t \\ \beta t & u - \alpha t \end{bmatrix}^* / H \right\}.$$

It is then seen that we have a flock of a hyperbolic quadric (See (6,8,3) and a flock of a quadratic cone, since the translation planes admit both a regulus-inducing elation group and a regulus-inducing homology group. Furthermore, there is also a regular hyperbolic fibration with constant back-half, and we see that the regulus nets disjoint from the components $x = 0$ and $y = 0$ admit the group H, which is also a group of central collineations with axis $x = 0$ and coaxis $y = 0$.

We note that the Andrè translation planes constructed (See part 10) also represent the same hyperbolic fibration. There are, therefore, connections with translation planes and regular hyperbolic fibrations with constant back half. It may not be completely apparent but there is an associated central collineation group G with axis $x = 0$ and co-axis $y = 0$ that acts regularly on each hyperbolic quadric. Consequently, multiple derivation does not change the hyperbolic fibration.

The main point for finite translation planes of order q^2 with spread in $PG(3, q)$ is that there is a cyclic homology group of order $q + 1$, and the hyperbolic quadrics become regulus nets of $q + 1$ components. Also, the algebraic connection between the translation plane and the hyperbolic fibration effectively maps

$$GF(q^2) \to Det \begin{bmatrix} u & t \\ ft & u + gt \end{bmatrix} = u(u + g) - ft^2,$$

which is a surjective mapping onto $GF(q)$. In the infinite case, with respect to a field K, there is also the same algebraic connection, however now the mapping takes

$$K(\theta) \to Det \begin{bmatrix} u & t \\ ft & u + gt \end{bmatrix},$$

but this mapping is no longer surjective. Hence, in the general case, over an arbitrary field K, the connection carries translation planes admitting an appropriate homology group to partial flocks of quadratic cones, indexed by the Det subset of elements of K.

Here are the main results of this theory with enough details to follow some applications to come. The reader is directed to Johnson [88], chapters 17 and 18, for complete details. Also, the reader might look into the chapter on infinite flocks of hyperbolic quadrics in (6,8,11) for a discussion on the Andrè planes over infinite fields K.

THEOREM 56. ([88] *Theorem 110 p. 257) Let π be a translation plane with spread in $PG(3, K)$, for K a field. Assume that π admits an affine homology group H so that some orbit of components is a regulus in $PG(3, K)$.*

(1) Then π produces a regular hyperbolic fibration with constant back half.

(2) Conversely, each translation plane obtained from a regular hyperbolic quadric with constant back half admits an affine homology group H one orbit of which is a regulus in $PG(3, K)$.

H is isomorphic to a subgroup of a Pappian spread Σ, coordinatized by a Galois quadratic extension $K(\theta)$. If

$H \simeq \langle h; h^{\sigma+1} = 1; h \in K(\theta)^* \rangle$, *for σ the involutory automorphism of $K(\theta)$,*

as

$$x = 0, y = 0, y = x \begin{bmatrix} u & t \\ ft & u + gt \end{bmatrix}; \delta_{u,t} = 1 : \forall t, u \in K - (0, 0).$$

(3) *Assume that the irreducible polynomial defining the Galois quadratic extension is $x^2 + gx + f$, for $\alpha, \beta \in K$. Then the collineation subgroup of the hyperbolic fibration that leaves each hyperbolic quadric invariant is $\langle \rho, H \rangle$, where*

$$\rho = \begin{bmatrix} I_2 & 0 \\ 0 & \begin{bmatrix} 1 & 0 \\ g & -1 \end{bmatrix} \end{bmatrix}.$$

H is regular on the sets of components and Baer subspaces (of each hyperbolic quadric as a regulus net) and ρ interchanges these two sets of 2-dimensional K-subspaces; regulus and opposite regulus.

The following theory indicates the nature of translation planes giving rise to regular hyperbolic fibrations with constant back half.

THEOREM 57. *(Johnson [88] Theorem 111, page 257) Let π be a translation plane with spread in $PG(3,q)$ that admits a cyclic affine homology group of order $q + 1$. Then π produces a regular hyperbolic fibration with constant backhalf.*

COROLLARY 16. *Finite flocks of quadratic cones are equivalent to translation planes with spreads in $PG(3,q)$ admitting cyclic affine homology groups of order $q + 1$. We shall call the translation planes 'cyclic $q + 1$-planes'.*

The following theorem connects spreads and partial flocks involved in the theory.

8. The correspondence theorem

To connect the previous notations of flocks of quadratic cones and spreads admitting certain affine homology groups over an arbitrary field K, we shall see that the connection is valid but produces partial flocks of quadratic cones in $PG(3,K)$.

THEOREM 58. *([88] see p. 269 for the slightly more complete version) The correspondence between any spread for a translation plane in $PG(3,K)$, for K a field that admits a Galois quadratic extension $K(\theta)$, and a regular hyperbolic fibration with constant back half, and the partial flock of a quadratic cone is as follows:*
If the spread for π is

$$x = 0, y = 0, y = x \begin{bmatrix} u & t \\ F(u,t) & G(u,t) \end{bmatrix} : \forall u, t \in K - (0,0);$$

$$F \text{ and } G \text{ functions on } K \times K, \ F(1,0) = 0 \text{ and } G(1,0) = 1,$$

then the partial spread for the partial flock of a quadratic cone is:

$$\begin{bmatrix} \delta_{u,t} & \mathcal{G}(\delta_{u,t}) \\ 0 & -\mathcal{F}(\delta_{u,t}) \end{bmatrix} : \forall \delta_{u,t},$$

where $\delta_{u,t} = \det \begin{bmatrix} u & t \\ ft & u + gt \end{bmatrix}$. And, where

$$\mathcal{G}(\delta_{u,t}) = g(uG(u,t) + tF(u,t)) + 2(u\,F(u,t) - tfG(u,t))$$
$$(gu - 2tf)G(u,t) + (gt + 2u)F(u,t)$$
$$-\mathcal{F}(\delta_{u,t}) = \delta_{F(u,t),G(u,t)} = F(u,t)^2 + gF(u,t)G(u,t) - fG(u,t)^2.$$

where

$$\delta_{F(u,t),G(u,t)} = \det \begin{bmatrix} F(u,t) & G(u,t) \\ fG(u,t) & F(u,t) + gG(u,t) \end{bmatrix} \in \det K(\theta)^*.$$

If there is a regular hyperbolic fibration with constant back half in $PG(3, K)$, there are corresponding functions

$$\phi_s : \phi_s(t) = s^2 t + \mathcal{G}(t) - \mathcal{F}(t)$$

which are injective for all $s \in K$ and for all $t \in \det K(\theta)^*$.

The converse that such functions and injective mappings produce a regular hyperbolic fibration with constant back half over K, and in turn, a set of corresponding translation planes is also valid.

In the construction, there is an embedding on a partial spread of the matrix spread of the cyclic $q + 1$-plane into an associated Desarguesian plane. Basically, when we have a $q + 1$-plane, there is a field extension of $K = GF(q)$ to $K(\theta)$, such that $\theta^2 + f\theta - g = 0$, and the automorphism σ fixes K and maps $\theta \to -\theta + g$. We adopt the vector space notation (x, y) so that $x, y \in K(\theta)$, representing $x = x_1 + x_2\theta$, and then $x^\sigma = x_1 + gx_2 - \theta x_2$. To be a spread, the matrices M and their differences $M - N$, for all $M \neq N$, $M, N \in \lambda$, and assuming that $I_2 \in \lambda$, are all non-singular. This implies that there is a corresponding spread for the translation plane:

$$x = 0, y = 0, y = xM : M \in \lambda,$$

for M a set of 2 by 2 matrices over K. There is also an associated Desarguesian spread:

$$x = 0, y = 0, y = x^\sigma M : M \in \lambda.$$

Consequently, it follows that

$$y = xM \begin{bmatrix} u & t \\ ft & u + gt \end{bmatrix} : \forall \delta_{u,t} = 1 \text{ for each } M \in \lambda,$$

is a derivable regulus net and the replacement net is

$$y = x^\sigma M \begin{bmatrix} u & t \\ ft & u + gt \end{bmatrix} : \forall \delta_{u,t} = 1 \text{ for each } M \in \lambda,$$

$$x^\sigma = (x_1, x_2) \begin{bmatrix} 1 & 0 \\ g & -1 \end{bmatrix}, \text{ when } (x_1, x_2) = x_1 + \theta x_2,$$

$$y = xM \begin{bmatrix} 1 & 0 \\ g & -1 \end{bmatrix} \begin{bmatrix} u & t \\ ft & u + gt \end{bmatrix} : \forall \delta_{u,t} = 1.$$

So, we represent the spread in the form

$$x = 0, y = 0, y = x : \forall u, t \in K$$

where this background notation is built into the coordinatization. Hence, there are $q - 1$ derivable regulus nets, choose any subset of k of these and denote the representatives to be $y = xM_i, i = 1, 2..., k$ and derive these. Now we have a spread of the following form (Correction-for readers of the Combinatorics book [**88**]: typo on page 262 on line -7: the mapping τ_{M_i} should be $(x, y) \to (xM_i, y)$)

$$x = 0, y = 0, y = xM_i \begin{bmatrix} 1 & 0 \\ g & -1 \end{bmatrix} \begin{bmatrix} u & t \\ ft & u + gt \end{bmatrix} :$$

$$\forall \delta_{u,t} = 1, i = 1, 2...k, \text{ for } u, t \in K$$

$$\cup y = xM_j \begin{bmatrix} u & t \\ ft & u + gt \end{bmatrix}, j = k + 1, ..., q - 1.$$

All of these translation planes produce the same flock of the quadratic cone, which makes an identification from the flock back to the cyclic $q+1$-plane problematic.

In the following, α-hyperbolic flocks are studied. At one point, j-planes will be considered, for which the theory of hyperbolic fibrations will be employed to realize certain interesting classes of finite α-hyperbolic planes.

COROLLARY 17. *Given a cyclic $q+1$-plane, with functions $F(u,t)$ and $G(u,t)$, the corresponding quadratic flock translation plane then has the following form:*
For

$$\mathcal{G}(\delta_{u,t}) = (gu - 2tf)G(u,t) + (gt + 2u)F(u,t)$$

and

$$\mathcal{F}(\delta_{u,t}) = -F(u,t)^2 - gF(u,t)G(u,t) + fG(u,t)^2$$

$$x = 0,$$

$$y = x \begin{bmatrix} w + \mathcal{G}(\delta_{u,t}) & \mathcal{F}(\delta_{u,t}) \\ u(u+gt) - ft^2 = \delta_{u,t} & w \end{bmatrix}$$

(7.1) $\forall u, t, w \in GF(q),$

9. Flocks to cyclic planes

Translation planes of order q^2 admitting a cyclic homology group of order $q+1$ are, indeed, equivalent to flocks of quadratic cones but there are so few known planes of this type, the use of such planes has not traditionally been an area of much research. It is known that j-planes and monomial flocks of quadratic cones are equivalent. And, later, we take this connection up again. We discuss t-nest planes in a later chapter and show that $q+1$-nest replaceable translation planes are cyclic homology planes and discuss the associated flocks. In this chapter, we shall discuss a few of the infinite flocks of quadratic cones and determine the associated cyclic homology planes. First, we lay out the procedure. Here again is the set up. The F, G functions refer to the cyclic homology planes and $\mathcal{G}$ and $\mathcal{F}$ functions correspond to the flocks of quadratic cones. Note that $\delta_{u,t}$ refers to the determinant of $\begin{bmatrix} u & t \\ ft & u + gt \end{bmatrix}$, for all $u, t \in GF(q) = K$, which defines a field $K(\theta)$ isomorphic to $GF(q^2)$, $f, g \in K$

$$\mathcal{G}(\delta_{u,t}) = (gu - 2tf)G(u,t) + (gt + 2u)F(u,t)$$

$$-\mathcal{F}(\delta_{u,t}) = \delta_{F(u,t),G(u,t)} = F(u,t)^2 + gF(u,t)G(u,t) - fG(u,t)^2.$$

We note that if $F(u,t)$ and $G(u,t)$ are given, then it is obvious how to determine $\mathcal{G}(\delta_{u,t})$ and $\mathcal{F}(\delta_{u,t})$. The reason for this is that the correspondence theorem was originally given from the cyclic $q+1$-plane to the flock. The correspondence also works the other way from the flock of a quadratic cone to the cyclic $q+1$-plane. While the direction from the cyclic $q+1$-plane to the flock is essentially immediate, the other direction, the flock to cyclic $q+1$-plane actually constructs not only the cyclic $q+1$-plane but also all of the multiply derived translation planes. Suppose that $\mathcal{G}$ and $\mathcal{F}$ are known. Then the idea to recover F and G is simply to solve the simultaneous equations. The requirements for g and f are obtained by the flock functions. But, one cannot actually recover the functions F and G uniquely. So, there are two conditions that must be observed:

The spread for the cyclic plane is

$$x = 0, y = 0, y = x \begin{bmatrix} u & t \\ F(u,t) & G(u,t) \end{bmatrix} : \forall u, t \in GF(q) = K.$$

The orbits under the cyclic homology group of order $q+1$ has elements $(x,y) \to (x,y \begin{bmatrix} u & t \\ ft & u+gt \end{bmatrix})$ where the determinant of the matrix $\delta_{u,t} = 1$. This is the cyclic group of order $q+1$. Since it is necessary to know what derivable nets look like, let $\theta^2 + \theta g - f = 0$ and realize that the set of matrices $\left\{ \begin{bmatrix} u & t \\ ft & u+gt \end{bmatrix} ; \forall t, u \in GF(q) \right\}$ as a field is isomorphic to the set of elements$\{\theta t + u; t, u \in GF(q)\}$, that is, $GF(q^2)$ or $K(\theta)$. In this setting, the replacement for $y = x(u+\theta t)$ is $y = x^\sigma(u+\theta t)$, where σ is the automorphism that maps $(u+\theta t) \to (u+(-\theta+g)t)$. Now write $x = x_1 + \theta x_2$, for $x_1, x_2 \in GF(q)$ and realize that

$$x^\sigma \begin{bmatrix} u & t \\ ft & u+gt \end{bmatrix} = x \begin{bmatrix} 1 & 0 \\ g & -1 \end{bmatrix} \begin{bmatrix} u & t \\ ft & u+gt \end{bmatrix},$$

for all $\delta_{u,t} = 1$. When the replacement net is

$$\left\{ y = x \begin{bmatrix} u^* & t^* \\ F(u^*,t^*) & G(u^*,t^*) \end{bmatrix} \begin{bmatrix} u & t \\ ft & u+gt \end{bmatrix} \right\}$$

for all $\delta_{u,t} = 1$, the derived net is

$$\left\{ y = x \begin{bmatrix} u^* & t^* \\ F(u^*,t^*) & G(u^*,t^*) \end{bmatrix} \begin{bmatrix} 1 & 0 \\ g & -1 \end{bmatrix} \begin{bmatrix} u & t \\ ft & u+gt \end{bmatrix} \right\},$$

for all $\delta_{u,t} = 1$.

Since we shall be ignoring these multiple derived planes, when determining functions F and G that produce a cyclic $q+1$-plane, we also provide the following identities:

$$\begin{bmatrix} u^* + t^* g & -t^* \\ F(u^*,t^*) + G(u^*,t^*)g & -G(u^*,t^*) \end{bmatrix}.$$

These identities are not necessary to be used for all values $\delta_{u,t}$, but for any particular value or values. If used for all $\delta_{u,t}$, the associated cyclic $q+1$-translation plane obtained would be the plane obtained by derivation of all of the $q-1$ derivable nets.

$$\begin{aligned} F_1(u^* + t^* g, -t^*) &= F(u^*,t^*) + G(u^*,t^*)g \\ G_1(u^* + t^* g, -t^*) &= -G(u^*,t^*) \end{aligned}$$

may be substituted for any (u^*,t^*). Also, since $\begin{bmatrix} u & t \\ ft & u+gt \end{bmatrix}$, for all u,t, such that $\delta_{u,t} = 1$. Note that if $g = 0$ then $G_1(u,-t) = -G(u,t)$ and $F_1(u,t) = F(u,t)$. Consequently, q must be odd, for otherwise, it would be not be possible to make derivations. Ultimately, we have a translation plane with spread

$$x = 0, y = x \begin{bmatrix} u & t \\ F(u,t) & G(u,t) \end{bmatrix}.$$

And, to be a spread, differences of different non-zero matrices must be non-singular matrices.

So, with the identities and the derivations (multiple), we solve for the cyclic $q + 1$-planes.

Therefore, we choose a set of $q - 1$ matrices as representative of the orbits of the homology group of order $q + 1$, indexed by (u_k^*, t_k^*) for $k = 1, 2, ..., q - 1$ and associated functions $F(u_k^*, t_k^*)$, $G(u_k^*, t_k^*)$, when considering the recovery of the cyclic plane.

Also, note that theoretically, every flock of a quadratic cone may be given in terms of functions.

We shall give the recovery in general and then apply this to various flocks of quadratic cones, $\mathcal{G}(\delta_{u,t})$ and $\mathcal{F}(\delta_{u,t})$. In all of the finite flocks of quadratic cones given in the handbook (See [94] p. 439, chapter 59), all of the infinite classes will have a variable s, as in $\begin{bmatrix} s & \mathcal{G}(s) \\ 0 & -\mathcal{F}(s) \end{bmatrix}$, for any of these classes, just let $s = \delta_{u,t}$ realizing that for a general term $\begin{bmatrix} u & t \\ ft & u + gt \end{bmatrix}$ of order $q^2 - 1$ in $GF(q^2)$, and the fact that $(u + \theta t)^{\sigma+1} = \delta_{u,t}$, we have that the set of determinants form a bijective group equal to $GF(q)^*$. Therefore, for any of these flocks, the process that will be shown will allow the reader to work out a nice representation for the cyclic $q + 1$-planes. On the other hand, the BLT-sets will, in general, not work directly, a transformation will need to made to place them in the correct form of $\begin{bmatrix} \delta_{u,t} & \mathcal{G}(\delta_{u,t}) \\ 0 & -\mathcal{F}(\delta_{u,t}) \end{bmatrix}$. These transformations will not be considered in this text for lack of space.

With these conditions in place, here is how to go from the flock of a quadratic cone to the cyclic $q + 1$-planes.

So, starting with

$$\begin{aligned} \mathcal{G}(\delta_{u,t}) &= g(uG(u,t) + tF(u,t)) + 2(u\,F(u,t) - tfG(u,t)) \\ &\quad (gu - 2tf)G(u,t) + (gt + 2u)F(u,t) \\ -\mathcal{F}(\delta_{u,t}) &= \delta_{F(u,t),G(u,t)} = F(u,t)^2 + gF(u,t)G(u,t) - fG(u,t)^2. \end{aligned}$$

I. Let $c = (gu - 2tf)$ and $d = (gt + 2u)$. It is possible that c and d could individually be 0, but they cannot both be zero.

So, we have

$$(*) : \mathcal{G}(\delta_{u,t}) = cG(u,t) + dF(u,t).$$

Resist the temptation to consider $c = 0$ sometimes.

II. Use

$$(**) : -\mathcal{F}(\delta_{u,t}) = F(u,t)^2 + gF(u,t)G(u,t) - fG(u,t)^2$$

to solve twice: Once for F and once for G. To solve for $G(u,t)$, form $d^2(**)$. That is, multiply equation $(**)$ by d^2 to realize

$$d^2(**) : -d^2\mathcal{F}(\delta_{u,t}) = (dF(u,t))^2 + gd(dF(u,t))G(u,t) - fd^2G(u,t)^2.$$

Substitute:

$$dF(u,t) = \mathcal{G}(\delta_{u,t}) - cG(u,t).$$

This will give you a quadratic in $G(u,t)$. In the odd order case, use the quadratic equation. In the even order case, that would be the final step.

III. Work out the quadratic in $F(u,t)$, then do the same by forming $c^2(**)$ and put in $cG(u,t) = \mathcal{G}(\delta_{u,t}) - dF(u,t)$. Here is what you will obtain:

Without reference to order:

$a)-$ The $F(u,t)-$Quadratic:

$$F(u,t)^2(\delta_{c,d}) + F(u,t)(gc+2df)\mathcal{G}(\delta_{u,t}) + c^2\mathcal{F}(\delta_{u,t}) - f\mathcal{G}(\delta_{u,t})^2 = 0$$

$b)-$ The $G(u,t)-$Quadratic:

$$G(u,t)^2\delta_{c,-d} + G(u,t)(gd-2c)\mathcal{G}(\delta_{u,t}) + d^2\mathcal{F}(\delta_{u,t}) + \mathcal{G}(\delta_{u,t})^2 = 0.$$

IV. Work out that $\delta_{c,d}$ and $\delta_{c,-d}$ to show that $\delta_{c,d} = \delta_{(gu-2ft),(gt+2u)}$, and $\delta_{c,-d} = \delta_{(gu-2ft),-(gt+2u)}$. Note the coefficient on $G(u,t)$ is $(g^2+4f)t$ and on $F(u,t)$ is $(g^2+4f)u$ (note the discriminant of the original quadratic equation defining the field $GF(q^2)$ is g^2+4f).

(a) Odd order: Then we have solutions for $F(u,t)$ and $G(u,t)$ using the quadratic equation noting that $\delta_{c,d}$ and $\delta_{c,-d}$ are never zero. Hence, we have the following solutions:

$$F(u,t) = \frac{(g^2+4f)u\mathcal{G}(\delta_{u,t})\pm \sqrt{((g^2+4f)u\mathcal{G}(\delta_{u,t}))^2 - 4((gu-2tf)^2\mathcal{F}(\delta_{u,t}) - f\mathcal{G}(\delta_{u,t}))}}{\delta_{(gu-2tf),(gt+2u)}}$$

and

$$G(u,t) = \frac{(g^2+4f)t\mathcal{G}(\delta_{u,t})\pm \sqrt{((g^2+4f)t\mathcal{G}(\delta_{u,t}))^2 - 4((gt+2u)^2\mathcal{F}(\delta_{u,t}) - f\mathcal{G}(\delta_{u,t}))}}{\delta_{(gu-2tf),-(gt+2u)}},$$

where the notation is to indicate $\delta_{(gu-2tf),-(gt+2u)}$ is a common denominator.

These functions will provide one of the cyclic $q+1$-planes obtained by multiple derivation.

(b) Even order: We may choose $g=1$.

$$F(u,t)^2(\delta_{u,t}) + F(u,t)(\delta_{u,t})\mathcal{G}(\delta_{u,t}) + u\mathcal{F}(\delta_{u,t}) + f\mathcal{G}(\delta_{u,t})^2 = 0$$

and

$$G(u,t)^2\delta_{u,t} + G(u,t)(\delta_{u,t})\mathcal{G}(\delta_{u,t}) + t\mathcal{F}(\delta_{u,t}) + \mathcal{G}(\delta_{u,t})^2 = 0.$$

Again, here the equations are defined modulo the

$$\mathcal{G}(\delta_{u,t}) = uG(u,t) + tF(u,t).$$

Use any of these functions to obtain one of the translation planes of the various multiple derivations of any cyclic $q+1$-planes. That is, the set of multiple derivations of any of these cyclic $q+1$-planes is equal to the set of the multiple derivations of any of the other cyclic $q+1$-planes.

One last comment on $g=0$. Recall

$$(*)\mathcal{G}(\delta_{u,t}) = -2tfG(u,t) + 2uF(u,t).$$

Note that $\delta_{u,-t} = \delta_{u,t}$, when $g = 0$. Since $G_1(u, -t) = -G(u, t)$ and $F_1(u, -t) = F(u, t)$.

REMARK 17. *Given any conical flock plane of finite order, there are the associated s-inverted and s-square conical flock spreads. All of these planes correspond to cyclic $q + 1$-planes. In general, the conical spreads should be in the following form:*

$$x = 0, y = x \begin{bmatrix} v & 0 \\ 0 & v \end{bmatrix}, \ y = x \left\{ \begin{bmatrix} u + h(t) & f(t) \\ t & u \end{bmatrix} \right\} ;$$

$$\forall t \neq 0, u, v \in K, \ h, f \ functions \ on \ K = GF(q).$$

REMARK 18. *(1) s-inversions: For conical flock spreads of odd order q, of the form*

$$x = 0, y = x \begin{bmatrix} v & 0 \\ 0 & v \end{bmatrix},$$

$$y = x \left\{ \begin{bmatrix} -(h(t) - h(s))/2 & f(t) - f(s) \\ t - s & (h(t) - h(s))/2 \end{bmatrix}^{-1} + \begin{bmatrix} u & 0 \\ 0 & u \end{bmatrix} \right\} ;$$

$$\forall t \neq s, u, v \in K, \ h, f \ functions \ on \ K = GF(q), \ q \ odd.$$

DEFINITION 66. *(2) If the order q is even, noting that h is bijective then there is a set of other flock spreads called the set of 's-square' spreads,' which have the following form: For any fixed $s \in K$,*

$$x = 0, y = x \begin{bmatrix} v & 0 \\ 0 & v \end{bmatrix},$$

$$y = x \left\{ \frac{1}{(h(t) + h(s))^2} \begin{bmatrix} h(t) + h(s)) & f(t) + f(s) \\ t + s & 0 \end{bmatrix} + \begin{bmatrix} u & 0 \\ 0 & u \end{bmatrix} \right\} ;$$

$$\forall t \neq s, u, v \in K, \ h, f \ functions \ on \ K = GF(2^r).$$

Here are four problems, two odd order, two even order.

PROBLEM 20. *For the Ganley semifield planes, with spread*

$$x = 0, y = x \begin{bmatrix} w + n\delta_{u,t}^3 & -n\delta_{u,t} - n^3\delta_{u,t}^9 \\ \delta_{u,t} & w \end{bmatrix}$$

$$: \ \forall w, \delta_{u,t} \in GF(3^r), \ n \ a \ fixed \ non\text{-}square,$$

determine the corresponding cyclic $q + 1$-planes. Note $\delta_{u,t} = u(u + gt) - ft^2$.

PROBLEM 21. *For any $s-$inverted of the Ganley semifield planes, determine the corresponding cyclic $q + 1$-planes.*

The variables now need to be in the form $\delta_{u,t}$ for the general variable and fix s as $\delta_{v,k} \neq \delta_{u,t}$.

Then the s-inverted conical flock spread will have

$$\begin{bmatrix} -(h(t) - h(s))/2 & f(t) - f(s) \\ t - s & (h(t) - h(s))/2 \end{bmatrix}^{-1}$$

$$= \begin{bmatrix} -(n\delta_{u,t}^3 - n\delta_{v,k}^3)/2 & -n\delta_{u,t} - n^3\delta_{u,t}^9 + n\delta_{v,k} + n^3\delta_{v,k}^9 \\ \delta_{u,t} - \delta_{v,k} & (n\delta_{u,t}^3 - n\delta_{v,k}^3)/2 \end{bmatrix}^{-1}.$$

Hint: Use the fact that $(r + m)^3 = r^3 + m^3$, in this field.

PROBLEM 22. *For the Walker/Betten/Denniston (W/B/D) even order version of a flock spread of order* $(q = 2^{2r+1})^2$, *with spread*

$$x = 0, y = x \begin{bmatrix} w + \delta_{u,t}^2 & \delta_{u,t}^3/3 \\ \delta_{u,t} & w \end{bmatrix} : \forall \delta_{u,t} \in GF(2^{2r+1}),$$

determine the corresponding cyclic $q + 1$*-planes.*

PROBLEM 23. *For any* s^2*-conical flock spread of the W/B/D flock spread, determine the corresponding cyclic* $q + 1$*-planes,* $q = 2^{2r+1}$.

Here $h(\delta_{u,t}) = \delta_{u,t}^2$, $f(\delta_{u,t}) = \delta_{u,t}^3/3$, *let* $s = \delta_{v,k} \neq \delta_{u,t}$.

Form $\dfrac{1}{(h(t)+h(s))^2} \begin{bmatrix} h(t) + h(s)) & f(t) + f(s) \\ t + s & 0 \end{bmatrix}$, *using this notation.*

Part 6

Twisted hyperbolic flocks

The work in this part appeared originally in [89] and [90] in Innovations in Incidence Geometries and Algebraic Combinatorics, respectively. The author gratefully acknowledges the assistance of the journals.

CHAPTER 8

Hyperbolic flocks and generalizations

A hyperbolic quadric H when viewed in an affine form is a classical regulus net within a 4-dimensional vector space V_4 over a finite field $GF(q)$. A flock of H is a covering of H by a set of $q + 1$ mutually disjoint planes in $PG(3, q)$. Associated with a hyperbolic flock is a translation plane in V_4, that admits an affine homology group of order $q - 1$, one of whose orbits union the axis and coaxis, becomes a regulus net and then all orbits that union the axis and coaxis are regulus nets. The union of these nets define a translation plane; the translation plane of the hyperbolic flock. These two geometries, the hyperbolic flock and the translation plane, are equivalent. There are exactly the following classes; the flocks are the linear flock, where the associated planes of $PG(3, q)$ share a line, and the Thas flocks, with a few exceptions. The Thas flocks correspond to the regular nearfield planes and the exceptional flocks correspond to certain of the irregular nearfield planes and are due to a number of mathematicians from various different points of view (Bader [5], Baker-Ebert [9], Bonisoli [17], Johnson [76]). There are three irregular nearfields of orders $11^2, 23^2, 59^2$. Bader, Bonisoli and Johnson independently determined the same results for all three orders. And, Baker and Ebert showed the same results but for orders $11^2, 23^2$. Each of the authors determined the flocks/translation planes by using essentially different methods. The main point here is that the associated translation planes are all Bol planes; which has been of considerable interest, and there is a complete classification due to Thas and Bader-Lunardon ([136], [8]). There are various possible formulations for this classification, depending on whether it is phrased in the associated translation plane or in the hyperbolic flock. Sometimes, names are used, sometimes the name of the algebra coordinatizing the structures is used to describe the structures. For uniformity, here we shall use the name of the coordinate structures for the translation plane version, and the names of the mathematicians finding the flocks in the flock version.

THEOREM 59. *Thas, Bader-Lunardon ([136].[8]) Classification of finite hyperbolic flocks/translation planes admitting regulus-inducing homology groups in* $PG(3, q)$.

Plane version: The translation planes are exactly the nearfield planes; based upon the regular nearfields of order q^2 *and the three irregular nearfields of orders* $11^2, 23^2$ *and* 59^2.

Flock version: The hyperbolic flocks are exactly the Thas flocks and the flocks of Bader, Baker-Ebert, Bonisoli, Johnson.

In the infinite case, the translation planes corresponding to hyperbolic flocks, it is not the case that all Bol planes need be nearfield planes, and there is a much richer set of examples. The interested reader is referred to the Handbook [94] for additional information.

There are also some interesting infinite classes of partial spreads related to hyperbolic flocks. In particular, there are partial spreads of degree $p^2 + 1$ that contain the irregular nearfields spreads of order p^2 by Bader, Durante, Law, and Lunardon [7]).

It is also an interesting question on how large a partial hyperbolic flock could be. For example, could there exist a deficiency-one partial hyperbolic flock; missing exactly one plane? There are geometric reasons for asking such a question. In [79], [77], the author showed any embeddable partial deficiency-one hyperbolic flock arises from a translation plane with spread in $PG(3, q)$ that admits a Baer group of order $q - 1$. The partial flock may be uniquely extended to a hyperbolic flock if and only if the net of degree $q + 1$ containing the Baer axis and coaxis is a regulus net.

Such partial geometries that are not extendable are extremely rare and occur in just a few known translation planes. These are as follows: orders 2^4 and 3^4. These translation planes are derivable by a twisted regulus net and are also transitive on the partial spread defining the partial hyperbolic flock. The most general result in this regard is the theorem of Johnson and Cordero [92] that classifies such translation planes of order p^4, where p is a prime. The result is that there are exactly two such partial extendable hyperbolic flocks of deficiency-one of order p^4 which are transitive and derivable, the Johnson partial flock of degree 4 in $PG(2, 4)$ and the Johnson-Pomareda partial flock of degree 9 in $PG(3, 9)$.

In Royle [128], there are four partial hyperbolic flocks of deficiency-one, two in $PG(3, 5)$ and two in $PG(3, 7)$.

Concerning Baer groups over infinite fields or over skewfields, these have been developed somewhat in the author's texts [86],[88].

Using an algebraic method, we are able to extend the theory to include the type 2 or twisted regulus nets and to show equivalence between flocks of α-reguli by planes of $PG(3, K)$, and translation planes using a twisted regulus net instead of the regulus hyperbolic quadric type plane. We will also use the term 'twisted hyperbolic flock', in this setting.

In particular, the equivalence is between twisted hyperbolic flocks in $PG(3, K)$, over fields K that admit a non-identity automorphism α, and translation planes with spreads in $PG(3, K)$, that admit an affine homology group, one of whose orbits together with the axis and coaxis is a twisted regulus net. Both finite and infinite examples are given in this text.

For the derivable nets of derivative type possibilities, we will construct what we shall call the 'classical derivative type spread', which is a translation plane that conceivably could correspond to a flock of a D-conic and a flock of a D-derivable net, but the connections with derivable type spreads have not yet been established.

1. Algebraic theory of twisted hyperbolic flocks

We begin with our definition of an α-twisted hyperbolic quadric in $PG(3, K)$, for K an arbitrary field, and α is an automorphism of K, possibly trivial. We use homogeneous coordinates for the definition.

DEFINITION 67. *An α-quadric Q^α is a non-degenerate variety of the following form:*

$$
\begin{aligned}
Q^\alpha(x_1, x_2, x_3, x_4) \;=\; & Ax_1^{\alpha+1} + Bx_2^{\alpha+1} + Cx_3^{\alpha+1} + Dx_4^{\alpha+1} + \\
& + Ex_1^\alpha x_2 + Fx_1^\alpha x_3 + Gx_1^\alpha x_4 + \\
& + Hx_2^\alpha x_1 + Zx_2^\alpha x_3 + Jx_2^\alpha x_4 + \\
& + Kx_3^\alpha x_1 + Lx_3^\alpha x_2 + Mx_3^\alpha x_4 + \\
& + Nx_4^\alpha x_1 + Rx_4^\alpha x_2 + Sx_4^\alpha x_3.
\end{aligned}
$$

An α-conic would then be any plane intersection of Q^α. By the plane in question to intersect H^α in a non-degenerate α-conic, it is meant that the plane does not contain a line of the α-regulus (or α-twisted regulus).

We have defined 'generic reducible derivable nets' as those containing Baer subplanes that force the representation of the derivable to be diagonal or partially diagonal. We are concerned with translation planes that contain derivable nets that contain either 1 or 2 Baer subplanes of an associated regulus derivable net, where the derivable net is not diagonal or partially diagonal. By results of part 2 [**91**], there are mappings in the collineation group of the associated regulus derivable net that map the types 1, 2, and 3 derivable nets into a generic net. When a translation plane with spread in $PG(3, F)$, the mappings are in $GL(2, F)_{left}$, where the associated vector spaces is a left 4-dimensional F-subspace.

THEOREM 60. *A translation plane π with spread in $PG(3, F)$, for F a skewfield such that the components are left subspaces may be represented in the form $x = 0$, $y = xM$, where $M \in \lambda$ is a subset of $GL(2, F)_{left}$.*

(1) Assume that there is a left derivable net D within λ containing $x = 0, y = x$ and $y = x$ and assume that D is reducible with respect to the standard right pseudo-regulus net R. Then there is an element g of $GL(2, F)_{left}$ such that g leaves the classical right R relative to F invariant and maps D to the associated generic pseudo-regulus net.

(2) Moreover, πg is isomorphic to π and the spreads for both are in $PG(3, F)$.

Why this theorem is important is that although an α-twisted derivable net is of type 2, it may not be in generic form. The above theorem states that any translation plane that corresponds to a reducible derivable net and admits an affine homology group, one orbit of which and the axis and coaxis, is a type 1, 2, or 3 derivable net that is isomorphic to one that admits a generic derivable net of type 1, 2, or 3, respectively.

In this section, we show the equivalence between certain spreads in $PG(3, K)$, for K a field and flocks of planes within $PG(3, K)$ that form a cover that we call an α-twisted hyperbolic quadric. This theory could be developed when K is a skewfield initially, except that when K is non-commutative, the connection with an α-flock is not completely clear. Some remarks of the non-commutative possibilities are given in (12,15,1-4).

The main result is the following:

THEOREM 61. *Let Σ be a translation plane with spread in $PG(3, K)$, for K an arbitrary field. Let α denote an automorphism of K, possibly trivial. Assume* *that Σ admits an affine homology group one orbit of which, together with the axis and coaxis, is a twisted regulus net. Then all orbits are twisted regulus nets and the spread may be coordinatized in the following form:*

(1) Let V_4 be the associated 4-dimensional vector space over K. Letting x and y denote 2-vectors, then the spread is:

$$x = 0, y = 0, y = x \begin{bmatrix} u^\alpha & 0 \\ 0 & u \end{bmatrix},$$

$$\text{and } y = x \begin{bmatrix} f(t) & g(t) \\ 1 & t \end{bmatrix} \begin{bmatrix} v^\alpha & 0 \\ 0 & v \end{bmatrix};$$

$$\forall u, t, v, uv \neq 0, \text{ of } K, \alpha \text{ an automorphism of } K.$$

and functions f, g on K. Furthermore, f is bijective.

(2) For $PG(3, K)$, and points written in homogeneous coordinates, the 'α-twisted hyperbolic quadric' has the form $\{(x_1, x_2, x_3, x_4), \text{ such that } x_1 x_4^\alpha = x_2^\alpha x_3\}$.

Then there is a flock of the α-twisted hyperbolic quadric with the flock of planes

$$\{\pi_t; \forall t \in K, \text{ and } \rho\}$$

of $PG(3, K)$, as follows:

$$\pi_t : -x_1 g(t)^\alpha + x_2 f(t) - x_3 t^\alpha + x_4 = 0, \text{ and } \rho : x_2 = x_3.$$

Furthermore, the planes intersect the twisted hyperbolic quadric in a non-degenerate α-conic.

(3) Conversely, a flock F of the α-twisted hyperbolic quadric by planes of the given form constructs a translation plane Σ as above.

The proof shall be given as a series of lemmas.

LEMMA 7. *The proof of (1).*

PROOF. A necessary condition for a spread is that the differences of the matrices involved in the components must be non-singular, which implies that f is injective. And to become a spread, the components must cover the points uniquely. To see that f is bijective, consider the vector $(1, -a, 0, 1)$ for $a \in K$. This point is on

$$y = x \begin{bmatrix} f(t) & g(t) \\ 1 & t \end{bmatrix} \begin{bmatrix} v^\alpha & 0 \\ 0 & v \end{bmatrix},$$

for some $v \neq 0$, so that $(f(t) - a)v^\alpha = 0$, and $(g(t) - at)v = 1$, so that $f(t) = a$, for some t, proving that f is bijective. Since the twisted regulus net may be assumed to be generic and then it is well known to have the given form, we have the proof. $\square$

LEMMA 8. *Let Σ denote the spread of the previous theorem.*

For $PG(3, K)$, and points written in homogeneous coordinates, the 'α-twisted hyperbolic quadric' has the form:

$$\{(x_1, x_2, x_3, x_4); \text{ such that } x_1 x_4^\alpha = x_2^\alpha x_3\}.$$

PROOF. We may use the affine form $x = 0, y = x \begin{bmatrix} u^\alpha & 0 \\ 0 & u \end{bmatrix}; \forall u \in K$, of the α-regulus net. Consider the set of vectors on $y = x \begin{bmatrix} u^\alpha & 0 \\ 0 & u \end{bmatrix}$, $(x_1, x_2, x_1 u^\alpha, x_2 u)$.

Note that if these vectors are also represented by (x_1, x_2, x_3, x_4), then $x_1 x_4^\alpha = x_1(x_2 u)^\alpha$ and $x_2^\alpha x_3 = x_2^\alpha(x_1 u^\alpha)$. On $x = 0$, $(0, 0, x_3, x_4)$, this relationship is still valid. This proves that the form is correct. □

LEMMA 9. *We claim that the α-twisted quadric H^α is*

$$\{(x_3 x_1^\alpha, x_4^\alpha x_1, x_3 x_2^\alpha, x_4^\alpha x_2); \text{ where}$$

$$(x_1, x_2, x_3, x_4) \text{ are the points of } PG(3, K)\}.$$

PROOF. We note that $(x_3 x_1^\alpha)(x_4^\alpha x_2)^\alpha = (x_4^\alpha x_1)^\alpha(x_3 x_2^\alpha)$. It remains to show the set is onto. For this, we use the affine form $x = 0, y = x \begin{bmatrix} u^\alpha & 0 \\ 0 & u \end{bmatrix}; \forall u \in K$. Take $x_1 = 0$ to obtain $x = 0$. So, now assume that $x_1 \neq 0$. Represent $(z_1, z_2) = (x_3 x_1^\alpha, x_4^\alpha x_1)$. Then $(x_3 x_2^\alpha, x_4^\alpha x_2) = (z_1(x_1^{-\alpha} x_2^\alpha), z_2(x_1^{-1} x_2))$. Therefore, for $u = x_1^{-1} x_2$, for all $x_1, x_2 \in K$, the set represents the affine form and then the set of homogeneous coordinates represents the projective form of the α-twisted regulus (twisted hyperbolic quadric). Note that $x_2 = 0$ is $y = 0$ so that the full affine form is obtained. The reader could determine that when $\alpha = 1$, this is also an alternative method of describing the hyperbolic quadric. This completes the proof of the assertion. □

LEMMA 10. *The planes π_t satisfy (2).*

PROOF. Continuing with the proof of the theorem, since Σ is a translation plane, given any vector (x_1, x_2, x_3, x_4), we know that there is a unique spread component containing this vector. Since, we are interested in connecting the spread components other than the affine form of the α-regulus, we assume the following: $(*)$ x_1 or x_2 is non-zero and $(x_3, x_4) \neq (x_1 v^\alpha, x_2 v)$, for any $v \in K$. What we are trying to do is find planes that cover the α-hyperbolic quadric, so the use of affine and projective α-regulus terms will be confusing here. What to look for is what sort of elements

$$(x_3 x_1^\alpha, x_4^\alpha x_1, x_3 x_2^\alpha, x_4^\alpha x_2)$$

are covered by the planes that arise from the components. These will not be sufficient for a complete cover and this is where the plane ρ comes in. So, for each vector (x_1, x_2, x_3, x_4) that satisfies $(*)$, there is a unique pair (t, v) and corresponding component such that

$$(x_3, x_4) = (x_1, x_2) \begin{bmatrix} f(t) & g(t) \\ 1 & t \end{bmatrix} \begin{bmatrix} v^\alpha & 0 \\ 0 & v \end{bmatrix}.$$

Hence, we obtain $x_3 = (x_1 f(t) + x_2)v^\alpha$, and $x_4 = (x_1 g(t) + x_2 t)v$. And, we have

$$x_4^\alpha = (x_1^\alpha g(t)^\alpha + x_2^\alpha t^\alpha)v^\alpha.$$

Therefore,

$$x_4^\alpha((x_1 f(t) + x_2)v^\alpha) = x_3((x_1^\alpha g(t)^\alpha + x_2^\alpha t^\alpha)v^\alpha).$$

Recalling that, here $v \neq 0$, we then have

$$x_4^\alpha(x_1 f(t) + x_2) = x_3(x_1^\alpha g(t)^\alpha + x_2^\alpha t^\alpha).$$

Recall, the points $(x_3 x_1^\alpha, x_4^\alpha x_1, x_3 x_2^\alpha, x_4^\alpha x_2)$, and note that we align the above equation, using this order, as follows:

$$(**) : \pi_t \cap H^\alpha : x_3 x_1^\alpha(-g(t)^\alpha) + x_4^\alpha x_1 f(t) - x_3 x_2^\alpha t^\alpha + x_4^\alpha x_2 = 0.$$

So, for all vectors (x_1, x_2, x_3, x_4) satisfying $(*)$, such that

$$(x_3, x_4) = (x_1, x_2) \begin{bmatrix} f(t) & g(t) \\ 1 & t \end{bmatrix} \begin{bmatrix} v^\alpha & 0 \\ 0 & v \end{bmatrix},$$

we obtain a plane of $PG(3, K)$, $\pi_t : x_1(-g(t)^\alpha) + x_2 f(t) - x_3 t^\alpha + x_4 = 0$, such that the intersection of π_t with H^α is $(**)$.   □

LEMMA 11.

$$\pi_t \cap H^\alpha : x_3 x_1^\alpha (-g(t)^\alpha) + x_4^\alpha x_1 f(t) - x_3 x_2^\alpha t^\alpha + x_4^\alpha x_2 = 0,$$

is a non-degenerate α-conic.

PROOF. So, the intersection of each π_t with H^α has the form

$$x_3 x_1^\alpha (-g(t)^\alpha) + x_4^\alpha x_1 f(t) - x_3 x_2^\alpha t^\alpha + x_4^\alpha x_2 = 0,$$

so that $-g(t)^\alpha = F$, $f(t) = N$, $-t^\alpha = I$, and $1 = Z$, in the above proposition and the definition of the α-hyperbolic quadric. Assume that the plane π_t contains a line of the α-regulus. In affine form, the lines are $x = 0$, and $y = x \begin{bmatrix} u^\alpha & 0 \\ 0 & u \end{bmatrix}$, for $u \in K$. Since $x = 0$ is $(0, 0, x_3, x_4)$, this has been excluded by $(*)$. Similarly the line $y = x \begin{bmatrix} u^\alpha & 0 \\ 0 & u \end{bmatrix}$ cannot be contained in π_t, again by $(*)$. Hence, the intersection is a non-degenerate α-conic.   □

LEMMA 12. *The remaining plane is* $\rho; x_2 = x_3$ *and the intersection with* H^α *is a non-degenerate* α-*conic.*

PROOF. So, we have used $(*)$ to find the planes π_t, from the translation plane. Then all elements of $\{(x_3 x_1^\alpha, x_4^\alpha x_1, x_3 x_2^\alpha, x_4^\alpha x_2);$ where (x_1, x_2, x_3, x_4) are the points of $PG(3, K)\}$, such that (x_1, x_2, x_3, x_4) satisfies $(*)$. The 'points', that do not satisfy $(*)$, are $x_1 = x_2 = 0$ and $(x_3, x_4) = (x_1 v^\alpha, x_2 v)$, for $v \in K$. First note that $x_1 = x_2$ produces the zero vector. For

$$\begin{aligned}
(x_3, x_4) &= (x_1 v^\alpha, x_2 v), \text{ then} \\
&\quad (x_3 x_1^\alpha, x_4^\alpha x_1, x_3 x_2^\alpha, x_4^\alpha x_2) \\
&= (x_1^{\alpha+1} v^\alpha, x_1 x_2^\alpha v^\alpha, x_1 x_2^\alpha v^\alpha, x_2^{\alpha+1} v^\alpha) \\
&= (x_1^{\alpha+1}, x_1 x_2^\alpha, x_1 x_2^\alpha, x_2^{\alpha+1}),
\end{aligned}$$

in homogeneous coordinates. Hence, this is $\rho \cap H^\alpha$, and so ρ has the form $x_2 = x_3$, and the intersection is $x_2^{\alpha+1} - x_1 x_4^\alpha = 0$, a non-degenerate α-conic, as it contains one point from each of the lines of the α-twisted conic. This set is the set containing $(1, 0, 0, 0)$, $(0, 0, 0, 1)$ and one point from each of the non-zero components $y = x \begin{bmatrix} u^\alpha & 0 \\ 0 & u \end{bmatrix}$, for $u = (x_1^{-1} x_2)$. So, we see that we have a set of planes mutually disjoint on the α-hyperbolic quadric and intersects the α-hyperbolic quadric in α-conics. It remains to show that we have a cover. So assume that some point

$$(x_3 x_1^\alpha, x_4^\alpha x_1, x_3 x_2^\alpha, x_4^\alpha x_2)$$

is not covered. We know that the points (x_1, x_2, x_3, x_4) that satisfy $(*)$ must be within this set. This only leaves the points on $x = 0$, since all of the rest have been considered and checked by looking at ρ. But, $(0, 0, 0, 1)$ is in ρ. So we are looking for

$(0, 0, 1, 0)$. Consider the question of whether $(0, 0, 1, 0)$ is on π_t, which would say that there exists a unique t_o such that $f(t_o) = 0$. Since f is bijective, there is such a unique plane π_{t_0}. Note that when $\alpha = 1$, the planes are $\pi_t : -x_1 g(t) + x_2 f(t) - x_3 t + x_4 = 0$, and $\rho : x_2 = x_3$. when using the form $x_1 x_4 = x_2 x_3$. For flocks of hyperbolic quadrics, the planes are

$$\pi_t^* : x_1 - x_2 t + x_3 f(t) - x_4 g(t) = 0, \rho : x_2 = x_3,$$

when using the form for the hyperbolic quadric of $x_1 x_4 = x_2 x_3$ and homogeneous coordinates (x_1, x_2, x_3, x_4), so we are using (x_4, x_3, x_2, x_1) in the form that works also with $\alpha \neq 1$. Note that this form is clearly equivalent to the standard form in the $\alpha = 1$ case. This completes the proof of the theorem.   □

LEMMA 13. *The proof of (3): Conversely, a flock F of the α-twisted hyperbolic quadric by planes of the given form constructs a translation plane Σ as above.*

PROOF. Assume that there is a flock

$$F : \pi_t : -x_1 g(t)^\alpha + x_2 f(t) - x_3 t^\alpha + x_4 = 0, \text{ and } \rho : x_2 = x_3.$$

We shall use the formulation of the α-hyperbolic quadric H^α, using homogeneous coordinates, as follows:

$$\{(x_3 x_1^\alpha, x_4^\alpha x_1, x_3 x_2^\alpha, x_4^\alpha x_2),$$
$$\text{where } (x_1, x_2, x_3, x_4) \text{ are the points of } PG(3, K)\}.$$

Furthermore, we claim that the following mappings

$$h_{u,v} : (x_1, x_2, x_3, x_4) \to (x_1 u^\alpha, x_2 u, x_3 v^\alpha, x_4 v); \forall u, v \in K^*,$$

preserve H^α. To see this, note that $(x_1 u^\alpha)(x_4 v)^\alpha = (x_2 u)^\alpha (x_3 v^\alpha)$. Also, note this is the mapping describing the Baer components, which must preserve the α-hyperbolic quadric (See above for the affine form for the α-regulus net and the Baer components $P(a, b)$). Then

$$-x_3 x_1^\alpha g(t)^\alpha + x_4^\alpha x_1 f(t) - x_3 x_2^\alpha t^\alpha + x_4^\alpha x_2 = 0,$$

is the set of 'points' of $PG(3, K)$, then clearly, by working the proof backwards, shows that this is equivalent to having a vector (x_1, x_2, x_3, x_4) being an element of

$$y = x \begin{bmatrix} f(t) & g(t) \\ 1 & t \end{bmatrix} \begin{bmatrix} v_t^\alpha & 0 \\ 0 & v_t \end{bmatrix},$$

for some element $v_t \in K^*$. So, the α-conic of intersection of π_t, then becomes the set of 1-dimensional subspaces of this particular component. This then shows that the set of planes π_t, for all $t \in K$, will directly reconstruct a set of elements $\{y = x \begin{bmatrix} f(t) & g(t) \\ 1 & t \end{bmatrix} \begin{bmatrix} v_t^\alpha & 0 \\ 0 & v_t \end{bmatrix} ; \forall t \in K\}$, where $v_t \in K^*$, depends on t. Note that we are representing the points of the twisted hyperbolic quadric H^α in the form $(x_3 x_1^\alpha, x_4^\alpha x_1, x_3 x_2^\alpha, x_4^\alpha x_2)$. Consider again:

$$h_{1,u}(x_1, x_2, x_3, x_4) = (x_1, x_2, x_3 u^\alpha, x_4 u),$$

and the effect on the corresponding elements on H^α, then becomes

$$(x_3 x_1^\alpha u^\alpha, x_4^\alpha x_1 u^\alpha, x_3 x_2^\alpha u^\alpha, x_4^\alpha x_2 u^\alpha),$$

which is, of course, the same point of $PG(3, K)$, but also shows that we may expand the point to belong to

$$\begin{bmatrix} f(t) & g(t) \\ 1 & t \end{bmatrix} \begin{bmatrix} v_t^\alpha & 0 \\ 0 & v_t \end{bmatrix} \begin{bmatrix} u^\alpha & 0 \\ 0 & u \end{bmatrix}.$$

In this way, from the set of planes $\pi_t; \forall t \in K$, and the intersections with H^α, we have reconstructed all components of the form of the target translation plane, with the exception of the set of elements $x = 0, y = x \begin{bmatrix} u^\alpha & 0 \\ 0 & u \end{bmatrix}$. We now turn to the plane ρ. We first note that the element of elements of H^α, that correspond to the elements vectors (x_1, x_2, x_3, x_4) satisfying (∗), produced the elements H^α, $(x_3 x_1^\alpha, x_4^\alpha x_1, x_3 x_2^\alpha, x_4^\alpha x_2)$, with the same (∗) condition on the elements (x_1, x_2, x_3, x_4). Hence, we are missing the possible points of H^α such that either $x_1 = x_2 = 0$ of $(x_3, x_4) = (x_1 u^\alpha, x_2 u)$, for $u \in K^*$. When $x_1 = x_2 = 0$, we see there are no corresponding elements of H^α. We shall come back for this technicality in a moment. Thus, assume $(x_3, x_4) = (x_1 u^\alpha, x_2 u)$, for $u \in K^*$. Then

$$(x_3 x_1^\alpha, x_4^\alpha x_1, x_3 x_2^\alpha, x_4^\alpha x_2) = (x_1^{\alpha+1} u^\alpha, x_2^\alpha x_1 u^\alpha, x_2^\alpha x_1 u^\alpha, x_2^{\alpha+1} u^\alpha),$$

by which is the intersection of ρ with H^α. Therefore, we see that we have recovered the components $y = x \begin{bmatrix} u^\alpha & 0 \\ 0 & u \end{bmatrix}$. While it may seem that we are missing the component $x = 0$, we note that a spread is the covering of V_4/K by mutually disjoint 2-dimensional K-subspaces. The partial spread is missing exactly the set of points $(0, 0, x_3, x_4)$. So, by the formal addition of $x = 0$, we have proved that we may recover the translation plane from the α-flock. The reader should not confuse the existence of the α-regulus in affine form with the cover of the projective version of the α-regulus. That is, since $x = 0$ in the affine form was not recovered, does not mean that the projective α-regulus was not covered. To make a point a little stronger, we consider where the $x = 0$ version of the projective version was covered. So, consider again, $(x_3 x_1^\alpha, x_4^\alpha x_1, x_3 x_2^\alpha, x_4^\alpha x_2)$. If $x_1 = 0$ then we obtain $(0, 0, x_3 x_2^\alpha, x_4^\alpha x_2)$. Thus, these points correspond to the vectors $(0, x_2, x_3, x_4)$, were covered by the π_t planes/or corresponding spread components. Similarly, $x_2 = 0$, produces $(x_3 x_1^\alpha, x_4^\alpha x_1, 0, 0)$, and then corresponds to the vector $(x_1, 0, x_3, x_4)$, which again is covered by the π_t. So, it is counter-intuitive but the projective $x = 0$ and $y = 0$, not the affine components $x = 0$, $y = 0$, are covered. So, the missing $x = 0$, the component, is recovered by realizing the covering required for a translation plane is uniquely extended by the partial spread consisting of the other components that correspond directly to the α-flock. Note that every element (x_1, x_2, x_3, x_4) is covered back in the vector space as this point produces $(x_1^\alpha x_3, x_4^\alpha x_1, x_2^\alpha x_3, x_4^\alpha x_2)$, a point on the α-hyperbolic quadric. If (x_1, x_2, x_3, x_4) is not covered in the putative spread, then the orbit of this element in the α-regulus-inducing group is not covered. But, this orbit is the complete set of points corresponding to $(x_1^\alpha x_3, x_4^\alpha x_1, x_2^\alpha x_3, x_4^\alpha x_2)$ which must covered by the α-flock. Hence, we have recovered a spread from an α-flock. This completes the proof of the theorem.   □

We have discussed the α-Klein quadric, K^α, for K a field. In part 12, we shall offer some ideas for extension of flocks of twisted hyperbolic flocks and α-flocks of non-commutative skewfields. Furthermore, there is a way to discuss non-commutative generalizations of the analogues of cyclic $q + 1$-spreads.

In section 5, we provide an alternative proof of equivalence with twisted hyperbolic flocks and translation planes admitting a twisted regulus-inducing homology group, which shows how the Baer subplanes of the derivable nets provide the 'points' for the twisted hyperbolic quadric.

In addition, there is similar research with a different focus in N. Durante [47], and also G. Donati, N. Durante [46], which is recommended to the reader.

2. Simultaneous α-flocks & twisted hyperbolic spreads

In this section, we look at translation planes that provide both α-flocks of α-conics and α-twisted hyperbolic flocks over a field K. This work originated in Johnson [89], where the planes are the Hughes-Kleinfeld semifield planes and their infinite generalizations. The α-twisted hyperbolic flocks are all 'star flocks', where the planes of the flock share a point.

We have noticed this form previously when $\alpha^2 = 1$, and $\alpha \neq 1$ and $g = 0$. In this setting, we have both an α-flock of an α-conic and an α-twisted hyperbolic flock. These flocks are both linear and occur also for the quaternion division rings.

In this more general setting, consider a spread of the following form:

$$x = 0, y = x \begin{bmatrix} u^\alpha + gt & ft^{\alpha^{-1}} \\ t & u \end{bmatrix} : \forall t, u \in K, \text{ for } K \text{ a field.}$$

which is an α-flock plane. And, by realizing that

$$\begin{aligned} x &= 0, y = x \begin{bmatrix} u^\alpha + gt & ft^{\alpha^{-1}} \\ t & u \end{bmatrix} = \begin{bmatrix} v^\alpha + g & f \\ 1 & v \end{bmatrix} \begin{bmatrix} s^\alpha & 0 \\ 0 & s \end{bmatrix} \\ &= \begin{bmatrix} (vs)^\alpha + gs^\alpha & fs \\ s^\alpha & vs \end{bmatrix} \end{aligned}$$

then letting $vs = u$ and $s^\alpha = t$, we see that $fs = ft^{\alpha^{-1}}$.

In the α-flock setting, we used

$$x = 0, \ y = x \begin{bmatrix} u + g(t) & f(t) \\ t & u^\alpha \end{bmatrix}; \ t, u \in K$$

and corresponding α-flock planes:

$$\rho_t: \ x_0 t - x_1 f(t)^\alpha + x_2 g(t)^\alpha - x_3 = 0.$$

Therefore, in our current situation the spread corresponds to an α^{-1}-flock plane so we obtain

$$\rho_t: \ x_0 t - x_1 f^{\alpha^{-1}} t^{\alpha^{-2}} + x_2 g^{\alpha^{-1}} t^{\alpha^{-1}} - x_3 = 0.$$

Hence, if $g = 0$ and $\alpha^2 = 1$, we obtain a linear α^{-1}-flock with line

$$\{(x_1 f^{\alpha^{-1}}, x_1, x_2, 0)\}.$$

Otherwise, the α^{-1}-flock is non-linear. If $g \neq 0$ and $\alpha^2 = 1$ then the intersection of all planes is

$$\{(x_1 f^{\alpha^{-1}}, x_1, 0, 0)\},$$

a point; a so-called 'star α^{-1}-flock'. If $g = 0$ and $\alpha^2 \neq 1$ then the intersection of all planes is $\{(0, 0, x_2, 0)\}$, another star α^{-1}-flock. Then if $g \neq 0$ and $\alpha^2 = 1$, the intersection of all planes is the empty set.

Then for the twisted hyperbolic flock, one uses the spread as

$$\begin{bmatrix} F(v) & G(v) \\ 1 & v \end{bmatrix} \begin{bmatrix} s^\alpha & 0 \\ 0 & s \end{bmatrix}$$

has the flock form as:

$$F : \pi_t : -x_1 G(t)^\alpha + x_2 F(t) - x_3 t^\alpha + x_4 = 0, \ and \ \rho : x_2 = x_3,$$

where $F(t) = t^\alpha + g$ and $G(t) = f$, becomes:

$$F : \pi_t : -x_1 f^\alpha + x_2(t^\alpha + g) - x_3 t^\alpha + x_4 = 0, \ and \ \rho : x_2 = x_3,$$

the intersection of the spreads shows that $-x_1 f^\alpha + x_2 g + x_4 = 0$ and $x_2 = x_3$ so that we obtain a linear α-hyperbolic flock with common line $\{(x_1, x_2, x_2, x_1 f^\alpha - x_2 g); x_1, x_2 \in K\}$.

THEOREM 62. *For spreads over a field K,*

$$x = 0, y = x \begin{bmatrix} u^\alpha + gt^\alpha & ft^{\alpha^{-1}} \\ t & u \end{bmatrix}$$

(1) The α^{-1}-flock is linear if and only if $g = 0$ and $\alpha^2 = 1$.
(2) If $g \neq 0$ and $\alpha^2 = 1$ then the α^{-1}-flock is a star flock.
(3) If $g = 0$ and $\alpha^2 \neq 1$ then the α^{-1}-flock is a star flock.
(4) If $g \neq 0$ and $\alpha^2 \neq 1$ then the α^{-1}-flock has planes that trivially intersect.
(5) In all cases, the α-hyperbolic flocks are linear with line

$$\{(x_1, x_2, x_2, x_1 f^\alpha - x_2 g); x_1, x_2 \in K\}.$$

COROLLARY 18. *All quaternion division ring planes give rise to linear α-flocks of cones, and linear flocks of α-hyperbolic flocks.*

3. Flocks of D-cones

Here are a few of the general problems that we consider in this section. Let D be any reducible derivable net in $PG(3, K)$, for K a field. When a derivable net is reducible and of type 1 or 2, there are either 1 or 2 associated Baer subspaces (incident with the zero vector) that are shared by a classical regulus net. We have also mentioned 'generic subspaces'. By this term, we mean that if there is type 1 derivable net then the Baer subplane has the short notation of $(0, 1)$ and if there are two such common subspace for type 2 derivable nets, these two Baer subplanes are $(0, 1)$ and $(1, 0)$, and, more generally for skewfields K, the generic type 3 derivable nets would include common $(0, 1), (1, 0)$, and $(1, 1)$ Baer subspaces.

(1) Are there translation planes with spreads in $PG(3, K)$, that admit an affine homology group such that together with the axis and coaxis and one orbit form a net isomorphic to D? If so, are there associated 'flocks of D-nets' by planes within $PG(3, K)$?

(2) Are there translation planes with spreads in $PG(3, K)$, that admit an affine elation group such that together with the axis and one orbit form a net isomorphic
to D? If so, are there associated 'flocks of D-cones' by planes within $PG(3, K)$?

When K is $GF(q)$, the Desarguesian planes satisfy both of the conditions
of (1) and (2), simultaneously; the associated flocks of the regulus and flocks of
quadratic cones are always 'linear', in that the covering planes share a line of
$PG(3, q)$.

(3) Suppose a translation plane with spread in $PG(3, K)$ admits both an affine homology group and an elation group corresponding to the same type of derivable
net? Are the translation planes known? If so, are there associated flocks and
then, if so, are the flocks linear?

We consider spreads of the following type:

$$x = 0, y = x(\begin{bmatrix} F(t) & G(t) \\ t & 0 \end{bmatrix} + \begin{bmatrix} u^\alpha & A(t) \\ 0 & u \end{bmatrix}); \forall t, u \in K,$$

where F and G are functions on K, for a skewfield K. We only consider the field
case in this section. However, in part 12, there are comments about such translation
planes in $PG(3, K)$, for a skewfield K. We have the following conditions:

- (1) Where if $\alpha = 1$, then A is a derivative function (type 1, A non-
  identically zero derivation); derivative net.
- (2) If $\alpha \neq 1$ then this is a type 2 or twisted regulus net (the derivative
  function occurs only when $\alpha = 1$, in the field case); this has been developed
  in Cherowitzo, Johnson and Vega [**25**]. The form for the flocks of α-cones
  is as follows:

$\gamma_t : x_1 t + x_2 G(t)^{\alpha^{-1}} - x_3 F(t)^{\alpha^{-1}} + x_4 = 0$, where (x_1, x_2, x_3, x_4) represent points
of $PG(3, K)$, using homogeneous coordinates.

- (3) If $\alpha = 1$ and A is identically zero, this is the conical flock type, which
  is known to exist in both finite and infinite versions.

The general form is then

$$x = 0, y = x \begin{bmatrix} v^\alpha & A(t) \\ 0 & v \end{bmatrix},$$

$$y = x(\begin{bmatrix} F(t) & G(t) \\ 1 & t \end{bmatrix} \begin{bmatrix} u^\alpha & A(t) \\ 0 & u \end{bmatrix});$$

$$\forall t, v, u \neq 0 \in K,$$

where F and G are functions on K.

In this section, we now consider spreads in $PG(3, K)$, for K a field which are
simultaneously equivalent to flocks of D-cones and flocks of D-homology nets. Many
of the results are valid for arbitrary skewfields, but we consider here only when K
is a field. We shall take up the situation over arbitrary skewfields in part 12.

Let Σ be a translation plane with spread in $PG(3, K)$, for K a field that admits
both an elation group whose axis L has an orbit Δ such that $L \cup \Delta$ is a D-affine
derivable net and a homology group whose coaxis L and axis M and some orbit is
an D-affine derivable net. We first point out that we may always deal with generic
forms of type 1 and type 2 nets by a previous theorem 59.

We first define a net of the derivable net flock types: (Note: the form given in the original article [**89**] has a misprint, which is corrected here. The $(2,2)$ entry in that matrix was shown as tv and should have been $A(v) + tv$.

$$x = 0, y = x \begin{bmatrix} u & A(u) \\ 0 & u \end{bmatrix}, \ y = x \begin{bmatrix} tv & A(u + tv) + bv \\ v & A(v) + tv \end{bmatrix},$$

for all $u, t \neq 0, v \in K$. Letting $w = u + tv + A(v)$, the putative spread has the form:

$$x = 0, \ y = x \begin{bmatrix} w + av - A(v) & A(w) + aA(v) + bv - A(A(v)) \\ v & w \end{bmatrix}, \ \text{for all } w, v$$

$\in K$. This is a semifield spread if and only if the determinants are all non-zero and we have the covering property. Note for K a quadratic extension $F(\theta)$ and $A(v) = \delta/\delta\theta \ v$, then $A(A(v)) = 0$ and characteristic 2.

THEOREM 63. *In the above example, let $a = 0$, let c be a non-square in F, F of characteristic 2 and $A(v) = (\delta/\delta\sqrt{c}) \ v$ then $A(A(v)) = 0$. Furthermore, the following is an additive quasifibration.*

$$x = 0, \ y = \begin{bmatrix} w + A(v) & A(w) + \sqrt{c}v \\ v & w \end{bmatrix}; \forall w, v \in F(\sqrt{c}).$$

PROOF. Let $v = \sqrt{c}v_1 + v_2$, where $v_i \in F$, for $i = 1, 2$. Then $(\delta/\delta\sqrt{c}) \ v = v_1$, and $A(v_1) = (\delta/\delta\sqrt{c}) \ v_1 = 0$. So, $A(A(v)) = 0$. Also, note that $v^2 = cv_1^2 + v_2^2 \in F$. The general determinant of the matrices is $w^2 + wA(v) + vA(w) + \sqrt{c}v^2 = w^2 + \sqrt{c}v^2 + A(wv) = 0$. Since, w^2, v^2, and $A(wu)$ are all elements of F, a determinant is 0 only if $v^2 = 0$, so that $w^2 + A(0) = 0$, implying $w^2 = 0$, so that $w = v = 0$. This shows that we have a quasifibration. □

PROBLEM 24. *Determine whether the above quasifibration is proper or is actually a spread.*

Reviewing the proof of the previous general result, there were no instances where commutativity was required in the set up. Therefore, the question arises whether there is a non-commutative skewfield that could replace $F(\sqrt{c})$ so as to obtain a quasifibration or spread in $PG(3, F(\sqrt{c}))$.

PROBLEM 25. *Let F be a skewfield of characteristic 2 and let $F(\sqrt{c})$ be a central quadratic extension, where $c \in Z(F)$ and $A(v) = (\delta/\delta\sqrt{c}) \ v$, then $A(A(v)) = 0$. Then determine if the following is an additive quasifibration.*

$$x = 0, \ y = \begin{bmatrix} w + A(v) & A(w) + \sqrt{c}v \\ v & w \end{bmatrix}; \forall w, v \in F(\sqrt{c}).$$

So, since we have additivity, it remains to show that

$$(w + A(v))v^{-1}w + A(w) + \sqrt{c}v)v \neq 0, \ for \ v \neq 0.$$

DEFINITION 68. *A spread/quasifibration of the above type shall be called a 'classical semifield spread with derivable net admitting a derivation'.*

DEFINITION 69. *A translation plane/quasifibration that admits both elation and homology groups of the type considered is called 'simultaneous' plane/quasifibration.*

THEOREM 64. *If Σ is a simultaneous translation plane over a field K then Σ is one of the following translation planes:*

(0) Σ is a Desarguesian spread

(1) K is finite and Σ is a Knuth/Hughes-Kleinfield semifield plane.

(2) K is an infinite field and Σ is a generalized Knuth/Hughes-Kleinfield semifield plane or

(3) K is an infinite field and Σ is a quaternion division ring plane.

(4) In all previous cases, there is both a corresponding α-flock of a twisted hyperbolic quadric, and a flock of an α-conic.

(5) K is infinite and Σ is a classical semifield spread with derivable net admitting a derivation.

PROOF. Let L be coordinatized by $x = 0$. If one of the orbits is an α-regulus derivable net, choose $y = 0$ to belong to that orbit. Then the form of the elation group E must be given by

$$(x, y) \to (x, x \begin{bmatrix} u^\alpha & 0 \\ 0 & u \end{bmatrix} + y),$$

where α is an automorphism of K, possibly trivial. This implies that all of the orbits of E are α-regulus nets that share $x = 0$. Therefore, we have a translation plane corresponding to a flock of an α-cone in $PG(3, K)$. Hence, the form for the translation plane is as follows:

$$x = 0, y = \begin{bmatrix} u^\alpha + f(t) & g(t) \\ t & u \end{bmatrix}; \forall t, u \in K.$$

Since there is also an affine homology group H with coaxis $x = 0$ and axis $y = 0$, we choose the orbit defining H to contain $y = x$. Hence, we have the homology group

$$(x, y) \to (x, y \begin{bmatrix} u^\alpha & 0 \\ 0 & u \end{bmatrix}); \forall u \in K^*.$$

It then follows that

$$\begin{bmatrix} u^\alpha + f(t) & g(t) \\ t & u \end{bmatrix} =$$
$$\begin{bmatrix} t^\alpha + f(1) & g(1) \\ 1 & t \end{bmatrix} \begin{bmatrix} v^\alpha & 0 \\ 0 & v \end{bmatrix}; \forall t, v \in K, \text{ for } v \neq 0.$$

Hence, $f(1)v^\alpha = f(v^\alpha)$, and $g(1)v = g(v^\alpha)$. Thus, the spread has the following form:

$$x = 0, y = \begin{bmatrix} u^\alpha + av^\alpha & bv \\ v^\alpha & u \end{bmatrix}; \forall u, v \in K, \text{ where } f(1) = a \text{ and } g(1) = b.$$

When K is $GF(q)$, this is the definition of a Knuth/Hughes-Kleinfield semifield plane. For the completion of the proof of this result, we need to consider the possible derivable nets with derivation and their associated spreads. Now assume that we have both of the types of cones and hyperbolic quadratic generalizations that we have been discussing in the case where we have a derivative type derivable net: First, we consider the flock of the D-cone, the spread components would be as follows. We have various functions F, A, G, f and g which shall be determined.

$$x = 0, y = x \begin{bmatrix} r + F(s) & A(r) + G(s) \\ s & r \end{bmatrix}; \forall s, r \in K,$$

for K a field. In general, to have a flock of a D-net, we would have the form:

$$x = 0, y = x \begin{bmatrix} u & A(u) \\ 0 & u \end{bmatrix}, y = x \begin{bmatrix} f(t) & g(t) \\ 1 & t \end{bmatrix} \begin{bmatrix} v & A(v) \\ 0 & v \end{bmatrix}; u, t, v \neq 0 \in K,$$

which is

$$\begin{bmatrix} f(t)v & f(t)A(v) + g(t)v \\ v & A(v) + tv \end{bmatrix}.$$

Let $A(v) + tv = r$, for $v \neq 0$, and $v = s$. Thus, $A(v) + tv + F(v) = f(t)v$, letting $t = 1$, we have $f(1)v = A(v) + v + F(v)$. So, $A(v) + F(v) = v(f(1) - 1)$, for all nonzero $v \in K$. Since $F(0) = 0 = A(0)$, this is also valid for all $v \in K$. Then $tv + v(f(1) - 1) = f(t)v$, and $f(t) = t + (f(1) - 1)$, noting that $A(1) = 0$ (as $A(1) = 2A(1)$). Let $f(1) - 1 = a$. Also,

$$f(t)A(v) + g(t)v = A(r) + G(s) = A(A(v) + tv) + G(v).$$

Thus, we have:

$$(t + a)A(v) + g(t)v = A(A(v) + tv) + G(v) = A(A(v)) + tA(v) + vA(t) + G(v),$$

if and only if

$$aA(v) + g(t)v = A(A(v)) + vA(t) + G(v).$$

Let $v = 1$, so that $g(t) = A(t) + G(1)$. Let $G(1) = b$. Hence,

$$aA(v) + bv = A(A(v)) + G(v),$$
$$\text{so } G(v) = aA(v) + bv - A(A(v)).$$

So, the functions are resolved as follows:

$$f(t) = t + a, \; g(t) = A(t) + b,$$
$$F(v) = av - A(v), \; G(v) = aA(v) + bv - A(A(v))$$

$$x - 0, y = x \begin{bmatrix} u & A(u) \\ 0 & u \end{bmatrix}, y = x \begin{bmatrix} t + a & A(t) + b \\ 1 & t \end{bmatrix} \begin{bmatrix} v & A(v) \\ 0 & v \end{bmatrix}$$

is

$$x = 0, y = x \begin{bmatrix} u + (t + a)v & A(u) + (t + a)A(v) + (A(t) + b)v \\ v & u + A(v) + tv \end{bmatrix} =$$
$$\begin{bmatrix} w + av - A(v) & A(w) + aA(v) + bv - A(A(v)) \\ v & w \end{bmatrix}$$

Thus, we have a semifield spread if and only if the determinants are $\neq 0$, for all $w, v \in K$, and the covering requirement is satisfied. This completes the proof of the theorem. $\qquad\square$

4. j–planes and twisted hyperbolic flocks

In the previous sections, we considered translation planes with spreads in $PG(3, K)$, for K a field, that admit an affine homology group one of whose orbits together with the axis and coaxis is a twisted-regulus that corresponds to a flock of a twisted hyperbolic quadric. However, there were only the Hughes-Kleinfeld semifield planes given as concrete examples, in the finite case, although every quaternion division ring plane provides an example. In this section, we show that there are infinitely many finite examples, provided by looking at certain j-planes, for $j = (p^s - 1)/2$ in order p^{2r}, p an odd prime and $r > 1$.

Actually, the associated translation planes were constructed 30 years by Johnson, Pomareda, and Wilke [**99**], and are actually j-planes. However, their significance was only just realized, basically after the proof that α-regulus-inducing homology groups on translation planes are equivalent to flocks of twisted hyperbolic quadrics, as we have seen in (6,8,2).

We have mentioned that finite regular hyperbolic fibrations with constant back half produce and are equivalent to flocks of quadratic cones. Moreover, by the correspondence theorem, (5,7,8), we see that translation planes of order q^2 with spreads in $PG(3, q)$ that admit a cyclic homology group of order $q + 1$ are equivalent to flocks of quadratic cones.

j-planes satisfy this hypothesis (are cyclic $q + 1$-planes) and therefore produce flocks of quadratic cones. In fact, the flocks are quite interesting, in that the associated flock translation planes have monomial spreads.

A few comments on the correspondence between translation planes with spreads in $PG(3, q)$ that admit a cyclic affine homology group of order $q + 1$, and flocks of quadratic cones are appropriate. While it is true that given such a translation plane, there is an associated flock of quadratic cone, the mapping is many to one. The reason for this is that given a translation plane with a cyclic homology group of order $q + 1$, the construction process uses the concept of a regular hyperbolic fibration with constant back half, conceived by Baker and Ebert and Penttila [**11**]. In this setting, there is always a constructed flock of a quadratic cone. Given any hyperbolic fibration; a set of $q - 1$ mutually disjoint hyperbolic quadrics plus two carrying lines, for each hyperbolic quadric, choose either the regulus or the opposite regulus, to produce a translation plane admitting a cyclic homology group of order $q + 1$. Conversely, if a translation plane of order q^2 admits a cyclic homology group of order $q + 1$, there is an associated hyperbolic fibration, it can then be shown that the hyperbolic fibration is regular with constant backhalf. For this process, all of the $q - 1$ possible regulus nets, induced by the affine homology group of order $q + 1$, are used in the construction of the flock of the quadratic cone.

The Desarguesian plane will then produce a linear flock of a quadratic cone by this process. With hyperbolic fibrations, all of the associated Andrè planes are then re-constructed. Consider any Andrè plane constructed from a Desarguesian affine plane by the replacement of k regulus nets of the $q - 1$ possible. Since the Andrè plane admits the affine homology group as well, there is a corresponding flock of the quadratic cone. The corresponding hyperbolic fibrations, reconstructs the Andrè plane, as well as all of the derivates of the plane, which include the Desarguesian plane. It follows that all Andrè planes construct exactly a linear quadric cone from the process. Since all Desarguesian planes are isomorphic, all constructed flocks are always linear in this setting.

However, for j-planes, the situation is somewhat different. While this many to one feature is still always valid, it turns out that the corresponding flock of the quadratic cone is always monomial.

The translation planes/flock planes constructed from j-planes have the following form:

$$x = 0, y = x \begin{bmatrix} u + gt^{j+1} & ft^{2j+1} \\ t & u \end{bmatrix} ; t, u \in GF(q^2), \text{ for } j \text{ an integer.}$$

For q even, the complete set of monomial j-planes were determined by Pentilla and Storme [124], who showed that the three known classes $j = 0, 1$ and 2 are the only possible classes.

THEOREM 65. *(See Johnson [88] Theorem 115 and Pentilla and Storme [124])* *Let π be a j-plane of even order q^2. Then $j = 0, 1, 2$ and the plane is one of the following types of planes:*
(1) Desarguesian ($j = 0$ and corresponding to the linear flock),
(2) Kantor-slice of a unitary ovoid ($j = 1$ and corresponding to the Betten flock),
(3) the Johnson-Pomareda-Wilke $j = 2$-planes (corresponding to the Payne flock).

Here we then consider q odd.

In the j-plane situation, the Abelian group trivially intersecting the kernel has the following form:

$$\left\{ \begin{bmatrix} 1 & 0 & 0 & 0 \\ 0 & \delta_{u,t}^{-j} & 0 & 0 \\ 0 & 0 & u & t \\ 0 & 0 & ft & u+gt \end{bmatrix} ; \forall u \in GF(p^r)^* \right\},$$

where $x^2 \pm gx - f$ is irreducible over $GF(p^r)$,

$$\delta_{t,u} = u^2 - gut - ft^2.$$

And the spread is of the form:

$$\begin{aligned} x &= 0, \, y = 0, \, y = x \begin{bmatrix} 1 & 0 \\ 0 & \delta_{u,t}^j \end{bmatrix} \begin{bmatrix} u & t \\ ft & u+gt \end{bmatrix} \\ &= \begin{bmatrix} u & t \\ \delta_{u,t}^j ft & \delta_{u,t}^j u \end{bmatrix} ; \forall u, t \in GF(p^r), \, (u,t) \neq (0,0). \end{aligned}$$

The spread may be written in the following form, using the group H_{q-1}.

$$x = 0, \, y = 0, \, y = x \begin{bmatrix} v & 0 \\ 0 & v^{2j+1} \end{bmatrix} \begin{bmatrix} u & t \\ \delta_{u,t}^j ft & \delta_{u,t}^j (u+gt) \end{bmatrix} ;$$

$$\forall v, t \in GF(p^r), \, vt \neq 0,$$

$$\text{and } y = x \begin{bmatrix} w & 0 \\ 0 & w^{2j+1} \end{bmatrix} ; \forall w \in GF(p^r)^*.$$

Using results of Dickson [43] on permutation polynomials, Johnson, Pomareda and Wilke [99], show that there are always j-planes of odd order p^{2r}, where every integer s, $1 \leq s < r$ and where $2j + 1 = p^s$, so that $j = (p^s - 1)/2$. Then the spreads have the following form:

$$x = 0, \, y = 0, \, y = x \begin{bmatrix} v & 0 \\ 0 & v^{p^s} \end{bmatrix} \begin{bmatrix} u & t \\ \delta_{u,t}^j ft & \delta_{u,t}^j u \end{bmatrix} ;$$

$$\forall v, t \in GF(p^r), \, vt \neq 0,$$

$$\text{and } y = x \begin{bmatrix} w & 0 \\ 0 & w^{p^s} \end{bmatrix} ; \forall w \in GF(p^r)^*.$$

If we could map the spread as follows $(x_1, x_2, x_3, x_4) \to (x_4, x_3, x_2, x_1)$, we would have the form used in this text. Hence, we have a set of of p^s-regulus-inducing groups and a set of corresponding twisted hyperbolic flocks.

$$x = 0, y = x \begin{bmatrix} u + gt^{j+1} & ft^{2j+1} \\ t & u \end{bmatrix} ; t, u \in GF(q).$$

Now we consider when $g = 0$ (or this is immediate under the conditions assumed). Since the group is transitive on the set of components other than $x = 0$ and $y = 0$, and the components have associated matrices that are non-singular, we see that a spread is obtained if and only if the difference of $y = x \begin{bmatrix} u & t \\ \delta^j_{u,t} ft & \delta^j_{u,t} u \end{bmatrix}$ and $y = x$ is non-singular. Letting $g = 0$, this becomes

$$y = x \begin{bmatrix} u - 1 & t \\ \delta^j_{u,t} ft & \delta^j_{u,t} u - 1 \end{bmatrix}.$$

The determinant of this matrix is

$$\delta^{j+1}_{u,t} - \delta^j_{u,t} u + (1 - u).$$

We shall now use the Kantor-Knuth flocks of quadratic cones to construct two infinite families of finite α-twisted hyperbolic flocks.

From (5,7,7&8) on hyperbolic fibrations and the correspondence between flocks of quadratic cones and finite translation planes of order q^2 that admit a cyclic affine homology group of order $q + 1$, our method uses the Kantor-Knuth flocks and the correspondence theorem to construct two infinite classes of j-planes of orders q^2, for $q = p^r$, and $j = \frac{p^s-1}{2}$, for $s = 0, 1, 2, ..., r$. and also for the $j = \frac{p^s-1}{2} + \frac{q-1}{2}$-planes. The correspondence will demonstrate the many to one aspect between the flocks of cones and the corresponding translation planes admitting the cyclic group.

We use the correspondence result. Note we will be working with three different types of spreads. Let π be any translation plane corresponding to a flock of a quadratic cone with associated flock spread of the following type. Let K be a field spread:

$$\text{Conical flock spread} \quad : \quad x = 0, y = x \begin{bmatrix} u + g(t) & f(t) \\ t & u \end{bmatrix} :$$

$$\forall t, u \in K, f, g \text{ functions on } K.$$

Now consider a translation plane of order q^2 that admits a cyclic homology group of order $q + 1$. This group will have the following form:

$$\left\langle \begin{bmatrix} I_2 & 0_2 \\ 0_2 & \begin{bmatrix} u & t \\ ft & u + gt \end{bmatrix} \end{bmatrix} : \forall t, u \in K \right\rangle,$$

$$\delta_{u,t} = u(u + gt) - ft^2 = 1 \text{ for all } u, t, (u, t) \neq (0, 0).$$

Note that the matrix for all t, u, represents a field extension of dimension 2 over K, $K(\theta)$, such that $\theta^2 = \theta g + f$. Using Andrè, in the associated Desarguesian translation plane, we have a set of $q - 1$ mutually disjoint derivable nets (See part 10). And the cyclic subgroup of order $q + 1$ acts transitively on all components of each net. It will become important to note that the spread has the form

$$x = 0, y = 0, \cup_{m_i \in \lambda} \left\{ y = xm_i \begin{bmatrix} u & t \\ ft & u + gt \end{bmatrix} : \delta_{u,t} = 1 \right\},$$

for a set λ of matrices m_i. Now when the order q^2 is odd, taking $g = 0$, there is an associated automorphism σ mapping $\theta t + u \to -\theta t + u$. The multiply derived spreads use the matrices $m_i \begin{bmatrix} 1 & 0 \\ 0 & -1 \end{bmatrix} \begin{bmatrix} u & t \\ ft & u + gt \end{bmatrix}$, if $x = \theta x_1 + x_2 = (x_1, x_2)$.

Cyclic Group plane:

$$x = 0, y = x \begin{bmatrix} u & t \\ F(u,t) & G(u,t) \end{bmatrix} :$$

$$\forall t, u \in K, \ f, g \text{ functions on } K = GF(q),$$

$$F \text{ and } G \text{ functions on } K \times K \text{ and } F(0,t) = ft, \ G(u,t) = u + gt,$$

$$\text{when } \delta_{u,t} = 1.$$

Then corresponding to the cyclic group plane turns out to be a translation plane of a quadratic cone, represented as follows. We start with a given flock plane of the

$$x = 0, y = \begin{bmatrix} u & t \\ \mathcal{F}(t) & u + \mathcal{G}(t) \end{bmatrix} : u, t \in K.$$

We have just conjugated the ordinary representation by $\begin{bmatrix} 0 & 1 \\ 1 & 0 \end{bmatrix}$. This mapping will interchange the matrix entries as follows: $(1,1) \longleftrightarrow (2,2)$ and $(1,2) \longleftrightarrow (2,1)$. Now the correspondence theorem uses the fact that over a finite field K and a quadratic Galois extension $K(\theta)$, with automorphism σ fixing K pointwise, we would have $K = K(\theta)^{\sigma+1}$. The fact this is almost never true for infinite fields and their quadratic Galois extensions is what makes extensions of this theory to the infinite case difficult. The correspondence theorem represents the flock spreads as follows:

$$\delta - form \text{ of a } \textbf{Flock Spread}$$

$$: \quad x = 0, y - \begin{bmatrix} \delta_{s,k} & \delta_{u,t} \\ \mathcal{F}(\delta_{u,t}) & \delta_{s,k} + \mathcal{G}(\delta_{u,t}) \end{bmatrix} : s, k, u, t \in K.$$

Here is the correspondence theorem:

$$(*) \quad : \quad \textbf{Cyclic Group Plane:}$$

$$\begin{bmatrix} u & t \\ F(u,t) & G(u,t) \end{bmatrix}$$

$$\Longleftrightarrow \quad \delta\text{-quadratic flock Form} \begin{bmatrix} \delta_{s,k} & \delta_{u,t} \\ \mathcal{F}(\delta_{u,t}) & \delta_{s,k} + \mathcal{G}(\delta_{u,t}) \end{bmatrix}$$

where

$$\mathcal{G}(\delta_{u,t}) = g(uG(u,t) + tF(u,t)) + 2(u \, F(u,t) - tfG(u,t))$$
$$= (gu - 2tf)(G(u,t)) + (gt + 2u)F(u,t) \text{ and}$$
$$-\mathcal{F}(\delta_{u,t}) = \delta_{F(u,t),G(u,t)} = \det \begin{bmatrix} F(u,t) & G(u,t) \\ fG(u,t) & F(u,t) + gG(u,t) \end{bmatrix}.$$

We now restrict to the δ-quadratic flock form of the set Kantor-Knuth spreads:

$$\delta \quad : \quad \text{Kantor-Knuth flock spreads}: \begin{bmatrix} \delta_{s,k} & \delta_{u,t} \\ f\delta_{u,t}^{p^s} & \delta_{s,k} \end{bmatrix}:$$

$$\delta_{s,k}, \delta_{u,t} \in K = GF(p^r), \ p \text{ odd}, \ f \text{ non-square in } K,$$

for each s properly dividing r.

Recall, we assume that $F(1,0) = G(1,0) = 1$, to have a component $y = x$. Then $(g)(G(1,0)) + (2)F(1,0) = g = 0$. We will now use the correspondence theory to determine the cyclic spreads. Note that we shall be looking for a set of cyclic spreads, as any multiple derivation of the cyclic spreads will correspond to the same δ-quadratic flock form. Once we find the set of cyclic spreads, we shall determine the set of these that correspond to the α-flocks of hyperbolic quadrics (or rather to the corresponding translation planes).

Using $(*)$, and since p is odd, we see that as $\mathcal{G}(\delta_{u,t}) = 0$, and the characteristic is not 2 then

$$u \ F(u,t) - ftG(u,t) = 0.$$

In this context, $u = 0$ and $t \neq 0$, implies $G(u,t) = 0$, a contradiction unless $t = 0$. Hence, $u = 0$ if and only if $t = 0$. Thus,

$$(**) : F(u,t) = f\frac{t}{u}G(u,t).$$

Also,

$$-\mathcal{F}(\delta_{u,t}) = \delta_{F(u,t),G(u,t)} = \det \begin{bmatrix} F(u,t) & G(u,t) \\ fG(u,t) & F(u,t) + gG(u,t) \end{bmatrix}$$

$$= \det \begin{bmatrix} f\frac{t}{u}G(u,t) & G(u,t) \\ fG(u,t) & f\frac{t}{u}G(u,t) \end{bmatrix}.$$

Since $\mathcal{F}(\delta_{u,t}) = f\delta_{u,t}^{p^s}$, we see that, after some reduction, that

$$\delta_{u,t}^{p^s} = ((\frac{t}{u})^2 f - 1)G(u,t)^2 = \delta_{u,t}u^{-2}G(u,t)^2.$$

Thus, we have:

$$G(u,t) = \pm\delta_{u,t}^{\frac{p^s-1}{2}} u$$

Then using $(**)$, we obtain:

$$F(u,t) = \pm ft\delta_{u,t}^{\frac{p^s-1}{2}}.$$

Notice that the $\pm$ are not arbitrary but only are applicable when their assignment determines a translation plane. Recalling the form of a multiply derived translation plane from a cyclic plane, it follows that $y = xm_i \begin{bmatrix} 1 & 0 \\ 0 & -1 \end{bmatrix} \begin{bmatrix} u & t \\ ft & u+gt \end{bmatrix}$ represents a derived net and the $\pm = -$ means the net has been replaced. Therefore, if we replace a subset i of the $q-1$ mutually disjoint derivable nets, then we put a $-$ on each of these nets. First, we see that if no nets are replaced then $\pm = 1$, for all nets which means that we have a j plane with $j = \frac{p^s-1}{2}$, for s properly dividing r. If all nets are replaced then $\pm = -1$ for all nets, which means that we have an isomorphic j-plane.

We now consider the nets determined by whether $\delta_{u,t}$ is a square or a non-square. Consider, $s = 0$, then the $j = 0$ plane is a Desarguesian plane. In this setting, the regular nearfield plane may be obtained by replacement of all non-square nets. In the more general case, a replacement of all nets when $\delta_{u,t}$ is a non-square also then provides a 'near-nearfield' translation plane. We have established that we are now working in a j-plane. The j-plane admits an affine homology group

$$H_{q-1} = \left\{ \begin{bmatrix} u & 0 & 0 & 0 \\ 0 & u^{2j+1} & 0 & 0 \\ 0 & 0 & 1 & 0 \\ 0 & 0 & 0 & 1 \end{bmatrix} ; \forall u \in GF(p^r)^* \right\},$$

where now $2j+1 = p^s$. If $s \neq 0$, we have an alternative version of a $\alpha = p^s$-hyperbolic quadric. In the ordinary version, the axis of the homology group is $x = 0$ and the coaxis is $y = 0$. Here the situation is changed to a homology group with axis $y = 0$ and coaxis $x = 0$. Thus, the $j = \frac{p^s-1}{2}$-plane is a translation plane corresponding to the p^s-twisted hyperbolic quadric. Now we assert that the near-nearfield translation plane, obtained by replacement of all nets where $\delta_{u,t}$ is non-square provides a $j+\frac{q-1}{2}$-plane. Note that $\delta_{u,t}^{j+\frac{q-1}{2}}$, for $\delta_{u,t}$ a square is $\delta_{u,t}^{j+\frac{q-1}{2}} = \delta_{u,t}^j$, and is $-\delta_{u,t}^j$, when $\delta_{u,t}$ is a non-square. Accordingly, we have proved the following theorem.

THEOREM 66. *Consider the δ-version of the Kantor-Knuth flock spreads.*

$$\delta \quad : \quad \text{Kantor-Knuth flock spreads} : \begin{bmatrix} \delta_{v,k} & \delta_{u,t} \\ f\delta_{u,t}^{p^s} & \delta_{v,k} \end{bmatrix} :$$

$$\delta_{v,k}, \delta_{u,t} \quad \in \quad K = GF(p^r), \ p \ odd, \ f \ non\text{-}square \ in \ K,$$

$$\text{for each } s \text{ properly dividing } r.$$

Then using the correspondence theorem between translation planes of order p^{2r} admitting cyclic homology groups of order $p^r + 1$, there are two classes of j-planes, where $j = \frac{p^s-1}{2}$ and $j = \frac{p^s-1}{2} + \frac{q-1}{2}$ admitting $\alpha = p^s$-twisted regulus-generating homology groups thus generating infinite classes of twisted hyperbolic flocks (these are more properly α^{-1}-twisted regulus-generating homology groups).

The j-planes and thus the infinite classes of twisted hyperbolic flocks are due to Johnson, Pomareda, and Wilke ([99], noting that s must divide r, which was not mentioned in the original article).

We shall make a few more remarks about how the construction of j-planes in question were made, thirty years before the development of the work on hyperbolic fibrations and translation planes of order q^2 admitting cyclic homology groups of order $q + 1$.

To create a j-plane, a spread is required. Moreover, in general, the differences of any distinct pair of the individual matrices are required to have non-determinants. What this means is that defining

$$\phi(u,t) = \delta_{u,t}^j ft = ft(u^2 - ft^2)^j.$$

Let $j = \frac{p^s-1}{2}$ so $2j + 1 = p^s$, where $q = p^r$, this function is required to be a permutation polynomial for each $u \in K$ to have a j-plane of this type.

THEOREM 67. (Dickson [**43**]*p. 63*) *If d is a divisor of $p^s - 1$ and v is not a d th power in $GF(p^r)$, then $\phi(\alpha) = \alpha(\alpha^d - v)^{\frac{p^s-1}{d}}$. is a permutation polynomial of $GF(p^r)$, for each $s = 0, 1, 2, ..., r$.*

Taking $d = 2$, u^2/f is not a square in $GF(p^r)$ then, we have have a permutation polynomial and therefore a j-plane. The $j + \frac{q-1}{2}$-plane was constructed exactly as above in the theorem arising from the associated flock of a quadratic cone.

To consider the type of twisted hyperbolic flock that is obtained, one may choose to invert the spread of the j-planes by $(x, y) \rightarrow (y, x)$, which would represent the spread and therefore the flock in what might be called 'standard form'. The group

$$H_{q-1} = \left\{ \begin{bmatrix} u & 0 & 0 & 0 \\ 0 & u^{2j+1} & 0 & 0 \\ 0 & 0 & 1 & 0 \\ 0 & 0 & 0 & 1 \end{bmatrix} ; \forall u \in GF(p^r)^* \right\},$$

will be mapped into the group

$$\widehat{H}_{q-1} = \left\{ \begin{bmatrix} 1 & 0 & 0 & 0 \\ 0 & 1 & 0 & 0 \\ 0 & 0 & u & 0 \\ 0 & 0 & 0 & u^{p^s} \end{bmatrix} ; \forall u \in GF(p^r)^* \right\},$$

And this mapping perseveres the kernel, so we have a translation plane admitting a α^{-1}-twisted regulus net and therefore, by the algebraic method above, we have a have an α^{-1}-twisted hyperbolic flock.

PROBLEM 26. *The spread for the α^{-1}-twisted flock plane is as follows: Let $\alpha = 2j + 1 = p^s$ or $p^s + \frac{q-1}{2}$, for any $s = 1, 2, ..., r$, then*

$$x = 0, y = 0, y = x \begin{bmatrix} u & t \\ \delta^j_{u,t} ft & \delta^j_{u,t} u \end{bmatrix}^{-1} : \forall u, t \in K = GF(q), (u, t) \neq (0, 0),$$

and admits an α^{-1}-twisted regulus inducing homology group.

Determine the form of the planes for the α^{-1}-twisted hyperbolic quadric. Hint: Determine the spread in the form of

$$x = 0, y = x \begin{bmatrix} f(k) & g(k) \\ 1 & k \end{bmatrix} \begin{bmatrix} u^{\alpha^{-1}} & 0 \\ 0 & u \end{bmatrix} : \forall k, u \neq 0 \text{ in } K,$$

f and g functions on K.

REMARK 19. *As mentioned, all finite j-planes correspond to monomial flocks of quadratic cones using the correspondence theorem. And, we know all possible even order j-planes. There are two other infinite classes of monomial flocks, for $j = 1, 2$. And, of course, there are another two infinite classes using the Kantor-Knuth flock planes (as the flock planes correspond to any multiple derived j-plane).*

*All of these monomial flocks were known and listed in the original article on j-planes, except these were listed as j-planes, as the correspondence theorem was not yet known. Moreover, in Johnson, Pomareda, Wilke [**99**], there is a set of 40 other j-planes of order p^2, p a prime < 100. The terms $p(2i)$ are the number $2i$ of* *j-planes of order p^2. The actual j's are given in a table on p. 283.*

$$p = 11(2), 17(4), 23(4), 37(2), 41(2), 43(2), 47(4),$$
$$53(4), 59(2), 67(2), 71(2), 73(2), 83(4), 89(2), 97(2).$$

The sums of the numbers in parentheses is 40. For any j-plane of odd order, the same near-nearfield construction is available to construct a $j + \frac{q-1}{2}$-plane. This is a multiple derived j-plane and therefore corresponds to the same monomial flock. This means, there are are 20 additional monomial flock planes that are perhaps not completely well known. For some of the primes listed, there are computer programs that must have found the same flocks, perhaps without the recognition that the flocks are monomial (See the Handbook on sporadic flocks). The construction of j-planes is relatively straight forward with a computer, due to the permutation polynomial that must be verified.

PROBLEM 27. *Previously, we have formed T-copies and twisted T-copies of Hughes-Kleinfeld look-alikes. Furthermore, in the notation of (7,9,2), if $\sigma\rho \neq 1$, then all T-copies and twisted T-copies from finite spreads are proper quasifibrations. The Kantor-Knuth flock planes are Hughes-Kleinfeld look-alikes and therefore, the T-copy and twisted T-copy are proper quasifibrations. Decide what happens to the associated $j = \frac{p^s-1}{2}$, j-planes. Since the twisted T-copy is still a flock proper quasifibration, see if the argument from the flock backwards still works for the twisted T-copy. Is there an associated j-plane type proper quasifibration?*

5. Joint theory of α-flocks

In this section, it is convenient to treat aspects of the theories of flocks of α-cones and twisted hyperbolic flocks jointly, due to the similarity of the ideas of each theory.

Our main results in this section also extend the above result for partial α-conical flocks and partial α-twisted hyperbolic flocks of deficiency-one over any field K that admits an automorphism α.

In Jha and Johnson [69], the analysis focused on understanding the structure of the hyperbolic flocks and of flocks of quadratic cones. The analysis is that from a translation plane with spread in $PG(3, K)$ admitting a regulus-inducing elation group E or a regulus-inducing homology group H actually involves partitioning the hyperbolic quadric or quadratic cone by the invariant Baer subplanes of the associated group H or E, respectively. For convenience, we list those results here.

THEOREM 68. *(Jha-Johnson ([69]) Let K be a field and let π be a translation plane with spread in $PG(3, K)$ that admits a full point-Baer elation group. Then there is a corresponding partial conical flock of deficiency-one in $PG(3, K)$. Conversely, any partial conical flock of deficiency-one constructs a translation plane with spread in $PG(3, K)$ that admits a full point-Baer elation group. The partial conical flock of deficiency-one may be extended to a flock of a quadratic cone if and only if the net containing the point-Baer axis is a regulus net sharing the axis.*

THEOREM 69. *(Jha-Johnson ([69])) Let K be a field and let Σ be a translation plane with spread in $PG(3, K)$ that admits a full point-Baer homology group. Then there is a corresponding partial hyperbolic flock of deficiency-one in $PG(3, K)$. Conversely, any partial hyperbolic flock of deficiency-one constructs a translation plane* *with spread in $PG(3, K)$ that admits a full point-Baer homology group. The partial hyperbolic flock may be extended to a hyperbolic flock if and only if the net containing the point-Baer axis and coaxis is a regulus net.*

REMARK 20. *Note that in the above two theorems there is the term 'point-Baer' in the hypotheses. A Baer subplane in an affine plane must have two properties: Every point of the plane must be on a line of the subplane ('point-Baer') and every line of the plane must be on a line of the subplane ('line-Baer'). These conditions are equivalent in the finite case but, in the infinite case, Barlotti [12] shows that they are not. We have mentioned that the author [84] constructs dual translation planes that admit a derivable net which are not derivable. So, when dealing with the infinite case, care must be taken (the interested reader might also look at (25.3) of [86] for more explanation). When we give our main extension results, we will show that we may remove the point-Baer hypothesis in the setting under consideration.*

We shall give the pertinent theory providing the structure of the set of invariant 2-dimensional K-subspaces in the associated 4-dimensional K-vector spaces. But, also, we have noted that instead of a partition of V_4 over K by regulus-inducing groups, we may also consider analogous theory by twisted regulus-inducing groups, either elation or homology groups. We now consider a partition of the twisted quadrics; the twisted conical flocks and the twisted hyperbolic flocks. After the preliminary material below, the extension of the theorem of Jha-Johnson [69] may be directly extended by using α-reguli instead of reguli. The idea is to work over K^{α}, when dealing with the Baer subplanes, then all of the arguments extend.

Our main results for the hyperbolic situation are as follows. However, 'translation plane' may be replaced by 'quasifibration' in the most general version of the theorem.

THEOREM 70. *Let Σ be a translation plane with blended kernel (K, K^{α}) that admits a full Baer α-homology group, where $\Sigma-$Baer axis (components containing the Baer subplane in question) is a set of K-subspaces. Then there is a corresponding twisted hyperbolic flock of deficiency-one in $PG(3, K)$. Conversely, any partial twisted hyperbolic flock of deficiency-one constructs a translation plane of blended kernel (K, K^{α}) that admits a full Baer α-homology group. The partial α-hyperbolic flock may be extended to an α-hyperbolic flock if and only if the Baer net is a derived α-regulus net.*

The theorem for Baer α-elation groups is then as follows:

THEOREM 71. *Let Σ be a translation plane with blended kernel (K, K^{α}) that admits a full Baer α-elation group, where $\Sigma-$Baer axis (components containing the Baer subplane in question) is a set of K-subspaces. Then there is a corresponding partial α-flock of an α-cone of deficiency-one in $PG(3, K)$. Conversely, any partial flock of deficiency-one of an α-cone constructs a translation plane of blended kernel (K, K^{α}) that admits a full Baer α-elation group. The partial α-flock may be extended to an α-flock if and only if the Baer net is a derived α-regulus net.*

The proofs will follow after some results on group orbits. We shall state the elation and homology Baer theorems separately, but will tend to combine the proofs, as there are definite similarities. First the Baer elation group situation:

THEOREM 72. *Let K be a field and let α be an automorphism of K. Let V_4 be a vector space over K and let $x = 0$ and $y = 0$ be disjoint 2-dimensional K and K^α subspaces, where x and y are 2-vectors. We note that V_4 may also be considered a 4-dimensional vector space over K^α. Consider the following group in $GL(4, K)$:*

$$E^\alpha = \left\{ \begin{bmatrix} I_2 & \begin{bmatrix} u^\alpha & 0 \\ 0 & u \end{bmatrix} \\ 0_2 & I_2 \end{bmatrix} ; u \in K \right\}.$$

We note that E^a fixes $x = 0$ pointwise.

(1)(a) Then the elements of the set of E^α invariant 2-dimensional subspaces not equal to $x = 0$ are each generated from a 1-dimensional K^α-subspace of $x = 0$ and another 1-dimensional space K^α. Two of these 2-dimensional subspaces over K^α are disjoint if and only if they are disjoint on $x = 0$.

(1)(b) Let $\{x_i; i \in \lambda\}$ denote the set of 1-dimensional K^α-subspaces of $x = 0$ and for each x_i, choose any other 1-dimensional subspace w_i not incident with $x = 0$ such that the generated 2-dimensional subspace is E^α-invariant. Then the set of all 2-dimensional K^α-subspaces of

$$\{\langle x_i, w_i \rangle, i \in \lambda\}$$

is a partial spread net covering $x = 0$. If we can form these subspaces into a derivable net, then the derived net is an α-regulus net over K.

For the Baer homology situation, we have:

THEOREM 73. *Let K be a field and let α be an automorphism of K. Let V_4 be a vector space over K and let $x = 0$ and $y = 0$ be disjoint 2-dimensional K and K^α-subspaces, where x and y are 2-vectors. We note that V_4 may also be considered a 4-dimensional vector space over K^α. Consider the following group in $GL(4, K)$,*

$$H^\alpha = \left\{ \begin{bmatrix} I_2 & 0_2 \\ 0_2 & \begin{bmatrix} u^\alpha & 0 \\ 0 & u \end{bmatrix} \end{bmatrix} ; u \in K^* \right\}.$$

(2)(a) Then all H^α-invariant 2-dimensional subspaces are generated by a 1-dimensional K^α-space from $x = 0$ and a 1-dimensional K^α-space of $y = 0$.

(2)(b) Let $\{x_i; i \in \lambda\}$ denote the set of 1-dimensional K^α subspaces of $x = 0$ and let $\{y_i; i \in \lambda\}$ be the set of all 1-dimensional K^α-subspaces of $y = 0$. Let Γ be any bijective mapping from $\{x_i; i \in \lambda\}$ onto $\{y_i; i \in \lambda\}$. Then the set of 2-dimensional K^α-subspaces

$$\{(x_i, \Gamma(x_i)); i \in \lambda\}$$

defines a partial spread net. If these subspaces can be formed into a derivable net then the derived net is an α-regulus net whose components are K-subspaces and K^α-subspaces, and whose 1-dimensional K^α-subspaces on $x = 0$ and $y = 0$ are completely covered.

PROOF. (1)(a). $E^\alpha = \left\{ \begin{bmatrix} I_2 & \begin{bmatrix} u^\alpha & 0 \\ 0 & u \end{bmatrix} \\ 0_2 & I_2 \end{bmatrix} ; u \in K \right\}.$ Consider any vector (x_1, x_2, x_3, x_4) and consider the E^α orbit of this vector. This is

$$\langle (x_1, x_2, x_1 u^\alpha + x_3, x_2 u + x_4); u \in K \rangle.$$

Then subtracting (x_1, x_2, x_3, x_4) from a general term, we see that we have the vectors $(0, 0, x_1 u^\alpha, x_2 u)$. Recall that V_4 is a K^α-vector space as well, and we have the scalar multiplication $\cdot$ given by

$$(x_1, x_2, x_3, x_4) \cdot u = (x_1 u^\alpha, x_2 u, x_3 u^\alpha, x_4 u),$$

we see that we must have $(0, 0, x_1, x_2)$ and the 1-dimensional K^α space within this subspace, in any two dimensional subspace that is invariant under E^α. This subspace is 2-dimensional over K if and only if $x_1 = x_2 = 0$, for $\alpha \neq 1$. We consider Ω^α as the set of 'points' that are either E^α or H^α 2-dimensional K^α orbits: The set of points has the following form

$$\{(x_1^\alpha x_4, x_1 x_3^\alpha, x_2^\alpha x_4, x_2 x_3^\alpha); x_i \in L.\}$$

Abstractly, this forms the basis for the 3-dimensional projective intersection with the K^α-Klein quadric, which will be discussed more completely in the subsequent material. We note the following: The E^α and H^α-orbits define 'points' and furthermore, if a vector

$$(x_1, x_2, x_3, x_4) \to (x_1^\alpha x_4, x_1 x_3^\alpha, x_2^\alpha x_4, x_2 x_3^\alpha)$$

then every K-space generated by this vector

$$(x_1 \beta, x_2 \beta, x_3 \beta, x_4 \beta)$$

maps is the same 'point'/modulo

$$\begin{aligned} K &\to (x_1^a x_4 \beta^{\alpha+1}, x_1 x_3^\alpha \beta^{\alpha+1}, x_2^\alpha x_4 \beta^{\alpha+1}, x_2 x_3^\alpha \beta^{\alpha+1}) \\ &\equiv (x_1^\alpha x_4, x_1 x_3^\alpha, x_2^\alpha x_4, x_2 x_3^\alpha). \end{aligned}$$

So, in the associated flocks, the plane intersections could be visualized as K-vectors whose corresponding mapped K^α-Klein quadric are objects that arise from either E^α or H^α-2-dimensional or K^α-subspace/orbits. In this manner, we may consider plane intersections as either covering Ω^α in the H^α-case or by covering an associated α-flock of an α-quadratic cone. Now assume that two of the E^α-subspaces are not disjoint

$$\langle (x_1, x_2, x_3, x_4) E^\alpha \rangle \cap \langle (y_1, y_2, y_3, y_4) E^\alpha \rangle .$$

Assume that

$$(x_1, x_2, x_1 u^\alpha + x_3, x_2 u + x_4) \cdot \delta = (y_1, y_2, y_1 v^\alpha + y_3, y_2 v + y_4),$$

for $u, v, \delta \in K$. Then $\langle (x_1, x_2) \rangle = \langle (y_1, y_2) \rangle$, and hence these subspaces share a 1-dimensional subspace on $x = 0$. It now follows directly that the two subspaces are equal. This proves (1)(a). Now it is clear that $\{\langle x_i, w_i \rangle, i \in \lambda\}$ is a set of E^α-invariant K^α-subspaces that are mutually disjoint and completely cover $x = 0$. Let $w_i = (x_{1,i}, x_{2,i}, x_{3,i}, x_{4,i})$, where the 2-vector over K^α on $x = 0$ is $x_i = (x_{1,i}, x_{2,i})$. The rest of the theorem will now follow directly once we have worked out the form of the derived nets. (2)(a) Now consider the group

$$H^\alpha = \left\{ \begin{bmatrix} I_2 & 0_2 \\ 0_2 & \begin{bmatrix} u^\alpha & 0 \\ 0 & u \end{bmatrix} \end{bmatrix} ; u \in K^* \right\}.$$

Consider any vector (x_1, x_2, x_3, x_4) not in either $x = 0$ or $y = 0$ and consider the H^α-orbit, $\langle (x_1, x_2, x_3 u^\alpha, x_4 u); u \in K^* \rangle$. By subtracting the generating vector, and then realizing that we have a K^α-subspace, it follows that $(0, 0, x_3, x_4)$ is a non-zero vector in the generated K^α-subspace. Now subtracting the original vector from this preceding vector, we also have $(x_1, x_2, 0, 0)$ a non-zero vector in the generated K^α-space. This says that the only H^α-invariant 2-dimensional subspaces are generated from two non-zero vectors on each of the two invariant K-subspaces $x = 0$ and $y = 0$. Moreover, it now follows that the only way that two of these invariant 2-dimensional K^α-subspaces can non-trivially intersect is if they intersect on either $x = 0$ or $y = 0$. Accordingly, $\{(x_i, \Gamma(x_i); i \in \lambda\}$ defines a partial spread net of K^α-invariant 2-dimensional subspaces that cover $x = 0$ and $y = 0$. This completes the proofs of both (2)(a) and (b). $\square$

6. The K^α-Klein quadric

The reader will notice that we are essentially using ideas of Thas and Walker, Thas [**136**], coupled with insights about derivable nets, and certain α–regulus orbits under elation or homology groups as 'points', for these extension notions.

Consider a 6-dimensional vector space V_6 over K in the standard manner and over K^α with special scalar multiplication $\cdot$ as follows:

$$(x_1, x_2, x_3, x_4, x_5, x_6) \cdot \delta = (x_1\delta^\alpha, x_2\delta, x_3\delta^\alpha, x_4\delta, x_5\delta^\alpha, x_6\delta),$$

and written over the associated standard basis, we obtain vectors as

$$(x_1^{\alpha^{-1}}, x_2, x_3^{\alpha^{-1}}, x_4, x_5^{\alpha^{-1}}, x_6),$$

then the standard hyperbolic form Ω_5 becomes

$$x_1^\alpha x_6 + x_2 x_5^\alpha + x_3^\alpha x_4 = 0.$$

Now embed $PG(3, K)$ as $x_3 = x_4 = 0$. We note that the group E^α now acts on $PG(3, K)$.

We consider the associated α-hyperbolic form, which can be given by $x_1^\alpha x_6 - x_2 x_5^\alpha$, without loss of generality. Now consider the set of points within $x_2 = x_5$, to obtain $x_1^\alpha x_6 = x_2^{\alpha+1}$, an α-conic. We realize that E^α leaves the associated homogeneous points on the α-regulus of Baer subspaces pointwise fixed (as 'points'). Identifying the α-quadric projectively as within the α-twisted hyperbolic quadric, we see that we may consider the α-conical flock as a set of points within the α-twisted quadric in $PG(3, K)$ as follows: Consider a point of $PG(3, K)$ say $v_0 = (0, 0, 0, 0, 0, 1)$ as the vertex of the cone, then a set of planes of $PG(3, K)$ that partition the points of the lines from v_0 to the α-conic $x_1^\alpha x_6 = x_2^{\alpha+1}$, then the planes correspond to the K-components of a translation plane with spread in $PG(3, K)$ that admit an α-regulus-inducing elation group E^α. The converse is that a translation plane with spread in $PG(3, K)$ with a central collineation group E^α produces an α-flock of the α-quadratic cone is now immediate.

Hence, this provides an alternative construction of the main theorem of Cherowitzo-Johnson-Vega [**26**]. We shall use this approach when considering the Baer groups and the deficiency-one α-conical flocks.

- This shows that α-flocks and the set of Baer components of the translation plane have a bijective correspondence.
- The results of the algebraic method show that we may identity the K^α-subspaces of the invariant H^α-2-dimensional subspaces with the elements

Ω_3^α, the twisted hyperbolic quadric in $PG(3,K)$ and the Baer components of the associated α-regulus nets associated with the translation plane admitting H^α.

- This also shows that twisted hyperbolic flocks and the set of Baer components of the translation plane have a bijective correspondence.

The algebraic approach to the connection between flocks of quadratic cones and translation planes was completed by Gevaert and Johnson[**56**], whereas the previous view was accomplished using the Klein quadric by Thas and Walker ([**135**]. Similarly, in Cherowitzo, Johnson, Vega [**26**], an algebraic method established the connection.

7. Baer theory

We now complete the proof of our Baer theorems. We begin with the translation plane admitting a Baer elation group. The reader should note that the group

$$E^\alpha = \left\{ \begin{bmatrix} I_2 & \begin{bmatrix} u^\alpha & 0 \\ 0 & u \end{bmatrix} \\ 0_2 & I_2 \end{bmatrix} ; u \in K \right\},$$

has this form when considering the axis as $x = 0$. It might be expected that any associated translation plane admitting E^α has $x = 0$ as a component. However, this may not be the case, and will not be the case when E^α is considered a Baer group and, in this case, $x = 0$ becomes a Baer subplane.

If the translation plane admits E^α as a Baer group then the components that lie on the Baer axis (i.e. $x = 0$) are E^α subspaces, so we have the situation of a translation plane with blended kernel (K, K^α). Now, we still have the orbits of K-components as before, we just have one less α-regulus derivable net than in the case when E^α is an elation group. For clarity, we now speak of a Baer elation group. Since the other α-regulus nets still correspond to sets of E^α orbits of K^α-subspaces, we clearly have a deficiency-one α-conical flock.

Now suppose that we have a deficiency-one α-conical flock. Then on each line of the α-cone, we are missing exactly one point. This point corresponds to an E^α-orbit and since no two of these are on the same line of the α-cone, then the points are mutually disjoint. Let Λ denote the set of corresponding 2-dimensional K^α-subspaces. Let Ω denote the partial spread consisting of the α-regulus nets corresponding to the deficiency-one α-conical flock. Since now $\Lambda \cup \{\Omega - (x = 0)\}$, covers all of the E^α-orbits in the associated vector space over K^α, we have a spread, but now with blended kernel (K, K^α), admitting E^α as a Baer elation group.

Now the question of when the deficiency-one α-conical flock situation occurs depends on whether the set Λ is a derivable net or not. This is valid, as this is the only method that can produce a set of K-components to return to a translation plane with kernel K admitting E^α as an elation group. Note that it is possible that the net in question is derivable without the associated Baer subplanes being K-subspaces, but it is only when this occurs that the deficiency-one partial α-conical flock may be extended to a flock.

This completes the proof of the deficiency-one α-conical flock theorem. Now assume that we have a translation plane with blended kernel (K, K^α) admitting

$$H^\alpha = \left\{ \begin{bmatrix} I_2 & 0_2 \\ 0_2 & \begin{bmatrix} u^\alpha & 0 \\ 0 & u \end{bmatrix} \end{bmatrix} ; u \in K^* \right\}$$

as a Baer homology group, where now $x = 0$ and $y = 0$ are K-Baer subplanes. In this setting, we are missing one α-regulus net of H^α-invariant subplanes and hence the Baer subplanes of the other α-regulus nets produce a deficiency-one twisted hyperbolic flock.

Now assume that we have a deficiency-one twisted hyperbolic flock. From the analysis of the author in [89], it is clear that we may construct a partial spread Π which consists of a set of K-2-dimensional subspaces that admit H^α and is a set of α-regulus nets sharing $x = 0$ and $y = 0$. Recall that our α-hyperbolic quadric is a hyperbolic quadric with respect to K^α (since the α-regulus components and Baer components are all K^α-subspaces). Hence, there are two sets of ruling lines with respect to K^α. It follows that on each K^α-ruling line of each of the two sets of ruling lines, there is exactly one missing point. These points correspond to a set of mutually disjoint K^α-subspaces that are disjoint H^α-orbits and thus completely cover $x = 0$ and $y = 0$. Let W denote the set of these K^α-subspaces in the associated vector space. Consider $W \cup \Pi$. This partial spread is a spread, since it covers all of the H^α-invariant 2-dimensional K^α-subspaces. This translation plane has blended kernel (K, K^α) and admits H^α as a Baer homology group.

For the question of when the deficiency-one partial twisted hyperbolic flock may be extended, it depends on whether W is a derivable net with Baer components being K-subspaces, similar to the argument for the Baer elation situation.

This completes the proofs of the Baer theory for α-twisted flocks–except for that question about 'point Baer' that is in the hypotheses of the corresponding Baer theory for flocks of quadratic cones and for hyperbolic quadrics. So, we have shown that with an α-flock of an α-quadratic cone, there is an associated translation plane admitting an α-regulus-inducing elation group E^α and, conversely, such a translation plane constructs the α-conical flock. And, we have shown that the set of flocks of an α-twisted hyperbolic quadric and the set of translation planes admitting α-regulus-inducing groups are equivalent. The real problem seems to come in the deficiency-one theory.

Suppose that we have a translation plane of either type. Choose any of the α-regulus nets and suppose that the subplanes involved are 'point Baer' but not 'line Baer'. Now derive the net, what is obtained? This cannot be a translation plane any longer, just a Sperner space with blended kernel (K, K^α) (which could be K, if $\alpha = 1$). But, this structure still admits a full 'point Baer' elation or full 'point Baer' homology group. Accordingly, we obtain a deficiency-one α-flock, quadratic or hyperbolic, now with a point missing on each of the relative lines of the cone or of the α-rulings. Our proof still shows that we can recover a spread-a 'translation plane' –not a Sperner space. So, all of the point Baer subplanes must be Baer subplanes, since all of the α-regulus nets can be derived to affine planes. This completes the proofs of the theorems on Baer theory.

8. Quasi-flocks

Quasifibrations have been mentioned previously. Much of the work on flocks and associated translation planes has been done algebraically. We have a certain form of a translation plane and using this form, we create an associated flock, using that algebraic representation. More generally, we may use the forms defining partial spreads to create what appear to be flocks but are actually quasifibration/ maximal partial flocks. If we define a 'quasi-flock' as a partial flock that has the basic form of a flock but does not satisfy the covering criterion, we have necessarily a maximal partial flock. Proper quasi-flocks like proper quasifibrations are infinite.

In any case, all our results may be phrased more generally in terms of quasifibrations. Without listing all of various theorems, here is an omnibus theorem:

THEOREM 74. *Quasifibrations with α-regulus inducing groups are equivalent to quasi α-flocks.*

9. The Baer forms

Here we indicate what the translation planes with blended kernel (K, K^α) look like that admit Baer groups.

- Baer Elation translation planes would have the following form:

$$x = 0, y = (x_1^{\alpha^{-1}}, x_2^{\alpha^{-1}}) \begin{bmatrix} v & z(v) \\ 0 & v \end{bmatrix}; \forall v \in K,$$

$$y = x \begin{bmatrix} 1 & -u^\alpha \\ 0 & 1 \end{bmatrix} \begin{bmatrix} F(t) & G(t) \\ t & 0 \end{bmatrix} \begin{bmatrix} 1 & u \\ 0 & 1 \end{bmatrix}$$

for all $u, t \neq 0 \in K$. z a function on K so that $z(0) = 0$, $z(1) = 1$, and functions F and G on K.
- A partial α-conical flock may be extended if and only if $z(v) = 0$ for all $v \in K$.
- Baer Homology translation planes would have the following form:

$$x = 0, y = (x_1^{\alpha^{-1}}, x_2^{\alpha^{-1}}) \begin{bmatrix} n(v) & 0 \\ 0 & v \end{bmatrix}; \forall v \in K,$$

n a function on K so that $n(0) = 0$, $n(1) = 1$ and

$$y = x \begin{bmatrix} 1 & 0 \\ 0 & u^{-\alpha} \end{bmatrix} \begin{bmatrix} g(t) & f(t) \\ 1 & t \end{bmatrix} \begin{bmatrix} 1 & 0 \\ 0 & u \end{bmatrix}; \forall u \neq 0, t \neq 0 \text{ in } K,$$

for functions g and f on K.
- The deficiency-one partial twisted hyperbolic flock may be extended if and only if $n(v) = v \; \forall v \in K$.

PROOF. The elation group changes form when considering $x = 0$ as $(0, x_2, 0, x_4)$ in the formulation of the axis of the elation group. The components incident with the Baer axis are K^α-subspaces that are left fixed by the Baer elation group. We leave, as an exercise, to check out the remaining parts of the form. Similarly, the homology group changes form when considering $x = 0$ as $(0, x_2, 0, x_4)$ and $y = 0$ as $(x_1, 0, x_3, 0)$. Again the components incident with the Baer axis and coaxis are K^α-subspaces that are left fixed by the Baer homology group. Here is another exercise to determine out the remaining parts of the form. Note that there are two distinct uses of the notation of $x = 0$ and $y = 0$; one use is when considering the group axis/coaxis and the other use is when considering the form of the translation plane, when $x = 0$ and $y = 0$ are Baer subplanes.   □

PROBLEM 28. *Supply the details to show that the Baer forms are as stated.*

10. Algebraic and α-Klein methods

When the ideas of flocks of quadratic cone and flocks of hyperbolic quadrics were introduced, there were a variety of new studies, in the infinite case, and also later with what we are calling α-flocks of α-quadratic cones and α-twisted hyperbolic flocks. Many of these were algebraic in nature. When this was done, there was essentially no connection between α-regulus-inducing groups, elation and homology, that really becomes the essence of understanding the α-Klein methods. Moreover, there is no uniformity with the notation of the α-cones of α-twisted quadrics.

In Cherowitzo, Johnson, and Vega [26], it was pointed out that whenever an α-conical flock is constructed, there is also an α^{-1}-conical flock which may be constructed and is isomorphic to the original. How these two examples may be understood using the α-Klein method is by a translation of vectors $(x_1, x_2, x_3, x_4) \rightarrow (x_4, x_3, x_2, x_1)$, which changes the α-conic that is used to the associated α^{-1}-conic. To see this, just note that

$$\left\{ x = 0, y = x \begin{bmatrix} u^\alpha & 0 \\ 0 & u \end{bmatrix} ; u \in K \right\} \rightarrow \left\{ x = 0, y = x \begin{bmatrix} u & 0 \\ 0 & u^\alpha \end{bmatrix} ; u \in K \right\}.$$

This mapping works for the α-twisted hyperbolic flocks. There is also a corresponding change of functions defining the α or α^{-1}-flocks. This is the mapping $(x_1, x_2, x_3, x_4) \rightarrow (x_2, x_1, x_4, x_3)$, which does the same thing, changing the α-twisted regulus net to the α^{-1}-twisted regulus net.

In the author's work on α-twisted hyperbolic quadrics (See [89]), the process of translating components of α-regulus-inducing group H^α to the associated α-twisted hyperbolic quadrics does not use the understanding of the 'points' being the invariant 2-dimensional subspaces over K^α (not equal to the two components $x = 0$ and $y = 0$, the axis and coaxis of H^α). In fact, in the algebraic method, it may be seen that the projective connection is accomplished with the invariant 2-dimensional subspaces over $K^{\alpha^{-1}}$. This does not cause any difficulty as, similar to the elation case, given an α-twisted hyperbolic flock, there is always an isomorphic α^{-1}-twisted hyperbolic flock.

The algebraic approach to the connection between flocks of quadratic cones and translation planes was done by Gevaert and Johnson[56], whereas the previous view was accomplished using the Klein quadric by Thas and Walker ([135], both of these restricted to the finite case. Similarly, in Cherowitzo, Johnson, Vega [26] and Johnson [89], the interconnections with α-flocks of α-cones and twisted hyperbolic flocks and translation planes was by algebraic methods. In Johnson [90], the α-Klein quadric was used. The question then is whether the translation planes and flocks obtained from the algebraic approaches are isomorphic to those obtain using the α-Klein quadric in the finite and infinite cases. The answer is yes for flocks of quadratic cones and the associated flock spreads by section 5 of Gevaert and Johnson [56] together with additional notes by J.A. Thas (also, see Gevaert, Johnson and Thas [57]). In general, we leave the remaining parts of this as an open problem/question.

PROBLEM 29. *Show that the algebraic methods and the α-Klein procedures give rise to isomorphic translation planes and α-flocks both of α-cones and of twisted hyperbolic quadrics.*

11. Infinite flocks of hyperbolic quadrics

In this text, the finite flocks of hyperbolic quadrics have previously been discussed. In this case, the associated translation planes are either Desarguesian or nearfield planes. Also, there are only the linear flock when the order is even.

The question then is whether there exist infinite flocks of hyperbolic quadrics other than these classes of planes. In this chapter, we consider the existence question. One of the main techniques for the proof in the finite case due to Thas ([135], [136]), who shows that for odd order, for each conic of the flock, there is an involutory homology fixing the conic pointwise and which leaves the flock invariant. This means that the corresponding translation is a 'Bol' translation plane (1,1, 2) by the following theorem of Burn [18].

THEOREM 75. *A translation plane is a Bol plane if and only if there exist components L and M such that for each component N distinct from L and M, there is an involutory perspectivity with axis N that inverts L and M.*

To re-emphasize the finite case, there are no non-linear hyperbolic flocks of even order and the only hyperbolic flock spreads of odd order are Bol planes, which are also nearfield Andrè planes.

We begin with a discussion of the Andrè planes, which form the largest class of hyperbolic flock planes.

In the Multiple Replacement Theorem part 10, given any Pappian spread Σ coordinatized by a Galois quadratic extension $K(\theta)$ of a field K, so equipped with an automorphism σ of order 2, there is a set λ of mutually disjoint K-regulus nets together with two components $x = 0$ and $y = 0$ that define the spread.

DEFINITION 70. *Choose any subset δ of λ and multiply replace the regulus nets of δ. The resulting translation plane with spread in $PG(3, K)$ is called an 'Andrè plane'.*

There are a number of examples of interesting Andrè planes, here we shall concentrate with a brief introduction to this subject by describing the Andrè planes that correspond to hyperbolic flocks. The interested reader is directed to Johnson [83] for further reading. In particular, all such hyperbolic flock spreads are Bol spreads.

To ensure that characteristic 2 Andrè planes are included and considered, consider any field K that admits an irreducible quadratic $x^2 - \alpha x - \beta$. Let $\theta^2 = \theta\alpha + \beta$ and the automorphism σ be defined by $\theta^\sigma = -\theta + \alpha$. In order that a hyperbolic flock plane is obtained, it is assumed that the constructed translation plane admits the Pappian collineation group

$$\{(x, y) \rightarrow (x, yu) : \forall u \in K^*\} = H_{x=0}.$$

Since, we have just come from the study of twisted hyperbolic flocks, it might be noted that when $\alpha = 1$, the case we are considering, a hyperbolic flock plane also admits

$$\{(x, y) \rightarrow (xv, y) : \forall \in K^*\} = H_{y=0}.$$

The subscript on the group symbol indicates the axis of the group (the two components $x = 0, y = 0$ are the axis and coaxis). This is not true for twisted hyperbolic flock spreads.

The mutually disjoint regulus nets D_γ are defined as follows:

$$\left\{ y = xm : m^{\sigma+1} = \gamma, \text{ for fixed } \gamma \in K \right\}, \text{for } m \in K(\theta)^*.$$

Note that it might not be the case that all elements of K are of the form $m^{\sigma+1}$, as this depends upon the field. Accordingly, by the Multiple Replacement Theorem, the spread for the Andrè plane π_δ when replacing the set of regulus nets indexed by the subset $\delta \subset \lambda$ is as follows. A regulus net D_γ for $\gamma \in \delta$ then produces the spread:

$$
\begin{aligned}
x &= 0, y = 0, \\
y &= x^\sigma m; m^{\sigma+1} = \gamma, \gamma \in \delta \\
y &= xn; n^{\sigma+1} = \rho, \text{ for } \rho \in \lambda - \delta.
\end{aligned}
$$

$$\left\{ m^{\sigma+1} : \forall m \in K^* \right\} = K^- \subseteq K^*.$$

In order that $H_{x=0}$ is a collineation group of π_δ, it follows that if $D_\gamma \in \delta$ then $D_{\gamma u^2} \in \delta$ for all $u \in K^*$. Since $u^{\sigma+1} = u^2$, then the set of squares S_2 is a subgroup of K^-, where K^- is a subgroup of K^*. Consider then K^-/S_2. Any mapping g from K^- to $GF(2)$ then defines a translation plane corresponding to a hyperbolic flock.

- For characteristic 2, K would need to be an inseparable field to avoid $u \to u^2$ being bijective, but examples exist here, as well.
- In order that there is an identity then $1g = 0$.

Accordingly, we may define an Andrè multiplication:

$$x \circ m = x^{\sigma(m^{\sigma+1}g)} m, \text{ where } g \text{ is any mapping from } K^- \text{ to } GF(2).$$

THEOREM 76. K^-/S_2 *is an elementary Abelian 2-group and thus may be considered a vector space over* $GF(2)$. *The nearfield Andrè planes correspond to homomorphisms* f *of* K^- *onto* $GF(2)$.

(1) The set of Andrè translation planes corresponding to hyperbolic quadrics is isomorphic to K^-/S_2, *such that 1 maps to 0.*

(2) Accordingly, there is a $1-1$ *correspondence between the dual space of* K^-/S_2 *and the set of nearfield Andrè planes.*

(3) Hence, the number of Andrè hyperbolic flock planes is $2^{|K^-/S_2|-1}$ *and the number of nearfield Andrè hyperbolic flock planes is* 2^d, *where the dimension d of* K^-/S_2 *is* $log_2 |K^-/S_2|$. *So, if* $2^{|K^-/S_2|-1} > |K^-/S_2|$ *there are Andrè hyperbolic flocks that are not nearfield hyperbolic flocks.*

(4) If K is an inseparable field of characteristic 2 that admits a Galois quadratic extension, that satisfies the inequality of part (3), there are characteristic 2 Andrè hyperbolic flocks that are not nearfield hyperbolic flocks.

PROOF. Any subset $\delta \subset \lambda$ that produces an Andrè hyperbolic flock spread has the property that $\gamma \in \delta$ implies that $u^2\gamma \in \delta$. So, if $\lambda = K^-$, it follows that $\gamma S_2 = u^2 \gamma S_2$, which is equivalent to K^-/S_2 being an elementary Abelian 2-group. The set of functions g from K^- to $GF(2)$, clearly defines the possible Andrè planes that give rise to hyperbolic flock spreads. Now consider that $(x \circ m) \circ n = (x \circ (m \circ n))$, $\forall m, n \in Q$. This will be valid if and only if the mapping g that maps the group

$(K^-/S_2, \cdot)$ as an elementary Abelian 2-group to $GF(2)$ is a linear mapping to $GF(2)$. For example, if $m^{\sigma+1} \to 1$, $n^{\sigma+1} \to 1$, then $(mn)^{\sigma+1} \to 1 + 1 = 0$. Note also that if $m^{\sigma+1} \to 1$, then $(m^\sigma)^{\sigma+1} \to 1$. For g then $(x \circ m) \circ n = (x^\sigma m)^\sigma n = x(m^\sigma n)$, and note that $x \circ (m \circ n) = x \circ (m^\sigma n) = x(m^\sigma n)$. If g maps $m^{\sigma+1} \to 1$, $n^{\sigma+1} \to 0$ and then $(mn)^{\sigma+1} \to 1$, we have $(x \circ m) \circ n = x^\sigma mn$, and $(x \circ (m \circ n)) = x \circ mn = x^\sigma mn$. If g maps $m^{\sigma+1} \to 0$, g maps $n^{\sigma+1} \to 1$ an $(mn)^{\sigma+1} \to 1$, $(x \circ m) \circ n = xm \circ n = x^\sigma m^\sigma n$ and $(x \circ (m \circ n)) = x \circ m^\sigma n = x^\sigma m^\sigma n$. Similarly, for all such elements m, n then g as a linear mapping to $GF(2)$, implies that the linear functionals from $K^-/S_2 \to GF(2)$ correspond to the nearfield Andrè hyperbolic flock spreads. Actually, a completely direct abstract argument may be given using $\sigma^2 = 1$, and $x \circ y = x^{\sigma(y^{\sigma+1}g)}$ that

$$(x \circ m) \circ n = x \circ (m \circ n) \text{ is equivalent to:}$$
$$(m^{\sigma+1}g) + (n^{\sigma+1}g) = (m^{\sigma+1}n^{\sigma+1}g).$$

That is, the dual space of K^-/S_2 is equivalent to the set of nearfield Andrè spreads. The proofs of parts (3) and (4) are now immediate. □

THEOREM 77. *All Andrè hyperbolic flock planes are Bol planes.*

PROOF. It is necessary to show that

$$\forall a, b, c \in Q : ((c \circ a) \circ b) \circ a = c \circ ((a \circ b) \circ a).$$

Write $x \circ y = x^{\sigma(y)}y$, where $\sigma(y) = \sigma^{(y^{\sigma+1}g)}$. Note that $\sigma(w) = \sigma(w^{\sigma(z)})$, for all z since $\sigma(z) = 1$ or σ and $(w^\sigma)^{\sigma+1} = w^{\sigma+1}$, as $\sigma^2 = 1$. Working out what is required to satisfy the Bol requirement, we obtain the equivalent:

$$c^{\sigma(a)\sigma(b)\sigma(a)} = c^{\sigma(abc)}.$$

Now write out the degree requirement on both side of the putative equation modulo 2 to obtain:

$$b^{\sigma+1}g \equiv (a^2b)^{\sigma+1}g \bmod 2.$$

Letting $\beta = b^{\sigma+1}$ and $\alpha = a^{\sigma+1}$, we see the original requirement

$$\beta g \equiv \alpha^2 \beta g \bmod 2.$$

This proves that the Andrè planes of hyperbolic flock type are Bol planes. □

- For an example of a field of the type of (4), see (3,5,12).
- The originator of Bol translation planes is R.H. Burn, who also discovered a class of Bol hyperbolic flock planes, which are of the Andrè type under discussion (Burn [18]).
- The only non-Andrè non-Bol hyperbolic flock planes that are known are the examples of Riesinger [127].

The most general result concerning Bol hyperbolic flock spreads that will be presented in this book is the following.

COROLLARY 19. *Let K be any field that admits a Galois quadratic extension and corresponding automorphism σ. Let K^- contain two non-squares m_0 and n_0, whose product m_0n_0 is a non-square. Then there is an Andrè Bol hyperbolic flock spread that is not a nearfield spread.*

PROOF. Let g map $m_0 \to 1$ and $n_0 \to 1$ and then it is possible to map $m_0 n_0 \to 1$. Choose g to map K^-/S_2 to $GF(2)$ by any other method. Then the constructed spread is a Bol Andrè hyperbolic flock spread and cannot be a nearfield spread, since otherwise the mapping g would have to map $m_0 n_0 \to 1 + 1 = 0$.   □

Consider the above result, and ask how to not have such a situation?

If $K(\theta)$ is the Galois quadratic extension and $\theta^\sigma = -\theta + \alpha$, where the associated irreducible polynomial is $x^2 - \alpha x - \beta$, for $\alpha, \beta \in K$. Then would be linear functionals as follows, let:

$$K^- = \left\{ c^2 - \beta b^2 + \alpha bc : \forall c, b \in K \right\}.$$

For example, take $\mathbb{Q}$ as the field of rational numbers and let d be any non-square in $\mathbb{Q}$ such that $1 - d$ is also non-square and $d(1 - d)$ is non-square and form the Galois quadratic extension $\mathbb{Q}(\sqrt{d})$. Then K^-/S_2 contains the set $\{(1 - d)S_2, \ -dS_2\}$. For example, take $-d = 7$, $1 - d = 8$ and since -42 is a non-square, there is such an Bol Andrè plane that is not a nearfield plane.

When the characteristic is $\neq 2$, then we see then α may be chosen to be 0 without loss of the structure of squares and non-squares (2,3,1).

In general, if only nearfield hyperbolic flock spreads are possible, then K^-/S_2 has order 1 or 2. Accordingly, there are usually infinitely many non-nearfield hyperbolic flocks spreads possible.

- For additional research in this area, there are very interesting results in T. Grundhoefer and M. Gruinger [59] and G. Cherlin, T. Grundhoefer, A. Nesin, H. Volklein, [22], which are recommended to the reader.

Part 7

Lifting

Quasifibrations were introduced in the literature to try to understand more completely the role that surjectivity has when dealing with injective maximal partial spreads. Quasifibrations look like spreads or fibrations, but it is somewhat mysterious that they exist. And, since injectivity is equivalent to bijectivity in the finite case (in the finite affine plane case), quasifibrations are necessarily infinite maximal partial spreads. . All of the known examples have been studied in the dimension 2 case in translation plane theory. However, it is certainly possibly to generalize the definition.

Quasifibrations also turn up when trying to generalize various theories in finite geometry to the infinite case. This occurs, for example, in the theory of α-conics over infinite fields, or in connections to flocks of quadratic cones using ideas of hyperbolic fibrations in the finite case. Moreover, going to the theory of finite geometries extended to analogous geometries over skewfields, indicates that the idea of a quasifibration could be central to general and complete theories that have their origins in the finite case.

The idea of 'lifting' a spread of dimension 2 originated with Hiramine, Matsumoto, and Oyama in [62]) finite translation planes of dimension 2. The ideas were then extended to infinite order and then to quasifibrations. In Biliotti, Jha and Johnson [16]) it is shown that every spread in $PG(3, K)$, for K a field, lifts to a spread in $PG(3, K(\theta))$, for $K(\theta)$ a Galois quadratic field extension of K. Quasifibrations lift to quasifibrations. The reader is further referred to Johnson [82], Jha and Johnson [70]) for additional details on lifting quasifibrations.

Further, the construction of lifted spreads of dimension 2 may be repeated, indefinitely, assuming that there is an arbitrary tower of Galois quadratic extensions. So, we consider quasifibrations possibly constructed using Galois chains.

The main point of this part is to ask if when there is a Galois chain of quadratic field extensions, based on a given quasifibration, which is not a spread, is it ever possible that somewhere in the chain, the chain of quasifibrations could turn into spreads? If the transformation from quasifibration to spread occurs, say at stage k, then, at all stages beyond k, the remaining chain is a chain of spreads.

CHAPTER 9

Chains & surjectivity of degree $\frac{1}{k}$

What we have noted that it does not take a quasifibration to be a spread to turn it into a spread upon lifting. Instead of a surjectivity requirement, what is actually necessary turns out to be what we call 'surjective of degree $\frac{1}{k}$'. where k is a positive integer or ∞, where surjective of degree 1 is simply surjective. Once this is established, then it is possible to provide an exact requirement on the initial quasifibration to ensure that at the k-stage in the quadratic chain, a spread would appear exactly at stage k. The larger the k, the theory would hope to say, the weaker the requirement on the initial quasifibration. We show that this condition is necessary and sufficient.

The main conjecture is as follows:

CONJECTURE 1. *If a quasifibration is not a spread (a proper quasifibration) then for any Galois chain of Galois lifted quasifibrations, there is never a link in the chain that turns into a spread; Any Galois chain of quasifibrations consist of proper quasifibrations if and only the initial quasifibration is proper.*

This chapter provides a structure for which it is suitable to comment on the conjecture.

Our main result says that if the k-th lift in a chain of quasifibrations is a spread, and all $k + i$-lifts are spreads for $i \geq 0$, if and only if only the base or initial quasifibration satisfies a restricted surjective condition.

The main theorem will be stated after some preliminaries:

Let Q be a quasifibration

$$x = 0, y = x \begin{bmatrix} f(u,t) & g(u,t) \\ t & u \end{bmatrix} ; u, t \in K; f, g \ functions \ on \ K \times K,$$

where K is a field. Consider a Galois chain of Galois quadratic extensions:

$$K = K_0 \subset K_0(\theta_1) = K_1 \subset K_1(\theta_2) = K_2 \subset K_2(\theta_3) = K_3 \subset ... \subset K_{j-1}(\theta_j)....$$

for $j = 1, 2, ..., k$ where k could $\to \infty$. Let elements of K_j be denoted by z_j in general, and writing $t = t_0$, $t_1 = \theta_1 t_0 + u_0$, $t_2 = \theta_2 t_1 + u_1$, ...$t_j = \theta_j t_{j-1} + u_{j-1}$. The induced automorphism of order 2 induced in K_i is denoted by σ_i. We then construct a chain of lifted quasifibrations from Q which then have the following form: We shall see that the j-th lift for $j \geq 2$ has a spread of the following form:

$$x = 0, y = x \begin{bmatrix} u_j^{\sigma_j} & \sum_{i=2}^{j}(\theta_i^{\sigma_i} u_{i-1}^{\sigma_{i-1}}) + \theta_1^{\sigma_1} f(u,t) + g(u,t) \\ t_j & u_j \end{bmatrix} ;$$

$\forall t_j, u_j \in K_j, u_{i-1} \in K_{i-1}, i = 2, ..., j.$

The 1-st lift is

$$x \;=\; 0, y = \begin{bmatrix} u_1^{\sigma_1} & \theta_1^{\sigma_1} f(u_0, t_0) + g(u_0, t_0) \\ t_1 & u_1 \end{bmatrix};$$

$$\forall t_1 \;=\; \theta_1 t_0 + u_0, u_1 \in K_1 = K_0(\theta_1).$$

DEFINITION 71. *The initial quasifibration Q is said to be of restricted surjectivity of degree $\frac{1}{k}$ if and only if*

$$\prod_{\tau \in E_{2k}} z_k^\tau \;:\; t - \Big(\prod_{\tau \in E_{2k}} z_k^\tau \Big) f(u, t) \text{ and}$$

$$u - \Big(\prod_{\tau \in E_{2k}} z_k^\tau \Big) g(u, t)$$

$$\forall z_k \;\in\; K_k, \text{ the functions are}$$

both independent and jointly surjective

$$\forall t, u \;\in\; K_0,$$

where E_{2k} is the elementary Abelian 2-group

of automorphisms

$$\text{of } K_k \;=\; K_0(\theta_1, \theta_2, ..., \theta_k) \text{ and}$$

$$\prod_{\tau \in E_{2k}} z_k^\tau \text{ the norm of } K_k \text{ over } E_{2k} \text{ to } K_0.$$

The term 'jointly surjective' shall mean that for all $\Big(\prod_{\tau \in E_{2k}} z_k^\tau \Big) \in K_0$ and for all elements $r, s \in K_0$, there is a pair $(t, u) \in K_0 \times K_0$ such that

$$t - \Big(\prod_{\tau \in E_{2k}} z_k^\tau \Big) f(u, t) = r \text{ and } u - \Big(\prod_{\tau \in E_{2k}} z_k^\tau \Big) g(u, t) = s.$$

THEOREM 78. *Given a Galois chain of Galois quadratic field extensions and a corresponding chain of k lifted quasifibrations from Q as the base quasifibration, then the k-th lifted quasifibration is a spread if and only if the base quasifibration is of restricted surjective degree $\frac{1}{k+1}$. Furthermore, all of the $k + i$ lifts for $i \geq 0$ are also spreads.*

On one end of the spectrum are chains whose corresponding fields are the type that the norm of the next field in the chain is the preceding field: $K_z^{\sigma_{z+1}} = K_{z-1}$.

COROLLARY 20. *If there is an infinite Galois chain of Galois quadratic extensions of a proper quasifibration then assume that for every quadratic extension K_z with the associated involutory automorphism σ_z, we have $K_z^{\sigma_{z+1}} = K_{z-1}$. Then all lifts in the chain are proper quasifibrations.*

However, there are a variety of fields that do not have the condition in the preceding corollary. In particular, every quaternion division ring is constructed from a Galois quadratic field extension that does not have this assumption (See (2,3,2)).

We take the first part of this section from Johnson and Jha [93].

A quasifibration is a maximal partial spread in $PG(3, K)$, for some field K. It satisfies all of the requirements for a spread except for the covering property, not all vectors are necessarily covered by the components (See (1,1,1)).

Given a quasifibration of the form

$$x = 0, y = x \begin{bmatrix} f(u,t) & g(u,t) \\ t & u \end{bmatrix}; u, t \in K,$$

where K is a field and f and g are functions of $(t, u) \in K \oplus K$ into K defining the quasifibration. Assume that K admits a quadratic extension $K(\theta)$, where $\theta^2 = \theta\alpha + a$, for $\alpha, a \in K$; $x^2 \pm x\alpha + a$. Then there is an automorphism σ defined as follows: $\theta^\sigma = -\theta + \alpha$. The theory of lifting quasifibrations then constructs the following quasifibration:

$$x = 0, y = x \begin{bmatrix} w^\sigma & (\theta f(u,t) + g(u,t))^\sigma \\ \theta t + u & w \end{bmatrix}; \forall \theta t + u, w \in K(\theta).$$

Let

$$F(\theta t + u = \tilde{t}) = (\theta f(u,t) + g(u,t))^\sigma$$

so that the quasifibration has the following form:

$$x = 0, y = x \begin{bmatrix} w^\sigma & F(\tilde{t}) \\ \tilde{t} & w \end{bmatrix}; \forall \tilde{t}, w \in K(\theta).$$

Considering that surjectivity of degree 1 of the quasifibration does not hold, what does it mean that a quasifibration could be surjective of degree $\frac{1}{2}$. This simply means that the first lift in a putative chain of lifted quasifibrations is a spread, but the quasifibration itself is not a spread.

For a lift, we need a Galois quadratic extension of K, $K(\theta)$ and K is infinite. Now consider what it means for the quasifibration to actually be a spread, we would then require that every vector (x_1, x_2, x_3, x_4), lies on a unique component of the partial spread. This takes the following form

$$x = 0, y = x \begin{bmatrix} f(u,t) & g(u,t) \\ t & u \end{bmatrix}; \forall u, t \in K.$$

Given $(x_1, x_2) \neq (0,0)$, there is a unique pair (t, u) such that

$$x_1 f(u,t) + x_2 t = x_3 \text{ and}$$
$$x_1 g(u,t) + x_2 u = x_4.$$

If $x_1 = 0$ then $x_2 \neq 0$ so that $t = \frac{x_3}{x_2}$ and $u = \frac{x_4}{x_2}$. Take $x_1 = -v$, $x_2 = 1$, so we see that

$$(*)^S \quad : \quad t - vf(u,t), \text{ and } u - vg(u,t)$$
$$\text{are jointly surjective functions on } K,$$

$$\text{for each } v \in K^*,$$

in the sense that for all elements $v \neq 0$, r, s of K, there is a pair (t, u) such that $t - vf(u,t) = r$ and $u - vg(u,t) = s$. Now let

$$K(\theta)^{\sigma+1} = K.^-$$

If the lifted quasifibration is a spread then we shall show that what is required of the quasifibration is that

$$(*)^{\sigma+1} \quad : \quad t - v^{\sigma+1} f(u, t) \text{ and}$$
$$u - v^{\sigma+1} g(u, t)$$

$$\forall v \quad \in \quad K(\theta), \text{ are both independent and jointly surjective.}$$

If $K^- = K$ then $(*)^S = (*)^{\sigma+1}$, would always define functions that construct a spread in both the base quasifibration and the lifted quasifibration. So, we assume that this is not the case, that K^- is a proper subset of K and note also that K^{*-} is a multiplicative subgroup of K^*. Note again that we are using the term 'jointly' surjective to indicate the condition on the pairs (t, u). Since K^- is $\subset K$, we shall see that we have a condition that will guarantee that the lifted spread is a spread, but will not show that the quasifibration is a spread. This is our condition of restrictive surjective of degree $\frac{1}{2}$.

We now establish that with this condition, we do, in fact, obtain a lifted 'spread'.

$$x \;=\; 0, y = x \begin{bmatrix} w^\sigma & (\theta f(u, t) + g(u, t))^\sigma \\ \theta t + u & w \end{bmatrix}$$

$$=\; \begin{bmatrix} w^\sigma & (\theta^\sigma f(u, t) + g(u, t)) \\ \theta t + u & w \end{bmatrix} ; \text{for all } \theta t + u, w \in K(\theta).$$

So, it remains to show that we have the jointly surjective covering condition for the quasifibration.

Now given a vector (x_1, x_2, x_3, x_4) over $K(\theta)$, such that $(x_1, x_2) \neq 0$, there is a unique element $(\theta t + u, \theta w_1 + w_2)$, where $w = \theta w_1 + w_2$, for $w_i \in K$, $i = 1, 2$ such that

$$(*)^+ : x_1(\theta w_1 + w_2) + x_2(\theta t + u) = x_3$$
$$x_1((\theta f(u, t) + g(u, t))^\sigma + x_2 w = x_4.$$

From part 6, we note the following condition: If a quasifibration is of the form:

$$x \;=\; 0, y = x \begin{bmatrix} w^\sigma & F(t) \\ t & w \end{bmatrix} ;$$

for all $t, u \in L$, σ an automorphism of L, such that $\sigma^2 = 1$,

for a field L, then the quasifibration is a spread if and only if for all $u \in L$ the function

$$(\#) : \phi_u : t \to t - u^{\sigma+1} F(t)^\sigma$$

is surjective on L.

Accordingly, consider again the jointly surjective equations

$$t - v^{\sigma+1} f(u, t) \;=\; r_0,$$
$$u - v^{\sigma+1} g(u, t) \;=\; l_0.$$

and note that we then obtain the surjective equations:

$$(**) : (\theta t + u) - v^{\sigma+1}(\theta f(u, t) + g(u, t)) = \theta r_0 + l_0.$$

In (#), change notation letting (#) : $\phi_u : t \to t - u^{\sigma+1} F(t)^{\sigma}$ be rewritten with $s = t$, $u = v$ and to first obtain

$$\phi_s : s - v^{\sigma+1} F(s)^{\sigma}.$$

Then letting $s = \theta t + u$ and $F(t) = (\theta f(u,t) + g(u,t))^{\sigma} = \theta^{\sigma} f(u,t) + g(u,t)$, and noting that $\sigma^2 = 1$, we have that (∗∗) is then

$$(\#)^1 : \phi_v : \theta t + u \to (\theta t + u) - v^{\sigma+1}(\theta f(u,t) + g(u,t))$$

is a surjective function for all elements v of $K(\theta)$. This proves that the lifted quasi-fibration is a spread. Conversely, if the lifted quasifibration is a spread, we now point out in this next section that this shows that the quasifibration is of restricted surjectivity of degree $\frac{1}{2}$. That is, we claim that $(\#)^1$ is surjective for all $v \in K(\theta)$, will show immediately that the two component functions for θ and 1 are surjective and split into the equations $(∗)^{\sigma+1}$.

In the next section, we shall show that this idea generalizes and provides a way to determine when the k-th lift in a Galois chain of quadratic extensions of K becomes a spread when beginning with an appropriate restricted surjectivity of degree $\frac{1}{k+1}$.

1. Restricted surjectivity

DEFINITION 72. *Let Q be a quasifibration*

$$x = 0, y = x \begin{bmatrix} f(u,t) & g(u,t) \\ t & u \end{bmatrix} ; u, t \in K,$$

where K is a field. Consider a Galois chain of Galois quadratic extensions:

$$K = K_0 \subset K_0(\theta_1) = K_1 \subset K_1(\theta_2) = K_2 \subset K_2(\theta_3) = K_3 \subset ... \subset K_{j-1}(\theta_j)....$$

for $j = 1, 2, ..., k$ where k could $\to \infty$. Let elements of K_j be denoted by z_j in general, and write $t = t_0$, $t_1 = \theta_1 t_0 + u_0$, $t_2 = \theta_2 t_1 + u_1$, ...$t_j = \theta_j t_{j-1} + u_{j-1}$. We then construct a chain of lifted quasifibrations from Q which then have the following form: The j-th lift for $j \geq 2$ is

$$x = 0, y = x \begin{bmatrix} u_j^{\sigma j} & \sum_{i=2}^{j}(\theta_i^{\sigma i} u_{i-1}^{\sigma_{i-1}}) + \theta_1^{\sigma_1} f(u,t) + g(u,t) \\ t_j & u_j \end{bmatrix};$$

$$\forall t_j, u_j \in K_j, u_{i-1} \in K_{i-1}, i = 2, ..., j.$$

The 1-st lift is

$$x = 0, y = \begin{bmatrix} u_1^{\sigma_1} & \theta_1^{\sigma_1} f(u_0, t_0) + g(u_0, t_0) \\ t_1 & u_1 \end{bmatrix};$$

$$\forall t_1 = \theta_1 t_0 + u_0, u_1 \in K_1 = K_0(\theta_1).$$

The initial quasifibration Q is said to be of restricted joint surjectivity of degree $\frac{1}{k+1}$ if and only if

$$\prod_{\tau \in E_{2^k}} z_k^\tau \quad : \quad t - \Big(\prod_{\tau \in E_{2^k}} z_k^\tau \Big) f(u,t) \text{ and}$$

$$u - \Big(\prod_{\tau \in E_{2^k}} z_k^\tau \Big) g(u,t)$$

$\forall z_k \in K_k$, are both independent and jointly surjective functions, where

E_{2^k} is the elementary Abelian 2-group of automorphisms

of $K_k = K_0(\theta_1, \theta_2, ..., \theta_k)$.

To understand the terms used, we note that

$$\prod_{\tau \in E_{2^k}} z_k^\tau = z_k^{(\sigma_k+1)(\sigma_{k-1}+1)....(\sigma_2+1)(\sigma_1+1)} = z_k^{\prod_{j=k}^{1}(\sigma_j+1)}$$

(the expression is written backwards), is just the norm of K_k relative to E_{2^k} over K_{k-1}. To see this, just note that we have k choices in the terms of the degrees (σ_i+1) and 2 possibilities σ_i or 1, in each, creating the elementary Abelian 2-group of order 2^k.

Our main theorem is as follows:

THEOREM 79. *Given a Galois chain of Galois field extensions and a corresponding chain of k lifted quasifibrations from Q as the base quasifibration. Then the k-th lifted quasifibration is a spread if and only if the base quasifibration is of restricted surjective degree $\frac{1}{k+1}$. Furthermore, all of the $k + i$ lifts for $i \geq 0$ are also spreads.*

- The notation suggests, but does not claim, that as $k \to \infty$, the requirement for the existence of a spread within the Galois chain would basically guarantee that there must be a spread somewhere in the chain. The problem of determining the restricted surjective degree is somehow tied up with non-commutative algebra. For example, suppose that $K_r^{*(\sigma_r+1)}$ is a proper subgroup of K_{r-1}^*, for some r, then there is a quaternion division ring within the chain. So, it would appear that the restricted surjectivity degree would then decrease, and for $j \geq r$, we must have $K_j^{*(\sigma_j+1)}$ a proper subgroup of K_{j-1}^*, and, by results of part II, the associated quaternion division ring then defines a chain of extension quaternion division rings.

The proof of restricted degree hinges on the following lemma:

LEMMA 14. *Consider the following set of functions for each element $u_k \in K_k^*$: Let $1 \leq j \leq k$.*

$$\phi_{u_k}^j \quad : \quad t_j \to t_j - u_k^{(\sigma_k+1)(\sigma_{k-1}+1)...(\sigma_j+1)}(\theta_j u_{j-1}^{\sigma_{i-1}}$$

$$+ \sum_{i=2}^{j-1} \theta_i^{\sigma_i} u_{i-1}^{\sigma_{i-1}} + \theta_1^{\sigma_1} f(u_0, t_0) + g(u_0, t_0)).$$

Then $\phi_{u_k}^j$ is surjective on K_j, for all $u_k \in K_k$, if and only if $\phi_{u_k}^{j+1}$ is surjective on K_{j+1}, for all $u_k \in K_k$. Moreover, when $j = 1$, $\phi_{u_k}^1$ is surjective on K_1, for all

$u_k \in K_k$. It is noted that the exponents on u_k converge to $\sigma_k + 1$ as $j \to k$, which will provide the requirement for a spread at the k-th link.

PROOF. When $j = 1$, we obtain

$$\phi_{u_k}^1 : t_1 \to t_1 - u_k^{(\sigma_k+1)(\sigma_{k-1}+1)\dots(\sigma_1+1)}(\theta_1^{\sigma_1} f(u_0, t_0) + g(u_0, t_0)).$$

Hence, this set of functions are all surjective on K_1, as we have seen in the previous section that this is equivalent to the original quasifibration assumed to be of restricted surjectivity of degree $\frac{1}{k}$. This is, in essence, our mathematical induction step 1. Now assume that

$$\phi_{u_k}^j \quad : \quad t_j \to t_j - u_k^{(\sigma_k+1)(\sigma_{k-1}+1)\dots(\sigma_j+1)}(\theta_j u_{j-1}^{\sigma_{i-1}}$$
$$+ \sum_{i=2}^{j-1} \theta_i^{\sigma_i} u_{i-1}^{\sigma_{i-1}} + \theta_1^{\sigma_1} f(u_0, t_0) + g(u_0, t_0)),$$

are surjective functions for each $u_k \in K_k$. Consider the following system of equations:

$$(*)_j(1) \quad : \quad t_j - u_k^{(\sigma_k+1)\dots(\sigma_{j+1}+1)} u_j^{\sigma_j},$$

$$(*)_j(2) \quad : \quad u_j - u_k^{(\sigma_k+1)\dots(\sigma_{j+1}+1)}(\theta_j u_{j-1}^{\sigma_{i-1}}$$
$$+ \sum_{i=2}^{j-1} \theta_i^{\sigma_i} u_{i-1}^{\sigma_{i-1}} + \theta_1^{\sigma_1} f(u_0, t_0) + g(u_0, t_0)).$$

It is claimed that both of these functions are independent and jointly surjective to K_j. It is pointed out that the joint surjectivity and independence of $(*)_j(1)$ and $(*)_j(2)$ then provides the surjectivity of $\phi_{u_k}^{j+1}$ acting on $\theta_{j+1} t_j + u_j$, since the mapping would then be $\theta_{j+1}(*)_j(1) + (*)_j(2)$. Indeed this would clearly establish the proof of the lemma that the function $\phi_{u_k}^j$ is surjective on K_j if and only if $\phi_{u_k}^{j+1}$ is surjective on K_{j+1}. Consider $(*)_j(1) = r_j$ and $(*)_j(2) = l_j$, for any choices of elements $r_j, l_j \in K_j$. Since $u_s^{\sigma_s+1}$ is in K_{s-1} for $u_s \in K_s$ then $u_k^{(\sigma_k+1)\dots(\sigma_{j+1}+1)} \in K_j$. Then the equations will test the surjectivity within K_j. Then form $((*)_j(1))^{\sigma_j} + u_k^{(\sigma_k+1)\dots(\sigma_{j+1}+1)\sigma_j}((*)_j(2))$, which means perform the indicated algebraic transformations to the elements of each of the corresponding rows. This procedure adds out the term u_j, the resultant equation is

$$t_j^{\sigma_j} - u_k^{(\sigma_k+1)(\sigma_{k-1}+1)\dots(\sigma_j+1)}(\theta_j^{\sigma_j} u_{j-1}^{\sigma_{i-1}}$$
$$+ \sum_{i=2}^{j-1} \theta_i^{\sigma_i} u_{i-1}^{\sigma_{i-1}} + \theta_1^{\sigma_1} f(u_0, t_0) + g(u_0, t_0))$$
$$= r_j^{\sigma_j} + u_k^{(\sigma_k+1)\dots(\sigma_{j+1}+1)\sigma_j} l_j.$$

Note that the actual required expression involving u_k is

$$u_k^{(\sigma_k+1)\dots(\sigma_{j+1}+1)\sigma_j} u_k^{(\sigma_k+1)\dots(\sigma_{j+1}+1)},$$

which is

$$u_k^{(\sigma_k+1)\dots(\sigma_{j+1}+1)(\sigma_j+1)}.$$

Now take the σ_j image of both sides, and observe that $\sigma_j^2 = 1$ and $(\sigma_j + 1)\sigma_j = (\sigma_j + 1)$. Note that this also means that the term $\theta_j^{\sigma_j}$ becomes θ_j and all other terms in the left side of the equation are fixed pointwise by the automorphism σ_j. Thus, we obtain

$$t_j - u_k^{(\sigma_k+1)(\sigma_{k-1}+1)\cdots(\sigma_j+1)}(\theta_j u_{j-1}^{\sigma_{i-1}}$$
$$+ \sum_{i=2}^{j-1} \theta_i^{\sigma_i} u_{i-1}^{\sigma_{i-1}} + \theta_1^{\sigma_1} f(u_0, t_0) + g(u_0, t_0))$$
$$= r_j + u_k^{(\sigma_k+1)\cdots(\sigma_{j+1}+1)} l_j.$$

Since the right side is an element of K_j and $\phi_{u_k}^j$ is surjective on K_j, there is a solution for t_j, noting that all expressions involve elements of fields K_i for $i \leq j$. Using this solution, there is clearly a solution for u_j, as the pair (t_j, u_j) is a solution connected to the equations. The solution is unique due to the fact that we always obtain quasifibrations in the chain. Hence, the claim that the functions $(*)_j(1)$ and $(*)_j(2)$ are independent and jointly surjective on K_j is proved. It now follows that this surjectivity is equivalent to the surjectivity involving $\theta_{j+1} t_j + u_j = t_{j+1}$. That is, we have now proved the lemma.   □

1.1. The proof of the theorem.

PROOF. We now complete the proof of the theorem by induction as follows. First assume that the base quasifibration has restricted surjectivity of degree $\frac{1}{k+1}$. Then this is equivalent to $\phi_{u_k}^1$ being surjective on K_1. This is part 1 of the induction argument. Assume that $\phi_{u_k}^j$ is surjective on K_j. This is equivalent to $\phi_{u_k}^{j+1}$ surjective on K_j. Finally, this implies that $\phi_{u_k}^k$ is surjective on K_k for all elements $u_k \in K_k$, because the exponents on u_k have converged to $\sigma_k + 1$. But, this is simply the requirement for complete surjectivity of a lifted spread, as we have seen in the previous section. Since all of these statements on surjectivity are equivalent by the previous lemma, we have the proof of the theorem, except for the $k \mid i$ -lifts being spreads. This last statement follows from a result in Biliotti, Jha and Johnson [16], in the sections on lifting. However, we may give a different proof using the ideas of this section. To continue the line of argument at the k-th lift, we have that

$$\phi_{u_k}^k \quad : \quad t_j \to t_j - u_k^{\sigma_k+1}(\theta_k u_{k-1}^{\sigma_{k-1}}$$
$$+ \sum_{i=2}^{k-1} \theta_i^{\sigma_i} u_{i-1}^{\sigma_{i-1}} + \theta_1^{\sigma_1} f(u_0, t_0) + g(u_0, t_0)),$$

is surjective for all $u_k \in K_k$. Assuming that the chain continues past the k-th lift, we have that

$$t_k - u_k u_k^{\sigma_k} \text{ and}$$

$$u_k - (\theta_k u_{k-1}^{\sigma_{k-1}} + \sum_{i=2}^{k-1} \theta_i^{\sigma_i} u_{i-1}^{\sigma_{i-1}} + \theta_1^{\sigma_1} f(u_0, t_0) + g(u_0, t_0)),$$

are both jointly surjective functions for all $u_k \in K_k$. The same analysis will now give

$$\phi_{u_k}^{k+1} : t_{k+1} \to t_{k+1} - u_k(\theta_{k+1} u_k^{\sigma_k} + \sum_{i=2}^{k} \theta_i^{\sigma_i} u_{i-1}^{\sigma_{i-1}} + \theta_1^{\sigma_1} f(u_0, t_0) + g(u_0, t_0)),$$

is surjective on K_{k+1}, for all $u_k \in K_k$. Since $u_{k+1}^{\sigma_{k+1}+1} \in K_k$, we have the condition for the $(k+1)$-st lift to be a spread. Hence, we may replace the leading u_k by $u_{k+1}^{\sigma_{k+1}+1}$ and repeat the argument to obtain the $(k+2)$-nd lift to be a spread. This concludes the proof of the main theorem.   □

COROLLARY 21. *If there is an infinite Galois chain of Galois quadratic extensions of a proper quasifibration then assume that for every quadratic extension K_z with the associated automorphism σ_z, we have $K_z^{\sigma_z+1} = K_{z-1}$. Then all lifts in the chain are proper quasifibrations.*

PROOF. If there is a lift at stage k which is a spread then $K_k^{\sigma_k+1} = K_{k-1}$. Notice that

$$K_k^{(\sigma_k+1)(\sigma_{k-1}+1)\ldots(\sigma_1+1)} = K_{k-1}^{(\sigma_{k-1}+1)(\sigma_{k-2}+1)\ldots(\sigma_1+1)} = K_1^{\sigma_1+1} = K_0.$$

This means that the base quasifibration satisfies

$$t_0 - v_0 f(u_0, t_0),$$
$$u_0 - v_0 g(u_0, t_0)$$

both representing functions that are independent and jointly surjective for all $v_0 \in K_0$. This means that the quasifibration satisfies the surjective property and hence is a spread, contrary to our assumptions. This completes the proof of the corollary.   □

Recalling that if $K_k^{\sigma_k+1} = K_{k-1}$, then it follows that there is no associated quaternion division ring with K_k, therefore, the lifts are all proper quasifibrations if there is never an associated division ring with any field of the chain. Note that once a field does have $K_j^{\sigma_j+1} \subset K_{j-1}$ then $K_{j+i}^{\sigma_{j+i}} \subset K_{j+i-1}$ for all $i \geq 0$. Therefore, we have proved the following.

COROLLARY 22. *If there are no quaternion division rings associated with any of the fields in the Galois chain of Galois quadratic extensions and the base quasifibration is proper (not a spread) then all of the quasifibrations in the chain are proper quasifibrations.*

REMARK 21. *Finally, in $\prod_{\tau \in E_{2k}} z_k^\tau$, since z_k is in K_k, we have that z_k may be written as an element as a 2^k-tuple over K, relative to the automorphism group of K_k over K. Hence, it follows that $\prod_{\tau \in E_{2k}} z_k^\tau$ is a multinomial of degree 2^k with variables and coefficients of K. This fact may enhance the understanding of whether a given quasifibration may satisfy the conditions to allow a k th. lifted quasifibration to turn into a spread.*

2. Hughes-Kleinfield look-alikes

Consider a quasifibration or spread of the following form

$$(*) \quad : \quad x = 0, y = x \begin{bmatrix} u & ft^\rho \\ t & u^\sigma \end{bmatrix} ; t, u \in K, \text{ a field},$$

where f an appropriate constant in K,

where $\sigma\rho = \rho\sigma$, σ, ρ automorphisms of K.

There is no restriction on the field K, and it could be infinite or finite. Examples are the Hughes-Kleinfeld planes and the infinite analogues and quaternion division ring planes. There are many other examples that can be obtained. We have mentioned T-copies previously, which are obtained from finite spreads by attempting to copy the spread over a rational function field $K(z)$. If it becomes a quasifibration, over the rational function field, we say that it is a T-copy. If K is an infinite field, attempting to copy a spread over the rational function field will also work if we allow that $\tau(z) = z$, for all automorphisms τ that are represented in the spreads. We call this a 'twisted T-copy'. When this process works, it produces a quasifibration which is usually a proper quasifibration but when $\sigma\rho = 1$, in the above quasifibration, a spread can be constructed, as the quaternion division rings as spreads are examples when $\sigma^2 = 1$. In part 7 on lifting, the question is whether a quasifibration Q is infinitely non-surjective, which is another way of saying that no quasifibration in any quadratic Galois chain based on Q can turn into a spread. Here we concentrate on what we call Hughes-Kleinfeld look-alikes.

DEFINITION 73. *A Hughes-Kleinfield look-alike is any quasifibration or spread that looks like/is a quasifibration or spread of type* (∗). *In the material on* α-*cones, we may use the functions describing a Hughes-Kleinfeld , with* $\alpha = \sigma$ *to see that the functions*

$$\phi_g(t) = t - g^{\sigma+1} f^\sigma t^{\sigma\rho}; g \in K$$

are all injective over K *if and only if the Hughes-Kleinfeld look-alike is a quasifibration. And, the quasifibration is a spread if and only all of the functions* ϕ_g *are surjective on* K.

If we have a quasifibration, then $(1 - g^{\sigma+1}f^\sigma) \neq 0$, *for all* $g \in K$. *Therefore, if* $\sigma\rho = 1$ *and we have a quasifibration then* $t(1 - g^{\sigma+1}f^\sigma) = w$, *for* $w \in K$ *has the obvious solution* $t = \frac{w}{1-g^{\sigma+1}f^\sigma}$ *so in this case, we obtain a spread.*

We will note the '1-test' is equivalent to $t - f^\sigma t^{\sigma\rho}$ *is surjective.*

Now consider the T-copy of the Hughes-Kleinfeld spread or the twisted T-copy in $K(z)$, the rational function field over K. First assume that K is finite. Then σ and ρ are endomorphisms but not automorphisms of $K(z)$ whenever they are non-trivial. Assume $\phi_u(t) = 0$ if and only if $t = 0$, for all $u \in K$. Then it is immediate that each ϕ_u is injective, since $\phi_u(t) = \phi_u(s)$ if and only if $(t - s) - u^{\sigma+1}f^\sigma(t - s)^{\sigma\rho} = 0$ so $t - s = 0$ by the assumption. Hence by the above theorem, we have a quasifibration provided $\phi_u(t) = 0$ if and only if $t = 0$. Furthermore, if σ or ρ is non-trivial, then ϕ_u is not subjective for any non-zero u, unless $\sigma\rho = 1$, as we show below.

So, for a T-copy of a Hughes-Kleinfeld look-alike (necessarily finite), we always obtain a proper quasifibration.

And, for a twisted T-copy of a Hughes-Kleinfeld look-alike whether finite or infinite, we always obtain a quasifibration which is always a spread if and only if $\sigma\rho = 1$. While we shall use a variation on the idea of showing that $\phi_u(t) = 0$ if and only if $t = 0$, the principle still applies.

THEOREM 80. *(**Hughes-Kleinfeld look-alikes theorem**) We consider the rational function field* $K(z)$ *over* K. *Assume that not both* σ *and* ρ *are trivial.*

(1) *If a quasifibration looks like/is a Hughes-Kleinfield quasifibration then the T-copy (if K is finite) or the twisted T-copy in the finite or infinite case is a quasifibration over $K(z)$.*

(2) *If $\sigma\rho \neq 1$ then the T-copy is a proper quasifibration and the twisted T-copy is a quasifibration or spread. By repeated application, the twisted T-copy over $K(z_1, z_2, ..., z_{k+1})$, for z_i for $i = 1, ..., k+1$ mutually commuting independent variables are quasifibrations or spreads.*

    (a) *If $\sigma\rho \neq 1$, then a twisted T-copy from a finite Hughes-Kleinfeld quasifibration is a proper quasifibration.*

    (b) *If $\sigma\rho \neq 1$, then a twisted T-copy is never a skewfield plane.*

(3) *If $\sigma\rho = 1$, and $\sigma \neq 1$, then the T-copy is a proper quasifibration and any twisted T-copy is a spread.*

    (a) *If $\sigma \neq 1$ and the twisted T-copy spread is a skewfield spread then it is a quaternion division ring spread. Then $\sigma^2 = 1$, $K = F(\theta)$, a Galois quadratic extension of a field F and σ is the unique element of the Galois group of $F(\theta)$. Furthermore, $f \in F - F(\theta)^{\sigma+1}$.*

    When the Desarguesian plane is coordinatized by a quaternion division ring center F, $(a, b)_F^\alpha$, then the twisted T-copy is a Desarguesian plane that is coordinated by a quaternion division ring $(a, b)_{F(z)}^\alpha$. Furthermore, there are quaternion division ring extensions $(a, b)_{F(z_1, z_2, ..., z_{k+1})}^\alpha$ for any set of mutually independent commuting variables z_i, for $i = 1, 2, ..., k+1$.

    (b) *If $\sigma\rho = 1$, $\sigma \neq 1$ and $F = GF(2)$ then the original plane produces a deficiency-one partial hyperbolic flock in $PG(3, 4)$.*

    The twisted T-copy produces a translation plane that contains a deficiency-one partial hyperbolic flock in $PG(3, GF(4)(z))$. However, the spread over $GF(4)(z)$ is not a deficiency-one partial hyperbolic flock as the required Baer group does not extend and act as a collineation group in $PG(3, GF(4)(z))$.

    (c) *If $\sigma\rho = 1$ and $\sigma \neq 1$, then there are infinitely many spreads obtained by iteration of the twisted T-copy procedure, where the spreads are in $PG(3, K(z_1, z_2, ..., z_{k+1}))$.*

PROOF. From our preliminaries, it remains to consider T-copies of spreads and to show that $\phi_u(t) = 0$ if and only if $t = 0$ for all $u^{\sigma+1} \neq 0$. We again note that if we obtain a quasifibration, it must be proper, if one of σ and ρ is not the identity. To see this for

$$(*) \quad : \quad x = 0, y = x \begin{bmatrix} u & ft^\rho \\ t & u^\sigma \end{bmatrix}; t, u \in K(z),$$

for K a finite field, f for an appropriate constant in K

to be a spread consider $(0, 1, 0, s)$ for $s \in K(z)$ and $s \neq 0$. Then $t = 0$ and $u^\sigma = s$, for the spread to cover this vector. Therefore, $u^\sigma = s$, means that σ is surjective. Therefore, if $s \in K[z]^*$, there is an element $\frac{u}{w}$, for $u, w \in K[z]$ with $wv \neq 0$ so that $u^\sigma = sw^\sigma$. If σ indicates a power in $K(z)$, it follows that s must be a σ power. If $\sigma \neq 1$, this is a contradiction. A similar argument for ρ, if $\rho \neq 1$, shows that if a quasifibration is obtained, it is a proper quasifibration.

That is, we need to show that the determinant $u^{\sigma+1} - ft^{\rho+1} \neq 0$, for all $u, t \in K(z)$. Assume that there exist u and t in $K(z)$ such that $u^{\sigma+1} - ft^{\rho+1} = 0$. By assumption, σ and ρ are not both automorphisms of $K(z)$, but they are endomorphisms and act as finite powers in $K(z)$. If $u^{\sigma+1} = ft^{\rho+1}$, then let $u = \frac{\Sigma_{i=0}^{n} u_i z^i}{\Sigma_{i=0}^{s} v_i z^i}$ such that $u_n v_s \neq 0$ with $u_i, v_j \in K$, for all relevant i and j. Similarly, let $t = \frac{\Sigma_{i=0}^{r} w_i z^i}{\Sigma_{i=0}^{k} y_i z^i}$ where $w_r y_k \neq 0$, and where all coefficients of the polynomials over z are in K. Hence, we have a polynomial identity

$$(\Sigma_{i=0}^{n} u_i z^i)^{\sigma+1}(\Sigma_{i=0}^{k} y_i z^i)^{\rho+1} = f(\Sigma_{i=0}^{s} v_i z^i)^{\sigma+1}(\Sigma_{i=0}^{r} w_i z^i)^{\rho+1}.$$

When multiplied out, we have a polynomial in z equal to 0 and hence all coefficients are zero. We consider the maximum term on the left side

$$u_n^{\sigma+1} y_k^{\rho+1} z^{n(\sigma+1)+k(\rho+1)}$$

and the maximum term on the right side

$$f(v_s^{\sigma+1} w_r^{\rho+1} z^{s(\sigma+1)+r(\rho+1)}).$$

Since the reduced polynomial must be the zero polynomial, in order for terms to cancel out, the maximum left term cannot cancel out any left term and the maximum right term cannot cancel out any of the right terms. This shows that these two elements of the reduced polynomial must be equal (that is, cancel out in the reduced zero polynomial). So, we obtain

$$u_n^{\sigma+1} y_k^{\rho+1} = f(v_s^{\sigma+1} w_r^{\rho+1}).$$

Therefore, we obtain

$$(\frac{u_n}{v_s})^{\sigma+1} = f(\frac{w_r}{y_k})^{\rho+1},$$

as $u_n y_k v_s w_r \neq 0$. However, this is a contradiction to the assumption that we have a finite spread initially. This proves that T-copies are always proper quasifibrations provided σ or ρ is non-trivial.

Now assume that we are considering twisted T-copies of either finite or infinite Hughes-Kleinfeld look-alikes. So, recall that σ and ρ now are automorphisms of $K(z)$ that fix z. Let $u = \frac{\Sigma_{i=0}^{n} u_i z^i}{\Sigma_{i=0}^{s} v_i z^i}$ such that $u_n v_s \neq 0$, to obtain the following polynomial equation:

$$(\Sigma_{i=0}^{n} u_i^{\sigma} z^i)(\Sigma_{i=0}^{n} u_i z^i)(\Sigma_{i=0}^{k} y_i^{\rho} z^i)\Sigma_{i=0}^{k} y_i z^i)$$
$$= f(\Sigma_{i=0}^{s} v_i^{\sigma} z^i)(\Sigma_{i=0}^{s} v_i z^i)(\Sigma_{i=0}^{r} w_i^{\rho} z^i)(\Sigma_{i=0}^{r} w_i z^i).$$

The maximum elements that are left are necessarily equal to the maximum elements on the right, so

$$u_n^{\sigma+1} z_k^{\rho+1} z^{2(n+k)} = f v_s^{\sigma+1} w_r^{\rho+1} z^{2(s+r)}.$$

Hence, we have the same contradiction. Now since σ and ρ are both automorphisms of $K(z)$, the question now arises whether we have a proper quasifibration or a spread.

$$t - g^{\sigma+1} f^{\sigma} t^{\sigma\rho} \text{ is surjective for all } g \neq 0 \text{ in } K(z).$$

If $\sigma\rho = 1$ then we obtain the question of whether

$$t(1 - g^{\sigma+1} f^{\sigma}) \text{ is surjective for all } g \neq 0 \text{ in } K(z).$$

If $1 - g^{\sigma+1}f^{\sigma} = 0$ or equivalently $f = (g^{\sigma+1})^{\sigma^{-1}} = (g^{\sigma^{-1}})^{\sigma+1}$, which is a contradiction as $u^{\sigma+1} - ft^{\rho+1} \neq 0$ for all $(t, u) \neq (0, 0)$, and choosing $t = 1$. Therefore, we have shown that if $\sigma\rho = 1$, then the twisted T-copy is a spread.

Now assume that $\sigma\rho \neq 1$. Consider the question of whether

$$t - g^{\sigma+1}f^{\sigma}t^{\sigma\rho} \text{ is surjective in } K(z) = w.$$

We will only complete this when the twisted T-copy is from a finite Hughes-Kleinfeld spread. The idea of the proof is to show that given any g and w in $K(z)$, there is a solution for t in $K(z)$ using the coefficients of the polynomials associated with t. Let $w = \frac{\Sigma_{i=0}^{r} w_i z^i}{\Sigma_{i=0}^{q} x_i z^i} = \frac{W(z)}{X(z)}$, let $t = \frac{\Sigma_{i=0}^{n} u_i z^i}{\Sigma_{i=0}^{s} v_i z^i} = \frac{U(z)}{V(z)}$ where $u_n v_s \neq 0$. Then the question is whether there is a solution to the following.

$$\frac{U(z)}{V(z)} - f^{\sigma}\left(\frac{U(z)}{V(z)}\right)^{\sigma\rho} = z$$

and therefore we have the following polynomial equation: This is the 1-test. We will show that the 1-test fails for all T-copies, and for all twisted T-copies that are constructed from finite Hughes-Kleinfeld spreads.

We obtain the following requirement:

$(**)$

$\qquad : \quad (\Sigma_{i=0}^{n}u_i z^i)(\Sigma_{i=0}^{s}v_i z^i)^{\sigma\rho-1} - f^{\sigma}((\Sigma_{i=0}^{n}u_i z^i)^{\sigma\rho} = z(\Sigma_{i=0}^{s}v_i z^i))^{\sigma\rho},$

for $u_n v_s \neq 0$.

Note that $\sigma\rho - 1 \neq 0$ makes sense as a power, if the automorphisms are of the form p^a, for p a prime.

The maximal degree set is $\{n + s(\sigma\rho - 1), n\sigma\rho, 1 + s\sigma\rho\}$. If there is exactly one absolute maximum, then the absolute maximum term cannot cancel out, a contradiction as $u_n v_n \neq 0$. So, two of the degrees must be equal, since $(n\sigma\rho, 1 + s\sigma\rho) = 1$. Assume that $n + s(\sigma\rho - 1) = 1 + s\sigma\rho$, then $n = s + 1$, but then $n\sigma\rho = (s+1)\sigma\rho > 1 + s\sigma\rho$, since $\sigma\rho > 1$, and we have an absolute maximum. Therefore, we must have $n + s(\sigma\rho - 1) = n\sigma\rho$, which implies that $n = s$, as $\sigma\rho - 1 > 0$. Hence, we have the maximal degree set as $\{s(\sigma\rho), s(\sigma\rho), 1 + s\sigma\rho\}$, a contradiction. This completes the proof. $\qquad\qquad\square$

COROLLARY 23. *If $\alpha\rho \neq 1$, then T-copies or twisted T-copies from Hughes-Kleinfeld finite spreads are proper quasifibrations that fail the 1-test.*

All T-copies of Hughes-Kleinfeld finite spreads when $\sigma\rho = 1$ pass the 1-test and any twisted T-copy when $\sigma\rho = 1$ is a spread.

DEFINITION 74. *If Q is a proper quasifibration of dimension 2 for which all Galois chains built from Q are comprised of proper quasifibrations, Q is said to be 'infinitely non-surjective'.*

REMARK 22. *The known examples of proper quasifibrations of dimension 2 are as follows:*

1. *The inversions of bilinear flocks.*
2. *T-copies or twisted T-copies of quasifibrations of finite Hughes-Kleinfeld look-alikes (when $\sigma\rho \neq 1$). All of these proper quasifibrations fail the 1-test and therefore are infinitely non-surjective.*

When $\sigma\rho = 1$, every T-copy of a Hughes-Kleinfeld look-alike is a proper quasifibration that passes the 1-test.
3. The Cherowitzo, Johnson, Vega quasifibration twisted by an automorphism α.
4. The Ganley additive quasifibrations.

REMARK 23. In addition, there is the following quasifibration. Let c be a non-square in F, F of characteristic 2 and $A(v) = (\delta/\delta\sqrt{c})\, v$, then $A(A(v)) = 0$. Furthermore, the following is an additive quasifibration or a spread.

$$x = 0, \ y = \left[\begin{array}{cc} w + A(v) & A(w) + \sqrt{c}v \\ v & w \end{array} \right]; \forall w, v \in F(\sqrt{c}).$$

PROBLEM 30. When $\sigma\rho \neq 1$, decide if the twisted T-copies of Hughes-Kleinfeld look-alikes are proper quasifibrations or spreads, when the Hughes-Kleinfeld look-alike is infinite.

3. The remaining quasifibrations of dimension 2

The remaining known quasifibrations of dimension 2 are the Cherowitzo, Johnson, Vega quasifibration twisted by an automorphism α, the Ganley Semifields and additive quasifibrations and the inversions of bilinear flocks. Finally, there is the additive quasifibration admitting the type 1 regulus-inducing elation group and homology group.

We begin with the Ganley additive quasifibrations, which are defined as follows ([16] 29.3.5, p. 444). Some of these are T-copies.

Let K be a field of characteristic 3 and n a fixed non-square in K. Then the following defines an additive quasifibration, and a spread corresponding to a flock of a quadratic cone when K is finite. For an example of a proper quasifibration, let $K = GF(3^r)(z)$, the rational function field over $GF(3^r)$; which is the T-copy of the finite flock spread.

Ganley Additive Quasifibrations

$$x \ = \ 0, \ y = x \left[\begin{array}{cc} u + nt^3 & nt^9 + n^3t \\ t & u \end{array} \right] : \forall u, t \in K.$$

For any such field K, since the partial spread is additive, it is only necessary to show that the determinants are all non-zero for any $(t, u) \neq (0, 0)$.

PROBLEM 31. Show the determinant is always non-zero. Hint: Look at the discriminant of the associated quadratic in u.

PROBLEM 32. If $K = GF(3^r)(z)$, and n any non-square in $GF(3^r)$, show that the T-copy is a proper quasifibration and the twisted T-copy is a quasifibration or a spread.

PROBLEM 33. Look at the T-copy of the Ganley quasifibrations to see if the quasifibration could cover $(1, 0, 0, z)$. Show that this question comes down to solving the following problem: Show that the following sets of positive integers has an absolute maximum, for all m, r, k, s positive or 0 integers:

$$\{2k + 10m, 3r + k + s + 7m, 2s + 10r, 2r + 2s + 8m, 1 + 2s + 10m\}.$$

PROBLEM 34. If the previous problem has a solution, show that the question of whether the twisted T-copy can cover $(1, 0, 0, z)$ also has a solution.

We next consider the Cherowitzo-Johnson-Vega quasifibration, which does not come from a finite spread.

(Cherowitzo, Johnson, Vega type)

$$: \qquad x = 0, y = x \begin{bmatrix} u^\alpha & -t^{3\alpha^{-1}} \\ t & u \end{bmatrix} ; \; K \text{ an ordered field}$$

and α an automorphism of K,

K not containing all cube roots

(*See* (Cherowitzo, Johnson, Vega [26](2.6)).

However, these quasifibrations for $\alpha = 1$, also can never support a Galois chain where there is a spread emerging at the k th. stage. To see this, we assume so and we again take the system of equations:

$$t - u^\alpha = r,$$
$$u + t^{3\alpha^{-1}} = m.$$

Since both of these equations must be independent and surjective, we see that we arrive at

$$t - (m - t^{3\alpha^{-1}})^a,$$

must be surjective. Let $\alpha = 1$, then we have a cubic equation of the form $t^3 + t + q = 0$, for all elements $q \in K$, and by Cardano's equations, all the cube roots all involve cube roots of q, a contradiction, by assumption. When $\alpha \neq 1$, let $m = 0$, to again a cubic equation of the same form $t^3 + t + q = 0$, and the same argument applies.

THEOREM 81. *The Cherowitzo-Johnson-Vega quasifibration is infinitely non-surjective.*

Finally, we have seen that there are inversions of certain flocks of quadratic cones, over the field of rational numbers $\mathbb{Q}$, of which many if not all are proper quasifibrations. For p any rational number, we have:

$$p > 0 : x = 0, y = x \begin{bmatrix} w & 0 \\ 0 & w \end{bmatrix} : w \in \mathbb{Q}; \; y = x \begin{bmatrix} u & \frac{1}{t-p} \\ \frac{1}{a_1(t-p)} & u \end{bmatrix}$$

$$; \forall, u, t > 0, t \neq p \in \mathbb{Q}$$

$$y = x \begin{bmatrix} v & \frac{1}{s-p} \\ \frac{1}{b_1 s - a_1 p} & v \end{bmatrix} ; \forall v, s < 0 \in \mathbb{Q}.$$

And

$$p < 0 : x = 0, y = x \begin{bmatrix} w & 0 \\ 0 & w \end{bmatrix} : w \in \mathbb{Q}; \; y = x \begin{bmatrix} u & \frac{1}{t-p} \\ \frac{1}{a_1 t - b_1 p} & u \end{bmatrix}$$

$$; \forall, u, t > 0, t \in \mathbb{Q}$$

$$y = x \begin{bmatrix} v & \frac{1}{s-p} \\ \frac{1}{b_1(s-p)} & v \end{bmatrix} ; \forall v, s < 0, s \neq p \in \mathbb{Q}.$$

In the proof that shows that is a proper quasifibration, it was shown that $(1, -1, c, d)$, for all c, d cannot be covered by a unique quasifibration component, so $(-1, 1, -c, -d)$ cannot be covered. But, this is

$$
\begin{aligned}
t - f(u, t) &= -c, \\
u - g(u, t) &= -d
\end{aligned}
$$

so the quasifibration fails the 1-test. Since this is part of the requirement for re-stricted surjectivity of type $\frac{1}{k+1}$ for a Galois chain based on this quasifibration to become a spread at state k, we see this cannot occur. Hence, any such Galois chain consists of proper quasifibrations. Hence, these quasifibrations are also infinitely non-surjective. Therefore, we have proved the following:

COROLLARY 24. *With the exception of the Hughes-Kleinfeld look-alikes when* $\sigma\rho = 1$, *the Ganley quasifibrations, and the type 1 additive quasifibrations admitting both elation and homology type 1-regulus inducing collineation groups (which may be spreads), of the remaining known proper quasifibrations of dimension 2, none can be the base quasifibration in a Galois chain of lifted quasifibrations, which eventually will turn into spreads; all such known proper quasifibrations of dimension 2 are infinitely non-surjective.*

4. Large dimension quasifibrations

We have seen that many of the known proper quasifibrations of dimension 2 all produce infinite chains of proper quasifibrations. Moreover, at each link in any chain, these quasifibrations are derivable and produce proper quasifibrations. However, the dimension changes from 2 to 4 for each derivation.

DEFINITION 75. *We define any quasifibration to be of 'large dimension' n provided $n > 2$.*

We will see in (10,12,5) that there is a set of additive quasifibrations arising from cyclic algebras of finite dimension n. The quasifibrations of dimension $n > 2$ are possibly spreads, but this is an open question.

PROBLEM 35. *Determine which of the additive quasifibrations are spreads.*

In addition to these spreads of dimension $n > 2$, quasifibrations of large dimension arise from T-copying the generalized twisted field planes (See the Handbook for reference to these finite semifield planes).

Finally, we ask the following question. Assume that a quasifibration Q of dimension 2 is infinitely non-surjective by the '1-test failure' at $(0, \theta)$', then given any quasifibration with functions $f(u, t)$ and $g(u, t)$,

$$
\begin{aligned}
t - f(u, t) &= 0, \\
u - g(u, t) &= \theta,
\end{aligned}
$$

where $t, u \in K(\theta)$, there is not a unique (t, u), which satisfies this system of equations. For many of the known quasifibrations, we had this 1-test failure. Now create a Galois chain and choose a proper quasifibration Q' in the chain. Is this quasifibration Q' infinitely non-surjective? If the original quasifibration has the 1-test failure, then Q' has the 1-test failure at θ_k, if Q' is the k th proper quasifibration. The quasifibrations are over $K(z)$, in the general case, we will assume that we have some proper quasifibration over a field extension $K(\theta_0)$.

THEOREM 82. *If Q_0 is a proper quasifibration over $K(\theta_0)$ that is infinitely non-surjective by failing the 1-test, for any Galois chain based at Q, and for any proper quasifibration Q' at any link k, then Q' fails the 1-test and is therefore infinitely non-surjective.*

PROOF. Assume the theorem is valid for the first link in any quadratic Galois chain. Then, by induction, using the first link, it is valid at link 2. Assume it is valid for link j, clearly this implies that it is valid at link $j+1$, and the theorem is proved by mathematical induction.

So, if we have the first quasifibration as

$$x = 0, y = x \begin{bmatrix} f(u_0, t_0) & g(u_0, t_0) \\ t_0 & u_0 \end{bmatrix}; u_0, t_0 \in K(\theta_0).$$

We then assume that this quasifibration fails the 1-test at $(0, \theta_0)$, which is equivalent to

$$t_0 - f(u_0, t_0) = r_0 = 0$$
$$u_0 - g(u_0, t_0) = s_0 = \theta_0$$

fails. Now let $K(\theta_0) \subset K(\theta_0)(\theta_1)$, a quadratic extension with automorphism σ_1 which fixes $K(\theta_0)$ pointwise. Then the lifted quasifibration is

$$x = 0, y = x \begin{bmatrix} u_1^{\sigma_1} & \theta_1^{\sigma_1} f(u_0, t_0) + g(u_0, t_0)) \\ t_1 & u_1 \end{bmatrix}; u_1, t_1 \in K(\theta_0)(\theta_1),$$

where $t_1 = \theta_1 t_0 + u_0$ and $u_1 \in K(\theta_0)(\theta_1)$. The 1-test at (r_1, s_1) has the following form:

$$t_1 - u_1^{\sigma_1} = r_1 = 0$$
$$(u_1 - (\theta_1^{\sigma_1} f(u_0, t_0) + g(u_0, t_0)) = s_1.$$

Therefore, we obtain the following test, noting that

$t_1^{\sigma_1} - (\theta_1^{\sigma_1} f(u_0, t_0) + g(u_0, t_0)) = s_1 = \theta_1(-r_0) + d_0$, and $t_1 = \theta_1 t_0 + u_0$.

The automorphism σ_1 maps $\theta_1^{\sigma_1} \to -\theta_1 + \alpha_0$, for some element α_0 of K_0.

Let $d_0 - \alpha_0(t_0 + f(u_0, t_0)) = s_0$, noting that s_0 is a general element if and only if d_0 is a general element of K_0. We then obtain

$$t_0 - f(u_0, t_0) = r_0$$
$$u_0 - g(u_0, t_0) = s_0.$$

So, if Q' passes the 1-test generally at (r_1, s_1) for all r_1, s_1 in K_1, then it follows that Q would pass the 1-test at (r_0, t_0) for all r_0, t_0 in K_0, which is contrary to our assumptions. Hence, by induction, Q' fails the 1-test.   □

COROLLARY 25. *For every T-copy/twisted T-copy of a Hughes-Kleinfeld look-alike quasifibration or spread, all of the proper quasifibrations generated are infinitely non-surjective. Hence, at any quadratic Galois extension, and at any link, that quasifibration is infinitely non-surjective.*

5. T-copies of generalized twisted field planes

In this section, we show that the finite generalized twisted field planes over a field $GF(q)$ may be extended to a quasifibration over the rational field over $GF(q)$, $GF(q)(t)$.

- If a finite spread is rationally extended/T-copied to a rational function field and satisfies the condition on non-singularity of differences of determinants then the extension is either a quasifibration or a spread. We note that the 'row-condition' is automatically valid, since the form is adopted for the extension.

Let $\alpha, \beta \neq 1$ be automorphisms of a field $GF(q)$. Then the generalized twisted field planes are defined as follows:

$$x = 0, y = 0, y = xm - cx^\alpha m^\beta; \forall\, x, m \in GF(q) \text{ and } c \neq x^{1-\alpha} m^{1-\beta}.$$

We claim that when extending this definition to $GF(q)(t)$, the rational field over $GF(q)$, a quasifibration is always constructed. Since we have an additive set, we only need to determine if, in $GF(q)(t)$,

$$xm \neq cx^\alpha m^\beta.$$

Accordingly, assume that

$$xm = cx^\alpha m^\beta,$$

for some $x, m \in GF(q)(t)$. Then $c^{-1} = x^{\alpha-1} m^{\beta-1}$. Let $x = \frac{P(t)}{Q(t)}$, and $m = \frac{R(t)}{S(t)}$, where all terms are polynomials in $GF(q)(t)$. Then we obtain

$$(*): c^{-1}(Q(t)^{\alpha-1} S(t)^{\beta-1}) = (P(t)^{\alpha-1} R(t)^{\beta-1}).$$

Let the degrees of $P(t)$, $Q(t)$, $R(t)$, $S(t)$ be p, z, r, s, non-negative integers, including 0. Then multiplying out the polynomials to reduced form, a polynomial that is identically zero is obtained so that all coefficients are equal to 0. Therefore, the left polynomial once reduced must be identical to the right polynomial. If we assume that the coefficients on the maximum degree left terms are non-zero before reduction, the product is non-zero after reduction. Working out the maximum degrees of the reduced terms, we have

$$z(\alpha - 1) + s(\beta - 1) = p(\alpha - 1) + r(\beta - 1).$$

Furthermore, letting the coefficients of the polynomials of $P(t)$, $Q(t)$, $R(t)$, $S(t)$ be denoted by α_i, β_i, δ_i, γ_i, respectively, we compute the maximum coefficient of the left and right sides of $(*)$.

Accordingly, noting that all of the coefficients on the maximum terms on each polynomial are necessarily non-zero, we see that

$$c^{-1} \beta_z^{\alpha-1} \gamma_s^{\beta-1} t^{z(\alpha-1)+s(\beta-1)} = \alpha_p^{\alpha-1} \delta_r^{\beta-1} t^{p(\alpha-1)+r(\beta-1)}.$$

Hence, we obtain

$$(\alpha_p \beta_z^{-1})(\delta_r \gamma_s^{-1}) = c(\alpha_p \beta_z^{-1})^\alpha (\delta_r \gamma_s^{-1})^\beta,$$

a contradiction, since all of the coefficients are all non-zero elements of $GF(q)$.

Therefore, we have the proof that we obtain a quasifibration or a spread. It then remains to show that

$$xm - cx^\alpha m^\beta$$

cannot be a surjective function on $GF(q)(t)$. To be a spread, for each (x, y), and $x \neq 0$, there exists a (unique) m in $GF(q)(t)$, so that $xm - cx^\alpha m^\beta = y$. Take $x = 1$, and $y = t$, then there must be an element m so that letting $m = \frac{u}{v}$ for u, v in $GF(q)[t]$.

$$m - cm^\beta = t.$$

Then

$$0 = tv^\beta - uv^{\beta-1} + cu^\beta,$$

is an identically zero polynomial. Let u_0 and v_0 be the integer degrees of u and v, respectively, assuming the corresponding coefficients of the maximum degree terms are non-zero. Then, we must have the set of three degrees is $\{u_0\beta, 1 + v_0\beta, u_0 + v_0(\beta - 1)\}$. If there is a degree that is largest, the coefficient must be 0, a contradiction. Therefore, at least two of these degrees must be identical. If $1 + v_0\beta = u_0 + v_0(\beta - 1)$ then $u_0 = v_0 + 1$ and $u_0\beta \leq 1 + (u_0 - 1)\beta$, implying that $\beta \leq 1$, so $\beta = 1$, since we may assume that β is positive, a contradiction to our assumptions. We see that

$$u_0\beta = \max(1 + v_0\beta, u_0 + v_0(\beta - 1)).$$

Since $\beta \neq 1$, if $u_0\beta = 1 + v_0\beta$, then we have the integer equation $(u_0 - v_0)\beta = 1$, which has no solution. So, $u_0\beta = u_0 + v_0(\beta - 1)$, so that $u_0(\beta - 1) = v_0(\beta - 1)$, and $v_0 = u_0$. Hence, we have the set of degrees is $\{u_0\beta, 1 + v_0\beta, u_0 + v_0(\beta - 1)\} = \{u_0\beta, 1 + u_0\beta, u_0(\beta - 1)\}$. However, the term with the degree $1 + u_0\beta$ cannot cancel out or have a non-zero coefficient in the zero polynomial. Hence, we have the proof that a spread is not obtained, and we obtain a proper quasifibration.

THEOREM 83. *The T-copies of the generalized twisted field planes are proper quasifibrations.*

Hence, here is the summary for the known quasifibrations of dimension > 2.

THEOREM 84.

(1) *Given any Galois quadric chain of quasifibrations based on a quasifibration that is infinitely non-surjective, all of the quasifibrations in the chains are proper quasifibrations of dimension 2. At any link k, each of these quasifibrations are derivable by a twisted regulus net. Since the covering conditions do not change, each of these derivable nets are necessary proper quasifibrations but the quasifibrations are now of dimension 4.*

(2) *Notice that any Hughes-Kleinfeld and any T-copy or twisted T-copy is a derivable quasifibration or derivable spread. Actually there are always at least two such derivable nets: These are*

$$x = 0, y = x \begin{bmatrix} u^\sigma & 0 \\ 0 & u \end{bmatrix} ; u \in F, \text{ for } F \text{ a field,}$$

and

$$x = 0, y = x \begin{bmatrix} 0 & ft^\rho \\ t & 0 \end{bmatrix} ; t \in F, \text{ for } F \text{ a field.}$$

In the second situation, a basis change must be made to see this in classical regulus form. These are either classical regulus nets or type 2 regulus nets. The derivable quasifibrations are still proper quasifibrations and have their spreads in $PG(4r - 1, Fix\sigma)$ or $PG(4s - 1, Fix\rho)$, if $F/Fix\sigma$ and $F/Fix\rho$ are finite-dimensional.

Can there be derivable nets of type 2, with an infinite automorphism σ?

(3) *The T-copies of generalized twisted field planes and their twisted T-copies (some of these are of dimension 2). These are all quasifibrations, of which some are spreads (the twisted T-copies when $\sigma\rho = 1$).*

(4) *The additive quasifibrations of the cyclic algebras of degree n by replacement of the non-norm element b by a generator of the associated Galois field of dimension n. The dimension 2 cases are spreads. It is an open question if these are proper quasifibrations, but if they are, the dimension is $n > 2$.*

(5) *Let c be a non-square in F, F of characteristic 2 and $A(v) = (\delta/\delta\sqrt{c})\,v$, then the following is an additive quasifibration.*

$$x = 0,\ y = \begin{bmatrix} w + A(v) & A(w) + \sqrt{c}v \\ v & w \end{bmatrix}; \forall w, v \in F(\sqrt{c}).$$

This partial spread admits a set of derivable nets containing $x = 0$ and partitions the partial spread. All of the derivable nets all of which are of type 1, with elation regulus-inducing groups. The partial spread also admits a set of derivable nets containing $x = 0$ and $y = 0$, all of which are of type 1, with homology regulus-inducing groups. Since all derivable nets are regulus nets, all derived translation planes are also quasifibrations of dimension 2.

Part 8

Lifting skewfields

CHAPTER 10

General theory

In this chapter, we ask if the lifting procedure over fields discussed in part 7 could be extended more generally to skewfields. The algebra is completely different and difficult. We have been able to lift certain skewfield spreads; where the skewfields are 'central' Galois quadratic extensions of a skewfield F, in that the extensions use irreducible polynomials $x^2 \pm \alpha x - \beta$, where α, β are in the center $Z(F)$ of F, see Cohn and Dicks [**30**] for the general definition of central skewfield extensions. In the work on derivable nets over skewfields in part 2, the analysis of certain derivable nets involved quaternion division rings, and how to extend the skewfields using central Galois quadratic extensions, in a manner particular to the study of derivable nets. Our work is more general than that of quaternion division rings, as it is applicable to any non-commutative (or commutative) skewfield that admits a central quadratic extension. In parts 11 and 12, we see how to find central Galois quadratic extensions of cyclic division rings of degree n, where n is even and of more general division rings.

In the process of lifting to arbitrary skewfields, there are a variety of new translation planes constructed. Everything that we consider and discuss in the text with quaternion division rings, generalized quaternion division rings, and cyclic division rings, later in the text, is valid in arbitrary characteristic.

The process uses a skewfield L that admits a central Galois quadratic extension $L(\theta)$. It follows that $L(\theta)$ may be viewed as a spread within $PG(3, L)$. When L is non-commutative, using an analogous lifting process, a semifield spread is constructed within $PG(3, L(\theta))$. These would be the first known class of semifield spreads of this type.

Non-commutative skewfields that admit a central Galois quadratic extension include most of the quaternion division rings. There is a related construction from the semifield spread analogous to the construction of a quaternion division ring, that could construct what might be called a 'generalized' quaternion division ring, provided $\{w^\sigma w; \forall w \in L(\theta)\}$ is a proper subset of L. This is the manner of construction of a quaternion division ring, when L is a field.

The lifting process constructs a semifield plane with spread in

$$PG(3, L(\theta)).$$

If L is a quaternion division ring $(a, b)_F$ of dimension 4 over a field F, then the semifield spread would be in $PG(7, (a, b)_F)$ and $PG(3, (a, b)_F(\theta))$.

Hence, in this case, we might also refer to these semifields as 'octonion-semifields', as the geometric process could be seen as similar to the Cayley-Dickson process for algebras.

In this preliminary section, we will review the idea of lifting quasifibrations, whose defining partial spreads lie in $PG(3, K)$, where K is a field. Hence, a quasifibration, in this setting, is a maximal partial spread in $PG(3, K)$ and has all of the requirements to be a spread, except the covering property.

Given a quasifibration of the form

$$x = 0, y = x \begin{bmatrix} f(u, t) & g(u, t) \\ t & u \end{bmatrix}; u, t \in K,$$

where K is a field and f and g are functions of $K \oplus K$, defining the quasifibration. Assume that K admits a Galois quadratic extension $K(\theta)$, where $\theta^2 = \theta\alpha + a$, for $\alpha, a \in K$. Then there is an automorphism σ defined as follows: $\theta^\sigma = -\theta + \alpha$. The theory of lifting quasifibrations then constructs the following quasifibration:

$$x = 0, y = x \begin{bmatrix} w^\sigma & (\theta f(u, t) + g(u, t))^\sigma \\ \theta t + u & w \end{bmatrix}; \text{for all } \theta t + u, w \in K(\theta).$$

To see how this works in the field case, to obtain a quasifibration, we need only to verify that the non-zero matrices and their distinct differences have non-zero determinant. Here is what one needs to determine to ensure the non-singularity of the matrices.

(i) $w^{\sigma+1} \in K$ and is equal to $m^2 + \alpha nm - an^2$, when $w = \theta n + m$ and note that $x^2 + \alpha x - a$ and $x^2 - \alpha x - a$ are irreducible over K. And then (ii) the θ-part of $(\theta t + u)(\theta f(u, t) + g(u, t))^\sigma = \theta(-(u(f(t, u) - tg(t, u)))$. Note that the determinant is zero if and only if both the θ and the 1-part of the element in $K(\theta)$ are zero. But the determinant of the original matrix in $PG(3, K)$ is $uf(t, u) - tg(u, t)$, which cannot be zero unless both u and t are zero, which means that for the lifted determinant to be zero, we are left with $w^{\sigma+1}$ must be zero, which by (i) shows that $n = m = 0$ so that $w = 0$.

A more general analysis, lifting over non-commutative skewfields, might be accomplished along the same lines as over a field, using right or left inverses of the associated matrices and their differences. This would be extremely difficult and perhaps computationally impossible, in the general case.

This theory initially appeared in Journal of Geometry by N.L. Johnson, V. Jha, Lifting Skewfields, **113: 5** (2022). The author is indebted to the Journal of Geometry for help with the article and for permission to adapt the work for this text.

1. Matrix forms and replacement

It is well known that for a quaternion division ring L, using the notation $(a, b)_F$, where F is a field and the center of L, there is a matrix version that gives rise to a Desarguesian affine plane with spread in $PG(3, F(\theta))$, where $F(\theta)$ is a Galois quadratic extension of F. This translation plane may be written as follows:

$$x = 0, y = \begin{bmatrix} w^\sigma & bs^\sigma \\ s & w \end{bmatrix}; \forall s, w \in F(\theta), \sigma \text{ induced from } F(\theta),$$
$$b \in F - F(\theta)^{\sigma+1}.$$

Although this translation plane looks very similar to a lifted plane, it cannot be a lifted translation plane. However, it can itself be used for lifting.

However, if we consider a translation plane lifted from a Pappian plane coordinatized by $F(\theta)$, we would obtain

$$x = 0, y = \begin{bmatrix} w^\sigma & \theta^\sigma s^\sigma \\ s & w \end{bmatrix}; \forall s, w \in F(\theta), \ \sigma \text{ induced from } F(\theta).$$

This means there is always a question of whether a lifted Pappian plane has the 'quaternion replacement'. We have seen, and will bring this out again in this series of sections, that there are conceivably 'generalized quaternion division ring planes', for which there is an analogous replacement possibility, when we lift central quadratic extensions of arbitrary skewfields.

There is also a lifted translation plane of the following form:

$$x = 0, y = \begin{bmatrix} w^\sigma & \theta s^\sigma \\ s & w \end{bmatrix}; \forall s, w \in F(\theta), \ \sigma \text{ induced from } F(\theta).$$

This translation plane arises from the previous Pappian plane obtained by net replacement of

$$\{y = xm; m \in F(\theta)\} \text{ by } \{y = x^\sigma m; m \in F(\theta)\}.$$

It then turns out that these two semifield planes, the lifted planes, have various net replacements that construct/reconstruct the set of Andrè planes, in the finite or infinite cases. In the non-commutative case, we may not obtain the set of Andrè planes, but we do obtain, analogously, the two semifield planes.

We now turn to the main study for this part; whether such a lifting procedure might work for skewfield spreads within $PG(3, K)$, where K is a non-commutative skewfield and the lifting process depends on central Galois quadratic skewfield extensions. When we do this, we recall that this is usually possible in quaternion division ring planes. Additionally, by using a different representation of a quadratic central skewfield extension of a skewfield, we will be able to extend any central Galois quadratic skewfield extension to a new class of semifield planes, but this time, the spreads lie in a 3-dimensional projective space over a non-commutative skewfield.

We are able to show that once there is a lifted skewfield producing a semifield plane of the form

$$x = 0, y = x \begin{bmatrix} w^\sigma & \theta^\sigma s^\sigma \\ s & w \end{bmatrix}; \forall s, w \in F(\theta), \ \sigma \text{ induced from } F(\theta),$$

there is an infinite set of semifield planes

$$x = 0, y = x \begin{bmatrix} w^\sigma & (\theta\delta + \gamma)s^\sigma \\ s & w \end{bmatrix}; \forall s, w \in F(\theta), \ \sigma \text{ induced from } F(\theta),$$

for all $\delta \neq 0$ and $\gamma \in F$. The isomorphism question for the skewfields has not been completely determined in the finite case. However, in the finite case, the reader is directed to I. Cardinali, O. Polverino, R. Trombetti [21], there is a complete determination of the isomorphism types. However, in general, it is shown that the pre-skewfields that produce these semifields correspond to the skewfields written over the other generators of $F(\theta)$, in particular when $F(\theta)$ is written over $\theta^* = \theta\delta + \gamma$, for $\delta \neq 0$; as in $\theta^* t + u$; $\forall t, u \in F$.

THEOREM 85. (**The Main Theorem on Lifting Skewfields**) *Let $F(\theta)$ be a central Galois quadratic extension of a skewfield F, generated by an irreducible quadratic polynomial $x^2 \pm \alpha x - \beta$, where $\alpha, \beta \in Z(F)$ is the center of F. The $\pm$ indicates that the form shows that both the plus and minus polynomials are irreducible if only one of them is irreducible.*

(1) *Then $Z(F(\theta)) = Z(F)(\theta)$, $\theta^2 = \theta\alpha + \beta$, and there is an induced involutory automorphism σ fixing F pointwise, where $\theta t + u$ represents the elements of $F(\theta)$ and $(\theta t + u)^\sigma = (-\theta + \alpha)t + u$.*

(2) *The spread associated with $F(\theta)$, denoted by $F_{\alpha,\beta}/F$, is given by the following set of matrices:*

$$x = 0, y = x \begin{bmatrix} u + \alpha t & \beta t \\ t & u \end{bmatrix} ; \forall t, u \in F.$$

F may be either a field or a non-commutative skewfield.

(3) *Then there is a semifield spread in $PG(3, F(\theta))$ of the following form:*

$$x = 0, y = x \begin{bmatrix} w^\sigma & \theta^\sigma s^\sigma \\ s & w \end{bmatrix} ; \forall s, w \in F(\theta),$$

denoted by $F(\theta)_{\theta^\sigma}/F(\theta)$ lifted from $F_{\alpha,\beta}/F$.

(4) *$F(\theta)_{\theta^\sigma}/F(\theta)$ implies that there is a set of semifield spreads*

$$x = 0, y = \begin{bmatrix} w^\sigma & (\theta\delta + \gamma)s^\sigma \\ s & w \end{bmatrix} ; \forall s, w \in F(\theta),$$

denoted by $F(\theta)_{\delta,\gamma}/F(\theta)$, for each pair of elements (δ, γ), for $\delta \neq 0$, $\gamma \in Z(F)$. These semifield planes are induced by the following set of pre-skewfield spreads $PF_{\delta,\gamma}/F$ of $PG(3, F)$ defined as follows:

$$x = 0, y = \begin{bmatrix} -\delta u + \gamma t & -\beta\delta t + (\alpha\delta + \gamma)u \\ t & u \end{bmatrix} ; \forall t, u \in F.$$

When $(\delta, \gamma) = (-1, \alpha)$, this is the original skewfield spread.

When $(\delta, \gamma) \neq (-1, \alpha)$, all of these are pre-skewfield spreads that do not contain a multiplicative unit element. These pre-skewfield spreads lift to the above semifield spreads. When $(\delta, \gamma) = (1, 0)$, this pre-skewfield spread is

$$\{y = x^\sigma m; m \in F(\theta)\}, \text{ represented in } PG(3, F).$$

(5) *The following set $F_{\delta,\gamma}/F$ for $\delta \neq 0$, defines a skewfield spread in $PG(3, F)$:*

$$x = 0, y = \begin{bmatrix} -\delta u + \gamma t & -\beta\delta t + (\alpha\delta + \gamma)u \\ t & u \end{bmatrix} \begin{bmatrix} -\frac{1}{\delta} & \frac{\alpha\delta + \gamma}{\delta} \\ 0 & 1 \end{bmatrix} ; \forall t, u \in F.$$

The skewfield spread represented as a central Galois quadratic extension $F(\theta_{\delta,\gamma})$ of F is as follows:

$$x = 0, y = \begin{bmatrix} u + (\alpha\delta + 2\gamma)t & (-\gamma(\alpha\delta + \gamma) + \beta\delta^2)t \\ t & u \end{bmatrix} ; \forall t, u \in F,$$

and

$$x^2 \pm (\alpha\delta + 2\gamma)x - (-\gamma(\alpha\delta + \gamma) + \beta\delta^2)$$

is irreducible. The corresponding generating element for the skewfield is denoted by $\theta_{\delta,\gamma}$, where

$$\theta_{\delta,\gamma}^2 = \theta_{\delta,\gamma}(\alpha\delta + 2\gamma) + (-\gamma(\alpha\delta + \gamma) + \beta\delta^2),$$

and the induced involutory automorphism $\sigma_{\delta,\gamma}$ is given by

$$(\theta_{\delta,\gamma}t + u)^{\sigma_{\delta,\gamma}} = (-\theta_{\delta,\gamma} + (\alpha\delta + 2\gamma))t + u,$$

for all $t, u \in F$.

(6) Each element in the set of these central Galois quadratic extensions $F(\theta_{\delta,\gamma})/F$ provides a Hughes-Kleinfeld spread that lifts to a semifield spread in $PG(3, F(\theta_{\delta,\gamma}))$,

$$x = 0, y = \begin{bmatrix} w^{\sigma_{\delta,\gamma}} & \theta^{\sigma_{\delta,\gamma}} s^{\sigma_{\delta,\gamma}} \\ s & w \end{bmatrix}; \forall s, w \in F(\theta_{\delta,\gamma}).$$

(7) The spreads in part (6) are all isomorphic and correspond to the skewfields written over the set of generators of $F(\theta)$. And, all of the previous semifield spreads are indexed by elements $\theta\delta + \gamma$, where $\delta \neq 0$. When $\delta = 0$, and a semifield spread is created in the above form and if F is a non-commutative skewfield then the semifield spread S becomes a non-commutative division ring.

REMARK 24. *In the finite case, there are Albert's isotopy results available, whereas there are no apparent methods in the infinite skewfield cases to determine isomorphisms. Additionally, there does not seem to be any methods at all to sort the semifield planes whose spreads lie in non-commutative projective spaces. When two lifting agents are non-isomorphic then the semifield planes are also non-isomorphic.*

We now study what might be considered the most fundamental version of our lifting problem in the non-commutative case. And, we provide the proofs to parts $1, 2, 3$ of the Main Theorem on Lifting Skewfields.

If a non-commutative skewfield F admits a central quadratic extension (an irreducible polynomial $x^2 - \alpha x - \beta$, such that $\alpha, \beta \in Z(F)$), can we lift to a spread in $PG(3, F(\theta))$, where $\theta^2 = \theta\alpha + \beta$, is there a corresponding lifted translation plane (a semifield plane, in this setting)?

There are basically three different versions of the basic lifting theorem. These are:

Case 1; when $\alpha = 0$, and β is a non-square in F. For characteristic $\neq 2$, we would require that $\beta r^2 \neq n^2$, for all $r, n \in F$.

Case 2, when $\alpha \neq 0$, part 1; characteristic 2. Case 3 is part 2 of case 2; characteristic $\neq 2$.

Quasifibrations are studied in (Biliotti, Jha, and Johnson [16]) and in that text, all are of dimension 2, where F is a field. We have given the more general definition in the preliminary sections.

LEMMA 15. *An F-quasifibration of dimension n which is a skewfield partial spread is a spread.*

PROOF. In the above representation, λ is then assumed to be a skewfield. Since there are homology groups whose elements are of the form $(x, y) \rightarrow (x, yM)$ and $(x, y) \rightarrow (xN, y)$ for all $M, N \in \lambda$, it follows that the group generated by these groups is transitive on both $x = 0$ and $y = 0$, and is transitive on the components of the partial spread. To see this, we may assume without loss of generality that the n th. row of the set of matrices representing the quasifibration is $[e_1, e_2, ..., e_n]$, for all $e_i \in F$, $i = 1, 2, ..., n$. In the first group, let y_1 be the n-vector $(0, 0, 0, ..., 1)$, so that the group maps y_1 onto $(e_1, e_2, ...e_n)$. Hence, these two groups act transitively on each of $x = 0$ and $y = 0$, respectively. Also, we have the group generated by $(x, y) \to (y, x)$. That is, just note that $x = 0$ and $y = 0$ would be interchanged and $y = xM$ mapped to (xM, x), which is $y = xM^{-1}$, where M^{-1} is the right inverse of M. Since λ^* is a multiplicative group, we see that we may interchange $x = 0$ and $y = 0$. Hence, given any vector (x, y), there is a unique $N \in \lambda^*$ such that $y = xN$. What this says is that we have a spread and since λ is a skewfield, we have a skewfield spread.   □

PROPOSITION 6. *Consider a left 4-dimensional vector space V over F, where F is a skewfield. Write vectors of V as (x, y), where x and y are left 2-vectors. Consider the set π of left 2-dimensional vector subspaces*

$$x = 0, y = x \begin{bmatrix} u + \alpha t & \beta t \\ t & u \end{bmatrix}; u, t \in F, \alpha, \beta \in Z(F),$$

$x^2 \pm \alpha x - \beta$ *is irreducible over F.*

Then, π is a spread for a Desarguesian translation plane.

PROOF. For $t = 0$, the right determinant of $\begin{bmatrix} u + \alpha t & \beta t \\ t & u \end{bmatrix}$ is u^2 and for $t \neq 0$,

is

$$((u + \alpha t)t^{-1}u - \beta t)t$$

(See (2,3,3). When $t \neq 0$, the right determinant is zero if and only if $(u + \alpha t)t^{-1}u - \beta t = 0$, if and only if $ut^{-1}ut^{-1} + \alpha ut^{-1} - \beta = 0$. Letting $ut^{-1} = x$, then $x^2 + \alpha x - \beta = 0$, a contradiction.

$$\begin{bmatrix} u + \alpha t & \beta t \\ t & u \end{bmatrix} \begin{bmatrix} v + \alpha s & \beta s \\ s & v \end{bmatrix}$$

$$= \begin{bmatrix} (u + \alpha t)(v + \alpha s) + \beta ts & (u + \alpha t)\beta s + \beta tv \\ t(v + \alpha s) + us & t\beta s + uv \end{bmatrix},$$

and since $\alpha, \beta \in Z(F)$, we see that

$$\begin{bmatrix} (u + \alpha t)(v + \alpha s) + \beta ts & (u + \alpha t)\beta s + \beta tv \\ t(v + \alpha s) + us & t\beta s + uv \end{bmatrix}$$

$$= \begin{bmatrix} uv + \beta ts + \alpha(t(v + \alpha s) + us) & \beta(t(v + \alpha s) + us) \\ t(v + \alpha s) + us & uv + \beta ts \end{bmatrix},$$

showing that we have closure of multiplication, additivity, 0, and 1, and clearly distributivity. We also have multiplicative inverses, as non-zero skewfield determinant implies right inverse matrices exist. Hence, the matrices form a skewfield which is a quadratic extension of F, which we call $F(\theta)$. It follows directly that $F(\theta)$ is isomorphic to $\{\theta t + u; t, u \in F\}$ and that θ commutes with F. Hence, we have a quasifibration and from the above theorem, we have a skewfield spread that is a central quadratic extension.   □

We now prove our lifting theorem for non-commutative skewfields. First we point out that we shall be forced to consider an equation apparently appearing ubiquitously when studying noncommutative skewfields.

- We recall, the so-called 'Metro-Equation' in a non-commutative skewfield L is

$$mn - nm = 1$$

for elements $m, n \in L$ (See, [112], [140] for the origin of this phrase).

In the following theorem, there are three cases, each involving particular metro-equations, which we show do not occur.

THEOREM 86. *Let F be a skewfield and let $\dot{x}^2 - \alpha x - \beta$ be an irreducible quadratic where $\alpha, \beta \in Z(F)$. Let $\theta^2 = \theta\alpha + \beta$, and let the automorphism $\sigma : \theta \to -\theta + \alpha$. If $\alpha \neq 0$, note that $(\frac{x}{\alpha})^2 - (\frac{x}{\alpha}) - (\frac{\beta}{\alpha^2})$ is also irreducible and produces a central extension. In this setting, we may take $\alpha = 1$ and consider the irreducible polynomial $x^2 - x - \delta$, In either case, we have:*
Then

$$x = 0, y = x \begin{bmatrix} w^\sigma & \theta^\sigma s^\sigma \\ s & w \end{bmatrix} ; \forall s, w \in F(\theta),$$

$$F(\theta) = \{\theta t + u; \forall t, u \in F\}, \quad \text{where } Z(F(\theta)) = Z(F)(\theta).$$

is a semifield spread within $PG(3, F(\theta))$, where $F(\theta)$ is a skewfield central quadratic extension of F.

PROOF. Note to mirror our lifting process in the field case, we would expect the following to occur, if we are able to extend the lifting process to skewfield quadratic extensions. Consider

$$x = 0, y = x \begin{bmatrix} u + \alpha t & \beta t \\ t & u \end{bmatrix} ; \forall u, t \in F, \alpha, \beta \in Z(F),$$

$$x^2 \pm \alpha x - \beta \text{ is irreducible over } F.$$

To obtain a convenient form, we use $\theta^2 = \theta\alpha + \beta$ to extend F. We would hope to lift to

$$x = 0, y = x \begin{bmatrix} w^\sigma & (\theta(u + \alpha t) + \beta t)^\sigma \\ \theta t + u & w \end{bmatrix} ; \forall u, t \in F, \ w \in F(\theta).$$

Note that now,

$$\theta(\theta t + u) = (\theta\alpha + \beta)t + \theta u = (\theta(u + \alpha t) + \beta t).$$

The change of representation of the extension allowed this, as now we would expect to have

$$x = 0, y = x \begin{bmatrix} w^\sigma & \theta^\sigma(\theta t + u)^\sigma \\ \theta t + u & w \end{bmatrix} ; \forall u, t \in F, \forall w \in F(\theta), \text{ that is}$$

$$x = 0, y = x \begin{bmatrix} w^\sigma & \theta^\sigma s^\sigma \\ s & w \end{bmatrix} ; \forall s, w \in F(\theta).$$

Due to the additivity, there is a transitive elation group of the form

$$\left\{ \left[\begin{array}{cc} I_2 & \left[\begin{array}{cc} w^\sigma & \theta^\sigma s^\sigma \\ s & w \end{array} \right] \\ 0_2 & I_2 \end{array} \right] ; \forall s, w \in F(\theta) \right\},$$

that fixes $x = 0$ and is regular on the remaining components. Hence, we need to show two properties: The associated matrices must be non-singular, which is equivalent to having the skewfield determinants non-zero for non-zero matrices, which produces a quasifibration, and to show we have a cover, which would complete the proof that we obtain a semifield spread. We will study the two parts to the proof and what is required for their proof and realize that, in all cases, we need to consider and verify the following inequality:

$$(*) : kk^\sigma \neq \theta^\sigma ; \forall k \in F(\theta).$$

First, we ask if the skewfield determinants are all non-zero: The skewfield determinant of

$$\left[\begin{array}{cc} w^\sigma & \theta^\sigma s^\sigma \\ s & w \end{array} \right]$$

is

$$w^\sigma w \text{ if } s = 0 \text{ and } (w^\sigma s^{-1} w - \theta^\sigma s^\sigma)s \text{ for } s \neq 0.$$

Hence, it remains to consider the case when $s \neq 0$. This determinant is 0 if and only if

$$(**) : w^\sigma s^{-1} w - \theta^\sigma s^\sigma = 0.$$

For skewfields L, and τ an automorphism of L, it is still the case that $(z^{-1})^\tau = (z^\tau)^{-1}$ and that $(zr)^{-1} = r^{-1}z^{-1}$. Hence, multiply $(**)$ on the right by $(s^\sigma)^{-1}$ to obtain the equivalent equation:

$$(**)' \ w^\sigma s^{-1} w (s^\sigma)^{-1} = \theta^\sigma.$$

Now write $w^\sigma s^{-1} w (s^\sigma)^{-1}$ as $(w^\sigma s^{-1})(w^\sigma s^{-1})^\sigma$, since $\sigma^2 = 1$.

Letting $w^\sigma s^{-1} = k$, we need to show that the inequality $(*)$ holds. Now, we turn to the question of cover: Given a vector (x_1, x_2, x_3, x_4) with elements $x_i \in F(\theta)$, for $i = 1, 2, 3, 4$, we need to find a unique pair (s, w) such that

$$(x_3, x_4) = (x_1, x_2) \left[\begin{array}{cc} w^\sigma & \theta^\sigma s^\sigma \\ s & w \end{array} \right],$$

where $(x_1, x_2) \neq (0, 0)$. Hence, we need to solve the system of equations

$$\begin{aligned} x_1 w^\sigma + x_2 s &= x_3 \text{ and} \\ x_1 \theta^\sigma s^\sigma + x_2 w &= x_4. \end{aligned}$$

Just as the proof of the previous preposition, clearly we need only consider the case, $x_1 x_2 \neq 0$. Since now

$$\begin{aligned} s &= x_2^{-1} x_3 - x_2^{-1} x_1 w^\sigma, \\ so \ s^\sigma &= (x_2^{-1} x_3 - x_2^{-1} x_1 w^\sigma)^\sigma = (x_2^{-1} x_3)^\sigma - (x_2^{-1} x_1)^\sigma w. \end{aligned}$$

and then, we collect w-terms and ask if the coefficient is non-zero, to solve uniquely for (s, w). Then $x_1 \theta^\sigma s^\sigma + x_2 w = x_4$, becomes

$$x_1 \theta^\sigma ((x_2^{-1} x_3)^\sigma - (x_2^{-1} x_1)^\sigma w) + x_2 w = x_4.$$

We then see that we obtain:

$$(-x_1\theta^\sigma(x_2^{-1}x_1)^\sigma + x_2)w = x_4 - x_1\theta^\sigma(x_2^{-1}x_3)^\sigma.$$

Now consider the coefficient on w:

$$-x_1\theta^\sigma(x_2^{-1}x_1)^\sigma + x_2 = 0$$

if and only if

$$(x_1^{-1}x_2)(x_2^{-1}x_1)^{-\sigma} = (x_1^{-1}x_2)(x_1^{-1}x_2)^\sigma = \theta^\sigma.$$

Now letting $k = x_1^{-1}x_2$, we see that we need to show that

$$kk^\sigma \neq \theta^\sigma; \forall k \in F(\theta),$$

which is inequality $(*)$. Thus, if we can prove $(*)$ then we obtain a semifield plane in $PG(3, F(\theta))$.

Again, the 'Metro-Equation' in a non-commutative skewfield L is

$$mn - nm = 1$$

for elements $m, n \in L$. In each of the three cases, equation $(*)$, will involve the metro-equation. So, we first consider case 1, when $x^2 - \alpha x - \beta$ is irreducible over F and $\alpha = 0$. Then letting $k = \sqrt{\beta}r + n$, for $r, n \in F$, we see that

$$kk^\sigma = (\sqrt{\beta}r + n)(-\sqrt{\beta}r + n) = -\beta r^2 + n^2 + \sqrt{\beta}(rn - nr) = \theta^\sigma = -\sqrt{\beta}.$$

Hence, we need to consider the following system of equations:

$$(i) \quad : \quad nr - rn = 1,$$
$$(ii) \quad : \quad \beta r^2 = n^2.$$

So, we left multiply (i) by r and right multiply (ii) by n to obtain:

$$(i)' \quad : \quad r^2 n = rnr - r,$$
$$(ii)' \quad : \quad \beta r^2 n = n^3.$$

Now we right multiply (i) by r and left multiply (ii) by n to obtain:

$$(i)'' \quad : \quad nr^2 = rnr + r,$$
$$(ii)'' \quad : \quad \beta nr^2 = n^3.$$

Hence, $nr^2 = r^2 n$, from $(ii)'$ and $(ii)''$, so that we have $-r = r$, from $(i)'$ and $(i)''$. If $r = 0$ then $1 = 0$, thus concluding case 1. We now consider case 2, parts 1 and 2. Again, we consider if

$$(*) : kk^\sigma \neq \theta^\sigma; \forall k \in F(\theta).$$

Here $\theta^\sigma = -\theta + 1$. Letting $k = \theta r + n$ then

$$kk^\sigma = (\theta r + n)(\theta r + n)^\sigma = (\theta r + n)((-\theta + 1)r + n) = (\theta r + n)(-\theta r + r + n).$$

We then obtain, using $\theta^2 = \theta + \beta$,

$$-(\theta + \beta)r^2 + \theta(rn - nr + r^2) + n(r + n)$$
$$= \theta(rn - nr) - \beta r^2 + n(r + n) \text{ and assume } = -\theta + 1.$$

Then we consider the system of equations:

$$nr - rn = 1,$$
$$-\beta r^2 + n(r + n) = 1.$$

Subtract the first equation from the second, to obtain:

$$(\#) \quad : \quad nr - rn = 1,$$
$$(\#\#) \quad : \quad \beta r^2 = n^2 + rn.$$

Left multiply $(\#)$ by r and then right multiply $(\#)$ by r to obtain

$$(\#)' \quad : \quad r^2 n = rnr - r, \text{ and}$$
$$(\#)'' \quad : \quad nr^2 = rnr + r.$$

Now right multiply $(\#\#)$ by n and then left multiply $(\#\#)$ by n to obtain

$$(\#\#)' \quad : \quad \beta r^2 n = n^3 + rn^2, \text{ and}$$
$$(\#\#)'' \quad : \quad \beta nr^2 = n^3 + nrn.$$

Subtract $(\#)'$ from $(\#)''$ to see that

$$r^2 n = nr^2 + 2r.$$

When the characteristic is 2 (part 1 of 2), then $r^2 n = nr^2$, and use $(\#\#)'$ and $(\#\#)''$ to see that

$$rn^2 = nrn, \text{ so that } rn = nr, \text{ a contradiction.}$$

So, $(*)$ is established for case 2 part 1. Now assume that the characteristic is not 2; case 2, part 2. Use

$$r^2 n = nr^2 + 2r.$$

in $(\#\#)' : \beta r^2 n = n^3 + rn^2$ to obtain $\beta(nr^2 + 2r) = n^3 + rn^2$, so that

$$\beta nr^2 \quad = \quad -2\beta r + n^3 + rn^2 \text{ and this is, from } (\#\#)'' = n^3 + nrn,$$
$$2\beta r \quad = \quad rn^2 - nrn.$$

Now use $(\#) : nr - rn = 1$, so that

$$2\beta r = rn^2 - (1 + rn)n = -n$$

Thus, we have

$$n = -2\beta r. \text{ But, then } nr - rn = 1 = -2\beta r^2 - r(-2\beta r) = 0,$$

yielding the contradiction $1 = 0$. This completes the proof of the theorem. $\qquad \square$

2. The general skewfield spread

In this section, we consider the form of the skewfield spread when representing over a generator other than θ, $\theta^* = \theta\delta + \gamma$, for $\delta \neq 0$. So, our original skewfield is represented by

$$x \quad = \quad 0, y = x \begin{bmatrix} u + \alpha t & \beta t \\ t & u \end{bmatrix}; \forall u, t \in F, \alpha, \beta \in Z(F),$$

$$x^2 \pm \alpha x - \beta \text{ is irreducible over } F,$$

where we use $\theta^2 = \theta\alpha + \beta$ to extend F. Let the representation be over $\theta\delta + \gamma = \theta^*$. We note that

$$\theta^* \delta^{-1} - \gamma \delta^{-1} = \theta.$$

We wish to determine θ^{*^2} in terms of θ^*.

$$\theta^{*^2} = (\theta\delta + \gamma)^2 = \theta^2\delta^2 + 2\delta\gamma\theta + \gamma^2 = (\theta\alpha + \beta)\delta^2 + 2\delta\gamma\theta + \gamma^2.$$

This expression is then

$$((\theta^*\delta^{-1} - \gamma\delta^{-1})\alpha + \beta)\delta^2 + 2\delta\gamma((\theta^*\delta^{-1} - \gamma\delta^{-1}) + \gamma^2,$$

which then reduces to

$$\theta^*(\delta\alpha + 2\gamma) + (-\gamma(\alpha\delta + \gamma) + \beta\delta^2.$$

Hence, the representation of the spread with respect to $\theta^* = \theta\delta + \gamma$, is

$$x = 0, y = \begin{bmatrix} u + (\alpha\delta + 2\gamma)t & (-\gamma(\alpha\delta + \gamma) + \beta\delta^2)t \\ t & u \end{bmatrix}; \forall t, u \in F,$$

In this representation there is a corresponding involutory automorphism σ^*, that will map

$$\theta^* \rightarrow -\theta^* + (\alpha\delta + 2\gamma)$$

and such that

$$\theta^{*^2} = \theta^*(\alpha\delta + 2\gamma) + (-\gamma(\alpha\delta + \gamma) + \beta\delta^2).$$

We note the following:

- Consider the equation $kk^\sigma = \theta^\sigma$, where k is represented in the form $\theta r + n$. We have shown that this cannot be valid for any $k \in F(\theta)$. Similarly, the equation $kk^\sigma = \theta$, can never hold. We note that $kk^\sigma = kk^{\sigma^*}$, when representing the first in the θ-form and the second in the $\theta^* = \theta\delta + \gamma$ form for $\delta \neq 0$. Since this condition is independent of the representation, it follows that $kk^\sigma \neq \theta^*$, as well.

THEOREM 87. *We shall give the proofs to parts* $4, 5, 6, 7$ *of the Main Theorem on Lifting Skewfields.*

PROOF. Here is a sketch of the proof: Recall we are trying to prove that from the lifting of a central Galois quadratic extension, and the proof of a semifield spread indexed by θ^σ, that there is a set of semifield spreads indexed by $\theta\delta + \gamma$, for all pairs $\delta \neq 0$ and $\gamma \in Z(F)$. Where in the previous, there were various metro-equations to solve, we here have some general equations similar to the metro equations to solve. Where before, everything hinged on showing that

$$kk^\sigma \neq \theta^\sigma = -\theta + \alpha,$$

we now will be forced to consider the analogous equations:

$$kk^\sigma \neq \theta\delta + \gamma,$$

where $\delta \neq 0$. We now provide the necessary pre-skewfield central extensions, that lift to these new semifield spreads. Consider

$$x = 0, y = \begin{bmatrix} -\delta u + \gamma t & -\beta\delta t + (\alpha\delta + \gamma)u \\ t & u \end{bmatrix}; \forall t, u \in F.$$

The right determinant, for $t = 0$, is clearly non-zero for $u \neq 0$ ad $\delta \neq 0$. When $t \neq 0$, the right determinant is

$$((-\delta u + \gamma t)t^{-1}u - (-\beta\delta t + (\alpha\delta + \gamma)u))t.$$

If one of these determinants is 0, then we would have

$$((-\delta u + \gamma t)t^{-1}u - (-\beta\delta t + (\alpha\delta + \gamma)u)) = 0,$$

so that

$$-\delta u t^{-1}u + \gamma u + \beta\delta t - \alpha\delta u - \gamma u = -\delta u t^{-1}u + \beta\delta t - \alpha\delta u = 0.$$

Since $\delta \neq 0$ and in $Z(L)$, we see that

$$-ut^{-1}u + \beta t - \alpha u = 0.$$

Then we obtain:

$$(ut^{-1})^2 + \alpha(ut^{-1}) - \beta = 0,$$

a contradiction. Since we have an additive set, it follows that we have a spread in $PG(3, F)$. We shall show that this is actually a pre-skewfield, but first we show that this spread lifts to

$$x = 0, y = \begin{bmatrix} w^\sigma & (\theta\delta + \gamma)s^\sigma \\ s & w \end{bmatrix} ; \forall s, w \in F(\theta).$$

The corresponding form lifted from the pre-skewfields would be

$$x = 0, y = \begin{bmatrix} w^\sigma & (\theta(-\delta u + \gamma t) + -\beta\delta t + (\alpha\delta + \gamma)u)^\sigma \\ \theta t + u & w \end{bmatrix} ;$$

$\forall \theta t + u, w \in F(\theta).$

It then remains to show that

$$(\theta\delta + \gamma)(\theta t + u)^\sigma = (\theta(-\delta u + \gamma t) + -\beta\delta t + (\alpha\delta + \gamma)u)^\sigma.$$

But, since $\sigma^2 = 1$, we notice that

$$\begin{aligned}
(\theta\delta + \gamma)(\theta t + u)^\sigma &= ((\theta\delta + \gamma)^\sigma(\theta t + u))^\sigma = ((\theta^\sigma\delta + \gamma)(\theta t + u))^\sigma \\
&= ((-\theta + \alpha)\delta + \gamma)(\theta t + u))^\sigma \\
&= -\theta^2\delta t + \theta((\alpha\delta + \gamma)t - \delta u) + (\alpha\delta + \gamma)u \\
&= (-(\theta\alpha + \beta)\delta t + \theta((\alpha\delta + \gamma)t - \delta u) + (\alpha\delta + \gamma)u)^\sigma \\
&= (\theta(-\delta u + \gamma t) + (\alpha\delta + \gamma)u - \beta\delta t)^\sigma.
\end{aligned}$$

This proves part 4.

Consider the spread for the pre-skewfield

$$x = 0, y = xM, \text{ for } M \in \lambda \leq GL(2, F).$$

Let N be any element of λ, so that there is a right inverse, as the right determinants are all non-zero, as we had additivity. We claim that

$$x = 0, y = xMN^{-1}, \text{ for } M \in \lambda \leq GL(2, F),$$

is a spread. This is straightforward and shows that we have a skewfield

$$x = 0, y = \begin{bmatrix} -\delta u + \gamma t & -\beta\delta t + (\alpha\delta + \gamma)u \\ t & u \end{bmatrix} \begin{bmatrix} -\frac{1}{\delta} & \frac{\alpha\delta+\gamma}{\delta} \\ 0 & 1 \end{bmatrix} ; \forall t, u \in F.$$

The skewfield spread represented a Galois central quadratic extension $F(\theta_{\delta,\gamma})$ of F as follows:

$$x = 0, y = \begin{bmatrix} u + (\alpha\delta + 2\gamma)t & (-\gamma(\alpha\delta + \gamma) + \beta\delta^2)t \\ t & u \end{bmatrix} ; \forall t, u \in F.$$

Now this is a quasifibration that becomes a spread by a previous theorem. Since we now have the original situation, we may lift to a semifield spread. This proves parts 5, and 6.   □

3. Generalized quaternion division rings

THEOREM 88. *The Proof of part (7).*

PROOF. It is pointed out in the introduction that (most) quaternion division rings may be extended using a central Galois quadratic extension which shows that there are basically infinitely many new semifield planes obtained using the lifting process. The basic structure of the matrix spreads obtained from quaternion division rings may be generalized. In (2,4,7) a definition of a 'generalized quaternion division ring' was given based on this matrix representation of a quaternion division ring. For quaternion division rings, the matrices may be realized within $GF(2, K)$, where K is a field. In the more general case, the matrices would be realized over a non-commutative skewfield L, instead of a field K. Given any non-commutative skewfield L that admits a Galois extension $L(\theta)$ with respect to an irreducible quadratic $x^2 - \alpha x - \beta$, $\alpha, \beta \in Z(L)$, we have shown that there is a lifted semifield

$$x = 0, y = x \begin{bmatrix} w^\sigma & \theta^\sigma s^\sigma \\ s & w \end{bmatrix}; \forall w, s \in L(\theta),$$

within $PG(3, L(\theta))$. Now to prove this, we were able to solve various Metro-equations. An interesting part of these lifted semifields is that they look almost identical to a quaternion division ring plane. In particular, if θ^σ would be replaced by an element $b \in L$, in the case that L is a field, this would be the matrix representation of a quaternion division ring plane. These ideas may be generalized. Now to consider the replacement ideas that can be developed to construct quaternion division rings. Recall that we were able to prove the following inequality:

$$(*) : kk^\sigma \neq \theta^\sigma; \forall k \in L(\theta).$$

So the question is for what replacements $\theta \delta + \gamma$ will work instead in place of θ^σ, so that the analogous equation may be solved. It was pointed out in the section on the General Skewfield that the pre-skewfields required to lift to more general $\theta\delta+\gamma = \theta^*$ questions correspond to the skewfield written over the variety of generators of $L(\theta)$. This proves the more general inequality

$$(*) : kk^\sigma \neq \theta \delta + \gamma; \forall k \in L(\theta), \delta \neq 0, \delta, \beta \in L.$$

This proves part 7.   □

REMARK 25.

(1) At this point, the question of the existence of a generalized quaternion division ring is completely open.

(2) In (2,3,1) we have seen that it is possible to form uniform matrix representations for quaternion division rings over a field, regardless of characteristic. It is not difficult to see that the analogous representation over a non-commutative skewfield, will automatically give a uniform representation for generalized quaternion division rings over a non-commutative skewfield, just by keeping the scalar elements on the left.

(3) *The other cases when $\theta^\sigma = -\theta + \alpha$, for $\alpha \neq 0$, produce analogous metro-equations, which would have to be eliminated using the replacement technique. Again, it is an open question whether examples of generalized quaternion division rings would exist in these cases. If there were examples, they would not appear to be like any of the examples in the literature.*

(4) *Note the interesting situation: if we represent a quaternion division ring plane $(a,b)_{F(\sqrt{c})}$, which extends $(a,b)_F$, using a central element c that is non-square in F, characteristic $\neq 2$, we obtain the following representation of the lifted plane, which we just proved is, in fact, a semifield translation plane with spread in $PG(3,(a,b)_{F(\sqrt{c})})$ of the following form:*

$$(S) \quad : \quad x = 0, y = x \begin{bmatrix} u^{\sigma_c} & -\sqrt{c}t^{\sigma_c} \\ t & u \end{bmatrix} ; u, t \in (a,b)_{F(\sqrt{c})},$$

$$\sigma_c \quad : \quad \sqrt{c} \to -\sqrt{c}.$$

But, if we represent the same division ring plane which is isomorphic to $(a,b)_{F(\sqrt{c})}$ with spread in $P(3, F(\sqrt{a}, \sqrt{c}))$ by

$$x = 0, y = x \begin{bmatrix} u^{\sigma_a} & bt^{\sigma_a} \\ t & u \end{bmatrix} ; u, t \in L = F(\sqrt{a}, \sqrt{c}),$$

$$\sigma_a \quad : \quad \sqrt{a} \to -\sqrt{a},$$

then we obtain a lifted plane from the Galois field extension $F(\sqrt{a}, \sqrt{c}, \sqrt{d})$ of the form

$$(F) \quad : \quad x = 0, y = x \begin{bmatrix} w^{\sigma_d} & (\sqrt{d}u^{\sigma_a} + u^{\sigma_a})^{\sigma_d} \\ \sqrt{d}t + u & w \end{bmatrix} ;$$

$$u, t \in F(\sqrt{a}, \sqrt{c}), w \in F(\sqrt{a}, \sqrt{c}, \sqrt{d}), \text{ where } \sigma_d : \sqrt{d} \to -\sqrt{d},$$

where σ_r maps $\sqrt{r}t+u$ to $-\sqrt{r}t+u$. It is pointed out in situation (S), t and u are in $(a,b)_{F(\sqrt{c})}$, the skewfield, whereas in situation (F) t and u are in $F(\sqrt{a}, \sqrt{c})$. Noting that $(a,b)_{F(\sqrt{c})}$ is a quadratic non-commutative skewfield extension of $(a,b)_F$, this makes both $(a,b)_{F(\sqrt{c})}$ and $F(\sqrt{a}, \sqrt{c}, \sqrt{d})$ 8-dimensional over F, but the projective spaces in which the defining spreads lie are non-isomorphic, one spread is in $PG(3,(a,b)_{F(\sqrt{c})})$ and the other spreads is in

$$PG(3, F(\sqrt{a}, \sqrt{c}, \sqrt{d})).$$

These two projective spaces, although both 8-dimensional over F, cannot be isomorphic, as one is defined over a non-commutative skewfield and one is defined over a field. Also of note is the fact that each of (S) and (F) have many derivable subnets. In particular, we have the following two derivable nets:

$$(S(D)) : x = 0, y = x \begin{bmatrix} u^{\sigma_c} & 0 \\ 0 & u \end{bmatrix} ; u \in (a,b)_{F(\sqrt{c})} \text{ in } (S).$$

This net is a twisted version of a left classical pseudo-regulus net with respect to $(a,b)_{F(\sqrt{c})}$, and is isomorphic to a left classical pseudo-regulus. There is also

$$(F(D)) : x = 0, y = x \begin{bmatrix} u^{\sigma_d} & 0 \\ 0 & u \end{bmatrix} ; u \in F(\sqrt{a}, \sqrt{c}, \sqrt{d}), \text{ in } (F).$$

This net is a twisted version of a classical regulus net with respect to

$$F(\sqrt{a}, \sqrt{c}, \sqrt{d}),$$

and is isomorphic to the classical regulus net. Note that all derivable nets are isomorphic to some pseudo-regulus net combinatorially embedded in $PG(3, Z)$, where Z is a skewfield and the net is a regulus net exactly when Z is a field.

(5) *Both of the lifted semifield spreads are derivable; one by a twisted regulus net and one by a twisted pseudo-regulus net. The twisted pseudo-regulus derivable plane within a non-commutative skewfield projective space is completely new.*

(6) *Every quaternion skewfield $(a, b)_F$ is derivable by a twisted regulus net within a field projective space. These derived planes are also completely new.*

(7) *For quaternion division rings $(a, b)_F = K$ of characteristic 2, then a non-square extension needs to be replaced by an extension that admits a non-trivial automorphism, using central extension. Then the lifting process still works. The theorem called 'Extending Skewfields', still is valid and shows that as long as $F(\theta)$ is a Galois extension producing $(a, b)_F$, any additional quadratic extensions may be Galois or non-Galois, so long as that they admit the involutory automorphism corresponding to θ.*

REMARK 26. *When using spreads over fields K, the form of the spread within the 3-dimensional space $PG(3, K)$ can change the isomorphism class of the lifted spread. For example, one may lift a Pappian matrix spread and obtain a lifted semifield plane. If one uses an associated prefield Pappian matrix spread set and obtains a lifted semifield plane, the two lifted planes are not always isomorphic, even though the original spread sets are isomorphic.*

We have seen that we may lift any skewfield L that admits a non-square c in the center $Z(L)$. In the field case, when the characteristic is not 2, one can turn to a Pappian matrix spread defined by an irreducible quadratic $x^2 + \alpha x + a$ into an isomorphic Pappian matrix spread set defined by a non-square, basically by using the quadratic formula in the associated field. There are then two different lifted spreads. To see what we are talking about, we shall determine the two spreads: First take the following spread over a field F of characteristic not 2 and such that $x^2 + \alpha x + a$ is irreducible over F. Then the following spread is a Pappian spread: $x = 0, y =$

$$x \begin{bmatrix} u + \alpha t & at \\ t & u \end{bmatrix}; u, t \in F.$$ *Let $\theta^2 = \theta\alpha + a$, and note that $\{\theta t + u; u, t \in F\}$ is isomorphic to the matrix spread over $PG(3, F)$. There is an automorphism $\sigma_\theta :$ $\theta^{\sigma_\theta} = -\theta + \alpha$. Using $F(\theta)$, we obtain an associated lifted spread of the following form:*

$$x = 0, y = x \begin{bmatrix} w^{\sigma_\theta} & (\theta(u + \alpha t) + at)^{\sigma_\theta} \\ \theta t + u & u \end{bmatrix}; u, t \in F, w \in F(\theta),$$

$$\theta^2 = \theta\alpha + a, \theta^{\sigma_\theta} = -\theta + \alpha.$$

Again, noting that $\theta(\theta t + u) = (\theta(u + \alpha t) + at)$, and letting $\theta t + u = s$, we have the following semifield spread:

$$x = 0, y = x \begin{bmatrix} w^{\sigma_\theta} & \theta^{\sigma_\theta} s^{\sigma_\theta} \\ s & w \end{bmatrix}; s, w \in F(\theta).$$

Now, using the quadratic equation, we see that a radical for the root for the irre-ducible quadratic is $\sqrt{\alpha^2 + 4a}$, which means that if you use $c = \alpha^2 + 4a$, this is a non-square and there is an isomorphic Pappian spread

$$x = 0, y = x \begin{bmatrix} u & ct \\ t & u \end{bmatrix} ; u, t \in F.$$

Now using the quadratic extension $F(\sqrt{c})$, we obtain the following semifield spread using the automorphism $\sigma_{\sqrt{c}} : \sqrt{c} \to -\sqrt{c}$:

$$x = 0, y = x \begin{bmatrix} r^{\sigma_{\sqrt{c}}} & \sqrt{c}^{\sigma_{\sqrt{c}}} k^{\sigma_{\sqrt{c}}} \\ k & r \end{bmatrix} ; k, r \in F\sqrt{c},$$

Are these two semifield spreads isomorphic?

REMARK 27. *Choose any field F of characteristic $\neq 2$. Assume that there is a quaternion division ring $(a, b)_F$ with center F. Assume that there exist Galois extensions $F(\sqrt{a}, \theta)$ and $F(\sqrt{a}, \tau)$ by quadratic extensions and neither field is equal to $F(\sqrt{a}, \sqrt{b})$ (note that here a could be b). Then there exist extensions of $(a, b)_F$, $(a, b)_{F(\theta)}$ and $(a, b)_{F(\tau)}$. Then both the spreads defined by $(a, b)_{F(\theta)}$ and $(a, b)_{F(\tau)}$ may be lifted to semifield spreads within $PG(3, (a, b)_{F(\theta)})$ or $PG(3, (a, b)_{F(\tau)})$, re-spectively. If $F(\theta)$ is not isomorphic to $F(\tau)$ then are the semifield spreads isomor-phic? That is, can a semifield spread be of dimension 2 with respect to different skewfields?*

We shall see that Pappian and Desarguesian spreads can be of dimension 2 with respect to a variety of different skewfields (See (11,13,3), Galois theory for division rings).

4. Retraction

We now return to the ideas that were raised in the section on Matrix Forms and Replacement, where it was mentioned that the matrix versions of any quaternion division ring cannot arise from a translation plane by lifting in the field case.

In the field case, there is a way to identify the type of spreads that have been lifted. That is, once identified, there is then the reverse construction, which is called 'retraction'. The condition that a spread may be retracted is a group theoretic one.

In the non-commutative skewfield case, we have shown that retraction exists in this setting as well, where some of the semifield spreads were retracted to pre-skewfield spreads. The retraction idea was essential in finding these semifield spreads and their origin.

However, asking that question in the non-commutative skewfield case is as fol-lows:

(1) Given a semifield translation plane over a 3-dimensional projective space $PG(3, L(\theta))$, where L is a non-commutative skewfield, when can it be certain that there is a retraction to a related translation plane within $PG(3, L)$?

(2) Given any skewfield L that admits a central quadratic extension with re-spect to a quadratic whose coefficients are in $Z(L)$, is there any related skewfield, in a different form, that also lifts to a related semifield plane?

(3) Actually, in our discussion previously, we discussed the idea of replacing θ^σ with other elements $\theta c + d$, where $\theta^2 = \theta\alpha + \beta$. Write the skewfield as $y = xm$, where $m \in L$. There is a corresponding automorphism, and consider $y = x^\sigma$. This is actually a 2-dimensional left L-space, in the associated 4-dimensional left vector space over L. However, it is not a component, which implies that it is a Baer subplane. Consider the intersection $(y = x^\sigma) \cap (y = xm)$, so we ask what are the components that lie on this Baer subplane by considering $x^\sigma = xm$. Raise the equation to σ, that is, $x^{\sigma^2} = x^\sigma m^\sigma$, which shows that $x = (xm)m^\sigma$, so for $x \neq 0$, we have $mm^\sigma = 1$. If L is taken to be a field, then $m^{\sigma+1} = 1$ defines a multiplicative subgroup of the associated field. Additionally, we would have found a derivable net. This idea is discussed further later in the field case in part 10. This is the key point of understanding infinite Andrè planes and their generalizations to analogous translation planes arising from non-commutative skewfields.

(4) In the non-commutative case, we have found another instance of the Metro equation $kk^\sigma = 1$, to solve. If $m = \theta r + n$, then we have found that

$$mm^\sigma = -\beta r^2 + n^2 + \alpha nr + \theta(rn - nr) = 1.$$

Hence, the set of element $y = xm$, such that $mm^\sigma = 1$ satisfies

$$n^2 + \alpha nr - \beta r^2 = 1, \text{ where } rn = nr.$$

Note that it might appear that $n^2 + \alpha nr - \beta r^2 = 1$ is simply a set of determinants equal to 1, and it would be if L is a field. In the non-commutative case, this merely represents a subset of components $y = x(\theta r + n)$, that intersect $y = x^\sigma$. However, we can see that there is a replacement of all of the components of the original non-commutative skewfield.

$$x = 0, y = xm; m \in L(\theta) \text{ by } x = 0, y = x^\sigma m; m \in L(\theta).$$

Now, x^σ is written as $(x_1, x_2) \begin{bmatrix} -1 & \alpha \\ 0 & 1 \end{bmatrix}$. Consider then

$$x = 0, y = x \begin{bmatrix} -1 & \alpha \\ 0 & 1 \end{bmatrix} \begin{bmatrix} u + \alpha t & \beta t \\ t & u \end{bmatrix}; u, t \in L.$$

Ask, does this pre-skewfield also lift to a semifield over $PG(3, L(\theta))$? Here is what would be expected to be obtained:

$$x = 0, y = x \begin{bmatrix} w^\sigma & (-\theta u + \alpha u - \beta t)^\sigma \\ \theta t + u & w \end{bmatrix}; \forall u, t \in L, \ \forall w \in L(\theta).$$

To connect this to our previous discussion, notice that

$$\theta^\sigma(\theta t + u) = (-\theta + \alpha)(\theta t + u) = -\theta u + \alpha u - \beta t.$$

So, the putative semifield would then become:

$$x = 0, y = x \begin{bmatrix} w^\sigma & \theta s^\sigma \\ s & w \end{bmatrix}; \forall u, t \in L, \ \forall w \in L(\theta).$$

So, we have then replaced θ^σ in the first constructed semifield with θ in the second. However, is it certain that this is actually a semifield 'spread'? If we hadn't noticed that Baer subplane connection, we would be forced into another Metro-equation argument to ensure that we obtain another semifield spread. However, we can now just observe that we have made that replacement of one semifield plane by another.

(5) The observation in number 3 forms the basis of showing that over fields, there are two semifield planes in which there are various replaceable nets that reconstruct the Andrè planes. Additionally, the replacements are inherited in chains of lifted spreads for the generalization to the infinite case of Cherowitzo and Johnson [**25**] in the finite case.

(6) We mentioned the idea of replacing θ^σ by an element b of L. This would construct a generalized quaternion plane, assuming that

$$b \in L - \{n^2 + \alpha nr - \beta r^2; n, r \in L\}$$

This is an ad hoc type of replacement, seemingly without theory attached. But, if this could occur, would this be a lifted translation plane? This would be the form:

$$x = 0, y = x \begin{bmatrix} w^\sigma & bs^\sigma \\ s & w \end{bmatrix} ; \forall s, w \in L(\theta),\ b \in L.$$

If there is a retraction, the retraction would produce:

$$x = 0, y = x \begin{bmatrix} bt & bu \\ t & u \end{bmatrix} ; \forall u, t \in L,\ b \in L,$$

which is not a spread. Just note the determinant, when $t = 0$ produces a non-singular matrix, but if $t \neq 0$, the determinant would be

$$bt(t^{-1})u - bu)t = 0.$$

So, even if there is a retraction process available, it would not work for any putative generalized quaternion plane.

REMARK 28. *We will come back to this subject using cyclic division algebras, once we have developed the Galois theory relative to extensions. When we do this, we shall also see that a Galois tower of quadratic extensions of $K(\theta_1, \theta_2, ..., \theta_{k+1})$, where $K(\theta_i)$ is also a quadratic Galois extension, for $k = 1, 2, .., k + 1$. If there is a quaternion division ring $(a, b)_K$ containing $K(\theta_1)$ then there is also a quaternion division ring $(a, b)_{K(\theta_2,...,\theta_{k+1})}$ containing $K(\theta_2, ..., \theta_{k+1})(\theta_1)$. Since the automorphism group is elementary Abelian of order 2^k, there are $2^k - 1$ mutually non-isomorphic sub-quaternion division rings of index 2, each of which is a central Galois quadratic extension.*

Part 9

Bilinearity

CHAPTER 11

General bilinear geometries

The concept of whether the planes of a flock of a quadratic cone can share a point has been answered by J.A. Thas [**137**]. We have seen the myriad connections that exist between flocks of a quadratic cone and generalized quadrangles, translation planes, herds of ovals, maximal partial spreads, hyperbolic fibrations and translation planes admitting cyclic homology groups whose orbits are regulus nets. This first chapter is obliquely related to bilinearity and is rather a novel group theoretic way of looking at the 'star flock theorem' of Thas, recalling that a 'star flock' is one where the planes of the flock share a point. However, for finite flocks of odd order, there can be no bilinear flocks as have been mentioned previously, due to the result of the author [**81**] that says that if there is a critical linear subflock which means at least $(q-1)/2$ planes sharing a line then the only non-linear possibility is the Fisher flocks, which shall be discussed later in this part.

1. Star flocks and rigidity

This work on rigidity in conical flocks first appeared in Jha and Johnson [**68**]. The impetus of this work was the following theorem of Thas.

THEOREM 89. *(Thas [**137**]* **The star flock theorem** *(1.5.6) Let F be a finite star flock of a quadratic cone Q with point P.*
(1) If q is even then F is linear (the planes share a line).
(2) (a) If q is odd and P is an interior point of Q then F is linear.
  (b) If q is odd an P is an exterior point, then F is a Kantor-Knuth flock.

Let the quadratic cone be represented using homogeneous coordinates (x_0, x_1, x_2, x_3) by $x_0 x_1 = x_2^2$.

Then the common point is $(0, 0, 1, 0)$, and the planes π_i of the flock have the form

$$\pi_i = a_i x_0 - m a_i^\sigma x_1 + x_3$$
$$\text{where } \{a_i : i = 1, 2, ..., q\} = GF(q),$$
$$\text{and } \sigma \text{ is an automorphism of } GF(q)$$
$$\text{and } m \text{ is a fixed non-square in } GF(q).$$

Now to discuss the concept of 'rigidity'.

Consider the mapping $\rho : (x_0, x_1, x_2, x_3) \to (x_0, x_1, -x_2, x_3)$. We notice that this mapping is a collineation of the flock fixing $x_0 x_1 = x_2^2$, which also fixes the common point $(0, 0, 1, 0)$. $\langle \rho \rangle$ is said to be an 'rigid group'. Note also that ρ must fix π_i for each $i = 1, 2..., q$.

DEFINITION 76. *A group acting on a partial flock G of a quadratic cone of s planes (conics) that fixes each plane of G is called a s-locally rigid group on G. The group is 'linear' if and only if it is in $PGL(4, q)$. If G is a flock of a quadratic cone, then we use the term 'rigid group' if the group is in $P\Gamma L(4, q)$ and 'rigid linear group' if the group is in $PGL(4, q)$.*

It turns out that there is a fundamental connection between s-locally rigid groups on G that makes G a partial star flock.

THEOREM 90. *(**The fundamental theorem on rigid groups**, Jha and Johnson [68]) A partial flock of s-planes is a partial star flock if and only if there exists a nontrivial linear s-locally rigid group.*

It is possible to overcome the linearity condition if s is large enough.

COROLLARY 26. *There exists an s-locally rigid group on a partial flock G and $s > 2\sqrt{q} + 1$ if and only if G is a star flock.*

Finally, then there is a characterization of the Kantor-Knuth flocks using rigid groups.

THEOREM 91. *(Jha and Johnson [68] (4.1) p. 61) A flock F of a quadratic cone is linear or a Kantor-Knuth flock if and only if F admits a non-trivial rigid automorphism group.*

DEFINITION 77. *A more general group is an 's-rigid group', which is a group of a flock that acts as an s-locally rigid group on a subflock of s planes. A group is a non-Desarguesian group if and only if the planes of the subflock do not share a line.*

Now in this part, we consider various instances of where the planes of a flock would share one of two lines. We discuss infinite bilinear flocks of quadratic cones and the k-linear flocks. Additionally, there are bilinear flocks of α-cones constructed by the ordinary technique of lifting.

The idea of s-rigidity is interesting, for example, there are nice results on the Fisher flocks, which we shall discuss further in section 10 of this part.

THEOREM 92. *((Jha and Johnson [68] (5.1) p. 62) If F is a nonlinear flock in $PG(3, q)$ that admits a s-rigid automorphism group of order $s \geq (q-1)/2 \geq 2$ then F is a Fisher flock.*

Concerning s-rigid groups, we consider what are called 'axial theorems' regarding when a subflock of planes could share a line.

THEOREM 93. *((Jha and Johnson [68] (5.2) p. 69, **The axial theorem for even order**). Let F be a flock of a quadratic cone in $PG(3, q)$ for q even. If F admits a non-trivial linear s-rigid automorphism group, then the group is Desarguesian (the s fixed planes share a common line).*

THEOREM 94. *((Jha and Johnson [68] (5.6) p. 73, **The axial theorem for odd order**). Let F be a flock of a quadratic cone in $PG(3, q)$ for q odd. If F admits a non-trivial linear s-rigid automorphism group of order > 2, then the group is Desarguesian (the s fixed planes share a common line.*

In (6,8,2&3), there was a discussion on simultaneous spreads, which are spreads that give rise to both α-conical flocks and also to α-twisted hyperbolic flocks. It was noted that

THEOREM 95. *For spreads over a field K,*

$$x = 0, y = x \begin{bmatrix} u^\alpha + gt^\alpha & ft^{\alpha^{-1}} \\ t & u \end{bmatrix}$$

(1) The α^{-1}-flock is linear if and only if $g = 0$ and $\alpha^2 = 1$.
(2) If $g \neq 0$ and $\alpha^2 = 1$ then the α^{-1}-flock is a star flock.
(3) If $g = 0$ and $\alpha^2 \neq 1$ then the α^{-1}-flock is a star flock.
(4) If $g \neq 0$ and $\alpha^2 \neq 1$ then the α^{-1}-flock has planes that trivially intersect.
(5) In all cases, the α-hyperbolic flock is linear with line

$$\{(x_1, x_2, x_2, x_1 f^\alpha - x_2 g); x_1, x_2 \in K\}.$$

The general α-flocks are

$$\rho_t : \quad x_0 t - x_1 f^{\alpha^{-1}} t^{\alpha^{-2}} + x_2 g^{\alpha^{-1}} t^{\alpha^{-1}} - x_3 = 0,$$

whereas the α-hyperbolic flocks are all linear. So, the situation is so much different for α-flocks. In particular, there are different star flocks.

2. Bilinear α-flocks

This material could have been included in part 4 on α-flocks. However, this work requires the understanding of how 'lifting' works, as well as how the multiple derivation/replacement process functions. The algebraic construction of α-flocks and their relationship to translation planes admitting twisted regulus-inducing elation groups is also mentioned in this chapter, and has been covered previously.

The idea of the construction of bilinear σ-flocks is as follows: Given a Pappian spread written in the form

$$(*) \quad : \quad x = 0, y = x \begin{bmatrix} u + \delta t & \beta t \\ t & u \end{bmatrix} ; t, u \in F;$$

$$x^2 \pm \delta x - \beta \text{ is irreducible over } F,$$

such that the matrix field is isomorphic to $F(\theta)$ but where the components are in $PG(3, F)$ and $F(\theta)$ is a Galois quadratic field extension. Then this spread lifts to a semifield plane that defines a linear σ-flock covered by planes in $PG(3, F(\theta))$. The lifted plane has the following form:

$$x = 0, y = x \begin{bmatrix} w^\sigma & \theta^\sigma(\theta t + u)^\sigma \\ \theta t + u & w \end{bmatrix} ; t, u \in F; w \in F(\theta),$$

$$\theta^\sigma = -\theta + \delta, \theta^2 = \theta\delta + \beta.$$

Note that

$$(\theta(u + \delta t) + \beta t) = \theta(\theta t + u).$$

The form then becomes:

$$x = 0, y = x \begin{bmatrix} w^\sigma & \theta^\sigma s^\sigma \\ s & w \end{bmatrix}; s, w \in F(\theta),$$

Using the method of the next part 10 on multiple replacement, we may construct a set of Andrè F-regulus nets (the subspaces defining the spread are 2-dimensional F-subspaces, thus constructing an F-regulus net, as described previously).

As we have seen, a flock of an α-cone is equivalent to a translation plane of the following form:

THEOREM 96. *Cherowitzo, Johnson, Vega* [26]. *Let K be any field and α an automorphism of K. Let (x_1, x_2, x_3, x_4) denote homogeneous coordinates of $PG(3, K)$. Define C_α an α-cone as $x_1^\alpha x_2 = x_3^{\alpha+1}$, with vertex $(0, 0, 0, 1)$. A set of planes which partition the non-vertex points of C_α will be called an α-flokki (also an α-conical flock). The plane intersections are called α-conics. A translation plane π with spread in $PG(3, K)$, for K a field, is an α-flokki plane if and only if there is an elation group E one of whose orbits is a derivable partial spread containing at least two Baer subplanes that are K-subspaces.*

Then the procedure to create bilinear α-flocks requires that $\alpha^2 = 1$, where there is a Pappian spread Σ admitting an automorphism α. Let Σ_α denote the corresponding Pappian spread (See the procedure in the multiple replacement in part 10).

More generally, if π is a translation plane that admits an elation group E^α, one component orbit of which, together with the axis is a α-regulus, then π has the following form

$$x = 0, y = x \begin{bmatrix} u + g(t) & f(t) \\ t & u^\alpha \end{bmatrix}; t, u \in K, f, g \text{ functions of } K$$

when writing the α-regulus in the form

$$x = 0, y = x \begin{bmatrix} u & 0 \\ 0 & u^\alpha \end{bmatrix}; t, u \in K,$$

then there is a corresponding α-flock with planes

$$\rho_t : x_1 t - x_2 f(t)^\alpha + x_3 g(t)^\alpha - x_4; t \in K.$$

Conversely, an α-flokki may be written in the form $\{\rho_t; t \in K\}$, for some functions $f(t)$ and $g(t)$ which constructs a translation plane with the above form. Hence, translation planes with spreads in $PG(3, K)$ that admit α-regulus-inducing elation groups are equivalent to flocks of an α-quadratic cone.

That being said, we see that we have a translation plane admitting a σ-elation group inducing a σ-regulus net. So, we obtain an associated σ-flock with planes

$$\rho_t : x_1 t - x_2 f(t)^\sigma + x_3 g(t)^\sigma - x_4; t \in K,$$

where $g(t) = 0$ and $f(t) = \theta^\sigma t^\sigma$. Hence, we obtain the linear σ-flock such that $\sigma^2 = 1$:

$$\rho_t : x_1 t - x_2 \theta t - x_4; t \in K,$$

so that

$$\bigcap \rho_t : \{(x_2\theta, x_2, x_3, 0)\}; x_2, x_3 \in K;$$

is a line.

Now if we write the Andrè F-regulus nets of the Pappian plane isomorphic to $F(\theta)$ in matrix form then

$$y = x^\sigma = (-\theta x_1 + (\delta x_1 + x_2)) = (x_1, x_2) \begin{bmatrix} -1 & \delta \\ 0 & 1 \end{bmatrix},$$

then the derived nets have the form:

$$y = x \begin{bmatrix} -1 & \delta \\ 0 & 1 \end{bmatrix} \begin{bmatrix} u + \delta t & \beta t \\ t & u \end{bmatrix};$$

$$\forall \det \begin{bmatrix} u + \delta t & \beta t \\ t & u \end{bmatrix} = \rho \in \{m^{\sigma+1}; m \in F(\theta)\} \subseteq F,$$

Call the corresponding derivable net D_ρ^F, for $\delta \in \{m^{\sigma+1}; m \in F(\theta)\} \subseteq F$.

Now we note, if we derive all Andrè F-regulus nets, we obtain the following spread:

$$\begin{bmatrix} -1 & \delta \\ 0 & 1 \end{bmatrix} \begin{bmatrix} u + \delta t & \beta t \\ t & u \end{bmatrix} = \begin{bmatrix} -u & -\beta t + \delta u \\ t & u \end{bmatrix}; \forall u, t \in F.$$

This spread set then also lifts, in the previous manner, to:

$$x = 0, y = x \begin{bmatrix} w^\sigma & (\theta(-u) - \beta t + \delta u)^\sigma \\ \theta t + u & w \end{bmatrix}; t, u \in F; w \in F(\theta),$$

$$\theta^\sigma = -\theta + \delta, \theta^2 = \theta \delta + \beta,$$

which may be written as follows:

$$x = 0, y = x \begin{bmatrix} w^\sigma & \theta(\theta t + u)^\sigma \\ \theta t + u & w \end{bmatrix}; t, u \in F; w \in F(\theta),$$

$$\theta^\sigma = -\theta + \delta, \theta^2 = \theta \delta + \beta.$$

To see this, we note that

$$\theta^\sigma(\theta t + u) = (-\theta + \alpha)(\theta t + u) = -\theta^2 t + \theta \alpha t - \theta u + \alpha u$$

$$= -(\theta \alpha + \beta)t + \theta \alpha t - \theta u + \alpha u = \theta(-u) - \beta t + \alpha u.$$

Since $\theta^{\sigma^2} = \theta$, we have the proof. What this says is that the spread obtained by deriving all of the Andrè F-regulus nets defines a semifield flock which is also linear of the following form:

$$\varepsilon_t : x_1 t - x_2 \theta^\sigma t - x_4; \ t \in K,$$

where

$$\bigcap \varepsilon_t : \{(x_2 \theta^\sigma, x_2, x_3, 0)\}; x_2, x_3 \in K$$

is the common line of the linear σ-flock.

We note that we have not yet verified that what we claimed was a Pappian 'spread' and a semifield 'spread' are actually spreads, in both cases. We will postpone this verification of spreads until we complete the analysis of the bilinear σ-flocks.

For the multiple derivation technique, we select any proper subset λ of the set of derivable nets

$$\{D_\rho^F; \rho \in \lambda \subset \{m^{\sigma+1}; m \in F(\theta)\}\}.$$

The union of the nets in λ when lifted partially will necessarily contain the line

$$\bigcap \varepsilon_t : \{(x_2 \theta^\sigma, x_2, x_3, 0)\}; x_2, x_3 \in K.$$

The union of the nets in

$$\left\{ D_\rho^F ; \rho \in \left\{ m^{\sigma+1}; m \in F(\theta) \right\} \right\} - \lambda$$

will contain the line

$$\bigcap \rho_t : \left\{ (x_2\theta, x_2, x_3, 0) \right\} ; x_2, x_3 \in K.$$

Note that the intersection of the two lines is a point $(0, 0, 1, 0)$.

THEOREM 97. *The field $F(\theta)$ listed above provides a spread in $PG(3, F)$ and the semifields listed above provide spreads in $PG(3, F(\theta))$.*

What is left to prove is the 'covering condition'. That is, given any vector (x_1, x_2, x_3, x_4) for $x_i \in F$, it is necessary to show that there is a unique component covering this vector. This turns out to be true for fields or non-commutative skewfields (with the appropriate skewfield quadratic extension), as we have seen in part 8.

THEOREM 98. *The two semifield spreads π_θ and π_{θ^σ} admit net replacements that reconstruct the F-Andrè translation planes. Additionally, the translation planes correspond to bilinear flocks of σ-cones, where the intersection of the lines is a point.*

PROOF. For each bilinear flock constructed by an Andrè translation plane may be thought of as a replacement set between π_{θ^σ} and π_θ, corresponding to that proper subset λ mentioned above. For more details in the finite case see Cherowitzo and Johnson [25]. The complete statement is given there, but for infinite planes, is identical to what we have in the finite case. ☐

PROBLEM 36. *This problem is completely open. Assume that π is a bilinear translation plane in $PG(3, q^2)$ that is lifted from a translation plane in $PG(3, q)$. Prove that π is an Andrè plane.*

3. Bilinear flocks of quadratic cones

This material on bilinear flocks of quadratic cones was inspired by the work of Thas [134], [136] on finite flocks whose planes contain a point. It is well known that the 'linear' flocks of quadratic cones, where the planes share a line, are those whose corresponding translation planes are Pappian (Desarguesian is the term used in the finite case). The star flock theorem of Thas has been previously discussed and shows that the only possible non-linear flocks whose planes share a point are those of Kantor-Knuth, which have associated translation planes. However, for infinite flocks, the situation is much different.

DEFINITION 78. *A 'bilinear flock' of any type in $PG(3, K)$, where K is a field, is a set of mutually disjoint planes covering a specific set S of points, such that planes share at least one of two lines L and M (at most one line can share both lines).*

In particular, there cannot be bilinear flocks of quadratic cones in the finite case, even though there can be bilinear α-flocks, as we have seen in the previous section. The question of the existence of bilinear flocks then became open only in the infinite case. This is considered here over possible flocks over ordered fields K.

DEFINITION 79. *An 'ordered field' is a field $(K,+,\cdot)$ together with a binary operation $<$ that satisfies the following properties:*

(1) $\forall x$, $\forall y$, exactly one of $x = y$, $x < y$, $y < x$ is satisfied.
(2) $\forall x$, $\forall y$, $\forall z$, $x < y$ and $y < z \Longrightarrow x < z$.
(3) $\forall x$, $\forall y$, $\forall z$, $x < y \Longrightarrow x + z < y + z$.
(4) $\forall x$, $\forall y$, $\forall z$, $0 < z$ and $x < y \Longrightarrow z \cdot x < z \cdot y$.

Using the definition of an ordered field, most of the customary properties may be satisfied, such as $c > 0 \Longrightarrow -c < 0$, $a^2 > 0$, for $a \in K$, $a > 0$, and $b < 0$ then $a(-b) > 0$.

PROBLEM 37. *Show that if $b^2 - 4a < 0$ then $x^2 + bx + a > 0$, for all $x \in K$. This will show that the quadratic is irreducible over K.*

Define a quadratic cone in the normal manner in $PG(3, K)$, where (x_1, x_2, x_3, x_4) are the points represented in homogeneous coordinates and let $x_1 x_2 = x_3^2$, define a quadratic cone with vertex $(0, 0, 0, 1)$. Recall that translation planes with spreads in $PG(3, K)$, for K a field, that admit regulus-inducing elation groups are equivalent to flocks of quadratic cones.

THEOREM 99. *Define $P(a_1, a_2, b_1, b_2)$, where $(a_1, a_2) \neq (b_1, b_2)$, $a_2^2 + 4a_1 < 0$ and $b_2^2 + 4b_1 < 0$. Then define the translation plane with spread in $PG(3, K)$ admitting a regulus-inducing elation group as follows:*

$$x = 0, \left\{ y = x \begin{bmatrix} u + a_2 t & a_1 t \\ t & u \end{bmatrix} ; \forall t, u \in K; \text{ where } t > 0 \right\};$$

denote the partial flock by N_t

$$\left\{ y = x \begin{bmatrix} u + b_2 t & b_1 t \\ t & u \end{bmatrix} ; \forall t, u \in K; \text{ where } t < 0 \right\};$$

denote the partial flock by S_t

$$\left\{ y = x \begin{bmatrix} u & 0 \\ 0 & u \end{bmatrix} \forall u \in K, \right\} \text{ denote the associated plane } I_0.$$

And,

$$(N_t; t > 0) \cup (S_s; s < 0) \cup I_0$$

is the associated flock of a quadratic cone.

PROOF. First note that the first set of matrices and including all values of t and not just the positive values, define a Pappian spread, since the determinant of this set of matrices is $u^2 + a_2 ut - a_1 t^2$. We note that the quadratic $x^2 + a_2 x - a_1$ is non-zero since $a_2^2 + 4a_1 < 0$, as we may use the quadratic equation in this setting. It is clear that this Pappian spread is equivalent to a linear flock of a quadratic cone. Similarly, we see that $x^2 + b_2 x - b_1$ is irreducible over K and also then is equivalent to a linear flock of a quadratic cone.

That being said, by taking differences of the matrices, to ensure that these differences form a non-singular matrix, we need only show that

$$\begin{bmatrix} u + a_2 t - b_2 s & a_1 t - b_1 s \\ t - s & u \end{bmatrix}$$

has a non-zero determinant, for $t > 0$ and $s < 0$. And, since we may use the quadratic formula for characteristic $\neq 2$ fields, we need to show that

$$(a_2t - b_2s)^2 + 4(t-s)(b_1s - a_1t) > 0.$$

Since $t - s > 0$, since $s < 0$, it remains to show that $(b_1s - a_1t) > 0$, but b_1, a_1 are both < 0 and $s > 0$, $t > 0$, so $b_1s > 0$ and $-a_1t > 0$. If we would change (b_1, b_2) to (a_1, a_2), we would obtain a spread, the Pappian translation plane $\Sigma_{(a_1,a_2)}$. So, effectively, the South S_t (for (a_1, a_2)) of the Pappian translation plane has been replaced by the South side S_t (for (b_1, b_2)) of the second Pappian translation plane $\Sigma_{(b_1,b_2)}$. This shows that we have a cover of the points by the union of N_t and S_t, $N_t \in \Sigma_{(a_1,a_2)}$ $S_t \in \Sigma_{(b_1,b_2)}$ and I_0. Since the two Pappian translation planes are linear with lines L_1 and L_2, we see that the constructed translation plane is bilinear. This completes the proof. □

DEFINITION 80. *The flocks of quadratic cones corresponding to the translation planes $P(a_1, a_2, b_1, b_2)$, are called the 'F(a_1, a_2, b_1, b_2)-flocks of quadratic cones'. We shall use the notation N_t for $t > 0$ and S_t for $t < 0$ for both the spread and the flock, and we call these sets the 'north and south partial spreads or partial flocks', respectively.*

Since the flocks each consist of two partial linear flocks with different lines, part of the following is noted in the previous proof.

THEOREM 100. *The $F(a_1, a_2, b_1, b_2)$-flocks are bilinear. The algebraic form of the flock is as follows:*

$$N_t : tx_1 - a_1tx_2 + a_2tx_3 + x_4; t > 0,$$
$$S_t : tx_1 - b_1tx_2 + b_2tx_2 + x_4; t > 0.$$
$$I_o : x_4 = 0.$$

Then $\cap_{t>0}N_t$ satisfies $x_1 - a_1x_2 + a_2x_3 = 0$, and $\cap_{t<0}S_t$ satisfies $x_1 - b_1x_2 + b_2x_3 = 0$. So, $I_o \cap \{\cap_{t>0}N_t\}$ is the line $x_1 = a_1x_2 - a_2x_3$, so that I_0 shares a line of each north and south partial flocks. The common point of the planes of the flock is as follows: If $a_1 = b_1$, the point is $(a_1, 1, 0, 0)$, if $a_2 = b_2$, the point if $(-a_2, 0, 1, 0)$ and if $a_1 \neq b_1$ and $a_2 \neq b_2$, the common point is $(-a_1\frac{a_2-b_2}{a_1-b_1} + a_2, \frac{a_2-b_2}{a_1-b_1}, 1, 0)$.

THEOREM 101. *Assume the hypothesis of the previous result. Any bilinear flock translation plane admits a group that fixes a regulus net, fixes the north net N_t and the south net S_t, and acts transitively on the regulus nets of each N_t and S_t.*

PROOF. Consider the following group $\left\langle \begin{bmatrix} I_2 & 0_2 \\ 0_2 & rI_2 \end{bmatrix}; \forall r > 0, r \in K \right\rangle$.
Note that

$$y = x \begin{bmatrix} u + a_2t & a_1t \\ t & u \end{bmatrix} \rightarrow \begin{bmatrix} ur + a_2tr & a_1tr \\ tr & ur \end{bmatrix},$$

for all $r > 0$, implies that group is transitive on the regulus nets of N_t. If $t < 0$, then since $tr < 0$, and all negative elements of K^* are $-r$ for all $r > 0$, it follows that the analogous argument on S_t, shows that the group is transitive on the regulus nets of each N_t and S_t. □

There are some bilinear flock spreads that admit groups that fix a regulus net and are transitive on the remaining regulus nets.

THEOREM 102. *Let $F(a_2, a_1, -a_2, a_1)$ be a bilinear flock spread. Then there is a collineation group that fixes a regulus net and is transitive on the remaining regulus nets.*

PROOF. Consider the group

$$\left\langle g = \begin{bmatrix} \begin{bmatrix} -1 & 0 \\ 0 & 1 \end{bmatrix} & 0_2 \\ 0_2 & \begin{bmatrix} -1 & 0 \\ 0 & 1 \end{bmatrix} \end{bmatrix} \right\rangle.$$

Then g fixes each component $y = x \begin{bmatrix} u & 0 \\ 0 & u \end{bmatrix}$, and $x = 0$ and maps

$$y = x \begin{bmatrix} a_2 t & a_1 t \\ t & 0 \end{bmatrix} \text{ to } y = x \begin{bmatrix} a_2 t & -a_1 t \\ -t & 0 \end{bmatrix} = \begin{bmatrix} -a_2(-1)t & a_1(-t) \\ -t & 0 \end{bmatrix};$$

it follows that N_t maps to S_t, so by the previous theorem, we have the proof to the theorem in question.   □

4. Translation planes admitting $SL(2, K)$

Using the Klein quadric, the collineation groups of the flock F of a quadratic cone and the associated translation plane π admitting a regulus-inducing elation group E are related as follows: The point orbits of E in the vector space associated with π, become the 'points' within the Klein quadric that defines the flock F. The normalizer $N_G(E)/EK^*$, where K^* is the kernel subgroup of the translation planes (the subgroup that fixes each component), where G_π is the collineation group of π, is the collineation group of the flock that permutes the planes of the flock. In the finite case, if the associated translation plane is not Desarguesian and/or the flock of the quadratic cone is not linear, then E is normal in the collineation group of π and the full collineation group of the flock G_F is $G_\pi/EK^* \simeq G_F$. However, in the infinite case, this is not the situation. And, for ordered fields K, there is a characterization of the corresponding groups of the flock and translation planes connected to bilinear spreads, which shall be presented in this section.

DEFINITION 81. *A 'partially transitive flock' of a quadratic cone is a flock that admits a group preserving the quadratic cone, fixes a plane and acts transitively on the remaining planes of the flock.*

PROBLEM 38. *Given any field K, it is an open problem to determine the set of partially transitive flocks of a quadratic cone in $PG(3, K)$.*

However, we see this problem may be solved for bilinear flocks over ordered fields such that the positive elements are always square.

COROLLARY 27. *All bilinear flocks corresponding to $F(a_2, a_1, -a_2, a_1)$ over ordered fields such that the positive elements are square are partial transitive.*

Now we show that all bilinear flock spreads admit the group $SL(2, K)$.

5. Double covers

DEFINITION 82. *Let π be a translation flock plane over $PG(3, K)$, for K a field (corresponds to a flock of a quadratic cone). A 'double cover' is a set of two sets of regulus nets $\mathcal{R}_1, \mathcal{R}_2$ and two components l_1, l_2 such that $\mathcal{R}_i$ is a set of regulus nets corresponding to regulus-inducing elation groups E_i, for $i = 1, 2$.*

DEFINITION 83. *A double cover is said to be a 'regular double cover' if and only if the set of regulus nets $\mathcal{R}_1, \mathcal{R}_2$ share a regulus net.*

In section 11 of this part, flocks of elliptic quadrics will be discussed where it will be noted that the two sets of reguli share at most one regulus or the translation plane is Pappian, and we have such a flock.

Assume that we have a regular double cover.

We may choose l_1 and l_2 to have coordinates $x = 0$ and $y = 0$, respectively. Isolating first on the group E_1, we may represent the components in the form

$$x = 0, y = x \begin{bmatrix} u + f(t) & g(t) \\ t & u \end{bmatrix} : \forall u, t \in K; f, g \text{ functions on } K,$$

where the group E_1 has the form:

$$\left\langle \begin{bmatrix} I_2 & uI_2 \\ 0_2 & I_2 \end{bmatrix} : \forall u \in K \right\rangle.$$

Since the regulus nets correspond to the orbits of $E_1 - \{x = 0\}$, it follows that l_2 may be chosen, as it was previously, as any component in any orbit of E_1, and so, as $y = 0$. This means that the group E_2 has the form

$$\left\langle \begin{bmatrix} I_2 & 0_2 \\ M_\lambda & I_2 \end{bmatrix} : \forall \lambda \in K \right\rangle.$$

Since we have a regular double cover, we may assume that that $M_\lambda = uI_2$.

THEOREM 103. *Let π be a translation plane admitting a regular double cover. Then π admits $SL(2, K)$ as a collineation group in the translation complement(fixes the zero vector) that is transitive on the components of the common regulus.*

PROOF. We have, from previous agreement, a translation plane that admits a collineation group G leaving invariant the regulus net

$$x = 0, y = x \begin{bmatrix} u & 0 \\ 0 & u \end{bmatrix} : \forall u \in K,$$

where

$$G = \left\langle E_1 = \begin{bmatrix} I_2 & uI_2 \\ 0_2 & I_2 \end{bmatrix}, E_2 = \begin{bmatrix} I_2 & 0 \\ vI_2 & I_2 \end{bmatrix} : \forall u, v, \in K \right\rangle.$$

Note that Baer subspaces on the regulus net are Pappian planes each admitting an isomorphic group induced upon each of them. The group induced is $SL(2, K)$, and generated by elation groups. For more information on the classical groups, the reader could consult Dieudonnè [45]. □

THEOREM 104. *All bilinear flock spreads $F(a_2, a_1, b_2, b_1)$ admit a regular double cover and therefore admit $SL(2, K)$ as a collineation group in the translation complement.*

PROOF. It remains to show that the translation planes admit

$$G = \left\langle \begin{bmatrix} I_2 & uI_2 \\ 0_2 & I_2 \end{bmatrix}, \begin{bmatrix} I_2 & 0 \\ vI_2 & I_2 \end{bmatrix} : \forall u, v, \in K \right\rangle.$$

So, we need only to prove that the collineation group

$$E_2 = \left\langle \begin{bmatrix} I_2 & 0 \\ vI_2 & I_2 \end{bmatrix} : \forall v, \in K \right\rangle$$

exists. It is easy to see that we need only show that the image of $y = xM$ is a component of the spread. In this case, we only need to consider that M is a matrix of the form $\begin{bmatrix} u + c_2 t & c_1 t \\ t & u \end{bmatrix}$, where $(c_2, c_1) = (a_2, a_1)$, when $t > 0$ and $= (b_2, b_1)$, when $t < 0$.

$$y = x \begin{bmatrix} u + c_2 t & c_1 t \\ t & u \end{bmatrix} \rightarrow$$

$$(x(I_2 + \begin{bmatrix} u + c_2 t & c_1 t \\ t & u \end{bmatrix} \begin{bmatrix} v & 0 \\ 0 & v \end{bmatrix}), x \begin{bmatrix} u + c_2 t & c_1 t \\ t & u \end{bmatrix}).$$

It follows immediately that we need to show that

$$\left[I_2 + \begin{bmatrix} u + c_2 t & c_1 t \\ t & u \end{bmatrix} \begin{bmatrix} v & 0 \\ 0 & v \end{bmatrix} \right]^{-1} \begin{bmatrix} u + c_2 t & c_1 t \\ t & u \end{bmatrix}$$

is a component for each $v \in K$. We note that since we have two Pappian spreads that admit the groups E_1 and E_2, the question is when the positive and negative parts of the bilinear spread are respected. A short calculation will produce the following matrix

$$\begin{bmatrix} w + c_2 \frac{t}{\Delta} & c_1 \frac{t}{\Delta} \\ \frac{t}{\Delta} & w \end{bmatrix},$$

where $\Delta = (1 + uv)(1 + uv + c_2 t^2 v) - c_1 t^2 v^2$, and $w = \frac{(1+uv)u+uvc_2 t)-c_1 t^2 v}{\Delta}$. If the reader wished to check this calculation, just pull out that part corresponding to the $(2, 1)$ entry, which is $\frac{t}{\Delta}$, to guide what w should be. When $(c_2, c_1) = (a_2, a_1)$ and $t > 0$, it remains to determine that $\Delta > 0$. To see this part, assume that $v \neq 0$ and let $z = \left(\frac{(1+uv)}{tv} \right)$, to end up asking if $z^2 + c_2 z - c_1$ is positive. Then complete the square to obtain $(z + \frac{c_2}{2}) - \frac{1}{4}(c_2^2 + 4c_1)$. And, since $a_2^2 + 4a_1 < 0$, this proves the assertion. When $(c_2, c_1) = (b_2, b_1)$, and $t < 0$, we still end up if $b_2^2 + 4b_1 < 0$, by analogy. We now obtain that a bilinear flock plane as a regulus double cover, so that $SL(2, K)$ is generated by the previous result. This completes the proof of the theorem. □

Finally, we close this section by characterizing translation planes admitting regular double covers over ordered fields, such that each positive element is a square, are exactly the bilinear spreads.

THEOREM 105. *Let π be a translation plane with spread in $PG(3, K)$, for K an ordered field, such that each positive element is a square. Assume that π admits a collineation group in the translation complement isomorphic to $SL(2, K)$, which is generated by regulus-inducing elation groups.*

(1) If we have a regular double cover then π is a bilinear flock spread.

(2) All double covers over such ordered fields are regular double covers.

PROOF. We have components of the form

$$x = 0, y = x \begin{bmatrix} u + f(t) & g(t) \\ t & u \end{bmatrix} : \forall t, u \in K,$$

for functions f and g on K. Let $t = 1, u = 0$ and define $f(1) = a_2$, and $g(1) = a_1$. And, let $f(-1) = b_2$ and $g(-1) = b_1$. By the proof of the previous theorem, we see that we have a collineation group of the following form:

$$G = \left\langle \begin{bmatrix} u & 0 & v & 0 \\ 0 & u & 0 & v \\ w & 0 & z & 0 \\ 0 & w & 0 & z \end{bmatrix} ; \forall u, v, w, z \in K : uz - vw = 1 \right\rangle.$$

Note that the group induces $SL(2, K)$ faithfully on two Baer subspaces of the regulus net so that $uz - vw = 1$ on each of these subspaces. Additionally, since the kernel group K^* acts as the subgroup $\left\langle Diag \begin{bmatrix} r & 0 \\ 0 & r \end{bmatrix} ; r \in K^* \right\rangle$, we see that GK^* is a group such that $uzr^2 - vwr^2 = r^2 > 0$. Choose $u = r^{-1}$, $z = r^2$ and $w = v = 0$. Hence, there exists a group of the form

$$\left\langle \begin{bmatrix} 1 & 0 & 0 & 0 \\ 0 & 1 & 0 & 0 \\ 0 & 0 & r^2 & 0 \\ 0 & 0 & 0 & r^2 \end{bmatrix} ; \forall u, v, w, z \in K : uz - vw = 1 \right\rangle.$$

Now the images of $y = x \begin{bmatrix} u + f(t) & g(t) \\ t & u \end{bmatrix}$ for $t = \pm 1$. We see that we have the spread components

$$y = x \begin{bmatrix} a_2 & a_1 \\ 1 & 0 \end{bmatrix} \rightarrow y = x \begin{bmatrix} a_2 r^2 & a_1 r^2 \\ r^2 & 0 \end{bmatrix}$$

and

$$y = x \begin{bmatrix} -b^2 & -b_1 \\ -1 & 0 \end{bmatrix} \rightarrow \begin{bmatrix} -b^2 r^2 & -b_1 r^2 \\ -r^2 & 0 \end{bmatrix}.$$

Since all positive elements are squares and all squares are positive, we have a partition of the spread into two parts plus the fixed regulus net. If $(a_2, a_1) = (b_2, b_1)$ then the plane is Pappian. Assume that $(a_2, a_1) \neq (b_2, b_1)$. Now consider the set of components

$$\left\{ y = x \begin{bmatrix} u + a_2 t & a_1 t \\ t & u \end{bmatrix} : \forall u, t > 0 \in K \right\}.$$

We know that the determinants $u^2 + a_2 tu - a_1 t^2 \neq 0$. This means that, again, by completing the square, $(u + \frac{a_2}{2} t) - ((\frac{a_2}{2})^2 + a_1)t^2 \neq 0$. This, in turn, means that

$$(\frac{a_2}{2})^2 + a_1 \neq (\frac{u + a_2 t}{t})^2.$$

But, since this is valid for all $u \in K$, then for t fixed and non-zero, $(\frac{u + a_2 t}{t})$ is bijective and the square of such elements is surjective on the squares. In order that we do not have an equality, we must have $a_2^2 + 4a_1 < 0$. The same argument shows that

$b_2^2 + 4b_1 < 0$. Hence, we have a bilinear flock of a quadratic cone, which completes the proof of part (1) of the theorem.

Now assume that we have a double cover but not necessarily a regular double cover. By suitable coordinate change, we still have the spread as

$$x = 0, y = x \begin{bmatrix} u + f(t) & g(t) \\ t & u \end{bmatrix} : \forall t, u \in K,$$

for functions f and g on K. There are two possibilities: there is a second regulus-inducing elation group with the same axis $x = 0$, or the second regulus-inducing elation group has an axis $L \neq (x = 0)$. In the first case, we obtain a second group

$$\left\langle \begin{bmatrix} I_2 & M_\alpha \\ 0 & I_2 \end{bmatrix} ; \alpha \in \lambda \right\rangle,$$

where $\{M_\alpha : \alpha \in \lambda \}$ is a set of additive matrices.

The image of $y = 0$ together with $x = 0$ is a regulus net and has the form

$$x = 0, y = x M_\alpha : \forall \alpha \in \lambda.$$

Choose some fixed matrix M_{α_0} and form the regulus net.

$$x = 0, y = x M_\alpha M_{\alpha_0}^{-1} : \forall \alpha \in \lambda.$$

These are the derivable nets studied in the main material on classifying derivable nets. It has been pointed out that there cannot be two distinct derivable nets in a translation plane of this type. Hence, the two regulus nets are identical. So

$$M_\alpha = \begin{bmatrix} v + f(t) & g(t) \\ t & v \end{bmatrix} \begin{bmatrix} u & 0 \\ 0 & u \end{bmatrix} : \forall u \in K, \ v, t \neq 0.$$

That being said, $f(t)u = f(tu)$ and $g(tu) = g(t)u$ for all $u \in K$. It follows easily that π is a Pappian plane. Assume then the plane is not Pappian and therefore, we must have the second case, that there is a regulus-inducing elation group with an axis $L \neq (x = 0)$. By a basis change, we then may assume that $L = (y = 0)$, since we may choose L as any component in a given E-orbit, where E denotes the regulus-inducing elation group with axis $x = 0$.

Therefore, we have a similar situation

$$\left\langle \begin{bmatrix} I_2 & 0 \\ M_\alpha & I_2 \end{bmatrix} ; \alpha \in \lambda \right\rangle,$$

where $\{M_\alpha : \alpha \in \lambda \}$ is a set of matrices.

In this case, $x = 0$ is mapped as $(0, y)$ to $x = y M_\alpha$. And this set together with $y = 0$ is a regulus net. If we form $x = y M_\alpha M_{\alpha_0}^{-1}$, then the regulus net is $y = 0$,
$x = y M_{\alpha_0}^{-1} \begin{bmatrix} u & 0 \\ 0 & u \end{bmatrix} : \forall u \in K$. We have the same situation as just above, implying
that since the plane is not Pappian, we have $y = 0, x = y \begin{bmatrix} u & 0 \\ 0 & u \end{bmatrix} : \forall u \in K$, and
equivalently that $x = 0, \ y = x \begin{bmatrix} u & 0 \\ 0 & u \end{bmatrix} : \forall u \in K$, is the regulus net with elation
group having axis $y = 0$. It then follows that we have a regulus double cover and part (2) now follows. $\qquad\square$

We now consider the question of the relationship with the group G_F of a flock of a quadratic cone and the group G_π of the associated translation plane. We resolve the relationship with flocks over ordered fields K where the positive elements are square.

THEOREM 106. *Let F be a flock of a quadratic cone over K, where K is an ordered field such that the positive elements are squares. Assume that the full collineation group is G_F. Let π denote the associated translation plane and let G_π denote the full collineation group. Let K^* denote the kernel homology group and E a regulus-inducing elation group.*

Then one of the following holds:

(1) π is Pappian and F is linear,

(2) π is a bilinear flock spread, there is a regular double cover and $SL(2, K)$ is generated by the elation groups.

(3) $G_\pi/EK^* \simeq G_F$.

6. nm-Linear flocks of quadratic cones

The concept of a bilinear flock of a quadratic cone became novel, since none can exist in the finite case. We have seen in the previous sections that there are a great variety of bilinear flocks of quadratic cones in the infinite case. In this section, we ask if there could be tri-linear, n-linear, nm-linear flocks of quadratic cones, again in the infinite case.

DEFINITION 84. *A k-linear flock of a quadratic cone is a flock whose planes contain one of k different lines of $PG(3, K)$, where K is an infinite field, for k an integer > 1.*

Here we take a less general construction method and consider ordered fields K, and translation planes of the following type:

$$x = 0, y = x \begin{bmatrix} u & at \\ t & u \end{bmatrix}; \text{ where } \alpha < 0; \forall t, u \in K.$$

Clearly, since the determinants have the form $u^2 - at^2 > 0$, for $(t, u) \neq (0, 0)$, these matrix spreads always give rise to linear flocks of a quadratic cone. We have seen that by choosing different Pappian spreads for the north and the south sections of the quadratic cone, there are a very large number of bilinear flocks of quadratic cones. Now we show how to partition the north and south sections, into n and m pieces, respectively, to create $nm-$linear flocks.

THEOREM 107. *Let K be an ordered field and choose sets*

$$\{a_i; i = 1, 2, ..., n\} \text{ and } \{b_i; i = 1, 2, ..., m\},$$

where n and m are positive integers and/or possibly $n, m \to \infty$, and

$$...a_n \quad < \quad a_{n-1} < ... < a_2 < a_1 < 0,$$
$$...b_m \quad < \quad b_{m-1} < .., < b_2 < b_1 < 0.$$

Choose partial spreads of the corresponding Pappian spreads with components

$$y = x \begin{bmatrix} u & a_i t_i \\ t_i & u \end{bmatrix}; \text{ where } \alpha_i < 0; \forall t, u \in K,$$

where

$$0 < t_1 \le d_1 < t_2 \le d_2 < t_3 \le d_3 < ... < t_n \to \infty$$

for an arbitrary but fixed set of positive elements $\{d_i; i = 1, 2, ..., n\}$ that forms a strictly increasing sequence of positive elements. Also, choose partial spreads of the corresponding Pappian spreads with components

$$y = x \begin{bmatrix} u & b_i s_i \\ s_i & u \end{bmatrix}; \; where \; b_i < 0,$$

where

$$0 > s_1 \ge e_1 > s_2 \ge e_2 > ...s_m \to -\infty$$

for arbitrary but fixed sets of negative elements $\{e_i; i = 1, 2, ..., m\}$, that forms a strictly decreasing sequence of negative elements.

   Then

$$x = 0, y = x \left(\begin{bmatrix} u & a_i t_i \\ t_i & u \end{bmatrix}; \forall t_i, u \in K; \\ where \; d_{i-1} < t_i \le d_i, \; i = 1, 2, ..., n, \; d_0 = 0 \right),$$

$$y = x \left(\begin{bmatrix} u & b_j s_j \\ s_j & u \end{bmatrix}; \forall s_j, u \in K; \\ where \; e_{j-1} > s_j \ge e_j, j = 1, 2, ,, m, e_0 = 0 \right)$$

$$y = x \left(\begin{bmatrix} u & 0 \\ 0 & u \end{bmatrix} \forall u \in K \right),$$

is a translation plane admitting a regulus-inducing elation group and constructs an nm-linear flock of a quadratic cone.

PROOF. The proof will be in three parts. First we show that

$$\begin{bmatrix} u & a_i t_i \\ t_i & u \end{bmatrix} - \begin{bmatrix} 0 & a_j t_j \\ t_j & 0 \end{bmatrix}$$

is a non-singular matrix, for the north part of the quadratic cone and further show that the north is completely covered. Then we show the analogous assertions for the south are also valid. Then it will be shown that the matrices corresponding to the north and the south have differences that are non-singular.
$\begin{bmatrix} u & a_i t_i - a_j t_j \\ t_i - t_j & u \end{bmatrix}$ is non-singular if and only if $-(t_i - t_j)(a_i t_i - a_j t_j) > 0$, for $t_i \ne t_j$. Assume that $i > j$. Then $t_i > t_j$ and so $(t_i - t_j) > 0$. Since $t_i > t_j$ then $a_i t_i < a_i t_j$, as $a_i < 0$. But, $a_i < a_j$, as $i > j$. For that reason, $a_i t_j < a_j t_j$ since $t_j > 0$. Thus $(a_i t_i - a_j t_j) < 0$, forcing $-(t_i - t_j)(a_i t_i - a_j t_j) > 0$. If $j > i$, then the same argument will work by interchanging all j terms with all i terms. In other words, $\begin{bmatrix} u & a_i t_i - a_j t_j \\ t_i - t_j & u \end{bmatrix}$ is nonsingular if and only if $-I_2 \begin{bmatrix} u & a_i t_i - a_j t_j \\ t_i - t_j & u \end{bmatrix}$ is non-singular.
   Now since the increasing sequence of terms

$$0 < t_1 \le d_1 < t_2 \le d_2 < t_3 \le d_3 < ... < t_n \to \infty$$

as a set is $t > 0$, we have a covering of the north side.

The argument for the south side is analogous, as consider

$$\begin{bmatrix} u & b_i s_i - b_j s_j \\ s_i - s_j & u \end{bmatrix}.$$

To show that we have a set of non-singular matrices, we need to show that $-(s_i - s_j)(b_i s_i - b_j s_j) > 0$. By similar remarks as for the north side argument, we may assume that $i > j$. For that reason $s_j > s_i$. Hence, $-(s_i - s_j) > 0$. For that reason, we need to show that $(b_i s_i - b_j s_j) > 0$. Since $s_j > s_i$ and $b_j > b_i$, therefore, $-s_j < -s_i$ and $-b_j < -b_i$, and since all terms are > 0, it follows that $b_j s_j < b_i s_i$, so that $b_i s_i - b_j s_j > 0$. It now follows that we have a cover for the south side. We now need to show that $-(t_i - s_j)(a_i t_i - b_j s_j) > 0$. Since $t_i > 0$ and $-s_j > 0$, then we need to show that $(t_i - s_j)(b_j s_j - a_i t_i) > 0$, or rather, that $a_i t_i < b_j s_j$. This, however, is clear since $a_i t_i < 0$ and $b_j s_j > 0$ as both b_j and $s_j < 0$. This completes the proof.   □

7. Nests of reguli

We have concentrated on the discussion of translation planes containing sets of derivable nets, where the nets are either mutually disjoint or share one or two components. In this section, we consider replaceable nets in a very different manner, by replacing 'nests of reguli'. This section concerns arbitrary and infinite nests of reguli. The concept of a nest or 'chain of reguli' was conceived in the finite case by Bruen [19] as a set of $(q+3)/2$ reguli in $PG(3,q)$, q odd, such that each line of the set is contained in exactly two reguli. For details, the reader is directed to Johnson [85], and the known chains of reguli are listed in the Handbook [94], pp. 486-489. The main importance of chains is there is a replacement procedure of certain sets of the Baer subplanes (or lines in the set of opposite reguli), creating important classes of translation planes. There are generalizations of chains of reguli that are called 't-nests', that were created and developed by Baker and Ebert in a series of articles (See [9] and references in [94]).

DEFINITION 85. *A 't-nest' is a set of t-reguli in $PG(3,q)$, for q odd, such that every line of the set is contained in exactly two of the reguli. The corresponding translation net has degree $t(q+1)/2$.*

- There are infinite classes of t-nests for $t = q, q-1, q+1$, or $2(q-1)$.
- All of the classes originate in a Desarguesian translation plane.

Since the arguments showing the existence of these nests use properties of squares and non-squares in odd order finite fields, it is possible to extend these translation planes to the infinite case, using fields K that admit a Galois quadratic extension. In this section, the focus shall be on the arbitrary case.

DEFINITION 86. *A 'nest of reguli' is a set of reguli in $PG(3,K)$, for K a field, such that every line of the set is contained in exactly two of the reguli and no two lines are contained in three reguli.*

The replacement procedure normally available involves a collineation group of a Pappian plane.

DEFINITION 87. *A 'full field' K of characteristic $\neq 2$, is a field such that the product of any two non-squares is a square and there exist non-squares. For that* *reason, in this setting, a field is full if and only if the multiplicative subgroup of non-zero squares is an index two subgroup in the multiplicative group.*

To describe the groups G of interest, let K of characteristic $\neq 2$ be a full field that is assumed to have a quadratic extension $K(\sqrt{\gamma})$, where γ is a non-square in K. Let σ denote the involutory automorphism of $K(\sqrt{\gamma})$ mapping $\sqrt{\gamma}t+u \to -\sqrt{\gamma}t+u$, for all $t, u \in K$. We consider the associated spread for the Pappian translation plane Σ to be

$$x = 0, y = 0, y = x \begin{bmatrix} u & \gamma t \\ t & u \end{bmatrix} : \forall t, u \in K - \{0\}.$$

We let G be a subgroup of $GL(4, K)$ and G will leave invariant a set of reguli called the 'base reguli', whose partial spreads are in the spread for Σ. The base reguli have one of the three following possible properties:

(1) the base reguli will be mutually disjoint and disjoint from $x = 0$ and $y = 0$,
(2) the base reguli will share exactly one component ($x = 0$) and otherwise be disjoint, or
(3) the base reguli will share exactly two components $\{x = 0, y = 0\}$ and otherwise be disjoint.

We have seen examples of such sets of base reguli in the construction of the Andrè planes, the flock (of quadratic cones) spreads, and the hyperbolic flock spreads. We have noted that there are central collineation groups that act so as to construct linear flocks in each of these three situations. We shall point this out more clearly when the description of G is fully complete.

The group required shall have the following requirements:

(i) G acts transitively on the components of each base regulus net, which are distinct from $\{x = 0, y = 0\}$(if any),
(ii) the kernel subgroup of Σ in G has two orbits on the 1-dimensional K-subspaces of any component and so we note under these conditions, that
(iii) the 1-dimensional subspaces of the base reguli which are not on any common component are in two orbits under G.

Suppose that there is a Baer subplane π_0 of Σ which is a 2-dimensional K and such that

(iv) π_0 shares either two or zero components with any base regulus and
(v) the two 1-dimensional K-subspaces of intersection on the components of a given base regulus of intersection are in distinct G-orbits.
(vi) We also assume that when the base reguli are disjoint from $x = 0$ and $y = 0$, G acts transitively on the set of Baer subplanes of the opposite regulus of each fixed base regulus.

Let H be the subgroup of squares of the full homology group

$$\langle (x, y) \to (xu^2, yu^2) : u \in K(\sqrt{\gamma})^* \rangle.$$

Let $A = \langle (x, y) \to (x, ym) : m \in K^* \text{ and } m^{\sigma+1} = 1 \rangle$ and let $G = AH$. This situation is tricky as the subplane π_0 needs to be taken so that the Andrè nearfield plane situation is avoided (See (6,8,11) to recall how such planes are constructed).

DEFINITION 88. *If $G/H \simeq A$, the associated nest is said to be an 'A-nest' and in the finite case, this becomes a '$(q+1)$-nest' as $|A| = q + 1$ in the finite case.*

REMARK 29. *$(q+1)$-nests and $(q+1)$-cyclic planes are connected in section 10.*

Let E denote the elation regulus-inducing group seen in the quadratic cone situation and $G = EH$.

DEFINITION 89. *This produces a '$G/H \simeq E$-nest of reguli'. In the finite case, this is a 'q-nest', as $|E| = q$ in the finite case.*

When G fixes two components $x = 0$ and $y = 0$, we use a group producing the regulus-inducing homology group of the form

$$\langle (x,y) \to (xa^{\sigma+2}, ya) : \forall a \in K^* \rangle\, H.$$

DEFINITION 90. *If G/H together with the K-kernel homology group produces a regulus-inducing homology group*

$$H_x = \langle (x,y) \to (x, ya^{\sigma+1}) : \forall a \in K^* \rangle.$$

This is denoted as an 'H_x-nest' and, is a '$(q-1)$-nest' in the finite case, as $|H_x| = q - 1$.

There is a set of classification theorems for the nests described above, the descriptions of which are beyond the scope of this text. The interested reader can find these results listed in Johnson [85]. The main classification theorem is listed in the following section.

DEFINITION 91. *A translation plane constructed from a Pappian plane using a group G satisfying $(i) - (vi)$ is said to be a 'group replaceable plane'.*

8. Group replaceable translation planes

Note that the assumption that the mapping $m \to m^{\sigma+1}$, where $m \in K(\theta)$, is not required to be surjective for the theorem to be valid, contrary to how it states in the original version. The reason for this is the set K^- is adequate to form a covering of base reguli, except for two components, without the surjective assumption in the infinite case (See Multiple Replacement Theorem in part 10).

THEOREM 108. *(See [85] 11.1 p. 261–should be stated without the surjective assumption) Let Σ denote the Pappian plane coordinatized by $K(\sqrt{\gamma})$, where K and $K(\sqrt{\gamma})$ are full fields of characteristic $\neq 2$.*

Let π be a group replaceable translation plane by use of a group of the Pappian plane Σ. Assume that G contains the K-kernel group. Let π_0 be an associated Baer subplane and let N_{π_0} denote the regulus net in Σ defined by the components of π_0. Then $\pi_0 G$ is a replacement for $N_{\pi_o} G$ and one of the following must occur:

(1) $N_{\pi_o} G$ is an H_x-nest (regulus-inducing homology group type), $q-1$-nest in the finite case,

(2) $N_{\pi_o} G$ is an A-nest (Andrè type homology group), $q+1$-nest in the finite case,

(3) $N_{\pi_0} G$ leads to an Andrè nearfield plane,

(4) $N_{\pi_0} G$ defines a flock of a hyperbolic quadratic in the infinite case only,

(5) $N_{\pi_0} G$ is an E-nest (regulus-inducing elation group type), q-nest in the finite case.

To completely cover the ideas of t-nests would require much more space than is available in this text. The reader might just note that for the E-nest constructions, for all of the translation planes constructed, there is a regulus-inducing elation group. This will show that there is a flock of a quadratic cone connected to each of the translation planes. These flocks are considered by Jha and Johnson [72], in both the infinite and finite case, where they are called the 'Fisher flocks'.

THEOREM 109. ([72], Theorem 1, p. 657) Let K be a full field of odd or 0 characteristic and let $K(\theta)$ be a quadratic extension of K that is also a full field. Let Σ be the Pappian affine plane coordinatized by $K(\theta)$ and let H be the kernel homology subgroup of squares in Σ. Let σ denote the automorphism of order 2 in $Gal_K K(\theta)$.

Let s be any element of $K(\theta)$ such that $s^{\sigma+1}$ is a nonsquare in K if -1 is a nonsquare in K, and $s^{\sigma+1}$ is square if -1 is square in K.

Let E denote the regulus-inducing elation group of Σ. Then the spread

$$x = 0, \{ EH(y = x^\sigma s)\} = \{y = x^\sigma s b^{2(1-\sigma)} + x\alpha : \forall b \in K(\theta)^*, \forall \alpha \in K\}$$
$$\cup\{y = xm; (m+\beta)^{\sigma+1} \neq s^{\sigma+1}, \forall \beta \in K.$$

is a generalized Fisher conical flock spread in $PG(3, K)$.

We shall return to both the generalized Fisher translation planes, the A-nest translation planes, and their constructions at the end of the following section. The A-nest translation planes in the finite case are translation planes of order q^2 and admit cyclic homology groups of order $q+1$ and thus by the correspondence theorem produce flock planes of quadratic cones; are $(q+1)$-cyclic planes. The Fisher planes using the correspondence theorem in the opposite order produce translation planes of order q^2 that admit homology groups of order $q+1$.

There are extension theorems concerning the finite Fisher planes that will be useful.

THEOREM 110. (Johnson [81]) Let $\mathcal{P}$ be a non-linear partial flock of a quadratic cone of odd order containing a linear flock of $(q-1)/2$ conics. Then $\mathcal{P}$ may be uniquely extended to the Fisher flock.

A linear subflock of $(q-1)/2$ is said to be 'critical' in the finite case. Since, the consideration is for generalized Fisher flocks, the idea of a 'critical linear subflock' will need to be defined without finiteness.

DEFINITION 92. Let $\mathcal{C}$ be a linear subflock of a quadratic cone in $PG(3, K)$, where K is a field. Assume that there is a flock $\mathcal{F}$ of a quadratic cone containing $\mathcal{C}$. Let the associated partial flock spreads containing $\mathcal{C}$ and $\mathcal{F}$, respectively be denoted by Π and Σ, so that $\Pi \subseteq \Sigma$. Then there is a regulus-inducing elation group with axis ℓ such that Σ is a union of reguli sharing ℓ and each regulus is induced from E. These reguli are called 'the base reguli'. Note that Π is invariant under E and is a union of base reguli.

We shall say that $\mathcal{C}$ is a 'critical partial subflock' if and only if the following two conditions hold:

(i) Every Baer subplane within the affine plane defined by Σ and disjoint from Π intersects each base regulus of $\Sigma - \Pi$ in two components and there is some Baer subplane which is disjoint from Π.

(ii) If $\mathcal{D}$ is a set of distinct base reguli that covers $\Sigma - \Pi$, then every Baer sub-plane within the affine plane defined by Σ and disjoint from Π and not in $\mathcal{D}$, intersects each base regulus of $\mathcal{D}$ in two components.

Now to see that a linear partial flock of $(q-1)/2$ in a flock of q reguli is critical, we use a counting argument. So in $\Sigma - \Pi$ there are $(q+1)/2$ remaining reguli that are disjoint from Π. If π_0 is a Baer subplane that is disjoint from Π then π_0 cannot share the axis ℓ but has $q+1$ components of intersection in $\Sigma - \Pi$. For that reason, π_0 cannot be a Baer subplane of any of the reguli of $\Sigma - \Pi$, and shares either 1 or 2 components with each. Letting k_1 and k_2 respectively denote the number of reguli that π_0 intersects in 1 and 2 components leads to the system of equations:

$$k_1 + 2k_2 = q + 1,$$
$$k_1 + k_2 = (q+1)/2,$$

showing $k_2 = (q+1)/2$ and $k_1 = 0$. That is, π_0 intersects each base regulus of $\Sigma - \Pi$ in 2 components. The same sort of argument shows that both conditions are valid for linear subflocks of $(q-1)/2$ components.

PROBLEM 39. *Show the second condition (ii) is valid for linear subflocks of $(q-1)/2$ components.*

The following theorem will be important in this later analysis.

THEOREM 111. *(Jha and Johnson* [**72**] *Theorem 3, p. 662)* (**Critical partial flock extension theorem**). *Let K be a full field of characteristic odd or 0 and let $K(\theta)$ be a full field quadratic extension of K. If there exists a linear critical partial flock $\mathcal{C}$ then any non-linear partial flock extension of $\mathcal{C}$ may be uniquely extended to a generalized Fisher flock.*

9. Circle geometry over $K(\sqrt{\gamma})$

Many of the arguments on nests of reguli use a model of the Pappian spread, based on the work of Orr [**116**], on the association of Miquellian inversive planes with $PG(1, q^2)$ that can define regulus nets using the circles of the geometry. Here we give an abbreviated version of the author's original article [**85**] in which the reader can find the details. In that article, there is a complete discussion of the varieties of t-nests and their infinite analogues. We shall be interested here in the $(q+1)$-nets and enough theory to consider flocks of elliptic quadrics. In the $(q+1)$-nest theory, we shall see that the associated translation planes are cyclic translation planes, in the sense that they are of order q^2 and admit a cyclic homology group of order $q+1$ and thus produce flocks of quadratic cones using the correspondence theorem. We consider first the theory of circle geometry and the affine version of it that we shall be using. For space considerations, proofs shall be given as problems for the reader.

DEFINITION 93. *An 'inversive plane' is an incidence geometry of 'points' and 'circles' such that*

(i) any three distinct points are incident with a unique circle,

(ii) given two points P and Q and circle C_P incident with P but not Q, there is a circle $T_{P,Q}$ incident with both P and Q such that $C_P \cap T_{P,Q} = \{P\}$,

(iii) there are at least four points, every circle is incident with at least one point and there is an point-circle pair (P, C), where P is not incident with C.

DEFINITION 94. *The inversive plane obtained from $PG(1, q^2)$, where the circles are reguli and the points are elements of $PG(1, q^2)$ is a 'Miquellian inversive plane', normally connected with elliptic quadrics.*

This model may also be used in the infinite case with $PG(1, K(\theta))$, where $K(\theta)$ is a quadratic extension field of the field K. As this is also useful in the study of flocks of elliptic quadrics in the following section, we include it here, by connecting the set of regulus nets in the Pappian spread over $K(\theta)$ to the Miquellian circle geometry.

PROBLEM 40. *Let K be a full field of characteristic $\neq 2$, with a full field quadratic extension $K(\sqrt{\gamma})$, where γ is a non-square in K, with unique involutory automorphism σ.*

Define the 'points' to be the elements of $PG(1, K(\sqrt{\gamma}))$ and 'circles' the set of 2 by 2 matrices of the following form:

$$(*): \left\{ \begin{array}{l} \begin{bmatrix} a & \alpha \\ \beta & -a^\sigma \end{bmatrix} \text{ for each } a \in K(\sqrt{\gamma}), \\ \text{and for each } \alpha, \beta \in K, \text{ such that } a^{\sigma+1} + \alpha\beta \neq 0 \end{array} \right\}.$$

With the basis $\{1, \theta\}$ and elements of $PG(1, K(\sqrt{\gamma})$ as ∞ and $\sqrt{\gamma}t + u$; elements of $K(\sqrt{\gamma}) \cup \{\infty\}$, where $\left(\sqrt{\gamma}t + u\right)^\sigma = -\sqrt{\gamma}t + u$ and $\sqrt{\gamma}^{\sigma+1} = -\gamma$. The elements z of a circle represented as above are

$$z = \frac{az^\sigma + \alpha}{\beta z^\sigma - a^\sigma}, \text{ and } \infty \text{ is a point of the circle if and only if } \beta = 0.$$

This particular model will identify regulus nets with circles.

PROBLEM 41. *Show that the classical regulus net $x = 0, y = x \begin{bmatrix} u & 0 \\ 0 & u \end{bmatrix} : \forall u \in K$ is the circle $\begin{bmatrix} \sqrt{\gamma} & 0 \\ 0 & (\sqrt{\gamma})^\sigma = -\sqrt{\gamma} \end{bmatrix}$.*

PROBLEM 42. *Show that the regulus net $x = 0, y = x \begin{bmatrix} 0 & \gamma t \\ t & 0 \end{bmatrix} : \forall t \in K$ is the circle $\begin{bmatrix} 1 & 0 \\ 0 & -1 \end{bmatrix}$*

PROBLEM 43. *Note that the Pappian translation plane Σ over $K(\sqrt{\gamma})$, is triply-transitive on the components. Show that two distinct regulus nets share at most two components. Give an argument that justifies the point-circle geometry as describing the set of regulus nets in Σ.*

PROBLEM 44. *Give a reason that different circles correspond to the same regulus net.*

PROBLEM 45. *Show that an André net $y = xm : m^{\sigma+1} = \rho$, a constant in K^* corresponds to the circle $\begin{bmatrix} 0 & \rho \\ 1 & 0 \end{bmatrix}$.*

To recognize how two circles intersect, a quadratic form $\mathbb{Q}$ is obtained which has the following associated bilinear form h:

$$h(C, D) = Det(C + D) - DetC - DetD, \text{ for circles } C \text{ and } D.$$

Then the following is obtained.

PROBLEM 46. *When K is a full field with full field quadratic extension $K(\sqrt{\gamma})$, γ non-square in K, prove the following result.*

THEOREM 112. *(**Orr's Theorem**, see Orr [116] for odd finite case). Two distinct circles C and D are tangent, disjoint or secant (share two points) if and only if $h(C, D)^2/4 - DetCDetD$ is zero, a non-zero square, or a non-square, respectively.*

PROBLEM 47. *Let π_0 be a Baer subplane incident with the origin of the Pappian translation plane Σ. Prove that π_0 is a 2-dimensional K-subspace. Let N_{π_0} denote the net defined by the components of Σ that non-trivially intersect π_0 (therefore, in a 1-dimensional K-subspace). Prove that N_{π_0} is a derivable net; a regulus net.*

So, with the problems listed above solved, the idea of the construction of a t-nest, once a particular set of base reguli is considered, is to find a regulus net that is disjoint from $x = 0$, and $y = 0$ secant or disjoint from each base regulus. For E-nests, once a Baer subplane defines a regulus net disjoint from $x = 0$ (the axis of the regulus-inducing elation group) N_{π_0}, the elation group E and the subgroup H of squares essentially solves the problem. A particular choice of a Baer subplane actually completes the theory.

THEOREM 113. *We consider the base nets relative to the group E acting on the Pappian plane Σ of regulus nets sharing the axis $x = 0$ of E. Then there is a Baer subplane π_0 (subspace) disjoint from $x = 0$ that has the following two properties:*

(1) π_0 intersects a base regulus net in exactly two components.

(2) In the regulus net N_{π_0} defined by π_0 are exactly two Baer subspaces of intersection with the base regulus net. Then π_0 may be be chosen so that these two Baer subplanes of intersection are in different H-orbits.

PROOF. Consider the base regulus $\mathcal{B} : x = 0, y = x \begin{bmatrix} u & 0 \\ 0 & u \end{bmatrix} : \forall u \in K$. It is necessary to avoid the Baer subspaces of this regulus $\mathcal{B}$. Take π_0 to share $y = 0$ and $y = x$. If $\pi_0 \cap y = 0$ is $\langle (1, 0) \rangle$ choose $\pi_0 \cap y = x$ to be (v, v) for $v \in K(\sqrt{\gamma})$, generating

$$(\alpha + \beta v, \beta v) ; \forall \alpha, \beta \in K.$$

We require that $\alpha + \beta v = 0$ if and only if $\alpha = \beta = 0$. For example, let v be any non-square in $K(\sqrt{\gamma})$. Consider the 1-dimensional $K(\sqrt{\gamma})$-subspaces on $x = 0$. We see that the Baer subspaces of intersection in the regulus net $\mathcal{B}$ are 2-dimensional K-subspaces $\langle (1, 0), (0, 1) \rangle$ and $\langle (v, 0), (0, v) \rangle$ which are in different H-orbits. $\square$

Then the following theorem is proved.

THEOREM 114. *(Jha and Johnson [67]) Let Σ be a Pappian plane over a full field K of characteristic $\neq 2$, that has a quadratic full field quadratic extension $K(\sqrt{\gamma})$. Let G be the group EH, where E is the regulus-inducing elation group and H is the subgroup of squares. We consider the base nets relative to the group E*

acting on the Pappian plane Σ of regulus nets sharing the axis $x = 0$ of E. Choose any Baer subplane π_0 (subspace) disjoint from $x = 0$ that intersects a base regulus net in exactly two components. In the regulus net N_{π_0} defined by π_0 there are exactly two Baer subspaces of intersection with the base regulus net. Assume that these two Baer subplanes of intersection in the base regulus are in different H-orbits.

(1) Then, whenever π_0 intersects a component of a base regulus, there are two components of intersection of this base regulus and the two Baer subspaces of intersection of this base regulus are in different H-orbits.

(2) There is a translation plane Σ_{π_0} obtained from the Pappian plane Σ by E-nest replacement.

In the same work, it is noted that although it is known by Ebert that there is a unique finite translation plane obtained by q-nest replacement, the Fisher quadratic cone flock spread, this is not true in the infinite cases, as there are a great variety of what are called 'generalized Fisher spreads and flocks' constructed (See Jha and Johnson [**72**]). In this work, there is also a variety of s-inversions that is studied where there are examples of both spreads/flocks and quasifibrations/quasiflocks. In this text, we have generalized those previous observations into a class of quasifibrations that are s-inversions that do not construct generalized quadrangles (5,7,5).

We now complete the connections with circles and regulus nets, leaving the proofs as problems.

PROBLEM 48. *Let Σ denote that Pappian translation plane coordinatized by a full field $K(\sqrt{\gamma})$ of characteristic $\neq 2$.*

$$(*): \left\{ \begin{array}{c} \begin{bmatrix} a & \alpha \\ \beta & -a^\sigma \end{bmatrix} \text{ for each } a \in K(\sqrt{\gamma}), \\ \text{and for each } \alpha, \beta \in K, \text{ such that } a^{\sigma+1} + \alpha\beta \neq 0 \end{array} \right\}.$$

Prove the following theorem:

THEOREM 115. *Then show that a circle corresponds to the following regulus net in Σ :*

$$(*) \quad : \quad \left\{ y = x \begin{bmatrix} u & \gamma t \\ t & u \end{bmatrix} : (u^2 - \gamma t^2)\beta + 2a_1\gamma t - 2a_2 u - \alpha = 0 \right\},$$

where $a = \sqrt{\gamma}a_1 + a_2, \; a_i \in K, i = 1, 2,$

where $x = 0$ *belongs to the net if and only if $\beta = 0$.*

PROBLEM 49. *For a base regulus net*

$$R_{(u,t)} = \left\{ x = 0, y = x \begin{bmatrix} u & \gamma t \\ t & u \end{bmatrix} \delta I_2 : \forall \delta \in K \right\}$$

then show that a regulus net of the previous form $(*)$ intersects a given base regulus net $R_{(v,s)}$ in zero, one, or two components if and only if

$$(a_1\gamma s - a_2 v)^2 + \alpha(v^2 - \gamma s^2)$$

is a non-square, zero, or a non-zero square, respectively.

10. $\langle a^{\sigma-1} \rangle$-nest planes

In this section, we consider again the correspondence theorem associating cyclic translation planes (translation planes of order q^2 admitting cyclic homology groups of order $q+1$). We have discussed this method with the creation of j-planes that then construct α-twisted hyperbolic quadrics. The $q+1$-nests planes are cyclic translation planes and thus correspond to certain translation planes of flocks of quadratic cones. This material is covered more completely in the Handbook ([**94**] chapter 63) and also in the author's work [**85**]. In order to work out the flock plane, the $q+1$-nest planes will need to be considered in the matrix form

$$(*) \quad : \quad \text{Cyclic Group Plane:} \begin{bmatrix} u & t \\ F(u,t) & G(u,t) \end{bmatrix}$$

$$\Longleftrightarrow \quad \delta\text{-quadratic flock Form} \begin{bmatrix} \delta_{s,k} & \delta_{u,t} \\ \mathcal{F}(\delta_{u,t}) & \delta_{s,k} + \mathcal{G}(\delta_{u,t}) \end{bmatrix}$$

where

$$\mathcal{G}(\delta_{u,t}) = g(uG(u,t) + tF(u,t)) + 2(u\,F(u,t) - tfG(u,t)) \text{ and}$$

$$-\mathcal{F}(\delta_{u,t}) = \delta_{F(u,t),G(u,t)} = \det \begin{bmatrix} F(u,t) & G(u,t) \\ fG(u,t) & F(u,t) + gG(u,t) \end{bmatrix}.$$

We will note that the $q+1$-nest planes admit two commuting affine cyclic homology groups of order $q+1$. There is a characterization of such planes as follows:

THEOREM 116. *(Johnson and Pomareda [**97**]* **Double Cyclic Theorem***) Let π be a translation plane of order q^2 that admits two commuting affine cyclic homology groups of order $q+1$. Then π is one of the following planes:*

(1) an Andrè plane,

(2) a Hall plane,

(3) q is odd and π is obtained from a Desarguesian plane by replacing a $(q+1)$-nest, or

(4) q is odd and π is obtained from a Desarguesian plane by replacing a $(q+1)$-nest and a set of mutually disjoint reguli.

The construction of a $q+1$-nest plane is as follows: Let π be a Desarguesian plane of odd order q^2 with spread as follows:

$$x = 0, y = 0, \ y = x \begin{bmatrix} u & t \\ \gamma t & u \end{bmatrix} \simeq$$

$$\sqrt{\gamma}t + u \quad : \quad \forall u, t \in GF(q); \gamma \text{ a non-square in } GF(q),$$

$$\text{and } (t, u) \ne (0, 0).$$

Let the collineation group be

$$G = \left\langle \begin{array}{c} (x,y) \to (xa^{\sigma-1}, yb^{\sigma-1}), \\ \sigma \text{ the inherited automorphism}, a, b \in GF(q^2) \end{array} \right\rangle.$$

This group has order $(q+1)^2$ and is the direct product of two commuting cyclic homology groups of order $(q+1)$. We consider the set of Andrè nets. From section 9 on circle geometry, it is possible to find a Baer subplane π_0 whose regulus net that contains it shares two components each from a set of $(q+1)/2$ Andrè nets. The main results on $\langle a^{\sigma-1}\rangle$-nests shows that the above group G works and provides a suitable Baer subplane π_0. There are exactly two G-orbits of 1-dimensional subspaces on any Andrè net and the Baer subplane intersected with the two common components has 1-spaces in different G-orbits. The group contains the square-kernel subgroup and thus exactly $(q+1)/2$ of the Andrè Baer subplanes (of intersection) are covered by $G\pi_0$. An appropriate choice for π_0 is given in the proof of the main result, which shall be listed after the statement of the theorem. Again, in the original statement of the following result, it was stated that $m \to m^{\sigma+1}$ is assumed to be surjective. This is not required and left out in the following statement.

THEOREM 117. *(Johnson 5.2 [85]). Let K be any full field of characteristic $\neq 2$, which has a quadratic extension $K(\sqrt{\gamma})$, which is also full, for $|K| > 3$. Let σ denote the unique involution in $Gal_K K(\sqrt{\gamma})$.*

Let G denote the collineation group in the Pappian plane Σ coordinatized by $K(\sqrt{\gamma})$ defined as

$$G = \left\langle \begin{array}{c} (x,y) \to (xa^{\sigma-1}, yb^{\sigma-1}), \ \sigma \text{ the inherited automorphism,} \\ \forall a,b \in GF(q^2)^* \end{array} \right\rangle.$$

The base regulus nets are the Andrè regulus nets.

(1) Then there are Baer subplanes π_0 which intersect the base regulus nets in 0 or two components and whose non-trivial G-images are mutually disjoint.

Let λ denote the set of base regulus nets that intersect π_0 in two components each.

Let N_{π_0} denote the regulus net defined by the components of π_0.

(2) Then $N_{\pi_0}G$ is a $\langle a^{\sigma-1}\rangle$-nest of reguli.

(3) There is a nest replacement of $\pi_0 G = \cup R_\beta$, for all $R_\beta \in \lambda$, and a corresponding translation plane constructed via $\langle a^{\sigma-1}\rangle$-nest replacement.

(4) In the finite case, all of the above assumptions are valid when the characteristic is odd and hence there is a corresponding $(q+1)$-nest in $PG(3,q)$ with associated translation plane.

Note that we have the components of Σ are $x = 0, y = 0, y = xm; m \in K(\sqrt{\gamma})^*$, the Andrè nets are

$$R_m = \{y = xm\delta : \forall \delta^{\sigma+1} = 1\},$$
$$\text{Baer subspaces} \quad : \quad \{y = x^\sigma m\delta : \forall \delta^{\sigma+1} = 1\}.$$

If $m^{\sigma+1} = \beta$, then the notation R_β is used as $(m\delta)^{\sigma+1} = \beta$.

We note that a circle

$$\begin{bmatrix} b & \alpha \\ \beta & -b^\sigma \end{bmatrix},$$

which has points m such that $m^{\sigma+1} - (mb^\sigma + m^\sigma b) - \alpha$ represent the regulus net:

$$\{y = xm; (m-b)^{\sigma+1} = (\alpha + b^{\sigma+1}) = \beta\}.$$

This is the notation that will initially represent the $(q+1)$-nest plane. In this setting, the Baer subplanes of this regulus net will have the form $y = x^\sigma n + xb$ such that $n^{\sigma+1} = 1$. When this form is obtained we need to transfer to the matrix form to discuss the associated flocks of quadratic cones.

In the proof of this theorem, π_0 is selected as follows: Let $(1,1) = (1,0,1,0)$.

$$Case\ I \quad : \quad \pi_0 = \left\langle \begin{array}{c} (1,1), (c, c^\sigma d), d^\sigma = d^{-1} \neq \pm 1, d \neq c^{\sigma-1}, \\ \text{if} - 1 \text{ is a non-square in } K \end{array} \right\rangle.$$

$$Case\ II \quad : \quad \pi_0 = \left\langle \begin{array}{c} (1,1), (e, -ed), d^\sigma = d^{-1} \neq \pm 1, \\ \text{if} - 1 \text{ is a square in } K \end{array} \right\rangle.$$

Here $(1,1)$ is in $y = x^\sigma$ and $(c, -cd)$ is in $y = x^\sigma$ if and only $d = 1$.

Then to represent π_0 as $y = x^\sigma n + xb$; $n^{\sigma+1} = \beta$. It follows that

$$\pi_0 \text{ is } y = x^\sigma n + x(1-n) : n^{\sigma+1} = \left(\frac{c^\sigma d - c}{c - c^\sigma} \right)^2 \text{ in case } I,$$

$$\pi_0 \text{ is } y = x^\sigma n + x(1-n) : n^{\sigma+1} = e^{\sigma+1} \left(\frac{d-1}{e - e^\sigma} \right)^2 \text{ in case } II.$$

Then π_0 is in $\{y = xm; m^{\sigma+1} = \beta\}$ if and only $(w, wm) = (w, w^\sigma n + w(1-n))$. This translation to π_0 is in that particular Andrè net if and only if there is a non-zero element w such that $m^{\sigma+1} = (w^{\sigma-1}n + (1-n))^{\sigma+1}$.

Then the remaining Andrè nets union $x = 0, y = 0$ together with $\pi_0 G$ define the translation plane.

Now apply the group G, written as HA, where H is the kernel subgroup of squares of Σ with elements $(x,y) \rightarrow (xb^2, yb^2)$, for all $b \in K(\sqrt{\gamma})^*$ and A is a homology group with elements $(x,y) \rightarrow (x, ya^{\sigma-1})$ for all $a \in K(\sqrt{\gamma})$. Then apply $G = HA$ to $y = x^\sigma n + x(1-n)$ to obtain $y = x^\sigma (b^{-2}a)^{\sigma-1} + x(1-n)a^{\sigma-1}$.

COROLLARY 28. *We see that the spread for the $\langle a^{\sigma-1} \rangle$-nest replaced translation plane has the following representation:*

$$x = 0, y = 0, \ \{y = x^\sigma (b^{-2}a)^{\sigma-1} + x(1-n)a^{\sigma-1};$$

$$n^{\sigma+1} = \left(\frac{c^\sigma d - c}{c - c^\sigma} \right)^2 \text{ in case } I, \text{ and}$$

$$n^{\sigma+1} = e^{\sigma+1} \left(\frac{d-1}{e - e^\sigma} \right)^2 \text{ in case } II\}$$

$$\cup \{y = xm; m^{\sigma+1} \neq (w^{\sigma-1}n + (1-n))^{\sigma+1},$$

$$\text{for any } w \neq 0, \text{ for all } b, a \in K(\sqrt{\gamma})^*.$$

Consider $y = x^\sigma n + x(1-n)$. To obtain a direct method for determining the associated flock of a quadratic cone, write $x = \sqrt{\gamma}x_2 + x_1 = (x_1, x_2)$, $n = \sqrt{\gamma}n_2 + n_1$ over $K(\sqrt{\gamma})$, for the subscripted elements in K. Then we may write π_0 as a matrix as

$$y = x \left\{ \begin{bmatrix} 1 & 0 \\ 0 & -1 \end{bmatrix} \begin{bmatrix} n_1 & \gamma n_2 \\ n_2 & n_1 \end{bmatrix} + \begin{bmatrix} 1 - n_1 & \gamma n_2 \\ n_2 & 1 - n_1 \end{bmatrix} \right\} = x \begin{bmatrix} 1 & 2\gamma n_2 \\ 0 & 1 - 2n_1 \end{bmatrix}.$$

Denote this matrix by M and consider the action on the group. We may write the group in terms of the squares of kernel group elements H and the affine homology group A with axis $x = 0$ and coaxis $y = 0$ over $K(\sqrt{\gamma})$ and note that all elements of K are squares.

Just so the reader knows what the direction is, in the finite case, this would be a set of $(q+1)/2$ distinct images of π_0, considered components in the nest replaced translation plane. And, the set of each of these images under the affine homology group A would be a derivable net. In the finite case, there would be $(q-3)/2$ remaining André nets not involved in the replacement procedure. Consequently, the images of these $q-1$ sets of components under the group A together with $x = 0$ and $y = 0$ constitute the translation plane. Then it would be a relatively easy process to write the plane in the arbitrary case in the form

$$x = 0, y = 0, y = x \left[\begin{array}{cc} u & \gamma t \\ \tilde{G}(u,t) & \tilde{F}(u,t) \end{array} \right].$$

Then write the element γt as s to obtain a corresponding representation in the form

$$x = 0, y = 0, y = x \left[\begin{array}{cc} u & t \\ G(u,t) & F(u,t) \end{array} \right].$$

At this point, the correspondence theorem can be used to construct a representation for the corresponding flock of a quadratic cone. In particular, any Pappian translation plane will correspond to a linear flock under the correspondence theorem. Furthermore, all Andrè planes correspond to the same linear flock. The replacement of the Andrè nets does not change the linear flock. So, if there is a net replacement procedure in the $q+1$-cyclic Pappian plane that leaves a set δ of of k Andrè nets invariant in the sense that some of these may have been replaced by their Andrè replacements, the correspondence theorem will correspond to a linear subnet of $k+1$ regulus nets that share the component $x = 0$, the axis of the elation regulus-inducing group of the translation plane of the quadratic cone. Therefore, we note that each of the regulus nets disjoint from $x = 0$ and $y = 0$ are orbits under the homology group A. Additionally, the correspondence theorem takes each such regulus net and converts the net to a component of the associated flock plane. Choose any subset ρ of the set of mutually disjoint reguli that are in a given Pappian translation plane. Then the correspondence theorem maps this set to a linear subset within the flock, or rather this is a Pappian linear subset. Consider the finite case: There are $q-1$ of these reguli. These map under the correspondence to obtain $q-1$ reguli in the flock translation plane that share a component $\ell = (x = 0)$. The q th. regulus is the orbit of $y = 0$ under the regulus-inducing elation group E. We have a set of $(q+1)/2$ Andrè regulus involved in the construction process of the $q+1$-plane, leaving $(q-3)/2$. In this particular case, the correspondence theorem will produce a flock of a quadratic cone that admits a linear subflock of $(q-3)/2+1 = (q-1)/2$; a critical linear subflock. The same idea will show that there is a critical linear subflock in the general case. But, the flock cannot be linear itself, so by the theory given at the end of the section on nests, we see that the flock plane must be generalized Fisher.

THEOREM 118. *Every* $\langle a^{\sigma-1} \rangle$*-nest (A-nest) replaced translation plane corresponds to a generalized Fisher plane under the correspondence theorem.*

Note the interesting result, valid in both the infinite and finite versions.

COROLLARY 29. *E-nest replaced translation planes are equivalent to A-nest replaced translation planes.*

Here is the finite situation:

COROLLARY 30. *Every q-nest plane constructs a Fisher flock plane. Every Fisher flock plane, by reversing the correspondence theorem, constructs a q + 1-nest plane.*

Hence, q-nest planes are equivalent to q + 1-nest planes.

REMARK 30. *Actually, the q+1-nest planes were originally constructed by Sherk and Pabst [132] in 1977 using indicator sets. For example, see the Handbook [94], chapter 50, especially Theorem 50.1.2 showing that indicator sets are equivalent to transversals to derivable nets in the finite case.*

11. Flocks of elliptic quadrics

In this section, elliptic flocks are considered. Ultimately, such flocks are connected with partitions of Pappian translation planes by disjoint regulus nets. We have discussed situations where the union of the regulus nets miss exactly two components L and M. This situation is what one's intuition would demand. But could the union of the disjoint regulus nets miss exactly one or perhaps no components? Thinking of an infinite setting with the image of $q^2 + 1$ components in a translation plane of order q^2 with components in $PG(3, q)$, partitioned in sets of $q + 1$ components for the regulus nets, it would appear impossible to not cover all but two components. However, in the infinite cases, it can be shown that there can be all three possibilities. This means that the set of translation planes obtained by multiple derivation of infinite Pappian planes is much more diverse than the set of Andrè planes discussed in the multiple replacement theorem in part 10.

The idea of the construction comes from a footnote in Dembowski [42]. This section is a variation of material first presented in Biliotti and Johnson [15] at Tedfest, T.G. Ostrom's 80th birthday conference. This work is presented in a fairly complete form due to the lack of wide availability of the notes 'Mostly Finite Geometries'.

EXAMPLE 4. **Dembowski's Example.** *Consider a sphere $\mathbb{Q}$ in 3-dimensional affine space over the field of real numbers. Let π denote the equatorial plane and let L and M be any two lines in π that are disjoint from $\mathbb{Q}$. Consider the northern and southern hemispheres $N_{\mathbb{Q}}$ and $S_{\mathbb{Q}}$ (strictly disjoint from $\pi \cap \mathbb{Q}$) and let $\mathfrak{N}$ be a point on $N_{\mathbb{Q}}$ such that the plane $\langle M, \mathfrak{N} \rangle$ is tangential to $\mathbb{Q}$ at $\mathfrak{N}$. Similarly, let a point $\mathfrak{S}$ be on $S_{\mathbb{Q}}$ such that the plane $\langle L, \mathfrak{S} \rangle$ is tangential to $\mathbb{Q}$ at $\mathfrak{S}$. Then consider the set of planes $\mathcal{N}_M = \{ \langle M, P \rangle : P \in N_{\mathbb{Q}} \}$ and the set of planes $\mathcal{S}_L = \{ \langle L, P \rangle : P \in S_{\mathbb{Q}} \}$*

$$\pi \cup \mathcal{N}_M \cup \mathcal{S}_L$$

is a bilinear flock of $\mathbb{Q}$ (an elliptic quadric).

DEFINITION 95. *For our purposes in this text, an 'elliptic quadric' $\mathbb{Q}$ in $PG(3, K)$, for K a field admitting a quadratic extension $K(\sqrt{\gamma})$, is defined as a set of points using homogeneous coordinates*

$$(x, y, z, w) : zw + y^2 = \gamma x^2.$$

DEFINITION 96. *Let $\mathbb{Q}$ be an elliptic quadric in $PG(3, K)$, where K is a field. 'A flock of an elliptic quadric' is a set of mutually disjoint conics that partition the quadric except for at most two points.*

DEFINITION 97. *A flock of an elliptic quadric* $\mathbb{Q}$ *in* $PG(3, K)$, *for* K *a field is an 'i-point flock' if and only if the union of conics together with exactly i-points cover* $\mathbb{Q}$, *with* $i = 0, 1, 2$.

When K is a finite field $GF(q)$, all flocks are linear (the planes containing the conics share a common line) by Thas [134] for q even, and by Orr [116], for q odd. However, there are elliptic flocks that are not linear over certain infinite fields, as in the example of Dembowski. There is interest in Pappian translation planes corresponding to the elliptic flock, and their multiply derived translation planes for this reason.

THEOREM 119. *Let* $\mathbb{Q}$ *be an elliptic quadric in* $PG(3, K)$, *for* K *a field, that has a quadratic extension field* $K(\theta)$. *Let* F *be a flock of* $\mathbb{Q}$. *Then there is a set of mutually disjoint regulus nets of the associated Pappian translation plane* Σ *that covers all but at most two components* L *and* M.

Derive all regulus nets to construct a translation plane $\widetilde{\Sigma}$. *The translation plane is non-Pappian if and only if the flock* F *is not linear.*

Conversely, any Pappian translation plane that admits a cover of the components with the exception of at most two components L *and* M, *by mutually disjoint regulus nets constructs a flock of an elliptic quadric.*

Given a Pappian translation plane with spread in $PG(3, K)$, coordinatized by a Galois quadratic field extension $K(\theta)$, we have seen that there is a partition of the spread by a union of mutually disjoint regulus nets, leaving exactly two components $x = 0, y = 0$. And, in this setting, there is an affine homology group acting regularly on the components of each of the regulus nets. And, this spread defines a linear elliptic flock.

Using the Klein quadric, a Pappian spread over field K, that has a Galois quadratic extension field $K(\theta)$ is equivalent to an elliptic quadric. Consequently, a partition of the spread produces a partition of the elliptic quadric. If the partition consists of regulus nets and at most two components, the flock of the elliptic quadric is partitioned by conics and at most two remaining points.

To prove the rest of this theorem, we require a general proposition first appearing in Biliotti, Johnson [15], which is a known result in finite Desarguesian planes (Ostrom [118], Lüneburg [110]) and also some additional preliminary work on elliptic quadrics.

PROPOSITION 7. *Let* V *be a vector space and let* Σ *and* Π *denote two Pappian affine planes defined on the points of* V *and coordinatized by fields* L *and* F *respectively.*

(1) *Then, if the two planes share at least three components, there is a net defined by common components which may be coordinatized by a subfield of* $L \cap F$.

(2) *Consider a regulus net* N *with components in* $PG(3, K)$, *for some field* K. *Then any 2-dimensional vector space* T *disjoint from the partial spread* N *of the associated 4-dimensional vector space may be embedded into a unique Pappian plane containing* N.

(3) *Assume that two Pappian spreads have components in the same* $PG(3, K)$, *for* K *a field. If the two spreads contain a regulus net* N, *and a component external to* N *then they are equal.*

PROOF. Choose a basis for the vector space so that vectors are of the form (x, y), for x, y in a common reference subspace W, and $V = W \oplus W$, and such that the three common components are of the form $x = 0, y = 0, y = x$, thereby arranging conditions so that the two fields L and F have a common zero and identity. Consequently, we have two coordinate systems $(L, +, \circ)$ and $(F, +, *)$ and the spreads have the following general form:

$$x = 0, y = 0, y = x, y = x \circ m \text{ for } m \in L \text{ for } \Sigma,$$
$$x = 0, y = 0, y = x, y = x * n \text{ for } n \in F \text{ for } \Pi.$$

Let the two Pappian spreads share a common net C. Then $y = x \circ m = x * n$, for all x, for all components in C. But, choosing $x = 1$, we see that $1 \circ m = m = 1 * n = n$. Hence, C is coordinatized by $L \cap F$. This proves part (1).

Consider part (2). Choose the regulus net N to have the standard partial spread:

$$x = 0, y = x \begin{bmatrix} u & 0 \\ 0 & u \end{bmatrix}; u \in K.$$

Let T have the general form $y = xM$, for $M = \begin{bmatrix} a & b \\ c & d \end{bmatrix}$. Form $y = MtI_2 + uI_2$, $\forall t, u \in K$, which is then

$$y = x \begin{bmatrix} u + at & bt \\ ct & u + dt \end{bmatrix}; \forall t, u \in K.$$

Write $w = u + dt$, $ct = s$, noting that c cannot be 0, as we have a partial spread and $M - \begin{bmatrix} u & 0 \\ 0 & u \end{bmatrix}$ must be non-singular. Then, the form is

$$x = 0, y = x \begin{bmatrix} w + \alpha s & \beta s \\ s & w \end{bmatrix}; \forall s, w \in K.$$

This set is additive and multiplicative, as a check will show, so we have a quadratic extension field of K. Consequently, there is a Pappian spread containing N and T, which is unique by part (1). This proves part (2) and also (3) is now immediate. □

12. Klein quadric and Pappian spreads

We recall the basic connection between the Klein quadric, circle geometry, and regulus nets in Pappian spreads.

PROPOSITION 8. *Consider the Klein quadric Ω_5 in $PG(5, K)$, for K an arbitrary field, admitting a Galois quadratic extension $K(\sqrt{\gamma})$, where points are defined by homogeneous coordinates*

$$(x_0, x_1, x_2, x_3); \ x_0 x_5 - x_1 x_4 + x_1 x_3 = 0.$$

Just as in the previous section, we represent the spread for the Pappian plane over $K(\sqrt{\gamma})$ as follows:

$$x = 0, y = x \begin{bmatrix} u & \gamma t \\ t & u \end{bmatrix} : \forall u, t \in K(\sqrt{\gamma}).$$

(1) $(1, t, u, -u, -t\gamma, u^2 - \gamma t^2) \in \Omega_5$.

(2) The circle/regulus

$$\begin{bmatrix} a & \alpha \\ \beta & -a^\sigma \end{bmatrix}$$

corresponds to the following regulus net in Σ:

$$(*)_{regulus} \quad : \quad \left\{ \begin{array}{c} y = x \begin{bmatrix} u & \gamma t \\ t & u \end{bmatrix} \\ : (u^2 - \gamma t^2)\beta + 2a_1\gamma t - 2a_2 u - \alpha; \ a^{\sigma+1} + \alpha\beta \neq 0 \end{array} \right\},$$

where $a = \sqrt{\gamma}a_1 + a_2$, $a_i \in K$, $i = 1, 2$,

where $x = 0$ belongs to the net if and only if $\beta = 0$.

To embed the elliptic quadric into Ω_5, consider the mapping

$$(x, y, z, w) \to^g (-z, x, y, -y, -\gamma x, w).$$

PROBLEM 50. *Show that if (x, y, z, w) satisfies $-\gamma x^2 + y^2 + zw = 0$, then the image point in $PG(5, K)$ is an element of the Klein quadric Ω_5. Hence, there is an embedding of the elliptic quadric of $PG(3, K)$ into the Klein quadric in $PG(5, K)$.*

PROBLEM 51. *Using the circle geometry connection with regulus nets of the Pappian translation plane coordinatized by $K(\sqrt{\gamma})$, prove the following:*

PROPOSITION 9.

(1) The regulus net given by $(*)_{regulus}$ above, intersects the elliptic quadric $\mathbb{Q}$ in a conic given by the points

$$(t, u, -1, u^2 - \gamma t^2) \ or \ (0, 0, 0, 1)$$

satisfying the same equation as does the regulus net.

(2) The plane containing the conic is

$$(2a_2\gamma)x - (2a_2)y - (u^2 - \gamma t^2)z = 0.$$

Hint: the general point is (x, y, z, w), the intersection with $\mathbb{Q}$ is

$$(t, u, -1, u^2 - \gamma t^2) \ or \ (0, 0, 0, 1))$$

(3) The tangent plane to $\mathbb{Q}$ at the point

$$(t, u, -1, u^2 - \gamma t^2) \ or \ (0, 0, 0, 1)$$

is

$$2\gamma tx - 2uy - (u^2 - \gamma t^2)z + w = 0 \ or \ z = 0, \ respectively.$$

Hint: write the tangent plane as

$$ax + by + cz + dw = 0.$$

The plane contains the point $(t, u, -1, u^2 - \gamma t^2)$ provided

$$at + by - c + d(u^2 - \gamma t^2) = 0.$$

Since the characteristic is assumed $\neq 2$, we may use the quadratic formula. There must be a unique solution, which then implies $a = 2\gamma t$ and $b = -2u$, since we may take $d = 1$ (why?)).

For the elliptic flocks, we take a convenient basis change.
Now change bases taking a new basis as

$$\{(1,0,0,0),(0,1,0,0),(0,0,1,1),(0,0,1,-1),$$

so that (x,y,z,w) in the old basis is $-\gamma x^2+y^2+zw$, then becomes in the new basis:

$$-\gamma x^2 + y^2 + (z^2 - w^2)/4 = 0.$$

So, we now realize the elliptic quadric in a 3-dimensional affine space so that the Dembowski concept may be considered. To do this, we now make the assumption that K is an ordered field such that all negative elements are non-squares. We may then choose $\gamma = -1$. That is, the points of the elliptic quadric (in the new basis) are then

$$(t, u, (-1 + u^2 + t^2)/2, (-1 - (u^2 + t^2))/2),$$

and

$$(0,0,-1,1) \equiv (0,0,-1/2,1/2).$$

Since $u^2+t^2 \neq -1$, then the elliptic quadric is in the affine space $AG(3,K)$, obtained by deleting the hyperplane $w = 0$.

In the $AG(3,K)$, and where the points of $\mathbb{Q}$ have the form

$$(x, y, (z + w)/2, (z - w)/2),$$

then the affine points are of the form

$$(\frac{2x}{z-w}, \frac{2y}{z-w}, \frac{z+w}{z-w}, 1), \ for \ z \neq w.$$

As the set of points $(x, y, z, 1)$, in generic form, then satisfy:

$$x^2 + y^2 + z^2 = 1.$$

To see this, we may then realize the points of the elliptic quadric in the form

$$(\frac{-2t}{1+(u^2+t^2)}, \frac{-2u}{1+(u^2+t^2)}, \frac{1-(u^2+t^2)}{1+(u^2+t^2)}, 1)$$
$$: \ \forall (t,u) \in K \times K, \ and \ (0,0,-1,1).$$

And,

$$\left(\frac{-2t}{1+(u^2+t^2)}\right)^2 + \left(\frac{-2u}{1+(u^2+t^2)}\right)^2 + \left(\frac{1-(u^2+t^2)}{1+(u^2+t^2)}\right)^2 = 1,$$

as can be immediately verified. We have then showed the first part of the following result.

THEOREM 120. *(**Elliptic Quadric Theorem**) Let K be an ordered field whose negative elements are all nonsquare. Let $K(\sqrt{-1})$ denote the quadratic extension of K and assume that Σ is the Pappian plane over $K(\sqrt{-1})$, with components*

$$x = 0, y = x \begin{bmatrix} u & -t \\ t & u \end{bmatrix} : \forall u, t \in K.$$

(1) *Then there is a basis for the 4-dimensional K-vector space such that the elliptic quadric $x^2 + y^2 + zw = 0$ in $PG(3,K)$ may be represented in an associated 3-dimensional affine space $AG(3,K)$ as*

$$x^2 + y^2 + z^2 = 1,$$

where the affine points are $(x, y, z, 1)$, *denoted by* (x, y, z) *and the projective points are written homogeneously as* (x, y, z, w). *In this representation, the elliptic quadric is represented by the following set of points:*

$$\left(\frac{-2t}{1 + u^2 + t^2}, \frac{-2u}{1 + u^2 + t^2}, \frac{1 - (u^2 + t^2)}{1 + u^2 + t^1}, 1\right) : \forall t, u \in K, \text{ and } (0, 0, -1, 1)$$

projectively and as affine points as

$$P_{t,u} : \left(\frac{-2t}{1 + u^2 + t^2}, \frac{-2u}{1 + u^2 + t^2}, \frac{1 - (u^2 + t^2)}{1 + u^2 + t^1}\right) : \forall t, u \in K \text{ and } (0, 0, -1).$$

(2) The hyperplane at infinity in this representation is $w = 0$.
(3) The tangent plane $Tan_{t,u}$ *to* $P_{t,u}$ *is*

$$-2tx - 2uy + (1 - (u^2 + t^2))z - (1 + u^2 + t^2) = 0.$$

DEFINITION 98. *In a 3-dimensional vector space over an ordered field* K, *define the 'open segment' to be*

$$oseg(R, T) = \{R + \lambda(T - R) : \forall 0 < \lambda < 1\}.$$

The following is now immediate, since $(x, y, z, 1)$ denotes the actual affine points, the open segment notation extends to the elliptic quadric.

THEOREM 121. *In the Elliptic Quadric Theorem, let* $R = (x, y, z)$ *and* $T = (x^*, y^*, z^*)$ *be affine points. Then*
(1)

$$oseg(R, T) = \{(x + \lambda(x^* - x), y + \lambda(y^* - y), z + \lambda(z^* - z)) : \forall 0 < \lambda < 1\}.$$

(2) Consider two points $P_{t,u}$ *and* $P_{s,v}$ *of* $\mathbb{Q}$, *as represented in* $AG(3, K)$. *Let* $L = Tan_{t,u} \cap Tan_{s,v}$ *denote the line of intersection of the tangent planes to the indicated points of* $\mathbb{Q}$. *Then, if*

$$P_{k,n} \in \mathbb{Q} - \{P_{t,u}, P_{s,v}\},$$

the plane $\langle P_{k,n}, L \rangle$ *intersects* $oseg(P_{t,u}, P_{s,v})$ *in a unique point.*

PROOF. (2): Without loss of generality, we may assume that the point $P_{t,u}$ is $(0, 0, 1)$ so the tangent plane is $z = 1$. The reason for this is that we may coordinatize the underlying Pappian translation plane by choosing any component to be $x = 0$. This will not change the form of the elliptic quadric or our particular representation. Recalling that there is an affine tangent plane and a projective tangent plane, the two projective tangent planes must intersect in a line. If $Tan_{s,v} \cap Tan_{t,u}$ is a line of the hyperplane at infinity, this means that the affine tangent planes are parallel, which forces $P_{s,v}$ to $(0, 0, -1)$ and the tangent plane must be $z = -1$. Now $oseg(P_{t,u}, P_{s,v}) = \{(0, 0, -1 + 2\lambda) : 0 < \lambda < 1\}$. Let $P_{k,n} = (x_1, y_1, z_1) + L$, where $L = \{(x, y, -1, 0) : \forall x, y \in K\}$. This shows that any plane $z = z_0$ since parallel to $z = 1$ will intersect the hyperplane at infinity in the same line L. Since $P_{k,n}$ is a point (x_0, y_0, z_0) and $z = z_0$ contains the point $P_{k,n}$ then in order that the plane $z = z_0$ intersects $\{(0, 0, -1 = 2\lambda) : 0 < \lambda < 1\}$, then $z_0 = -1 + 2\lambda$, exactly when $\lambda = \frac{z_0 + 1}{2}$. Since $-1 < z_0 < 1$ which is equivalent to $0 < \frac{z_0 + 1}{2} < 1$, it follows that this is the unique intersection point of $oseg(P_{t,u}, P_{s,v})$.

Now assume that $P_{t,u} = (0, 0, 1)$ with tangent plane $z = 1$ and $P_{s,v} = (x_0, y_0, z_0, 1) \neq (0, 0, 1, 1)$, with tangent plane $x_0 x + y_0 y + z_0 z = 1$ (the line of intersection of the tangent planes is an affine line). Then the points of $z = 1$ are $(x, y, 1, 1)$ and therefore there is a common line L with equation $x_0 x + y_0 y + z_0 = 1$. Since not both x_0 and y_0 are zero, assume without loss of generality that $y_0 \neq 0$. Then the set of points on the line is $\{(x, 1 - z_0 - x_0 x)/y_0 1, 1) : \forall x \in K\}$ or just $(x, 1 - z_0 - x_0 x)/y_o, 1)$ in $AG(3, K)$. We now work out the plane $\langle (P_{k,n} L \rangle$. Let $P_{k,n} = (x_1, y_1, z_1, 1)$. Let this plane be initially represented as $ax + by + cz = d$. First assume that $d = 0$. Let $x = 0$ and then 1 to obtain the following set of equations:

$$b(1 - z_0)/y_0 + c = 0,$$
$$a + b(1 - z_0 - x_0)/y_o + c = 0.$$

Solving these equations and canceling b, obtains the plane as

$$x_0 x + y_0 y + (z_0 - 1)z = 0.$$

Now $oseg((0, 0, 1), P_{s,v}) = \{(\lambda x_0, \lambda y_0, \lambda(z_0 - 1) + 1) : 0 < \lambda < 1\}$. There is a point of intersection with the segment on this plane if and only if

$$\lambda x_0^2 + \lambda y_0^2 + \lambda z_0^2 + 2\lambda(1 - z_0) + (z_0 - 1) = 0.$$

Since $z_0 \neq 1$, this reduces to $\lambda = 1/2$. Conversely, if a plane through $(0, 0, 0)$ contains the line of intersection of the two tangent planes, then the plane is exactly as above, and contains $(-x_0, -y_0, -z_0)$, and intersects the interval in $1/2$, then will also intersect (x_0, y_0, z_0).

Now assume that $d \neq 0$. Since (x_1, y_1, z_1) is on the plane, it follows that the plane has the following equation:

$$b x_0 x + b y_0 y + b y_0 y + (y_0 - b(1 - z_0))z = y_0, \text{ for}$$
$$b = y_0(1 - z_1)/(x_0 x_1 + y_0 y_1 + (z_0 - 1)z_1).$$

Note that $y_0 \neq 0$, $z_1 \neq 1$, so $b \neq 0$, and is well defined, as $d \neq 0$. The plane has an intersection within the open interval if and only if

$$b\lambda x_0^2 + b\lambda y_0^2 + (y_0 - b(1 - z_0))(\lambda(z_0 - 1) + 1) = y_0.$$

Rearranging, we have:

$$b\lambda((x_0^2 + y_0^2 + z_0^2) - 2z_0 + 1) + (y_0\lambda + b)(z_0 - 1) = 0,$$

which implies that

$$\lambda = b/(2b - y_0),$$

provided $2b - y_0 \neq 0$, where $b = y_0(1 - z_1)/(x_0 x_1 + y_0 y_1 + (z_0 - 1)z_1)$. If $2b = y_0$ then

$$2(1 - z_1) = (x_0 x_1 + y_0 y_1 + (z_0 - 1)z_1) = x_0 x_1 + y_0 y_1 + z_0 z_1 - z_1,$$

and

$$\begin{aligned} z_1 - 1 &= 1 - (x_0 x_1 + y_0 y_1 + z_0 z_1) \\ &= ((x_0 - x_1)^2 + (y_0 - y_1)^2 + (z_0 - z_1)^2)/2, \end{aligned}$$

since $x_0^2 + y_0^2 + z_0^2 = 1 = x_1^2 + y_1^2 + z_1^2$. And, we note that $-1 < z_1 < 1$, so $z_1 - 1 < 0$, a contradiction to the above equation. Consequently, $2b - y_0 \neq 0$. And, note for future reference that

$$1 - (x_0 x_1 + y_0 y_1 + z_0 z_1) > 0, \text{ so that } 1 - z_1 > x_0 x_1 + y_0 y_1 + (z_0 - 1)z_1,$$

which will appear in the next inequality. Let $m = 1/(1 - z_1)/(x_0 x_1 + y_0 y_1 + (z_0 - 1)z_1))$

Hence, we need to show that

$$0 < 1/(2 - y_0/b) = 1/(2 - y_0/(y_0(1 - z_1)/(x_0x_1 + y_0y_1 + (z_0 - 1)z_1)) < 1$$

if and only if

$$0 < 1/(2 - y_0/b) = 1/(2 - 1/(1 - z_1)/(x_0x_1 + y_0y_1 + (z_0 - 1)z_1)) < 1$$

Then $0 < 1/(2 - m) < 1$ if and only if $2 - m > 0$ and $m < 1$. So, the question is whether

$$m = (x_0x_1 + y_0y_1 + (z_0 - 1)z_1))/(1 - z_1) < 1$$

But, this is valid since $1 - (x_0x_1 + y_0y_1 + z_0z_1) > 0$. This completes the proof. $\square$

- In the finite case, an elliptic flock is always linear and consists of $q + 1$ planes, of which two are tangential planes and $q - 1$ planes intersect the quadric in a non-degenerate conic. Since there are $q+1$ planes on an exterior line, all planes containing L necessarily intersect the quadric.

  In an ordered field K where the negative elements are all non-square, we have a representation of the quadric in $AG(3, K)$ in the form $x^2 + y^2 + z^2 = 1$, where (x, y, z) are the points of the affine space. Consequently, for a plane, say $z = 2$, could not intersect the quadric. For this reason, we may use an intersection technique on $oseg(R, T)$ similar to what is suggested by Dembowski's example. This will provide n-linear elliptic flocks.

13. n-Linear elliptic flocks

In this section, we construct n-linear elliptic flocks, by which it is meant that the planes of the flock share one of n distinct lines. Let $\mathbb{Q}$ be an elliptic quadric over an ordered field K such that all negative elements are non-squares.

DEFINITION 99. *A 'T-line' is a line L exterior to the quadric $\mathbb{Q}$ which is the intersection of two tangent planes from points S_L and N_L of $\mathbb{Q}$.*

REMARK 31. *We have seen that a linear elliptic flock minus two points may be obtained as the planes sharing a T-line L and defined by the planes $\langle W, L \rangle$, where W is a point of $\mathbb{Q} - \{S_L, N_L\}$. Note that we have seen that there is a unique point of intersection of $\langle W, L \rangle$ and $oseg(S_L, N_L)$.*

We begin by a discussion of binary-elliptic flocks and north/south side of $\mathbb{Q}$ relative to a plane π and a line L.

DEFINITION 100. *Let L be a T-line and π a plane, which assumes π contains L and intersects $\mathbb{Q}$ in a point I_π on $oseg(S_L, N_L) \cup \{S_L, N_L\}$. Form the partial flock $F(\pi, L)^n$, called the 'north partial flock relative to π and L'*

$$\{\cup \langle U, L \rangle\} \cap \mathbb{Q} : U \in oseg(S_L, N_L) \text{ and } I_\pi < U < N_L \text{ and } \langle U, L \rangle \cap \mathbb{Q} \neq \phi,$$

which is called the 'north side of $\mathbb{Q} - \{N_L\}$, relative to π and L'. The set of north $\mathbb{Q}$-intersection points, together with N_L is called the 'north side of $\mathbb{Q}$, relative to π and L'. The 'north section' excludes N_L, and is denoted by $\mathbb{Q}^n(\pi, L)$, and $\mathbb{Q}^{n+}(\pi, L) = \mathbb{Q}^n(\pi, L) \cup \{N_L\}$.
Similarly, the 'south side of $\mathbb{Q} - \{S_L\}$, relative to π and L' is

$$\{\cup \langle U, L \rangle\} \cap \mathbb{Q} : U \in oseg(S_L, N_L) \text{ and } S_L < U < I_\pi \text{ and } \langle U, L \rangle \cap \mathbb{Q} \neq \phi,$$

and the 'south side of $\mathbb{Q}$, *relative to* π *and* L' *is the set of south* $\mathbb{Q}$-*intersection points together with* S_L, *and is denoted by* $\mathbb{Q}^s(\pi, L)$, *and* $\mathbb{Q}^{s+}(\pi, L)^+ = \mathbb{Q}^n(\pi, L) \cup \{S_L\}$. *The partial flock* $F(\pi, L)^s$ *is called the 'south partial flock relative to* π *and* L'. *And, the 'south section' excludes* S_L.

DEFINITION 101. *Let* $L_1 \neq L_2$ *be* T-*lines, relative to* π. *Note that it is not assumed that the points of tangency* $\{N_{L_1}, S_{L_1}\}$ *for* L_1 *are necessarily the same as the point of tangency* $\{N_{L_2}, S_{L_2}\}$ *for* L_2. *Also, we note that the terminology of north and south has no intrinsic meaning in this context. In any case, then* $\mathbb{Q}^{n+}(\pi, L_1)$ *is either* $\mathbb{Q}^{s+}(\pi, L_2)$ *or* $\mathbb{Q}^{n+}(\pi, L_2)$. *We note that we have two linear flocks* $F(\pi, L_1)$ *and* $F(\pi, L_2)$, *of the elliptic quadric, one of* $\mathbb{Q} - \{N_{L_1}, S_{L_1}\}$ *and one of* $\mathbb{Q} - \{N_{L_2}, S_{L_2}\}$.

Define the side of $\mathbb{Q}$ *not covered by* $\mathbb{Q}^{n+}(\pi, L_1)$ *relative to* L_2 *the 'opposite side of* $\mathbb{Q}^{n+}(\pi, L_1)$', *and denoted by* $opp(\mathbb{Q}^{n+}(\pi, L_1))$, *and use the terminology for the partial spread as* '$oppF(\pi, L_1), L_2)$'.

If K is the field of real numbers, then the idea of intersections of north and south make intrinsic sense and a bilinear flock is obtained using Dembowski, but suppose that K is a subfield of the real numbers. Is this still true?

REMARK 32. *Assume that* K *is a subfield of the field of real numbers and assume* π *is not a tangent plane and let* $C = \pi \cap \mathbb{Q}$. *Then the north section of* $\mathbb{Q}$ *relative to* π *and* L *is the set* $\mathbb{Q}^n(\pi, L)$ *of all points of* $\mathbb{Q} - C$ *such that* $oseg(Z, N_L)$ *does not intersect* π, *for all* $Z \in \mathbb{Q} - C$.

PROOF. Let R denote the field of real numbers. Then assume that a K-point W of $\mathbb{Q} - C$ does not intersect π, but is not in $\mathbb{Q}^n(\pi, L)$, where the point as an R-point does intersect π.

$$ax + by + cz + dw = 0$$

for $a, b, c, d \in K$. Since $oseg(Z, N_L)$ is

$$\{Z + \lambda(N_L - Z) : 0 < \lambda < 1\}.$$

So, there is an intersection for some $\lambda \in R$ but no intersection for all $\lambda \in K$. However all intersections of K-points and with all of the coefficients in K will force λ to be in K, since $Z \neq N_L$ and some element of the row vector of $Z - N_L$ is non-zero v_0, where $\lambda v_0 = \rho_0$, where ρ_0 non-zero due to the range of λ. □

Consequently, we see that Dembowski's example may be extended to subfields K of the field of real numbers.

COROLLARY 31. *Let* $\{L_1, L_2\}$ *be distinct exterior lines of an elliptic quadric* $\mathbb{Q}$ *over a subfield* K *of the real numbers. Let* π *be a plane that intersects* $\mathbb{Q}$ *in at least two points, and assume that* L_1 *and* L_2 *are* T-*lines relative to* π_1.

Then $\pi_1 \cup F(\pi_1, L_1)^n \cup opp(F(\pi_1, L_1), L_2)$ *is a flock of* $\mathbb{Q} - \{N_{L_1}, S_{L_2}\}$; *hence is a bilinear flock of a quadratic cone, where the planes of the flock contain either* L_1 *or* L_2.

There is a generalization of the procedure that produces bilinear or 2-linear flocks, that can create n-linear flocks of elliptic quadrics. The idea is to observe with two T-lines relative to a plane π_1, we can partition $\mathbb{Q}$ into two slabs, so to speak, with two extra points. To create L_1, take any two points W and Z of $\mathbb{Q}$ and construct the tangent planes and define L_1 as the intersection line of these planes. Then choose any point I_1 on $oseg(W, Z)$, and define $W = N_{L_1}$ and $Z = S_{L_1}$. Form $\langle I_1, L_1 \rangle = \pi_1$. Choose any T-line L_2 on π_1, thus creating N_{L_2} and S_{L_2}. By use of the bilinear construction, we have the north side of π_1 covered (that is, the N_{L_1} side). So, we have covered either $oseg(I_1, N_{L_1})$ or $seg(S_{L_1}, I_1)$. Assume, without loss of generality, that we have covered $oseg(I_1, N_{L_1})$ and with π_1, we have the north side covered except for N_{L_1}. We have also covered the north or the south side of $oseg(S_{L_2}, N_{L_2}) \cap \pi_1 = I_1^{L_2}$. Assume, for notational purposes, that the uncovered part is denoted by $oseg(S_{L_2}, I_1^{L_2})$ (it could, of course, be $oseg(I_1^{L_2}, N_{L_2})$). Choose any point I_2 of $oseg(S_{L_2}, I_1^{L_2})$, and again, assume that the north slab from π_2 to π_1 is covered by the partial spread $\left\{ \langle I_2, L_2 \rangle ; I_2 < U < I_1^{L_2} \right\}$ (again, it could be the south slab). So, by agreement on notation, we keep partitioning the elliptic quadric and cover the north by disjoint partial spreads of the form $F(oseg(A, B))$, together with planes π_1, π_2 which would now, if we cover the uncovered south, have a 3-linear flock. Continuing in this manner, we may use n lines $L_1, ..., L_n$ creating an n-linear flock of an elliptic quadric.

PROBLEM 52. *Look at the notation of the following theorem and create a 4-linear flock of an elliptic quadric. By doing this, one will see how the notation is essentially forced for clarity. Noting the arbitrary choice of elements from intervals shows that there are cardK ways to create such flocks.*

THEOREM 122. *(Theorem 20 page 150 [15]) For any subfield K of the field of real numbers, and for any finite positive integer n, there are cardK flocks of an elliptic quadric $\mathbb{Q}$ in $PG(3, K)$ such that the planes of the flock share one of n lines. We have*

$$\left\{ \cup \pi_i \cup F(oseg(I_{i-1}^{L_i}, I_i)) : 1 \leq i \leq n \right\},$$

where

$$N_{L_1} = I_0^{L_i}, L_{i-1}^{L_i} = oseg(S_{L_i}, N_{L_i}) \cap \pi_{i-1},$$

and I_i is any point on

$$oseg(S_i, I_{i-1}^{L_i}), for 1 \leq i \leq n - 1$$

and $S_{L_n} = I_n$
    is a flock of $\mathbb{Q} - \{N_{L_1}, S'_{L_n}\}$, where $S'_{L_n} \in \{N_{L_1}, S_{L_n}\}$.
    The planes share a line of the set of n-lines $\{L_1, L_2, ...L_n\}$.

REMARK 33. *There is much more in the original article that we are not able to cover here due to space considerations. These include a characterization of the flocks obtained by the above construction technique. Also, there is discussion and construction of 0-n-linear flocks, and 1-n-linear flocks (of $\mathbb{Q}$ missing either 0 or 1 points). The interested reader is directed to sections 4 and 5 of Biliotti, Johnson [14] for the theory and examples.*

COROLLARY 32. *Analogous to the Andrè planes are the multiple derived planes constructed by choosing any subset δ of the set λ of regulus nets, corresponding to the variety of elliptic flocks and deriving the regulus nets of the set δ. We note that, contrary to the Andrè planes, the replacement of all of the nets will not produce a Pappian plane, due to the non-linearity of the elliptic flocks.*

Additionally, for all of the i-n-linear flocks, for $i = 0, 1, 2$, in contrast to the Andrè planes, there will not be an associated collineation group of the Pappian plane acting transitively on the regulus nets of the Pappian spreads.

COROLLARY 33. *Any of the multiply derived planes will have spreads in $PG(3, K)$ and K admits a quadratic extension $K(\theta)$.*

(1) *Consequently, all of these spreads may be lifted to σ-spreads of σ-cones in $PG(3, K(\theta))$. If $x = 0$ is not involved in the multiple derivation process, coordinates must be changed to find the appropriate representation for lifting.*

(2) *If K is the set of rational numbers, there are infinite towers of quadratic field extensions, each of which will produce another class of lifted spreads.*

(3) *Each of the lifted spreads are further derivable and may be derived to translation planes with blended kernels.*

14. Tangential packings of ovoids

Here we are able to connect the material on bilinear flocks of quadratic cones with partial tangential packings of ovoids and elliptic flocks using the Klein quadric.

DEFINITION 102. *Let $\mathbb{Q}$ be an ovoid of points of the Klein quadric in $PG(5, K)$, where K is a field. Assume that there is a conic C in $\mathbb{Q}$ such that, for each point P of the conic, there is a partition of $\mathbb{Q}$ into a set of conics containing C which mutually share P. In this case, the set of such partitions is called a 'tangential packing' of the ovoid with respect to C.*

All of the following results have been proven in the section on bilinear spreads (9,11,3&4).

THEOREM 123. *Let π be a translation plane with spread in $PG(3, K)$, where K is a field that admits a regular double cover. Then, there is a collineation group isomorphic to $SL(2, K)$ generated by regulus-inducing elation groups. And, for each component L of the invariant regulus net, there is a set of regulus nets which share L and are orbits under a regulus-inducing elation group with axis L.*

THEOREM 124. *Under the assumptions of the previous theorem, for each component L of the invariant regulus net, there is a flock of a quadratic cone F_L in some 3-dimensional projective space Σ_L^3, isomorphic to $PG(3, K)$, which has as its basic points, the Baer subplanes of the regulus nets sharing the component L.*

Hence, each of these flocks of quadratic cones share a conic.

Each of these projective spaces Σ_L^3 lie in the same projective space Σ^5 isomorphic to $PG(5, K)$ and there is a collineation group which fixes the common conic and acts transitively on the cones and on the flocks.

Additionally, the ovoid Σ^5 corresponding to the spread of π admits a tangential packing with respect to the conic corresponding to the invariant regulus net.

THEOREM 125. *Assume that there is a flock of a quadratic cone F in a 3-dimensional projective space Σ^3 so that the cone is embedded in the Klein quadric. Take the intersection with the Klein quadric of the perpendicular planes of the planes which contains the conics of the flock. This forms an ovoid $\mathbb{Q}$ of points consisting of conics which share a point.*

For a fixed conic C, assume that there is a tangential packing of the ovoid with respect to C. Then there is a corresponding translation plane with spread in $PG(3, K)$ which admits $SL(2, K)$ generated by regulus-inducing elation groups with an axis on an invariant regulus net corresponding to C.

Part 10

Multiple replacement theorem

In this part, we discuss a very general net replacement procedure that works for any skewfield plane that admits an automorphism. Then we apply this result in several ways extending procedures of Andrè [4] and Ostrom. Along the way, we consider how this works in general quaternion division ring planes, as well as in cyclic division ring planes. What is necessary to apply the general skewfield lifting theorem, in these last mentioned skewfield planes, is that the skewfields be central cyclic Galois quadratic skewfield extensions. In part 12, we shall see that any quaternion division ring, regardless of characteristic, has an extension which has a central Galois quadratic extension. So, this procedure will produce a set of mutually disjoint pseudo-regulus or regulus-net derivable nets, the union of which covers the skewfield spread minus two components, which we denote by $x = 0$, $y = 0$. Note that a skewfield spread is also called a Desarguesian plane or Desarguesian translation plane. We also use the terminology Pappian plane or Pappian translation plane, when the skewfield is a field.

CHAPTER 12

The general theorem

THEOREM 126. *(**The Multiple Replacement Theorem**) Let Σ denote a Desarguesian translation plane coordinatized by a skewfield S that admits a non-trivial automorphism σ, which could be either inner or outer, finite or infinite.*

(1) Writing the spread of Σ as

$$x = 0, y = xm; m \in S, \forall\ x, y \in S,$$

defines a spread where all components are left vector spaces over S and where the elements of the set of kernel mappings S_L have the following form

$$S_L : \{(x, y) \to (\beta x, \beta y); \forall \beta \in S^*\}.$$

Then we form a corresponding spread Σ_σ, of the form

$$x = 0, y = mx^\sigma; m \in S,\ \forall x, y \in S,$$

where all components are right vector spaces over S and where the elements of the set of kernel mappings S_R^σ have the following form:

$$S_R^\sigma : \{(x, y) \to (xe, ye^\sigma); \forall e \in S^*\}.$$

(2) $S_L S_R^\sigma = S_R^\sigma S_L$ is a collineation group of both Σ and Σ_σ.

(3) Σ_σ is also a Desarguesian plane and Σ and Σ_σ cover each other.

(4) The intersections of $y = xm$ and $y = mx^\sigma$ are isomorphic subspaces over $Fix\sigma \cap Z(S)$, where $Z(S)$ is the center of S. As a result, we have a vector space that is a left/right bimodule (S, S^σ), where the components are isomorphic vector spaces over the field $Fix\sigma \cap Z(S)$.

(5) Let the subskewfield $Fix\sigma$ be denoted by F and note that Σ is a set of left F-subspaces and Σ_σ is a set of right F-subspaces.

We emphasize that the skewfield $Fix\sigma$ appears/acts in $S_L S_R^\sigma$ as the set of left mappings of the form $(x, y) \to (fx, fy)$, for all $f \in Fix\sigma^*$ that fixes each component of Σ and as the set of right mappings of the form $(x, y) \to (xf, yf)$ for all $f \in Fix\sigma^*$.

(a) The set of components of Σ that are images of $y = x$ under S_R^σ is

$$\{y = xbb^{-\sigma}; b \in S^*\}$$

and the set of components of Σ_σ that are images of $y = x^\sigma$ under S_L is

$$\{y = bb^{-\sigma}x^\sigma; b \in S^*\},$$

Furthermore,

$$\{y = xbb^{-\sigma}; b \in S^*\} = \{y = bb^{-\sigma}x^\sigma; b \in S^*\},$$

in the sense that the sets are replacements of each other.

(b) *Similarly, $y = xm$ and $y = mx^\sigma$ intersect non-trivially and have the following sets of images under $S_L S_R^\sigma$:*

$$(*) : \{y = bmb^{-\sigma}x^\sigma; b \in S^*\} \quad and \quad \{y = xbmb^{-\sigma}; b \in S^*\},$$

as replacement sets.

(6) *There is an index set λ such that*

$$\Sigma = \{x = 0, y = 0\} \cup_{i \in \lambda} \{\{y = xbm_ib^{-\sigma}; b \in S^*\} = \Gamma_i\},$$

and,

$$\Sigma_\sigma = \{x = 0, y = 0\} \cup_{i \in \lambda} \{\{y = bm_ib^{-\sigma}x^\sigma; b \in S^*\} = \Gamma_i^\sigma\},$$

where the unions are disjoint unions.

(7) *Choose any proper subset δ of λ (not the empty set or λ), then the multiply replaced translation plane has blended kernel $(S_L, S_R^\sigma) \cup \{0\}$ and formal kernel $Z(S) \cap Fix\sigma = (S_L \cap S_R^\sigma) \cup \{0\}$.*

PROOF. To prove (1), first consider

$$S_L : \{(x, y) \to (\beta x, \beta y); \forall \beta \in S^*\},$$

and choose any component $y = xm$ of Σ. Then an element of S_L, will map $(x, xm) = (\beta x, \beta(xm) = (\beta x)m)$ so setwise, $y = xm \to y = xm$. Consider S_R^σ : $\{(x, y) \to (xe, ye^\sigma); \forall e \in S^*\}$ and choose any component $y = nx^\sigma$. An element of S_R^σ will map $(x, nx^\sigma) \to (xe, (nx^\sigma)e^\sigma)$. If $xe = x_1$, then $n(x^\sigma e^\sigma) = n(x_1 e^{-1})^\sigma e^\sigma = nx_1^\sigma$, and therefore, as a set, $y = nx^\sigma \to y = nx^\sigma$.

Now consider the set $S_R^\sigma S_L : (x, y) \to (\beta x, \beta y) \to ((\beta x)e, (\beta y)e^\sigma) = (\beta(xe), \beta(ye^\sigma))$, which shows that $S_R^\sigma S_L = S_L S_R^\sigma$, implying that the product is a group. Note that the inverse of $(x, y) \to (\beta xe, \beta ye^\sigma)$ is $(x, y) \to (\beta^{-1}xe^{-1}, \beta^{-1}ye^{-\sigma})$, as $(e^\sigma)^{-1} = (e^{-1})^\sigma$. Consider a component $y = xm$ of Σ, then the image under $S_L S_R^\sigma$ is $(x, xm) \to (\beta xe, \beta xme^\sigma)$, and note that if $\beta xe = x_1$ then $\beta xme^\sigma = x_1 e^{-1}me^\sigma$, which implies that $y = xm \to y = xe^{-1}me^\sigma$. And the set of images of $y = xm$ under $S_L S_R^\sigma$ is $\{y = xe^{-1}me^\sigma : e \in S^*\}$.

Similarly, following the same argument, the image of $y = mx^\sigma$ under $S_L S_R^\sigma$ is $\{y = \beta m\beta^{-\sigma}x^\sigma : \beta \in S^*\}$.

Letting $\beta = e^{-1}$, we see that the two sets are $\{y = xe^{-1}me^\sigma : e \in S^*\}$ and $\{y = e^{-1}me^\sigma x^\sigma : e \in S^*\}$. This proves (1) and (2). Leaving (3) for the reader, we will now consider (4). There are two points to make. The group S_L is the kernel collineation group of Σ, which means S_L leaves invariant each component $y = xm$ and is transitive on the non-zero vectors of each component. And, S_R^σ leaves invariant each component $y = nx^\sigma$ and is transitive on the non-zero vectors of each component of Σ_σ. We note that $(y = xm) \cap (y = mx^\sigma)$ non-trivially intersect at $Z(S) \cap Fix\sigma \supseteq \{0, 1\}$. Since the vector space is a left-right (S_L, S_R^σ)-bimodule, we see that the components of both of the two Desarguesian planes are isomorphic vector spaces over $Z(S) \cap Fix\sigma$, which is easy to determine is a field. We now show that the point sets $\{y = xe^{-1}me^\sigma : e \in S^*\} = \{y = e^{-1}me^\sigma x^\sigma : e \in S^*\}$. Consider the image of $(y = xm) \cap (y = mx^\sigma)$ under $S_L S_R^\sigma$. The intersection is a $Z(S) \cap Fix\sigma$ subspace that lies on $y = xm$, Since S_L is transitive on $y = xm$, the image of this intersection covers $y = xm$. Continuing to follow the path of the intersection, then carries $y = xm$ onto $y = xe^{-1}me^\sigma$, which then shows that image of the intersection also completely covers $y = xe^{-1}me^\sigma$, for each $e \in S^*$. To see this, fix e as b and apply the argument again, which shows that the image of $y = xb^{-1}mb^\sigma$ under the path of the intersection is $y = xe^{-1}b^{-1}mb^\sigma e^\sigma$, and since $(e^{-1}b^{-1}) = (be)^{-1}$, and $b^\sigma e^\sigma = (be)^\sigma$, we have $y = x(be)^{-1}m(be)^\sigma$. To see that $\{y = xe^{-1}me^\sigma : e \in S^*\} = \{y = e^{-1}me^\sigma x^\sigma : e \in S^*\}$, now follow $y = mx^\sigma$ under the path of the intersection. This time the image of the intersection starts with S_R^σ and the image of the intersection covers the nonzero vectors of $y = mx^\sigma$, and maps $y = mx^\sigma$ onto $y = eme^{-\sigma}x^\sigma$ (notice e instead of e^{-1} in the previous case). This shows that the two sets are equal/cover each other. Therefore, we have that the two sets are replacement sets of each other. This proves all parts of the theorem except for the representation of the spread. So, considering the orbits of the spread of Σ under the group $S_L S_R^\sigma$, we may use the Axiom of Choice to choose one element component from each orbit. This shows that we have a set λ of mutually disjoint replaceable sets together with $x = 0$ and $y = 0$. Choose any nontrivial subset δ of λ and replace all of the replaceable nets corresponding to δ. Now we have a translation plane with a blended kernel $(S_L \cup \{0\}, S_R^\sigma \cup \{0\})$, and the formal kernel is $Z(S) \cap Fix\sigma$. Note that this is clearly $(S_L \cap S_R^\sigma) \cup \{0\}$. This completes parts (6) and (7) and therefore finishes the proof.   □

DEFINITION 103. *Let π be a translation plane whose components consist of a union of left subspaces over a skewfield F and right subspaces over the skewfield F, we shall then use the term 'blended left-right kernel F'. In the situation in the theorem, the standard kernel for π would then be $Z(F) \cap Fix\sigma$, where $Z(F)$ is the center of F. If the skewfield is a field, the term used is 'blended kernel'.*

COROLLARY 34. *Any collineation g of Σ that leaves $\{x = 0, y = 0\}$ invariant maps the set $\cup_{i \in \lambda} \{y = xbm_i b^{-\sigma}; b \in S^*\}$ to another mutually disjoint set of replaceable nets. If Σ is a Desarguesian but not Pappian spread, the group $S_L^* S_R^{*\sigma}$ is not normal in the collineation group that leaves $\{x = 0, y = 0\}$ invariant and there would be many such sets of mutually disjoint replaceable nets.*

For example, the group of homologies with axis $x = 0$ with elements $(x, y) \to (x, yf); f \in F^*$ is a collineation group of Σ that maps $y = xm$ to $y = x(mf)$ but is not a collineation group of Σ_σ, for non-Pappian spreads, although for $f \in Z(F)^*$, the associated subgroup

$$\langle (x, y) \to (x, yh^{\sigma-1}); h \in Z(F)^* \rangle,$$

leaves invariant each element of the set of mutually disjoint replacement sets.

COROLLARY 35. *There are no restrictions on the automorphism σ or the skew-field. So, note that for any element σ^j of the group $\langle \sigma \rangle$, there are translation planes*

$$\Sigma^j(\rho) = \{x = 0, y = 0\} \cup \left\{ \cup_{i \in \rho} \Gamma_i^{\sigma^j} \right\} \cup \left\{ \cup_{i \in \lambda - \rho} \Gamma_i^j \right\}.$$

If the group is an infinite cyclic group, there are infinitely many translation planes, and there are infinitely many choices of subsets of λ and for all nonzero positive and negative integers j and where $\Gamma_i^{\sigma^i}$ and Γ_i^j, σ is replaced by σ^j.

COROLLARY 36. *Let σ be any inner automorphism: $\sigma_k : \sigma_k(t) = ktk^{-1}$, when S is a non-commutative skewfield. There are infinitely many automorphism groups*

$\left\langle \sigma_k^j \right\rangle$, *for all k not in $Z(S)$ and for all nonzero positive and negative integers j. We shall see below that some inner automorphism groups can have finite order.*

So, there are always sets of mutually disjoint replaceable nets for any non-commutative skewfield.

REMARK 34. *Important note: We note that $(a,b)_K$ defines a Desarguesian spread. And, we will show in the following two sections that $(a,b)_K$ has two important automorphisms τ that define sets of replaceable nets covering all components but $x = 0$, $y = 0$. The same remark is valid for cyclic division rings of exponent n defining a Desarguesian spread.*

In the first case, τ is an inner automorphism of $(a,b)_K$ and the replaceable nets are 4-dimensional over K.

In the second case, τ is an automorphism of $(a,b)_{K(\rho)}^{K(\rho)(\theta)}$ fixing $(a,b)_K$ point-wise. Here derivable nets are obtained. The derivable nets are in the 1-dimension representation $y = xm$ where x, y, m are in $(a,b)_K$ and the multiply derived translation spreads are in $PG(15, K)$. For example, the line $y = x$ has dimension $[(a,b)_{K(\rho)}^{K(\rho)(\theta)} : Fix\tau][Fix\tau : Fix\tau \cap K(\rho)] = 8$. Since the components of the replacement net are 2-dimensional $(a,b)_K$-subspaces, the derivable nets are isomorphic to the classical pseudo-regulus nets relative to $(a,b)_K$.

In $(11,13,3\&4)$, this last example is generalized and applications are given.

REMARK 35. *In $(2,4,2)$, it was mentioned that for every quaternion division ring Desarguesian plane, there are at least three derivable nets. In the next sections 3 and 4, we will use the multiple replacement theorem to show that are many other derivable nets in $(a,b)_{K(\rho)}^{\alpha}$, actually a set of such derivable nets that cover the spread except for $x = 0, y = 0$. The question is, what automorphism τ is available? It turns out that we use an inner automorphism τ, which has order 2 if the characteristic $\neq 2$, and of order 3 in certain cases, finite or possibly infinite if the characteristic $= 2$. Each of these derivable nets are isomorphic to the type 2 twisted regulus net with respect to the classical regulus net over $K(\theta)$.*

Consider the quaternion division ring $(a,b)_{K(\tau)}^{\alpha}$, a 2-dimensional extension of $(a,b)_K^{\alpha}$. The associated field is $K(\theta)$ for $(a,b)_K^{\alpha}$, and $K(\theta,\rho)$ for $(a,b)_{K(\rho)}^{\alpha}$. In the proof of the extending skewfields theorem, it was shown that the Desarguesian plane coordinatized by $(a,b)_{K(\rho)}^{\alpha}$ has a derivable net coordinatized by $(a,b)_K$.

(1) *Using the fact that the $L = (a,b)_K$ has a central Galois quadratic extension $L(\rho)$, it then follows that there is an automorphism τ of $K(\theta,\rho)$ that fixes L pointwise. The multiple replacement theorem then shows that there is a set of mutually disjoint derivable nets that together with $x = 0, y = 0$ cover the Desarguesian plane coordinatized by $L(\rho)$. These derivable nets are pseudo-regulus nets isomorphic to the classical pseudo-regulus net over $(a,b)_K$.*

(2) *Using the previous remark, there is an inner automorphism of $L(\rho)$ that applies to show that the derivable nets are each isomorphic to the type 2 twisted regulus net with respect to the classical regulus net over $K(\rho)(\theta) = K(\theta,\rho)$.*

1. Skewfields of finite dimension/$Fix\sigma$

Let S be a skewfield that admits an involutory automorphism σ. Then, applying the multiple replacement theorem, we have a set of mutually disjoint replaceable nets in Σ, where the replacement nets are in Σ_σ.

THEOREM 127. *For each $i \in \lambda$,*

$$\left\{ y = xbm_ib^{-\sigma}; b \in S^* \right\} = \left\{ y = bm_ib^{-\sigma}x^\sigma; b \in S^* \right\}.$$

2. $(a,b)_K$–inner automorphisms.

All of the theory in this section may be generalized to arbitrary division rings of finite degree.

It might be expected when σ has order 2 and the intersection of the groups S_L^* and $S_R^{*\sigma}$ is $Z(S) \cap Fix\sigma$ is the kernel of the translation planes obtained by multiple replacement, that S has dimension 2 over $Fix\sigma$ as a vector space, which would be true if L is a field. The components $y = xbm_ib^{-\sigma}$, as elements of $\Sigma = \{y = xn; n \in S^*\}$ are therefore left 2-dimension vector spaces over $Fix\sigma$. Then $y = x^\sigma$ is a right/left 2-dimensional vector space over $Fix\sigma$, which implies that $y = nx^\sigma$ is a right 2-dimensional $Fix\sigma$ subspace.

THEOREM 128. *Let S be a non-commutative skewfield that admits an automorphism τ and further assume that S has dimension 2 over $Fix\tau$ as a vector space. Hence, the spread is in $PG(3, Fix\tau)$. Then the Desarguesian plane coordinatized by S admits a set of mutually disjoint derivable nets where the components are left 2-dimensional $Fix\sigma$ subspaces and the replacement components are right 2-dimensional $Fix\sigma$ subspaces. Therefore, we have a set of mutually disjoint pseudo-regulus nets that together with $x = 0$ and $y = 0$ form the spread. Since the components of the nets are left 2-dimensional $Fix\tau$ subspaces and the Baer subspaces are right 2-dimensional $Fix\tau$ subspaces, the derivable nets are all isomorphic to $x = 0, y = 0$, $y = xFix\tau$. So, if $Fix\tau$ is a field, we have regulus nets relative to $Fix\tau$ of type 2 and if $Fix\tau$ is a non-commutative skewfield, we have pseudo-regulus nets.*

The non-trivial translation planes obtained by multiple derivation has spreads with kernel $Fix\tau \cap Z(S)$.

PROOF. There will be a Desarguesian affine plane written using a set of 2 by 2 matrices over $Fix\tau$. And, recall that we now have a skewfield spread Σ in $PG(3, Fix\tau)$. Also, note that, by the structure of the multiple replacement theorem, Σ is a left spread over S, and Σ_τ is isomorphic to the dual translation plane of Σ. Furthermore, the components of Σ are left 2-spaces over $Fix\tau$ and the components of Σ_τ are right 2-subspaces over $Fix\tau$. Rereading the proof of the multiple replacement theorem, we see that the replacement nets $\{y = xbmb^{-\tau}; b \in S^*\}$ are subplane covered nets, where $y = bmb^{-\tau}x^\tau$ are subplanes that are Baer since any two such subplanes of right dimension 2 over $Fix\tau$ generate the 4-dimensional $Fix\tau$-space. This completes the proof except for the nature of the derivable nets. Suppose that $Fix\tau$ is a field. Consider the proof of the embedding theorem: We have a derivable net which may be transformed into a pseudo-regulus net. The Baer subplanes incident with the zero vector are Desarguesian 2-dimension vector spaces over a skewfield W, and are Pappian 2-dimensional over a field $Fix\tau$. This then means that W is $Fix\tau$. The fact that we have components that are left 2-dimension $Fix\tau$ subspaces and the Baer subspaces are right 2-dimensional $Fix\tau$ can only mean that the derivable nets are twisted regulus nets relative to an associated isomorphic field. The same proof shows that using embedding to $PG(3, W)$, W is isomorphic to $Fix\tau$. $\square$

Now we show that for any quaternion division ring plane coordinatized by $(a, b)_K^\alpha$, there exist appropriate automorphisms (inner automorphisms) and for any extension $(a, b)_{K(\rho)}^\alpha$, the corresponding Desarguesian plane admits two types of automorphisms that provide both types of derivable nets.

THEOREM 129. *(Quaternion division ring 4-dimensional replacement theorem for inner automorphisms). Let S be a quaternion division ring $(a, b)_K^\alpha$. Let the matrix representation be*

$$(a, b)_K^\alpha \simeq \left\{ \begin{bmatrix} u^\sigma & bt^\sigma \\ t & u \end{bmatrix} : \forall t, u \in K(\theta) \right\},$$

σ *the automorphism induced by $K(\theta)$.*

If the characteristic is 2, let $\theta^2 = \theta + \beta$, for $\beta \in K$ (choosing $\alpha = 1$). If the characteristic is $\neq 2$, let $\theta = \sqrt{a}$, where a is a non-square. Define an inner automorphism τ by

$$\tau\left(\begin{bmatrix} u^\sigma & bt^\sigma \\ t & u \end{bmatrix} \right) = \begin{bmatrix} \theta^\sigma & 0 \\ 0 & \theta \end{bmatrix} \begin{bmatrix} u^\sigma & bt^\sigma \\ t & u \end{bmatrix} \begin{bmatrix} \theta^{-\sigma} & 0 \\ 0 & \theta^{-1} \end{bmatrix}$$

$$= \begin{bmatrix} u^\sigma & -b(t\theta^{\sigma-1})^\sigma \\ t\theta^{\sigma-1} & u \end{bmatrix}.$$

(1) *Then for characteristic 2, the order of τ is 3 if $\beta = 1$, and the order is $k + 1$ if $1 + \beta + \beta^2 + ... + \beta^k = 0$, or possibly infinite. For characteristic $\neq 2$, τ has order 2.*

In all cases, $Fix\tau = \left(\begin{bmatrix} u^\sigma & 0 \\ 0 & u \end{bmatrix} ; \forall u \in K(\theta) \right)$ and $(a, b)_K^\alpha$ is of dimension 2 over $Fix\tau$.

(2) *The standard replacement net with components*

$$\{ y = xd^{-1}d^\tau : \forall d \in (a, b)_K^{\alpha*} \}$$

has replacement subspaces as

$$\{ y = d^{-1}d^\tau x^\tau : \forall d \in (a, b)_K^{\alpha*} \}.$$

The components are 2-dimensional left $Fix\tau$-subspaces and the replacement subspaces are 2-dimensional right $Fix\tau$-subspaces. Hence, we have a set of mutually disjoint pseudo-regulus derivable nets that are type 2 twisted regulus nets relative to $K(\theta)$.

(3) *Matrix representations are as follows for characteristic $\neq 2$*

$$\left\{ \begin{bmatrix} y_2^\sigma & by_1^\sigma \\ y_1 & y_2 \end{bmatrix} = \begin{bmatrix} x_2^\sigma & bx_1^\sigma \\ x_1 & x_2 \end{bmatrix} \begin{bmatrix} \frac{u^{\sigma+1}+bt^{\sigma+1}}{u^{\sigma+1}-bt^{\sigma+1}} & \frac{-2but^\sigma}{u^{\sigma+1}-b^{\sigma+1}} \\ \frac{-2u^\sigma t}{u^{\sigma+1}-bt^{\sigma+1}} & \frac{u^{\sigma+1}+bt^{\sigma+1}}{u^{\sigma+1}-bt^{\sigma+1}} \end{bmatrix} \right\}$$

$\quad : \ \forall u, t \in K(\theta),$

with replacement subspaces

$$\left\{ \begin{bmatrix} y_2^\sigma & by_1^\sigma \\ y_1 & y_2 \end{bmatrix} = \begin{bmatrix} \frac{u^{\sigma+1}+bt^{\sigma+1}}{u^{\sigma+1}-bt^{\sigma+1}} & \frac{-2but^\sigma}{u^{\sigma+1}-b^{\sigma+1}} \\ \frac{-2u^\sigma t}{u^{\sigma+1}-bt^{\sigma+1}} & \frac{u^{\sigma+1}+bt^{\sigma+1}}{u^{\sigma+1}-bt^{\sigma+1}} \end{bmatrix} \begin{bmatrix} x_2^\sigma & -b(x_1)^\sigma \\ -x_1 & x_2 \end{bmatrix} \right\}$$

$\quad : \ \forall u, t \in K(\theta).$

Note that x and y are in $(a, b)_K^\alpha$ not 2-vectors over $K(\theta)$, as it would appear. As a result, with $x = \begin{bmatrix} x_2^\sigma & bx_1^\sigma \\ x_1 & x_2 \end{bmatrix}$ and $x^\tau = \begin{bmatrix} x_2^\sigma & b(x_1\theta^{\sigma-1})^\sigma \\ \theta^{\sigma-1}x_1 & x_2 \end{bmatrix}$, and

$y = \begin{bmatrix} y_2^\sigma & by_1^\sigma \\ y_1 & y_2 \end{bmatrix}$, *we would then have the components in the form given.*

(4) *Matrix representations are as follows for characteristic 2:.*

$$\left\{ \begin{bmatrix} y_2^\sigma & by_1^\sigma \\ y_1 & y_2 \end{bmatrix} = \begin{bmatrix} x_2^\sigma & bx_1^\sigma \\ x_1 & x_2 \end{bmatrix} \begin{bmatrix} \frac{u^{\sigma+1}+bt^{\sigma+1}\left(\frac{\theta+\beta}{\theta}\right)}{u^{\sigma+1}+bt^{\sigma+1}} & \frac{but^\sigma\left(\frac{\theta}{\beta}\right)}{u^{\sigma+1}+b^{\sigma+1}} \\ \frac{u^\sigma t\left(\frac{\beta}{\theta}\right)}{u^{\sigma+1}+bt^{\sigma+1}} & \frac{u^{\sigma+1}+bt^{\sigma+1}\left(\frac{\theta}{\theta+\beta}\right)}{u^{\sigma+1}+bt^{\sigma+1}} \end{bmatrix} \right\}$$

$\quad : \ \forall u, t \in K(\theta),$

and the matrices for the replacement subspaces are

$$\left\{ \begin{bmatrix} \frac{u^{\sigma+1}+bt^{\sigma+1}\left(\frac{\theta+\beta}{\theta}\right)}{u^{\sigma+1}+bt^{\sigma+1}} & \frac{but^\sigma\left(\frac{\theta}{\beta}\right)}{u^{\sigma+1}+b^{\sigma+1}} \\ \frac{u^\sigma t\left(\frac{\beta}{\theta}\right)}{u^{\sigma+1}+bt^{\sigma+1}} & \frac{u^{\sigma+1}+bt^{\sigma+1}\left(\frac{\theta}{\theta+\beta}\right)}{u^{\sigma+1}+bt^{\sigma+1}} \end{bmatrix} \begin{bmatrix} x_2^\sigma & b(x_1\theta^{\sigma-1})^\sigma \\ \theta^{\sigma-1}x_1 & x_2 \end{bmatrix} \right.$$

$$\left. \begin{bmatrix} y_2^\sigma & by_1^\sigma \\ y_1 & y_2 \end{bmatrix} = \right\}$$

$\quad : \ \forall u, t \in K(\theta).$

(5) *There are matrix representations for the other derivable nets, which are of the following form: For the index set λ describing the derivable nets, choose any proper subset δ and let $m_i = \begin{bmatrix} v_i^\sigma & bs_i^\sigma \\ s_i & v_i \end{bmatrix}$ for $i \in \lambda$. Now replace all replacement nets associated with δ. Then the following describes the associated translation plane π_δ: The components are $x = 0, y = 0$ union the following: writing $x =$*

$$\left[\begin{array}{cc} x_2^\sigma & bx_1^\sigma \\ x_1 & x_2 \end{array}\right],\ x^\tau = \left[\begin{array}{cc} x_2^\sigma & b(x_1\theta^{\sigma-1})^\sigma \\ \theta^{\sigma-1}x_1 & x_2 \end{array}\right],\ \text{and } y = \left[\begin{array}{cc} y_2^\sigma & by_1^\sigma \\ y_1 & y_2 \end{array}\right].$$

$$\bigcup_{i\in\delta}\left\{\begin{array}{c} y = \\ x\left[\begin{array}{cc} u^\sigma & bt^\sigma \\ t & u \end{array}\right]^{-1}\left[\begin{array}{cc} v_i^\sigma & bs_i^\sigma \\ s_i & v_i \end{array}\right]\left[\begin{array}{cc} u^\sigma & bt^\sigma\theta^{1-\sigma} \\ \theta^{\sigma-1}t & u \end{array}\right]; \\ \forall (t,u)\in K(\theta)\times K(\theta)-\{(0,0)\} \\ \text{where } \theta = \theta + \beta \text{ for characteristic 2} \\ \text{and } \theta = \sqrt{a} \text{ for char. } \neq 2. \end{array}\right\} \bigcup$$

$$\bigcup_{i\in\lambda-\delta}\left\{\begin{array}{c} y = \\ \left[\begin{array}{cc} u^\sigma & bt^\sigma \\ t & u \end{array}\right]^{-1}\left[\begin{array}{cc} v_i^\sigma & bs_i^\sigma \\ s_i & v_i \end{array}\right]\left[\begin{array}{cc} u^\sigma & bt^\sigma\theta^{1-\sigma} \\ \theta^{\sigma-1}t & u \end{array}\right]x^\tau; \\ \forall (t,u)\in K(\theta)\times K(\theta)-\{(0,0)\} \\ \text{where } \theta = \theta + \beta \text{ for characteristic 2} \\ \text{and } \theta = \sqrt{a} \text{ for char. } \neq 2. \end{array}\right\}.$$

(6) *The components are left 2–dimensional Fixσ subspaces and hence right and left 4-dimensional K- subspaces. The replacement components are right 2-dimensional subspaces and hence left right 4-dimensional K-subspaces. The replacement nets are clearly derivable nets by the main embedding theorem. As a result, all translation planes obtained by multiple net replacement have spreads with kernel $Fix\tau \cap Z((a,b)_K^\alpha)$, in $PG(7,K)$.*

PROOF. Let $(a,b)_K$ be a quaternion division ring. So, we may assume that a and b are non-squares in K, in the characteristic $\neq 2$ case and the more general representation in the characteristic 2 case. Generally, in the matrix representation, we have $(a,b)_K \simeq \left[\begin{array}{cc} u^\sigma & bt^\sigma \\ t & u \end{array}\right] : \forall t, u \in K(\theta)$. Let τ denote the inner automorphism of $(a,b)_K$ defined by $\tau(r) = \theta r\theta^{-1}$, represented in matrix form. As a result, we consider

$$\tau\left(\left[\begin{array}{cc} u^\sigma & bt^\sigma \\ t & u \end{array}\right]\right) = \left[\begin{array}{cc} \theta^\sigma & 0 \\ 0 & \theta \end{array}\right]\left[\begin{array}{cc} u^\sigma & bt^\sigma \\ t & u \end{array}\right]\left[\begin{array}{cc} \theta^{-\sigma} & 0 \\ 0 & \theta^{-1} \end{array}\right],$$

which after a short computation is

$$\tau\left(\left[\begin{array}{cc} u^\sigma & bt^\sigma \\ t & u \end{array}\right]\right) = \left[\begin{array}{cc} u^\sigma & b(t\theta^{\sigma-1})^\sigma \\ t\theta^{\sigma-1} & u \end{array}\right] = \left[\begin{array}{cc} u^\sigma & -bt^\sigma \\ -t & u \end{array}\right],$$

in characteristic $\neq 2$. This implies that

$$\tau^2\left(\left[\begin{array}{cc} u^\sigma & bt^\sigma \\ t & u \end{array}\right]\right) = \left[\begin{array}{cc} u^\sigma & bt^\sigma \\ t & u \end{array}\right].$$

That is, τ has order 2 and $Fix\sigma = \left(\left[\begin{array}{cc} u^\sigma & 0 \\ 0 & u \end{array}\right]; u \in K(\theta)\right)$. Then note that

$$\left[\begin{array}{cc} u^\sigma & bt^\sigma \\ t & u \end{array}\right] = \left[\begin{array}{cc} u^\sigma & 0 \\ 0 & u \end{array}\right] + \left[\begin{array}{cc} t^\sigma & 0 \\ 0 & t \end{array}\right]\left[\begin{array}{cc} 0 & b \\ 1 & 0 \end{array}\right].$$

As a result, $S = (a,b)_K$ has a basis $\left\{ I_2, \begin{bmatrix} 0 & b \\ 1 & 0 \end{bmatrix} \right\}$ over

$$Fix\sigma = \left(\begin{bmatrix} u^\sigma & 0 \\ 0 & u \end{bmatrix} ; u \in K(\theta) \right),$$

and therefore, satisfies the assumptions of the previous result. We then consider

$$d^{-1}d^\tau, \ for \ d \in \left\{ \begin{bmatrix} u^\sigma & bt^\sigma \\ t & u \end{bmatrix} : t, u \in K(\theta) \right\}.$$

The computation shows that, since d is a matrix over a field, the inverse is immediate, which shows that

$$\begin{bmatrix} u^\sigma & bt^\sigma \\ t & u \end{bmatrix}^{-1} \begin{bmatrix} u^\sigma & bt^\sigma \\ t & u \end{bmatrix}^\tau = \begin{bmatrix} \frac{u^{\sigma+1}+bt^{\sigma+1}}{u^{\sigma+1}-bt^{\sigma+1}} & \frac{-2but^\sigma}{u^{\sigma+1}-b^{\sigma+1}} \\ \frac{-2u^\sigma t}{u^{\sigma+1}-bt^{\sigma+1}} & \frac{u^{\sigma+1}+bt^{\sigma+1}}{u^{\sigma+1}-bt^{\sigma+1}} \end{bmatrix};$$

$$d^{-1}d^\tau, \ for \ d \ \in \ \left\{ \begin{bmatrix} u^\sigma & bt^\sigma \\ t & u \end{bmatrix} : t, u \in K(\theta) \right\}$$

$$= \left\{ \begin{array}{c} \begin{bmatrix} \frac{u^{\sigma+1}+bt^{\sigma+1}}{u^{\sigma+1}-bt^{\sigma+1}} & \frac{-2but^\sigma}{u^{\sigma+1}-b^{\sigma+1}} \\ \frac{-2u^\sigma t}{u^{\sigma+1}-bt^{\sigma+1}} & \frac{u^{\sigma+1}+bt^{\sigma+1}}{u^{\sigma+1}-bt^{\sigma+1}} \end{bmatrix} \\ : t, u \in K(\theta) \end{array} \right\}.$$

Now $y = x^\tau = \begin{bmatrix} x_2^\sigma & bx_1^\sigma \\ x_1 & x_2 \end{bmatrix}^\tau = \begin{bmatrix} x_2^\sigma & b(x_1\theta^{\sigma-1})^\sigma \\ \theta^{\sigma-1}x_1 & x_2 \end{bmatrix}$, in characteristic $\neq 2$ we have $\theta^{\sigma-1} = -1$. We are considering then $y = d^{-1}d^\sigma x^\sigma$, and $y = xd^{-1}d^\sigma$, setwise. Since the matrices are over a field $K(\theta)$, we have the following: every quaternion division ring $(a,b)_K^\alpha$ over a field of characteristic $\neq 2$ admits a set of mutually disjoint derivable nets which are τ-twisted regulus nets over the field K. The fundamental τ-twisted regulus net is written as follows: The components are

$$\left\{ \begin{bmatrix} y_2^\sigma & by_1^\sigma \\ y_1 & y_2 \end{bmatrix} = \begin{bmatrix} x_2^\sigma & bx_1^\sigma \\ x_1 & x_2 \end{bmatrix} \begin{bmatrix} \frac{u^{\sigma+1}+bt^{\sigma+1}}{u^{\sigma+1}-bt^{\sigma+1}} & \frac{-2but^\sigma}{u^{\sigma+1}-b^{\sigma+1}} \\ \frac{-2u^\sigma t}{u^{\sigma+1}-bt^{\sigma+1}} & \frac{u^{\sigma+1}+bt^{\sigma+1}}{u^{\sigma+1}-bt^{\sigma+1}} \end{bmatrix} \right\}$$
$$: \ \forall u, t \in K(\theta),$$

and are left subspaces over $\left\{ (a,b)_{K_{Left}}^\alpha : (x,y) \to (dx, dy), \forall d \in (a,b)_L \right\}$ with Baer subspaces are right subspaces over $\left\{ (a,b)_{K_{Right}}^{\alpha\,\tau} : (x,y) \to (xe, ye^\tau) \right\}$.

$$\left\{ \begin{bmatrix} y_2^\sigma & by_1^\sigma \\ y_1 & y_2 \end{bmatrix} = \begin{bmatrix} \frac{u^{\sigma+1}+bt^{\sigma+1}}{u^{\sigma+1}-bt^{\sigma+1}} & \frac{-2but^\sigma}{u^{\sigma+1}-b^{\sigma+1}} \\ \frac{-2u^\sigma t}{u^{\sigma+1}-bt^{\sigma+1}} & \frac{u^{\sigma+1}+bt^{\sigma+1}}{u^{\sigma+1}-bt^{\sigma+1}} \end{bmatrix} \begin{bmatrix} x_2^\sigma & -bx_1^\sigma \\ -x_1 & x_2 \end{bmatrix} \right\}$$
$$: \ \forall u, t \in K(\theta).$$

The components and Baer subspaces are right 2-dimensional over the field $K(\theta)^\sigma$. And thus the derivable nets are σ-twisted regulus nets, as the other derivable nets are isomorphic to the fundamental net and hence, by the main embedding-contraction theory and classification theory, are isomorphic. The kernel of these translation planes is K. When the characteristic is 2, the analogous computation provides the same field $Fix\tau$ and shows that matrices for components are

$$\left\{ y = x \begin{bmatrix} \dfrac{u^{\sigma+1}+bt^{\sigma+1}\left(\frac{\theta+\beta}{\theta}\right)}{u^{\sigma+1}+bt^{\sigma+1}} & \dfrac{but^{\sigma}\left(\frac{\theta}{\beta}\right)}{u^{\sigma+1}+b^{\sigma+1}} \\ \dfrac{u^{\sigma}t\left(\frac{\beta}{\theta}\right)}{u^{\sigma+1}+bt^{\sigma+1}} & \dfrac{u^{\sigma+1}+bt^{\sigma+1}\left(\frac{\theta}{\theta+\beta}\right)}{u^{\sigma+1}+bt^{\sigma+1}} \end{bmatrix} \right\}$$

$$: \ \forall u, t \in K(\theta),$$

and the matrices for the Baer subspaces are

$$\left\{ y = \begin{bmatrix} \dfrac{u^{\sigma+1}+bt^{\sigma+1}\left(\frac{\theta+\beta}{\theta}\right)}{u^{\sigma+1}+bt^{\sigma+1}} & \dfrac{but^{\sigma}\left(\frac{\theta}{\beta}\right)}{u^{\sigma+1}+b^{\sigma+1}} \\ \dfrac{u^{\sigma}t\left(\frac{\beta}{\theta}\right)}{u^{\sigma+1}+bt^{\sigma+1}} & \dfrac{u^{\sigma+1}+bt^{\sigma+1}\left(\frac{\theta}{\theta+\beta}\right)}{u^{\sigma+1}+bt^{\sigma+1}} \end{bmatrix} x^{\tau} \right\}$$

$$: \ \forall u, t \in K(\theta).$$

The translation planes obtained by non-trivial multiple derivation then have spreads in $PG(7, K)$, since $Fix\tau \cap Z((a,b)_K^{\alpha}) = K$. For characteristic 2, things are more complicated. In this case, $\theta^{\sigma} = \theta + 1$, (with $\alpha = 1$), so the inner automorphism ρ maps $t \rightarrow t\frac{\theta+1}{\theta}$, where the associated irreducible polynomial is $x^2 + x + \beta$. Then ρ^2 would map $t \rightarrow t\left(\frac{\theta+1}{\theta}\right)^2 = \frac{\theta+\beta+1}{\theta+\beta}$. If $\beta = 1$ then ρ^3 maps $t \rightarrow t\frac{\theta}{\theta+1}(\frac{\theta^{\sigma}}{\theta}) = t$, and ρ would have order 3. Since, in general $\beta \in F$, the inner automorphism could be of infinite order. If $1 + \beta + \beta^2 + ... + \beta^k = 0$, then the order of the inner automorphism would be $k + 1$. This completes the proof.   □

We shall illustrate another inner automorphism of $(a,b)_K^{\alpha}$, leaving most of the details to the reader as homework, as these are quite similar. The reader might glance at (2,3,1&2) to notice the three basic derivable nets that share $x = 0$ and $y = 0$. These derivable nets correspond to the three fields which are of index 2 in $(a,b)_K^{\alpha}$.

THEOREM 130. *There are other inner automorphisms of $(a,b)_K^{\alpha}$, for example, if we would take* $\begin{bmatrix} 0 & b \\ 1 & 0 \end{bmatrix}$, *define τ as follows: If the characteristic is 2, let $\theta^2 = \theta + \beta$, for $\beta \in K$ (choosing $\alpha = 1$). If the characteristic is $\neq 2$, let $\theta = \sqrt{a}$, where a is a non-square. Define an inner automorphism τ by*

$$\tau\left(\begin{bmatrix} u^{\sigma} & bt^{\sigma} \\ t & u \end{bmatrix} \right) = \begin{bmatrix} 0 & b \\ 1 & 0 \end{bmatrix} \begin{bmatrix} u^{\sigma} & bt^{\sigma} \\ t & u \end{bmatrix} \begin{bmatrix} 0 & 1 \\ \frac{1}{b} & 0 \end{bmatrix}$$

$$= \begin{bmatrix} u & bt \\ t^{\sigma} & u^{\sigma} \end{bmatrix}.$$

$Fix\tau$ is then the set $\left\{ \begin{bmatrix} u & bt \\ t^{\sigma} & u^{\sigma} \end{bmatrix}; t^{\sigma} = t \text{ and } u^{\sigma} = u \right\}$. *The order in this case of τ is 2 in all characteristics. Also, $(a,b)_K^{\alpha}$ is a 2-dimensional vector space over $Fix\tau$. Therefore, there will be a set of mutually disjoint derivable nets which are isomorphic to the open form regulus derivable net which is are of type 2 relative to $K(\theta)$. Using the same notation as in the previous theorem, we may write the fundamental derivable net as follows (recalling that x, y and Σ, and Σ_{τ}) as written over $(a,b)_K^{\alpha}$ in matrix form, where τ is now given as above. We shall write here*

this derivable net in explicit form, leaving the following descriptions to the reader as homework.

$$\begin{bmatrix} y_2^\sigma & by_1^\sigma \\ y_1 & y_2 \end{bmatrix} = \begin{bmatrix} x_2^\sigma & bx_1^\sigma \\ x_1 & x_2 \end{bmatrix} \begin{bmatrix} \frac{u^2-bt^2}{u^{\sigma+1}-bt^{\sigma+1}} & b\frac{(tu^\sigma-ut^\sigma)}{u^{\sigma+1}-bt^{\sigma+1}} \\ \frac{ut^\sigma-tu^\sigma}{u^{\sigma+1}-bt^{\sigma+1}} & \frac{u^{2\sigma}-bt^{2\sigma}}{u^{\sigma+1}-bt^{\sigma+1}} \end{bmatrix}$$

$;\forall u,t \ \in \ K(\theta), (u,t) \neq (0,0).$

And the net of Baer subspaces is

$$\begin{bmatrix} y_2^\sigma & by_1^\sigma \\ y_1 & y_2 \end{bmatrix} = \begin{bmatrix} \frac{u^2-bt^2}{u^{\sigma+1}-bt^{\sigma+1}} & b\frac{(tu^\sigma-ut^\sigma)}{u^{\sigma+1}-bt^{\sigma+1}} \\ \frac{ut^\sigma-tu^\sigma}{u^{\sigma+1}-bt^{\sigma+1}} & \frac{u^{2\sigma}-bt^{2\sigma}}{u^{\sigma+1}-bt^{\sigma+1}} \end{bmatrix} \begin{bmatrix} x_2 & bx_1 \\ x_1^\sigma & x_2^\sigma \end{bmatrix}$$

$;\forall u,t \ \in \ K(\theta), (u,t) \neq (0,0).$

The multiply replaced translation planes are in $PG(7,K)$.

PROBLEM 53. *Use the multiple replacement theorem to write out the fundamental derivable net, given above.*

PROBLEM 54. *Write out the description of the full spread.*

PROBLEM 55. *Take a proper subset δ of the set λ of mutually disjoint derivable nets. Multiply derive this set and write out the description of the new translation plane.*

PROBLEM 56. *How do you see that the multiply derived spreads are in $PG(7,K)$?*

PROBLEM 57. *In (2,3,1&2), when the derivable nets are of type 0 relative to the classical regulus net corresponding to $K(\theta)$, there was a special requirement that the associated vector space be a bimodule for $K(\sqrt{b})$ and $K(\theta)$. How does this change the kernel of the new translation planes?*

3. Automorphisms of infinite order

There are fields K with automorphisms of infinite order, such as Hilbert's example of the rational function field $\mathbb{R}(t)$, where t is an independent variable, and $\mathbb{R}$ is the field of real numbers with the automorphism σ mapping $t \to 2t$, so elements $f(t) \to f(2t)$ (See e.g. Cohn [29] notes on page 45, where there is also discussion of this example using twisted/skew formal Laurent series creating a non-commutative skewfield admitting a non-Archimedean ordering). We noted in the previous section, for quaternion division rings of characteristic 2, there are inner automorphisms of various orders, some of which are perhaps infinite.

There is also the interesting example of the p-adic field $\mathbb{Q}_p$ admitting the Frobenius automorphism τ of infinite order. For reasons of space, this material using the applications of the multiple replacement theorem is left as a potentially important and interesting area for research.

The reader is also directed to (12,14,5), for a short discussion of twisted formal Laurent series by which there is a construction of a non-commutative division ring $\mathbb{Q}_p((t,\tau))$ with center $GF(p)$.

PROBLEM 58. *For either of these examples, show that it is then possible to construct extensions that admit finite inner automorphisms and create Desarguesian spreads admitting sets of mutually disjoint derivable nets. What are the dimensions over the kernels of the multiple derived translation planes?*

PROBLEM 59. *Using the original fields, the translation planes constructed by multiple replacement would then also appear to be of infinite dimension over their kernels. What can be said about such translation planes?*

4. $(a, b)_K^{K(\theta)} \otimes_K K(\rho)$-outer automorphisms

We shall also use some alternative notation identifying $(a, b)_K^\alpha$ with $(a, b)_K^{K(\theta)}$, if the defining Galois quadratic field extension $K(\theta) \subset (a, b)_K^\alpha$, that shall also be adopted when considering cyclic division rings of exponent n, where then the Galois extensions will be of dimension n.

In this section, we will consider central extensions of quaternion division rings $(a, b)_{K(\rho)}^{K(\rho)(\theta)} = (a, b)_K^{K(\theta)} \otimes_K K(\rho)$ of $(a, b)_K^{K(\theta)}$.

NOTATION 8. *For a quaternion division ring $(a, b)_K^\alpha$, we shall use the notation $(a, b)_K^{K(\theta_1)}$, where the α indicates the polynomial used to construct the field extension $K(\theta_1)$.*

DEFINITION 104. *Consider a central quadratic Galois tower of a field $K(\theta_1, \theta_2, \theta_3, ...\theta_{k+1})$, of length $k + 1$. Using the extension theorem for quaternion division rings, there is a related quaternion division ring*

$$(a, b)_{K(\theta_2,...,\theta_{k+1})}^{K(\theta_2,...,\theta_{k+1})(\theta_1)} = (a, b)_K^{K(\theta_1)} \otimes_K K(\theta_2, \theta_3, ...\theta_{k+1}).$$

where $(a, b)_K^{K(\theta_1)}$ is the original quaternion division ring. For each link j in the tower, we may associate a quaternion division ring $(a, b)_{K(\theta_2,...,\theta_j)}^{K(\theta_2,...,\theta_j)(\theta_1)}$.

The same notation shall be used for cyclic division rings of exponent n using the extra symbol ς_n to denote a primitive n-th root of unity. In this setting, the field extensions are cyclic Galois extensions of degree n over the previous field, and K contains ς_n and is of characteristic that does not divide n, then the symbolism becomes $(a, b)_{\varsigma_n K(\theta_2,...,\theta_j)}^{K(\theta_2,...,\theta_j)(\theta_1)}$. We normally consider $K(\theta_i)$ as a cyclic Galois extension of degree n, for all $i = 1, ..., k + 1$ and form the composition $K(\theta_1, \theta_2, \theta_3, ...\theta_{k+1})$.

That is, in part 8, on lifting skewfields, we have noted that if $L = (a, b)_K^{K(\theta)}$, if $K(\rho)$ is a Galois extension, and $K(\theta, \rho)$ is Galois over K then $L(\rho)$ is a central Galois quadratic extension of L. It was proven that $(a, b)_{K(\rho)}^{K(\rho)(\theta)}$ becomes a derivable net of type 0 relative to the classical regulus net over $K(\theta, \rho)$. The proof shows that $(a, b)_{K(\rho)}^{K(\rho)(\theta)}$ can define a spread in both $PG(3, K(\theta, \rho))$ and $PG(3, (a, b)_K^{K(\theta)})$. We use the same expression for both of these embeddings, as it was proved that the two different centralizer skewfields were isomorphic (while this is true for dimension 2, it is not valid for dimension $n > 2$, using Brauer groups). In this setting, there is an automorphism τ obtained from $K(\rho)$ that fixes $(a, b)_K^{K(\theta)}$ pointwise. As a result, using the multiple replacement theorem, and working in $PG(3, (a, b)_K^{K(\theta)})$, we see that the dimension of $(a, b)_{K(\rho)}^{K(\rho)(\theta)}$ over $Fix\tau$ is 2.

So, we consider the spread where x, y, m are in $(a, b)_{K(\rho)}^{K(\rho)(\theta)}$. For the spread we consider

$$(a, b)_{K(\rho)}^{K(\rho)(\theta)} \oplus (a, b)_{K(\rho)}^{K(\rho)(\theta)}$$

to represent the points (x, y). Consider the fundamental replacement

$$\left\{y = xd^{-1}d^\tau : d \in (a, b)_{K(\rho)}^{K(\rho)(\theta)*}\right\} = \left\{y = d^{-1}d^\tau x^\tau : d \in (a, b)_{K(\rho)}^{K(\rho)(\theta)*}\right\}.$$

Consider $(y = x) \cap (y = x^\tau)$. Since $Fix\tau$ is $(a, b)_K^{K(\theta)}$, we see $y = x$ and $y = x^\tau$ are both 2-dimensional $(a, b)_K^{K(\theta)}$-subspaces whose intersection is a 1-dimensional $(a, b)_K^{K(\theta)}$-subspace. The argument to the multiple replacement theorem using the group $S_L^* S_R^{*\tau}$ then shows that the components are left 2-dimensional $(a, b)_K^{K(\theta)}$-subspaces and the replacement subspaces are right 2-dimension $(a, b)_K^{K(\theta)}$-subspaces. With an appropriate basis change, this is a mirror for the procedure used to realize that $(a, b)_K^{K(\theta)}$ may be extended to $(a, b)_{K(\rho)}^{K(\rho)(\theta)}$ when using $(a, b)_K^{K(\theta)}$ as a derivable net relative to the classical regulus net $PG(3, K(\theta, \rho))$. Again, that proof showed that there is an isomorphic copy of $(a, b)_{K(\rho)}^{K(\rho)(\theta)}$ coordinatizing a spread in $PG(3, (a, b)_K^{K(\theta)})$. Assume the derivable net is written as a 2 by 2 matrix net over $(a, b)_K^{K(\theta)}$. Recall that a derivable net written in open form and isomorphic to the classical derivable net relative to $(a, b)_K^{K(\theta)}$ is actually as a type 2 derivable net relative to the classical $(a, b)_K^{K(\theta)}$-pseudo-regulus net. As a result, the nets that we have are actually type 2. However, type 2 derivable nets are isomorphic to the classical $(a, b)_K^{K(\theta)}$ pseudo-regulus net.

As a result, we obtain a set of mutually disjoint derivable nets, which are isomorphic to the classical pseudo-regulus net relative to $(a, b)_K^{K(\theta)}$.

THEOREM 131. *Let $(a, b)_K^{K(\theta)}$ have a central Galois extension $(a, b)_{K(\rho)}^{K(\rho)(\theta)}$. Consider the Desarguesian spread for the translation plane Σ in*

$$PG(3, (a, b)_K^{K(\theta)}).$$

The Desarguesian spread is initially in

$$PG(1, (a, b)_{K(\rho)}^{K(\rho)(\theta)}).$$

Then the multiple replacement theorem shows that there is a set of mutually disjoint derivable nets such that together with $x = 0, y = 0$, for $x, y \in (a, b)_{K(\rho)}^{K(\rho)(\theta)}$ constitutes the spread for the Desarguesian translation plane.

Each derivable net is isomorphic to the left classical pseudo-regulus net with respect to $(a, b)_K^{K(\theta)}$.

It is interesting to work out the fundamental derivable net with components

$$\left\{y = xd^{-1}d^\sigma; \forall d \in (a, b)_{K(\rho)}^{K(\rho)(\theta)*}\right\},$$

which has the representation as

$$(a, b)_{K(\rho)}^{K(\rho)(\theta)} \simeq \left\{\begin{bmatrix} u^\sigma & bt^\sigma \\ t & u \end{bmatrix} : \forall t, u \in K(\theta, \rho), (t, u) \neq (0, 0)\right\},$$

which follows:

THEOREM 132. *(Quaternion division ring derivable net construction for outer automorphisms)* We write $(a,b)_{K(\rho)}^{K(\rho)(\theta)}$ as $w+r\rho$ where $w,r \in (a,b)_K^{K(\theta)}$ and $\tau(w+r\rho) = w + r\rho^\tau$. Then

$$\tau\left(\begin{bmatrix} u^\sigma & bt^\sigma \\ t & u \end{bmatrix}\right) = \begin{bmatrix} u^{\tau\sigma} & bt^{\tau\sigma} \\ t^\tau & u^\tau \end{bmatrix} : \forall t, u \in K(\theta, \rho).$$

Hence,

$$\begin{bmatrix} u^\sigma & bt^\sigma \\ t & u \end{bmatrix}^{-1} \begin{bmatrix} u^\sigma & bt^\sigma \\ t & u \end{bmatrix}^\tau = \begin{bmatrix} \frac{u^{1+\tau\sigma}-bt^{\sigma+\tau}}{\Delta} & \frac{b(t^{\tau\sigma}u-t^\sigma u^\tau)}{\Delta} \\ \frac{t^\tau u^\sigma - tu^{\tau\sigma}}{\Delta} & \frac{u^{\sigma+\tau}-bt^{1+\tau\sigma}}{\Delta} \end{bmatrix},$$

$$\Delta = u^{\sigma+1} - bt^{\sigma+1}.$$

(1) The set of components of the standard derivable net are

$$\{y = xd^{-1}d^\tau\} = \left\{ = \begin{bmatrix} x_2^\sigma & bx_1^\sigma \\ x_1 & x_2 \end{bmatrix} \begin{bmatrix} \dfrac{u^{1+\tau\sigma}-bt^{\sigma+\tau}}{\Delta} & \dfrac{b(t^{\tau\sigma}u-t^\sigma u^\tau)}{\Delta} \\ \dfrac{t^\tau u^\sigma-tu^{\tau\sigma}}{\Delta} & \dfrac{u^{\sigma+\tau}-bt^{1+\tau\sigma}}{\Delta} \end{bmatrix} : \right\}.$$
$$\forall (t,u) \in K(\theta,\rho) \times K(\theta,\rho) - \{0,0)\}.$$

LEMMA 16. *(2) The set of Baer subspaces is*

$$\{y = d^{-1}d^\tau x^\tau\} = \left\{ = \begin{bmatrix} \dfrac{u^{1+\tau\sigma}-bt^{\sigma+\tau}}{\Delta} & \dfrac{b(t^{\tau\sigma}u-t^\sigma u^\tau)}{\Delta} \\ \dfrac{t^\tau u^\sigma-tu^{\tau\sigma}}{\Delta} & \dfrac{u^{\sigma+\tau}-bt^{1+\tau\sigma}}{\Delta} \end{bmatrix} \begin{bmatrix} x_2^{\tau\sigma} & bx_1^{\tau\sigma} \\ x_1^\tau & x_2^\tau \end{bmatrix} : \right\}.$$
$$\forall (t,u) \in K(\theta,\rho) \times K(\theta,\rho) - \{(0,0)\}.$$

(3) All of the derivable nets are isomorphic to a classical pseudo-regulus net over $(a,b)_K^{K(\theta)}$. The translation planes obtained by choosing any proper subset δ of the index set λ is a translation plane with spread in $Fix\tau \cap Z(\ (a,b)_{K(\rho)}^{K(\rho)(\theta)}) = K$. That being the case, the translation planes are in $PG(15, K)$.

5. Cyclic algebras and additive quasifibrations

For cyclic algebras of dimension n, we show here that there are always associated additive quasifibrations which might be called 'lifted Pappian quasifibrations of dimension n'. The construction is simple but seems appropriate to include. These additive geometries are similar to the finite cyclic semifield planes of Jha and Johnson (See, e.g. [**16**] chapter 27). In the following construction of a cyclic algebra of dimension n, we focus on the associated additive matrix set of n by n matrices.

DEFINITION 105. *Let K be a field that contains a primitive n-th root of unity ς_n and assume that the characteristic of K does not divide n, and n is invertible, let $a,b \in K$, for $ab \neq 0$. Let u and v be elements and generate a K-algebra as $F = K\langle u,v \rangle$ such that $u^n = a$ and $v^n = b$, such that $uv = \varsigma_n vu$. Then F is said to be a 'symbol algebra'.*

When $n = 2$ and K has characteristic $\neq 2$, the symbol algebra is a 'quaternion algebra'. When $n = 2$, the representation of quaternion algebras have been discussed in (2, 3, 1&2).

Note that it is possible that $a = b$, for example, $(-1, -1)_R$, where R, the field of real numbers, defines the real Hamiltonians. If the symbol algebra is a division ring, we use the term 'symbol division ring'., or 'symbol division algebra'.

Consider $n = 2$. We have seen that using the definition above, we require that characteristic $\neq 2$ for cyclic algebras and a matrix version of $(a, b)_K^{K(\theta)}$ is $\begin{bmatrix} u^\sigma & bt^\sigma \\ t & u \end{bmatrix}$ for all $t, u \in K(\sqrt{a})$, where $b \in K$, which is a division ring if and only if $b \in K - K(\sqrt{a})^{\sigma+1}$, where σ is the automorphism mapping $\sqrt{a} \to -\sqrt{a}$ and fixes K elementwise. Recall that $\left\{ I_2, \begin{bmatrix} -\sqrt{a} & 0 \\ 0 & \sqrt{a} \end{bmatrix}, \begin{bmatrix} 0 & b \\ 1 & 0 \end{bmatrix}, \begin{bmatrix} 0 & -b\sqrt{a} \\ 1 & 0 \end{bmatrix} \right\}$ is a basis for a quaternion algebra over K, which is a cyclic symbol algebra. For characteristic 2, modifications need to be made for the quaternion algebras.

DEFINITION 106. *Let $F = K(\theta)$ be a cyclic Galois field extension over a field K of degree n. Let $\{1, d, d^2, ..., d^{n-1}\}$ for $d \in F$ be a set of elements, where d is considered a symbol, such that $d^n = b \in K^*$. Form the algebra:*

$$K \oplus dK \oplus d^2 K \oplus ... \oplus d^{n-1} K;$$

$$fd \ = \ df^\sigma, \text{ for } \sigma \text{ a generator of the Galois group of } K(\theta)/K.$$

Note this will make the center of the algebra K and the dimension over K is n^2. The requirement that a division ring is obtained is that $b^i \in K - K(\sqrt[n]{a})^{\frac{\sigma^n - 1}{\sigma - 1}}$ for $i = 1, 2, ..., n - 1$. That is, the elements b^i satisfy the 'non-norm condition'. The algebras are called 'cyclic algebras of finite degree n'. We shall use the notation $(F/K, \sigma, b)$ to denote a cyclic division algebra of finite degree.

Note if d is considered an irreducible semilinear mapping of $\Gamma L(F, K)$, we obtain an additive quasifibration and in the finite case, this is a cyclic semifield plane, as defined in Biliotti, Jha and Johnson [16] chapter 27. The examples of Jha and Johnson were usually defined over F as $xd = \gamma x^\sigma$, for some γ in F. Note the obvious similarity of ideas. That being the case, if we now replace the element b as above, we will see a connection between cyclic algebras, which may or may not be cyclic division algebras or cyclic semifield planes or simply 'additive quasifibrations'.

Now to see an illustration of how we obtain the specific additive quasifibrations, we form a matrix set corresponding to a cyclic algebra of dimension n. This follows the construction used by [108] and others.

DEFINITION 107. *Given a cyclic algebra of degree n, the associated n by n matrix representation is given as follows: This is the matrix representation of $(K(\theta)/K, \sigma, b)$ where $K(\theta)/K$ is a cyclic Galois field extension of dimension n).*

$$\begin{bmatrix} a_o & ba_{n-1}^\sigma & ba_{n-2}^{\sigma^2} & \cdot & \cdot & \cdot & ba_1^{\sigma^{n-1}} \\ a_1 & a_0^\sigma & ba_{n-1}^{\sigma^2} & \cdot & \cdot & \cdot & ba_2^{\sigma^{n-1}} \\ a_3 & a_2^\sigma & a_0^{\sigma^2} & \cdot & \cdot & \cdot & \cdot \\ \cdot & \cdot & \cdot & \cdot & \cdot & \cdot & \cdot \\ \cdot & \cdot & \cdot & \cdot & \cdot & \cdot & \cdot \\ \cdot & \cdot & \cdot & \cdot & \cdot & \cdot & \cdot \\ a_{n-1} & a_{n-2}^\sigma & a_{n-2}^{\sigma^2} & \cdot & \cdot & \cdot & a_o^{\sigma^{n-1}} \end{bmatrix} : \forall a_i \in K(\theta), i = 0, 1, ..., n - 1.$$

In this setting, there is a cyclic division ring provided b^i for $0 < i < n$, are non-norm elements; that is $b^i \in K - K(\theta)^{1+\sigma+\sigma^2+...+\sigma^{n-1}}$.

PROBLEM 60. *Show that $K(u,v)$ for $u^n = a$, $v^n = b$ and $uv = \varsigma_n vu$, has a matrix representation analogous to that of the cyclic division ring in the case that K contains a primitive n-th root of unity, the characteristic of K does not divide n and that n is invertible and $K(\theta) = K(\sqrt[n]{a})$ for $a \in K$. Note that now for $K(\theta)$, such that $\theta = \sqrt[n]{a}$, $\sigma : \sqrt[n]{a} \to \varsigma_n \sqrt[n]{a}$ (See (11, 13, 1)).*

REMARK 36. *In 1932, A.A. Albert [1] showed that there are division algebras that are not cyclic in the sense that the algebras do not contain a cyclic field $K(\theta)$, where K is the center. These division algebras are of degree 4 (16-dimensional over K), where K is a rational function field of two independent variables) and arises as the direct product of two quaternion division algebras.*

REMARK 37. *Recently, C. D'Elbée [38] created interesting cyclic division algebras and revisited the examples of Albert of dp-rank for some analogous results on non-cyclic division rings.*

REMARK 38. *When considering automorphisms of division rings, there is inevitable use of the 'Skolem-Noether theorem', which states the following: (Skolem-Noether) If A is a central simple algebra over a field F and B is a simple F-algebra, assume that f and g are algebra mappings from $B \to A$. Then there exists a unit u in A such that $g(b) = uf(b)u^{-1}$ for all $b \in B$.*

In particular, for $B = A$ if σ is an automorphism $: A \to A$ then there is a unit in A such that $\sigma(b) = ubu^{-1}$. So, every automorphism of a central simple algebra over a field F, fixing F pointwise, is an inner automorphism.

For example, let D be a central division ring with center K such that $D \otimes K(\tau)$ is a central division ring with center $K(\tau)$. Then the Galois group of $K(\tau)$ is an automorphism group of $D \otimes K(\tau)$ that admits elements that do not pointwise fix the center, so would not necessarily be inner automorphisms.

We now show that there are additive quasifibrations that may be associated with any cyclic algebra. These additive quasifibrations could be spreads in certain situations, for example, in the dimension 2 situation.

THEOREM 133. *Let θ be any generator of F over K and in the above matrix replace b by θ then the matrix set consists of non-singular elements. Hence, this defines an additive quasifibration. When $n = 2$, the additive quasifibration is a Hughes-Kleinfeld semifield plane.*

PROOF. Assume that some matrix has determinant 0. Notice that the product of the elements in the main diagonal is $a_0^{1+\sigma+...+\sigma^{n-1}}$ and that all elements may be represented as linear sets with respect to the linearly independent set/basis $\{1, \theta, ...\theta^{n-1}\}$ over K, and furthermore note that there are no other coefficients of 1. This element must be equal to 0. That being the case $a_0^{(\sigma^n-1)/(\sigma-1)} = 0$, implying that $a_0 = 0$. Now similarly, work out the coefficient of θ^{n-1} to obtain the element

$$\theta^{n-1}(a_{n-1}^{1+\sigma+...+\sigma^{n-1}}),$$

which implies similarly that $a_{n-1} = 0$. Working inductively, then cut out the elements of each succeeding term that are not strictly elements of the indicated form to obtain

$$\theta^{n-2}(a_{n-2}^{1+\sigma+...+\sigma^{n-2}}),$$

so that $a_{n-2} = 0$, and hence that $a_i = 0$ for $i = 0, 1, 2, ..., n-1$. Note that when $n = 2$, we have the Hughes-Kleinfeld planes. For an explicit diagram of the proof, let $n = 3$. The determinant is as follows

$$a_0^{\sigma^3-1/\sigma-1} + \theta(a_1^{\sigma^3-1/\sigma-1} - a_0 a_1^{\sigma} a_2^{\sigma^2} - a_0^{\sigma} a_1^{\sigma^2} a - a_0^{\sigma^2} a_1 a_2^{\sigma})) + \theta^2(a_2^{\sigma^3-1/\sigma-1}) = 0$$

then implies that $a_0 = a_2 = 0$, which reduces the determinant to $\theta a_1^{\sigma^3-1/\sigma-1} = 0$, so $a_1 = 0$. In general, in the expansion of the determinants using Cramer's rule over the first column, once a_0 and a_n are shown to be zero, look at the $(1, n-1)$-entry with product $a_{n-1}M$, where M is the corresponding $n-1$ by $n-1$ matrix. The only way to obtain an θ^{n-2} coefficient is as $\theta^{n-2} a_{n-2}^{\sigma n-1/\sigma-1}$ and there will be no other such θ^{n-2}-terms in the other parts of the determinant expansion. This implies $a_{n-2} = 0$. Induction then completes the proof showing all $a_i = 0$. $\square$

6. Cyclic division ring automorphisms

The ideas in the last two sections generalize in two ways. First, we assume that we have a division ring of the form $K\langle u, v \rangle$, where K is a field that contains a primitive n-th root of unity ς_n, and the characteristic of K does not divide n, or a cyclic division ring $(K(\theta), K, \sigma, b)$. In the first case, $u^n = a$, $v^n = b$ and $uv = \varsigma_n vu$. We can use an inner automorphism τ in exactly the same way as when $n = 2$, for characteristic $\neq 2$. We have mentioned a matrix representation and the same would be true for cyclic division rings. In this same matrix representation, the idea will work when the elements of the matrix are elements of $K(\theta)$ such that $\theta^n = a$ and the non-norm element b in the matrix need not be a. The non-norm conditions relative to $K(\theta)$ now still apply; $b^i \in K - K(\theta)^{\frac{\sigma^n-1}{\sigma-1}}$, for $i = 1, ..., n-1$, where the automorphism acts as $\theta \to \varsigma_n \theta$. This will show that τ has order n and $Fix\tau$ is a field (the field on the main diagonal of the matrix representation isomorphic to $K(\theta)$) and there is a set of replaceable nets where the components are left n-dimension over $Fix\tau$ and the replacement nets are right n-dimension over $Fix\tau$.

THEOREM 134. (*Cyclic division ring of degree n, inner automorphism net replacement theorem*). *Let D_K be any general division ring, with center K. Assume that K contains a primitive n-th root of unity ς_n and the characteristic of K does not divide n. And let $K(\theta)$ be a cyclic Galois extension of K of dimension n, where an automorphism σ generating the Galois group is $\sigma : \theta \to \varsigma_n \theta$.*

Then there is an inner automorphism τ of D defined by $\tau(r) = \theta r \theta^{-1}$ and τ has order n. The corresponding multiple replaceable nets are left n-dimensional over a subfield $K(\theta)^{\sigma}$, isomorphic to $K(\theta)$ and the corresponding replacement nets are right n-dimensional over $K(\theta)^{\sigma}$.

The translation planes obtained by non-trivial multiple replacement have kernel $Fix\sigma \cap Z(D) = K$. That being the case, the spreads lie in $PG(2 \cdot n^2 - 1, K)$.

The second way that the ideas generalize is by realizing a cyclic division ring geometrically as a subplane covered net relative to a classical pseudo-regulus net (of degree n) and extend the skewfield replacement theorem to higher dimensions. This will mean certain cyclic division rings $(a, b)_{\varsigma_n F(\varsigma_n)=K}^{K(\theta)}$ extend to $(a, b)_{\varsigma_n K(\rho)}^{K(\rho)(\theta)}$, where $K(\rho)$ is also a cyclic Galois extension of K. Then the underlying field $K(\theta, \rho)$ is a Galois extension with Galois group $z_n \times z_n$, the Abelian group which is the direct product of two cyclic groups of order n. Then there is an automorphism τ of the extension, which fixes $(a, b)_{\varsigma_n K}^{K(\theta)}$ pointwise. Applying the multiple replacement theorem shows that the replacement nets are of left degree n over $(a, b)_{\varsigma_n K}^{K(\theta)}$ and the right replacement nets are of right degree n over $(a, b)_{\varsigma_n K}^{K(\theta)}$. That being the case, we obtain the following result.

THEOREM 135. *(Cyclic division ring n-outer automorphism subspace replacement construction)* If D_K is a general cyclic division ring of degree n and center K with associated cyclic Galois extension $K(\theta)$, has a extension $D_{K(\rho)}$ of degree n, with associated Abelian Galois field $K(\theta, \rho)$ of degree n^2 over K, then the associated replacement nets of degree n over $K(\theta)^\sigma$, are replaceable by nets isomorphic to D_K, and so the non-trivial multiple replacements give rise to translation planes with spreads in $PG(2 \cdot n^3 - 1, K)$.

In (11,13,3&4), we form Galois towers of cyclic Galois extension of dimension n and develop the possible applications with a continued study of the extensions of cyclic division rings. For the most part, it is easier symbolically to use the general division rings $(a, b)_{\varsigma_n K}$ but the results apply for every cyclic division ring as well.

PROBLEM 61. *Open problems: For any given non-commutative skewfield K, study the translation planes obtained from this theory, using inner automorphisms.*

Choose any inner automorphism $\sigma_k : c \to kck^{-1}$ for all $c \in K$. In particular, there is a replaceable net with partial spread defined by

$$\left\{ y = xcc^{-\sigma_k} = xckc^{-1}k^{-1} : \forall c \in K^* \right\},$$

with replacement set

$$\left\{ y = cc^{-\sigma_k} x^{\sigma_k} = ckc^{-1}k^{-1}(kxk^{-1}) = ckc^{-1}xk^{-1} : \forall c \in K^* \right\}.$$

Hence, the replaceable nets are related to the commutators of K. It is a difficult problem to determine the set of orbits of this 'commutator replaceable net', under $S_L^* S_R^{*\sigma}$. The reader interested in work is this direction might consult P. Gvozdevsky [60] for additional information.

COROLLARY 37. **Extending the Theorem of Andrè:** *When the Desarguesian plane is finite and σ has order 2, the nets and their replacements are regulus nets and the corresponding translation planes constructed are called 'Andrè planes', due to Andre in 1954, [4]. The same 'Andrè plane' terminology is used for Pappian infinite planes and where σ has order 2.*

In (11,13,4), we shall continue the discussion of applications of the multiple replacement theorem. Here are some immediate problems to consider.

PROBLEM 62. *When the Desarguesian plane is finite and σ has finite order n, $n > 2$, the nets and their replacements are called 'hyper-reguli', coined by Ostrom [120]. In this setting, the replacements are also called Andrè replacements, and there are $n - 1$ variations/different replacements, previously mentioned. There are other replacements in the finite case, where the original nets have non-Andrè replacements other than those given in the theory. When the Desarguesian plane is not Pappian,* *there has been no studies of such planes. The interested reader is directed to chapter 19 of the handbook* [**94**] *for constructions of hyper-reguli that are not of the Andrè type.*

PROBLEM 63. *Open Problem: Study infinite Desarguesian planes admitting cyclic automorphism groups of finite order > 2 and determine the possible replacements, using the above theorem. When are there non-Andrè type replacements? That is, replacements of $\{y = xb^{-1}b^{\sigma}; b \in S\}$ that are not of the form $\{y = b^{-1}b^{\sigma}x^{\sigma}; b \in S\}$ (or the various replacements mentioned above).*

PROBLEM 64. *No mention has been made about Galois groups in the field plane cases. Are the translation planes different in distinguishable ways if the extension of the fixed point subfield is Galois or not Galois, in the necessarily infinite Pappian planes and the order of the automorphism group is > 2?*

PROBLEM 65. *For Pappian planes admitting infinite cyclic automorphism groups, there have been no studies on the translation planes obtained, except for what we have described above. Determine the planes and their isomorphisms.*

PROBLEM 66. *For non-commutative skewfields admitting cyclic automorphism groups, determine the replacement nets and analyze the constructed translation planes.*

COROLLARY 38. *Let S be a non-commutative skewfield Desarguesian plane admitting a finite cyclic outer automorphism group of order n. If n = 2, then there is a multiple derivation set, where the derivable nets are pseudo-regulus sets. The blended left-right kernel is left-right $Fix\sigma$, in the case where $Fix\sigma$ is a non-commutative skewfield.*

We have previously noted the extension of skewfield theorem in part 2 and also noted that it is not required for extensions of skewfields to be Galois extensions in the characteristic 2 case.

In the following we consider central extensions of skewfields.

DEFINITION 108. *Consider a non-commutative Desarguesian spread $\sum_n$ of dimension n in $PG(2n-1, K)$, for K a field, and let F denote the associated skewfield coordinatizing $\sum_n$. Assume that F admits a cyclic automorphism of finite order n, obtained from a central extension $F(\theta)$ of F of finite dimension n, written over K as a matrix group. Additionally, there are sets of isomorphic mutually disjoint replaceable nets. Then there is a set of multiply replaced translation planes admitting the blended left-right F-kernel. We shall call the corresponding multiple replaceable nets 'hyper pseudo-regulus nets' in keeping with and extending the terminology of Ostrom.*

REMARK 39. *In the finite cases and the Pappian finite cyclic group cases, the original arguments of Andrè and Ostrom work much differently. Hilbert's theorem 90 (See e.g. Dummit and Foote* [**49**] *583, 814) is used to specify the representation of the replacement sets and the general derivable net when the cyclic group is of order 2. The argument given here could not use Hilbert's theorem in the general cases, for inner automorphism groups, for infinite automorphism groups or for non-commutative skewfields. The new argument shows that Hilbert's theorem is not actually necessary for any of these cases.*

However, when an automorphism σ has finite order n, consider

$$\left\{ bb^{-\sigma}; b \in S^* \right\},$$

for S the associated skewfield, and note that

$$(bb^{-\sigma})^{1+\sigma+\sigma^2+\dots+\sigma^{n-1}} = 1 = bb^{-\sigma}b^\sigma b^{-\sigma^2} \dots b^{-\sigma^n}.$$

We also have a proof of Theorem 90 for skewfields that admit automorphisms of finite order.

In part 9 on Bilinearity, there is a discussion on elliptic flocks, or rather flocks of elliptic quadrics. There, it is shown that it is possible to determine Pappian translation planes in $PG(3, K)$, for K a field, that admit a partition by mutually disjoint regulus nets of the components missing 0, 1, or 2 components; called 'elliptic i-flocks', for $i = 0, 1, 2$ respectively. Moreover, when the set of regulus nets cover all but two components, this cannot be an Andrè situation, in the non-linear examples, and the planes obtained by multiple replacement are never Andrè planes (9,11,11).

In this part, we have shown that given any skewfield that admits an involutory automorphism, the associated Desarguesian plane may be partitioned by isomorphic derivable nets except for two components $x = 0$ and $y = 0$. When the skewfield is a field, the replacement of all of the derivable (regulus nets in this case) produces the Pappian spread Σ_ρ. This is also the situation for any skewfield planes (Desarguesian plane) using the automorphism technique.

So, using the new ideas in the argument, we know what the nets are but not the cardinality. In the following, we provide some alternative ideas for the field case, and the automorphism group is cyclic of finite order n. We shall restrict the situation to the field case, since now we have a better understanding of what the replacement nets actually look like.

THEOREM 136. *Consider a Pappian spread Σ_n of dimension n in $PG(2n - 1, K)$, for K a field, and let F denote the associated field coordinatizing Σ_n. Assume that F admits a cyclic Galois central extension $F(\theta)$ of F of finite dimension n, written over K as a matrix group. Let σ denote an automorphism of order n. Represent the associated Desarguesian translation plane over an F-vector space in the form:*

$$x = 0, y = x, y = xm; m \in F(\theta)^*.$$

The left $F(\theta)^*$ kernel mappings $F_L(\theta)^*$: fixing each component are given by

$$(x, y) \to (dx, dy); d \in F(\theta)^*.$$

So, the components are then $F(\theta)$-subspaces, and are then n-dimensional F-subspaces. We consider the following Pappian translation plane:

$$x = 0, y = x, y = nx^\sigma; m \in F(\theta)^*,$$

The $\sigma - F(\theta)^*$ kernel mappings $F_R(\theta)^{\sigma*}$ as given by

$$(x, y) \to (xe, ye^\sigma); e \in F(\theta)^*.$$

This makes the components $y = rx^\sigma$ $F(\theta)^*$-subspaces and also n-dimensional F-subspaces. The $F_R(\theta)^{\sigma*}$-mappings are collineations of Σ_n. Then the following defines a hyper-regulus net $R_{\sigma,1}$: With n-dimensional replacement components as

$$\left\{ y = d^{1-\sigma}x^\sigma; d \in F(\theta)^* \right\},$$

and n-components as
$$\left\{ y = xd^{1-\sigma}; d \in F(\theta)^* \right\}.$$
For all $r \in F(\theta)^*$ then there is set of mutually disjoint hyper-regulus nets
$$\left\{ R_{\sigma,r} \right\}, \; where$$
$$\left\{ y = xrd^{1-\sigma}; d \in F(\theta)^* \right\}.$$
and n-dimensional replacement sets as
$$\left\{ y = rd^{1-\sigma}x^\sigma; d \in F(\theta)^*. \right\}$$
(1) Let H denote the subgroup of $F(\theta)^*$ such that $h \in H$, then
$$h^{(\sigma^n-1)/(\sigma-1)} = 1,$$
then
$$\left\{ r_\delta H; r_\delta \in F(\theta)^*; \; r_\delta^{(\sigma^n-1)/(\sigma-1)} = \delta \in F^* \right\} = F(\theta)^*/H.$$
Therefore,
$$\cup_{r_\delta} R_{\alpha, r_\delta}$$
is a disjoint union of hyper-regulus nets. The set of intersection subspaces D_1 with $y = x^\sigma$ is the set of components D_1' $y = xm$, such that
$$mm^\sigma m^{\sigma^2}...m^{\sigma^{n-1}} = m^{(\sigma^n-1)/(\sigma-1)} = 1.$$

LEMMA 17. *(2) Let d be a proper division of n and let τ be an automorphism of order d. Let H_d be the subgroup of H such that $h \in H_d$ then $h^{\frac{\tau^d-1}{\tau-1}} = 1$. Decompose $\{y = xbb^{-\sigma}; b \in F(\theta)^*\}$ relative to H/H_d. Then there is a replacement set of the form $\cup\{y = bb^{-\tau}r; b \in F(\theta)^*\}$ such that $r \in \lambda$, a representation set for H/H_d.*

When the order of an automorphism σ^k is n, then $\{y = xbb^{-\sigma}; b \in F(\theta)^*\}$ has a replacement of the form $\{y = bb^{-\sigma^k}x^{\sigma^k}; b \in F(\theta)^*\}$. In this manner, there are $n-1$ possible replacements of $\{y = xbb^{-\sigma}; b \in F(\theta)^*\}$, called Andrè replacements. So there are $n-1$ Andrè replacements.

PROOF. $xm = x^\sigma$, for some non-zero vector, x implies that $x^\sigma m^\sigma = x = xm^{1+\sigma}$, and inductively, $m^{(\sigma^\sigma-1)/(\sigma-1)} = 1$. We know this to be the case by Hilbert's Theorem 90, since $bb^{-\sigma} = b^{1-\sigma}$, in the commutative case. And the order of $b^{1-\sigma}$ is $(\sigma^n - 1)/(\sigma - 1)$. $\qquad\square$

Generally, in the various parts, we show what the replacement theorem means in terms of specific skewfield planes, the quaternion division ring planes and cyclic division ring planes. That latter study will require some preparation in setting up an ambient space of a subplane covered net which is analogous to the derivable net analysis.

We shall return to applications of the multiple replacement theorem after we have discussed Galois theory for skewfields (11, 13, 3&4).

PROBLEM 67. *For any field $F(\theta)$, determine if there are replacements for $\{y = xbb^{-\sigma}; b \in F(\theta)^*\}$ that are not Andrè replacements.*

Part 11

Classification of subplane covered nets

The classification of derivable nets may be studied more generally over subplane covered nets, which we shall discuss in this part. In the derivable nets cases, there are four types $0, 1, 2, 3$, based on the number of Baer subspaces shared by a suspect derivable net and a classical pseudo-regulus net. Noting that any two Baer subspaces are linearly independent, for types 1 and 2, we are saying that either one of two elements of a basis for the set of Baer subspaces are considered. In particular, for type 2, we are, in effect, asking for two linearly independent Baer subspaces that are shared by the two left and right derivable nets. If we use that same technique for subplane covered nets of finite dimension n, we may classify subplane covered nets as a more general theory, which contains the classification of derivable nets as a corollary.

Recall that a projective space of dimension $n + 1$ over a skewfield F, $PG(n + 1, F)_C$ (C for combinatorial embedding), constructs a classical right or left subplane covered net in a $2n$-dimension vector space V_{2n}/F. Then the partial spread for this subplane covered net is taken to be a right classical pseudo-regulus subplane covered net in the $PG(2n - 1, F)_S$ (S for projective space of subspaces). Note when $n = 2$, we have the setting of investigation of the classification of derivable nets, and only in that case could there be confusion between the projective space of the embedding and the projective space of the 'suspect subplane covered net'.

CHAPTER 13

Suspect subplane covered nets

Here is the general definition.

DEFINITION 109. *Consider any skewfield F and the projective space $PG(n + 1, F)$, for n a positive integer. Let R_{Right}^F denote the right classical pseudo-regulus subplane covered net. Recall the form for R_{Right}^F is*

$$\left\{ \begin{array}{c} x = 0, y = \delta x; \forall \delta \in F; \\ components\ are\ right\ n\text{-}dimensional\ F\text{-}subspaces \end{array} \right\},$$

and the set $B(R^F)_{left}$ of 'little subplanes' incident with the zero vector $B(R^F)_{left}(\overrightarrow{C})$, where $\overrightarrow{C} \in F^n - \{(0, 0, ..., 0)\}$ (the zero element as an n-tuple). Let $\overrightarrow{C}$ denote a non-zero n-vector over the standard basis with respect to F.

$$B(R^F)_{left} = \left\{ \begin{array}{c} B(R^F)_{left}(\overrightarrow{C}) \\ = \left\{ (c\overrightarrow{C}, d\overrightarrow{C}); \forall c, d \in F \right\} \end{array} \right\};$$

$$little\ subspaces/subplane$$

$$are\ left\ 2\text{-}dimensional\ F\text{-}subspaces,$$

$$\forall \overrightarrow{C} \in F^n - \{(0, 0, ..., 0)\}.$$

Note that when $n = 2$, the little subspaces are now the Baer subspaces.

Let the following suspect subplane covered net S_{Left}^K with partial spread in $PG(2n - 1, F)_S$ be defined as follows:

$$\left\{ \begin{array}{c} x = 0, y = 0, y = xM; M \in \lambda \subset GL(n, F)_{left} \\ ; for\ \lambda \cup \{0\} =\ a\ skewfield\ K\ ; \\ components\ are\ left\ n\text{-}dimension\ F\text{-}subspaces \end{array} \right\}.$$

and the set $B(S^K)_{right}$ of 'little subplanes' incident with the zero vector denoted by $B(S^K)_{left}(\overrightarrow{C})$, where $\overrightarrow{C} \in F^n - \{(0, 0, ..., 0)\}$, and $\overrightarrow{C}$ denotes an n-vector now written over the basis with respect to K.

$$B(S^K)_{right} = \left\{ \begin{array}{c} B(S^K)_{right}(\overrightarrow{C}) \\ = \left\{ (\overrightarrow{C}M, \overrightarrow{C}N); \forall M, N \in K \right\} \\ \forall \overrightarrow{C}\ non\text{-}zero\ n\text{-}vectors\ written\ over\ K \end{array} \right\};$$

$$little\ subspaces\ are\ right\ 2\text{-}dimensional\ K\text{-}subspaces.$$

In order for this suspect subplane covered net to actually be a subplane covered net, we need that V_{2n}/F is also a 2n-dimensional vector space over K. So we have paired V_{2n}/F as a left vector space with V_{2n}/K as a right vector space.

To see that we have a subplane covered net with the above condition, just re-alize that now the elements $M \in K$ become right scalar mappings acting on $\overrightarrow{C}$. Considered in this manner, it is clear that we have a subplane covered net that is isomorphic to the right classical pseudo-regulus subplane covered net over F and a suspect subplane covered net that is a left pseudo-regulus subplane covered net.

The definition shall be given in two parts, first for the types $i = 0, 1, ..., n$ and then second for the types $n + 1, ..., 2n - 1$.

DEFINITION 110. *The types will now be considered. First let $n = 2$. There are types $0, 1, 2$ and then at least 3. What 'at least 3' means is $1 + card Z(F)$, and this means that for a given type, one of entries of the n-tuple representing the common Baer subspace/subplane may be taken as any element over $Z(F)$, where F corresponds to/coordinatizes the classical right pseudo-regulus. Then the 'at least 3', is the end of the type system when $n = 2$. A 'left suspect subplane covered net of dimension n' relative to the right classical pseudo-regulus net with respect to F in $PG(2n-1, F)_S$' is said to be of type i, where $i = 0, 1, ..., 2n-1$, if and only if for $i = 0$ there are 0 shared little subplane subspaces in common with the classical left pseudo-regulus subplane covered net over F and of type $i = 1, 2, ..., n$ if and only if there are i shared little subplane subspaces that are F-linearly independent. At this point, taking the 'generic' set of n shared little subplanes means that the unordered short notation is just that of the standard basis over F, $\{(1, 0, 0, ...0), (0, 1, ..., 0), ..., (0, 0, ..., 0, 1)\}$, and then for type n the matrix is a diagonal n by n matrix when the (n, n) entry is w for all $w \in F$ and the (i, i) entry is w^{σ_i}, for $\sigma_i \neq \sigma_j$, for all $i \neq j$, $i = 1, 2, ..., n-1$, and $\sigma_i \neq 1$, and all automorphisms are outer.*

Any other distinct set of n little subspaces that make the subplane covered net a type n will form a basis for the set of all little subspaces, but the type would not be generic.

Below, we give the definition of the $n + i, ..., 2n - 1$ types, but first we look at an example. First note that every subplane covered net has a set of derivable subnets. For every pair of little subspaces, the direct sum of these subspaces generates a subspace of dimension 4 over F, which becomes a derivable subnet. In the example below, the subspace generated by the two little subplanes in the 2 by 2 submatrix $\begin{bmatrix} w^{\sigma_2} & 0 \\ 0 & w \end{bmatrix}$, will determine a type 2 derivable subnet. Let $n = 3$ and consider

$$x = 0, y = x \begin{bmatrix} w^{\sigma_1} & 0 & 0 \\ 0 & w^{\sigma_2} & 0 \\ 0 & 0 & w \end{bmatrix}; w \in F$$

where x, y and 3-vectors over F,

where $\sigma_1 \neq \sigma_2$ are non-identity automorphisms of F fixing L pointwise.

This would be a typical generic type $n = 3$ subplane covered net. Choose (e, f, g) as a $(n + 1)$-st common little subplane. One of the entries must be 1, say e, and one of f or g is not 0, say g. Then $(he, hf, hg) = (ew^{\sigma_1}, fw^{\sigma_2}, gw)$. With $e = 1$, then $h = w^{\sigma_1}$, so that $wf = fw^{\sigma_2}$ and $w^{\sigma_1}g = gw$. Since $g \neq 0$, and all automorphisms are outer, we now choose $\sigma_1 = 1$, to create a type $n + 1$, so that $g \in Z(F)$. Therefore, we would have $(1, 0, g)$ for all $g \in Z(F) - \{0\}$, is a common little subplane. Since there are n other common subplanes, we would have $n + CardZ(F) - 1$ common subplanes, once we have $n + 1$. So we consider the two entries $(1, 1)$ and $(3, 3)$ and construct the derivable subnet relative to these little subplanes. This would be the 'at least 3' type in the derivable subnet. So the type $n + 1$ will create a type 3 derivable net, relative to a particular derivable subnet.

DEFINITION 111. *A type $n + i$ subplane covered net is obtained if and only there is a derivable subnet that becomes a type 3 derivable net, such that the derivable subnet was not of type 3 in previous types $n + j$, for $j < i$ for $i, j = 1, 2, ..., n - 1$.*

The definition we have chosen is easy to apply once we go through an example for $n = 3$, as we shall do after our understanding of an ambient space.

1. Ambient theory for subplane covered nets

To get an idea why the definition makes sense, let F be a field that contains a primitive n th. root of unity ς_n, such that the characteristic does not divide n. Consider two cyclic Galois extensions of dimension n/F, $F(\sqrt[n]{a})$ and $F(\sqrt[n]{c})$ that are not isomorphic, so that the composition $J = F(\sqrt[n]{a}, \sqrt[n]{c})$ is an Abelian Galois extension with Galois group isomorphic to $z_n \times z_n$, where z_n is the cyclic group of order n, as J is the splitting field of a product of separable polynomials. In the section on cyclic division rings of degree n, $(a, b)_{\varsigma_n K}$, we noted the notation is just to remind the reader that we are dealing with cyclic division rings/or symbol division rings, instead of quaternion division rings $(c, d)_K^\alpha$.

REMARK 40. *Assuming the above notation, we note that a suspect left subplane covered net has partial spread in $PG(2n - 1, F(\sqrt[n]{a}, \sqrt[n]{c}))$. We note that every cyclic division ring L with matrix representation containing $F(\sqrt[n]{a})$ will become a right subplane covered net that shares $x = 0, y = 0, y = x$ with the (left) classical regulus net with respect to $F(\sqrt[n]{a}, \sqrt[n]{c})$. We note that the n-th. row of the matrix set has the form $[t_1, ..., t_n]$ for all $t_i \in F(\sqrt[n]{a})$, and if the skewfield L is non-commutative, then L will be a cyclic division ring $(a, b)_{\varsigma_n F}$ or $(F(\sqrt[n]{a})/F, \sigma, b)$, σ a generator of the Galois group of $F(\sqrt[n]{a}))$ of dimension n^2 over the center F, for some element $b \in F$ such that b^i is a non-norm element for all $i = 1, ..., n - 1$.*

We now develop an ambient theory of subplane covered nets that will provide examples of central extensions of cyclic division rings of finite degree in a manner similar to the theory of the classification of derivable nets. These provide new translation planes by multiple hyper pseudo-regulus replacements, which is the subject of part 10.

We consider the following situation: Consider a field $F(\theta, \tau)$ obtained by taking a Galois cyclic field extension $F(\theta)$ of degree n followed by a cyclic extension of degree n. Then assume that we have a field extension of degree n^2 admitting the cyclic automorphism group induced by θ (normally, we would assume that that there is an Abelian Galois group which is the direct product of two cyclic groups of order n). Consider a $n + 1$-dimensional projective space $PG(n + 1, F(\theta, \tau))$ over $F(\theta, \tau)$ and form the associated subplane covered net as a right subplane covered net of dimension n over $F(\theta, \tau)$, using the construction in part 1, by contracting this back into a $2n$-dimensional vector space over $F(\theta, \tau)$. Now assume that there is a non-commutative central division ring K of dimension n^2 over F; that is, of degree

n, with center F, which has a matrix form of n by n matrices over $F(\theta)$; thereby coordinatizing a left subplane covered net of the same dimension n over $F(\theta)$.

Then the set of matrices form the open form of K with a matrix skewfield $(F(\theta), F, \sigma, b)$, σ of order n, in $GL(n, F)_{left}$, contained in $GL(n, F(\theta))_{left}$, and clearly K is non-isomorphic to $F(\theta, \tau)$. The right/left theory developed for derivable nets is valid here and acting on a common n-dimensional subspace C in $\{x = 0, y = 0, y = x\}$, then K^*_{right} and $F(\theta, \tau)^*$ commute as scalar mappings; acting on the bimodule C induced by the sets of subplanes of each of the subplane covered nets. We claim that the action of each group is irreducible in the type 0 situation, that is, the minimal invariant subspace of C is C. Since the little subplanes restricted to C are 1-dimensional subspaces, then assume not, so there is a minimal invariant subspace of 1-dimensional subspaces, corresponding to a set of little subspaces. Hence, there is a sub-subplane covered net left invariant by both of the groups. Assume this is a maximal invariant sub-subplane covered net. Then, there is an extension of the cyclic division ring K_{right}, which acts as a matrix group of dimension k for $k < n$, a contradiction, since the degree n is minimal. Therefore, there are groups that are mutually irreducible and faithful acting on Com, just as in the derivable net case.

For that reason, there is a centralizer skewfield $C_{GL(n, F(\theta, \tau))_{left}}(K^*_{right})$ and a centralizer skewfield $C_{GL(n, K)_{right}}(F(\theta, \tau)^*_{left})$. Since K^*_{right} and $F(\theta, \tau)^*$ centralize each other and are both in $C_{GL(n, F(\theta, \tau))_{left}}(K^*_{right})$, and note that K^*_{right} is contained in $GL(n, F(\theta))$, it follows that the centralizer skewfield extends both K and $F(\theta, \tau)$ as a vector space of dimension n over both field/skewfield. Then $C_{GL(n, F(\theta, \tau))_{left}}(K^*_{right}) = K_{right} \otimes_F F(\tau)$. For that reason, we obtain an extension of the subplane covered net of dimension n over $F(\theta)$ to a net of dimension n over $F(\theta, \tau)$, with matrix net within $GL(n, F(\theta, \tau))$. This then makes $K_{right} \otimes_F F(\tau)$ a spread. So K_{right} with degree n over F, is extended to a division ring of degree n over $F(\tau)$. See the note in the formal proof showing that a spread is obtained.

THEOREM 137. **Skewfield extension theorem for division rings of degree n.** *Let F be a field that contains a primitive n-th root of unity and so that the characteristic of F does not divide n. Let S be an division ring of degree n with center F that contains a field $F(\theta)$, which is a Galois cyclic field extension of dimension n over F. Assume S induces a subplane covered net on $PG(2n - 1, F(\theta, \tau))$, where $F(\theta, \tau)$ is an Abelian central field Galois extension of dimension n^2, the composition of two non-isomorphic cyclic extensions of dimension n, containing $F(\theta)$ and $F(\tau)$:*

(1) Then S coordinatizes a matrix skewfield of n by n matrices over $F(\theta)$, defining a left subplane covered net as a suspect subplane covered net relative to the classical (left/right) regulus net over $F(\theta, \tau)$.

(2) Also, there is a central division ring extension of S over F, which is a division ring $S(\tau)$ such that $Z(S(\tau)) = F(\tau)$, that coordinatizes a spread in $PG(2n - 1, F(\theta, \tau))$ and a spread in $PG(2n - 1, S)$.

(3) This $S(\tau)$ central division ring extension defines a Desarguesian translation plane with left spread in $PG(2n - 1, F(\theta, \tau))$. Additionally, the dual translation plane may be written in $PG(2n - 1, S)$.

PROOF. It remains to show that the extension defines a spread. A spread of dimension n arises from a vector space V_{2n} of dimension $2n$ over a skewfield. The subplane covered net in $PG(2n-1, F(\theta, \tau))$ defined by the cyclic division ring K as a matrix group over $GL(n, F(\theta, \tau))_{left}$ has matrices defined by elements of $F(\theta)$, a cyclic Galois n-dimensional field extension of F, where F is a center of the skewfield. $F(\theta, \tau)$ is a field extension of dimension n^2 over F. It is noted, just as in the derivable net case with the quaternion division ring situation, that the degrees of the classical pseudo-regulus net and the suspect subplane covered net in the same ambient space must have the same dimension n^2 over a field F. In the extension, the skewfield is still of dimension n^2 (the degree does not change) but the center is now $F(\tau)$, and the representation in the matrix set is for all elements in $F(\theta, \tau)$. For that reason, the extended skewfield $S(\tau)$ is a central Galois extension of dimension n of S and defines a Desarguesian spread in $PG(2n-1, F(\theta, \tau))$. Since the vector space $S(\tau)$ is of dimension n over S and over $F(\theta, \tau)$, we see that $S(\tau)$ defines a Desarguesian plane defined by a vector space with spread in $PG(2n-1, S)$ and the translation plane defines a spread in $PG(2n-1, F(\theta, \tau))$. To conceive of the spread in $PG(2n-1, F(\theta, \tau))$, as a matrix spread, the n-th row would be $[t_1, .., t_n]$, $\forall t_i \in F(\theta, \tau)$. The suspect partial spread, the subplane covered net representing S, has $t_i \in F(\theta)$, but when S is extended to $S(\tau)$, we see the $t_i \in F(\theta, \tau)$, so a spread is obtained. This completes the proof of the theorem.   □

REMARK 41. *Assume that F is a field of characteristic not dividing n and contains ς_n. If, for some element $a \in F$, $x^n - a$ is an irreducible polynomial then the polynomial has roots $\varsigma_n^i \sqrt[n]{a}$. for $i = 0, 1, ..., n-1$. If, for some divisor d of n, $x^d - a$ factors $x^n - a$ then the roots in the extension are $\varsigma_n^i \sqrt[d]{a}$ and become as a set*

$$\varsigma_{\frac{n}{d}}^j \sqrt[\frac{n}{d}]{a}, \ for \ j = 0, 1, 2, ..., d.$$

Assuming a is not an e-th power in F, for all e dividing n, then a cyclic Galois extension is obtained.

In particular, for F an algebraic extension of the field $\mathbb{Q}$ of rational numbers, that contains ς_n, assume there are $k+1$ mutually non-isomorphic cyclic Galois extensions of F, of dimension n, $F(\theta_i = \sqrt[n]{c_i})$, for $c_1, ..., c_k \in \mathbb{Q}$. Then there would be Galois extensions $F(\theta_1, ..., \theta_{k+1})$ of degree n^{k+1}, admitting an Abelian Galois group isomorphic to $\prod_{k+1} z_n$ which contains an Abelian subgroup isomorphic to $\prod_{k} z_n$ that fixes $F(\theta_1)$ pointwise. We shall extend these ideas further in section 3.

To remark about the specific constructions of type 0 subplane covered nets, we add the following corollary.

COROLLARY 39. *Under the assumptions of the previous remark, $L = (a, b)_{\varsigma_n F}$ or $(F(\sqrt[n]{a})/F, \sigma, b)$, provides a type 0 subplane covered net. And, on $x = 0$ there is a generated cyclic division ring $L(\sqrt[n]{c}) = (a, b)_{\varsigma_n F(\sqrt[n]{c})}$, which is a cyclic Galois extension of $(a, b)_{\varsigma_n F}$. That being the case, we have the cyclic division rings*

$$C_{GL(2,K)_{right}}(L^*_{left}) \simeq ((a, b)_{\varsigma_n F})_{left} \bigotimes_F F(\sqrt[n]{c}).$$

Noting that $K^{opp} = K_{right}$,

$$C_{GL(2,L)_{Left}}(K^*_{right}) \simeq F(\sqrt[n]{c}) \bigotimes_F ((a, b)_{\varsigma_n F})_{right},$$

and these division rings are the right and left versions of the cyclic division ring and isomorphic to $(a, b)_{\varsigma_n F(\sqrt[n]{c})}$, *and each other if and only if* $n = 2$.

THEOREM 138. *Since* $(a, b)_{\varsigma_n F(\sqrt[n]{c})}$ *coordinatizes a Desarguesian plane with a left spread, this spread may be considered in* $PG(2n - 1, (a, b)_{\varsigma_n F})$ *and the right version Desarguesian plane as a right spread may be considered in* $PG(2n - 1, F(\sqrt[n]{a}, \sqrt[n]{c}))$.

PROOF. The tensor product assertion is due to results of Cohn and Dicks [**30**]. The right and left versions of a skewfield are inverses in the Brauer group, so the two skewfields are the right and left versions of $(a, b)_{\varsigma_n F(\sqrt[n]{c})}$, which are isomorphic exactly when $n = 2$.   □

We have indicated the idea of type 0 subplane covered nets. So, we shall continue with types $1, 2, ..., 2n - 1$.

THEOREM 139. ***Justification of the types.*** *Assume that we have a left suspect subplane covered net of dimension* n. *It is assumed that the spread is in matrix form with matrices in* $GL(n, F)_{left}$, *and that the spread is in* $PG(2n - 1, F)$, *for* F *a skewfield, and relative to the* F-*right pseudo-regulus.*

Just as in the derivable net case, there are irreducible cases and reducible cases, respectively, type 0 *and type* $\neq 0$. *In the reducible cases, all corresponding coordinate skewfields for the suspect covered nets are isomorphic to* F.

(1) *If the suspect subplane covered net is of type* 1, *assume that the common subplane induced on* $x = 0$ *is* $(0, 0, 0, ..., 0, 1)$. *This would imply that the nth row is* $[0, 0, 0, ..., 0, t_n]$, *for all* $t_n \in F$.

(2) *Type* 2 *will place the integer* 0 *in elements of a row except for some diagonal entry for example,* $(0, 0, ..., 0, 1, 0)$ *will have the* $(n-1)$-*st row as* $[0, 0, ..., 0, 1, 0]$. *If we keep choosing type* i *for* $i = 1, , , , .n$ *in this generic manner, the type* n *will be a diagonal matrix. And, it follows that diagonal elements will be of the form* $t_{n-1}^{\sigma_i}$, *where* $\sigma_i \neq \sigma_j$, *for* $i = 1, 2, ..., n$ *is an outer automorphism in the automorphism group of* F, *one of which is* 1. *Therefore, there must be such automorphisms for a type* 2 *subplane covered net to exist.*

(5) *Types* $n + 1$ *to* $2n - 1$ *will remove all automorphisms (make all automorphisms* $\sigma = 1$) *so that type* $2n - 1$ *is the right classical pseudo-regulus net over* F. *Recall that all of these types are 'at least' types.*

PROBLEM 68. *The standard choices for representations of common subplanes would lead to a definition of 'generic types' and there would be the issue of other representations of common subplanes to complete. This would an interesting project to take on as an on-going problem.*

- We now go through examples for the types when $n = 3$.

So, the types i would be $i = 0, 1, 2, 3, 4, 5$, again with the 'at least' built into the system for types 4 and 5, relative to the derivable subnets. The examples given below should clarify the meaning.

Consider some examples of suspect subplane covered nets over

$$PG(5, F(\sqrt[3]{a}, \sqrt[3]{c})),$$

where F is a field and $F(\sqrt[3]{a}, \sqrt[3]{c})$ is a Galois extension of F admitting a Galois group of order 3^2. The suspect subplane covered net will have the general form

$$x = 0, y = xM,$$

where $M \in K$, for a skewfield K, as a suspect subplane covered net in $PG(5, F(\sqrt[3]{a}, \sqrt[3]{c}))$, a type 0 example would either be a degree 3 skewfield, which is known to be cyclic (See [**64**]), or a field extension

$$F(\sqrt[3]{a}, \sqrt[3]{d}) \not\simeq F(\sqrt[3]{a}, \sqrt[3]{c})).$$

Type 0.

$$x = 0, y = x \begin{bmatrix} t & bv^{\sigma} & bu^{\sigma} \\ u & t^{\sigma} & bv^{\sigma^2} \\ v & u^{\sigma} & t^{\sigma^2} \end{bmatrix} ; t, u, v \in F(\sqrt[3]{a}),\ b^i \in F - F(\sqrt[3]{a}); i = 1, 2,$$

σ an automorphism of order 3.

The extension of skewfields theorem, gives a Desarguesian spread in

$$PG(5, F(\sqrt[3]{a}, \sqrt[3]{c})),$$

coordinatized by a cyclic division ring over $K(\sqrt[3]{c})$.

Type 1. Let L be a skewfield with center $Z(L)$ and let $L(\theta)$ be a central Galois quadratic skewfield extension, using an irreducible quadratic $x^2 \pm \alpha x - \beta$, for $\alpha, \beta \in Z(L)$. Then the following provides a type 1 example with respect to the classical right pseudo-regulus with respect to $L(\theta)$, which is in $PG(5, L(\theta))$ To see this, note that

$$\left\{ \begin{bmatrix} u + \alpha t & \beta t \\ t & u \end{bmatrix} ; t, u \in L \right\} \simeq L(\theta).$$

The following is a suspect subplane covered net with partial spread in $PG(5, L(\theta))$.

$$x = 0, y = x \begin{bmatrix} u + \alpha t & \beta t & 0 \\ t & u & 0 \\ 0 & 0 & \theta t + u \end{bmatrix} ; t, u \in L,$$

where x, y and 3-vectors over $L(\theta)$.

The determinant of the matrix is $((u + \delta t)t^{-1}u - \gamma t)t)(\theta t + u)$, when $t \neq 0$. The matrices are in $GL(3, F(\theta))_{left}$ and form (together with 0_3) an isomorphic copy of $L(\theta)$. Using the short notation for little subplanes, we see that $(0, 0, 1)$ is a common little subplane. If there is another little common subplane (e, f, g), then for every element of $h \in L(\theta)$, $(he, hf, hg) = (e(u + \alpha t) + ft, e\beta t + fu, g(\theta t + u))$. From here, $g = 1$ without loss of generality and $h = \theta t + u$. It follows by taking $t = 0$ that e and $f \in Z(L)$, from the first two entries. But this implies either $e = f = 0$, contrary to assumptions. Therefore, assume $g = 0$, and we may assume that $e = 1$. A similar argument shows that $f \in Z(L)$ and $h = (u + \alpha t) + fu \in L$, a contradiction. Therefore, we have a type 1 subplane covered net.

Type 2. For a generic type 2, consider $L(\theta, \tau)$, a Galois extension of a skewfield L, where $L(\theta)$ and $L(\tau)$ are non-isomorphic central Galois quadratic extensions of a skewfield L. Let A be a σ-derivation that is not a σ-inner derivation. Consider

$$x \;=\; 0, y = x \begin{bmatrix} w^\sigma & A(w) & 0 \\ 0 & w & 0 \\ 0 & 0 & w^\rho \end{bmatrix} ; w \in L(\theta, \tau),$$

where x, y and 3-vectors over $L(\theta, \tau)$,

where $\sigma \;\neq\; \rho$ are outer automorphisms of $L(\theta, \tau)$ fixing L pointwise.

Noting that $L(\theta, \tau)$ is a Galois extension, there is an elementary Abelian 2-group of order 4 that fixes L pointwise.

Let σ and ρ be non-identity automorphisms. Then there are two common little subplane covered nets, with short notation $(0, 0, 1)$ and $(0, 1, 0)$. First assume that there is another common little subplane (e, f, g), where $e \neq 0$, so without loss of generality, we may let $e = 1$. So (h, hf, hg). Therefore, this forces $h = w^\sigma$. Then $hf = A(w) + fw = w^\sigma f$, so that $A(w) = w^\sigma f - fw$, which is an inner σ-derivation. Therefore, $f = 0$. Then it follows that $w^\sigma g = gw^\rho$. Also, $(1, 0, g) = (w^\sigma, A(w), gw^\rho) = (w^\sigma, 0, w^\sigma g)$, a contradiction. Therefore, we have a type 2 subplane covered net. Hence, assume that $e = 0$. So that $(0, f, g) = (0, hf, hg) = (0, fw, gw^\rho)$. We may assume that either f or $g = 1$. If $f = 1$ then $h = w$, so that $wg = gw^\rho$, for all $w \in L(\theta, \tau)$. Since ρ is an outer automorphism, ρ fixes L pointwise, and shows that $g \in Z(L) \subset Z(L(\theta, \tau)) = Z(L)(\theta, \tau)$, which implies that $w = w^\rho$, a contradiction. A similar argument, when $g = 1$, shows that we have a type 2 subplane covered net.

Type 3. Assume generic with $(0, 0, 1), (0, 1, 0)$ and $(1, 0, 0)$, and note we have a set of linearly independent little subplanes. And, almost the same argument since $\{1, \sigma, \rho\}$ is a set of mutually distinct automorphisms shows that we have a type 3 subplane covered net (with $A \equiv 0$). So, we have the diagonal matrix type.

Type 4. This is the $n + 1$ type, which is an 'at least' type.

Choose $\rho = 1$. Let (e, f, g) be a common little subplane. If $e \neq 0$, choose $e = 1$, which then shows that $f = g = 0$, a contradiction. Therefore, $e = 0$. Then $fg \neq 0$, let $f = 1$. Then $(0, h, hg) = (0, w, gw)$, so that $wg = gw$, so that $g \in Z(L(\theta, \tau))$. Therefore, we have then $n + CardZ(L(\theta, \tau))$, to indicate at the $n + 1 = 4$ type, there are $3 - 1 + CardZ(L(\theta, \tau))$ common little subplanes corresponding to that $(n+1)$-st common little subplane.

Type 5. This is the $n + 2 = 2n - 1 = 5$ for $n = 3$, type, again which is an 'at least' type, where all automorphisms are trivial.

$$x \;=\; 0, y = x \begin{bmatrix} w & 0 & 0 \\ 0 & w & 0 \\ 0 & 0 & w \end{bmatrix} ; w \in L(\theta, \tau),$$

where x, y and 3-vectors over $L(\theta, \tau)$,

where $\sigma \;\neq\; \rho$ are automorphisms of $L(\theta, \tau)$ fixing L pointwise.

In the derivable net case, a set of 3 common subplanes was always the $n + 1$ case and every three subplanes were in an orbit under the collineation group of the associated classical derivable net. In general, the first n common little subplanes form a basis for the set of little subplanes. So, all types n will involve a basis and a change in basis should carry one set of n common little subplanes to the generic set of common little subplanes. How far the analogy may be carried involves the nature of $PG(n + 1, F)_N$ where N is a co-dimension 2 projective subspace. In any case, there are interesting types $0, 1, ..., 2n - 1$, as has been shown for $n = 3$. Various of the types might be of general interest. And, types 0 are always of special interest.

PROBLEM 69. *Push the analogy described above. Try this for $n = 3$, 4, and 5. At some point, the technical requirements will probably become insurmountable.*

Also, in the examples, that we have given, the type 0 is coordinatized by a cyclic division ring of degree 3, and the type 1 could be coordinatized by a quaternion division ring or a sub-division ring of a cyclic division ring of degree n when 2 divides n. This sort of example is interesting as it may be generalized as follows. Let $L(\theta)$ be a central Galois extension of degree $n - 1$ of a skewfield. Write $L(\theta)$ as an $n - 1$ by $n - 1$ matrix determined by an irreducible polynomial $f(x)$ of degree $n - 1$ over $Z(L)$. Then consider the following suspect subplane covered net with respect to the classical right subplane covered net with partial spread in $PG(2n - 1, L(\theta))$:

$$x = 0, y = x \begin{bmatrix} M_w & 0 \\ 0 & w \end{bmatrix}; w \in L(\theta);$$

where M_w indicates the $n - 1$ by $n - 1$ matrix and w is in the last row, with the 0 matrices the appropriate matrices to fill out the n by n matrix. Then this would/could be a type 1 subplane covered net.

PROBLEM 70. *Suppose that there is a division ring of finite degree k that has an open form representation. Find a subplane covered net of degree $k + 1$ with a type 1 subplane covered net.*

This idea then may be extended in a more or less obvious manner. Therefore, there could be a series of type $1, 2, ...$ types of subplane covered nets that are relative type 0 nets. Prove this.

PROBLEM 71. *For $n = 4$, find a type 1 subplane covered net that is coordinatized by a cyclic division ring of degree 3.*

Also, for $n = 4$, find a type 2 subplane covered net that is coordinatized by a quaternion division ring.

PROBLEM 72. *For $n = 5$, find a type 1 subplane covered net that is coordinatized by a cyclic division ring of degree 4.*

2. Fundamentals of Kummer theory

In this section, we discuss a bit of Kummer Theory, which is related to the construction of Abelian Galois groups that are direct sums of finite cyclic groups z_n of order n. In order to accomplish this, it is required to begin with a field F that admits a primitive n-th root of unity, denoted by ς_n, where the characteristic of F does not divide n and n is invertible. Here, we shall further sometimes restrict to the case where Q is the field of rational numbers and $F = \mathbb{Q}(\varsigma_n)$, a 'cyclotomic extension' of $\mathbb{Q}$, which is a Galois extension of $\mathbb{Q}$ of degree $\varphi(n)$, the 'Euler phi-function'. This function indicates the number of integers a, for $1 \leq a < n$, relatively prime to n. For example, $\varphi(8) = 4$, the relatively prime integers being $1, 3, 5, 7$.

In particular, we are interested in the construction of extensions of $\mathbb{Q}(\varsigma_n) = F$ of the form $F(\theta, \tau)$, where the non-isomorphic extensions $F(\theta)$ and $F(\tau)$ are cyclic

Galois extensions of dimension n and $F(\theta, \tau)$ is an Abelian Galois extension with Galois group the direct product of two cyclic groups of order n, $z_n \times z_n$, so the dimension of $F(\theta, \tau)/F$ is n^2.

In the skewfield classification of derivable nets, it was shown that there is a derivable net within a Galois extension of the form $F(\theta, \tau)$. For $n = 2$, it is possible to find extensions of $\mathbb{Q}$, say $\mathbb{Q}(\sqrt[2]{-3})$, that support a quaternion division ring $(-3, -1)_\mathbb{Q}$, a modified form of a Hamiltonian division ring. Note that $\varsigma_2 = -1$ so that representing a classical regulus net over $(\mathbb{Q}(\varsigma_2)(\sqrt[2]{-3}))(\sqrt[2]{5})$, then the quaternion division ring $(-3, -1)_{\mathbb{Q}(\varsigma_2)}$ represents a type 0 derivable net and the extension $(-3, -1)_{\mathbb{Q}(\varsigma_2)(\sqrt[2]{5})}$ coordinatizes a spread, which is a central Galois extension of a quaternion division ring. In a quaternion division ring $(a, b)_K^{K(\theta)}$, the matrix representation requires an automorphism σ of order 2 of the associated spread $K(\sqrt{a})$, in the characteristic $\neq 2$ cases. And the element b must be in $K - K(\sqrt{a})^{\sigma+1}$, where

$$K(\sqrt{a})^{\sigma+1} = \{c^2 - ad^2 : \forall c, d \in K\}.$$

If the above case $a = \sqrt{-3}$ and $b = -1$, we see that $c^2 - ad^2 = c^2 + 3d^2 > 0$, so the requirements for a quaternion division ring are fulfilled.

In (10,12,5), the definition of a symbol division ring of dimension n^2 over the center K is defined as an algebra $K\langle u, v \rangle$, where $u^n = a$, $v^n = b$ and $uv = \varsigma_n vu$. We also shall use the notation $(a, b)_{\varsigma_n}$ to denote cyclic division rings of this type, where there is an underlying field K containing the primitive n-th root of unity ς_n. To emphasize this, the notation used here is $(a, b)_{\varsigma_n \mathbb{Q}(\varsigma_n)}$. Taking the previous example when $n = 2$, the quaternion division ring $(\sqrt[2]{-3}, \varsigma_2)_{\mathbb{Q}(\varsigma_2)}$, which has a central Galois quadratic extension $(\sqrt[2]{-3}, \varsigma_2)_{\mathbb{Q}(\varsigma_2)(\sqrt[2]{5})}$. In the following section, it is shown that for every $n = p^r$, for p a prime, there is a cyclic division ring of degree n of the form $(a, \varsigma_n)_{\varsigma_n \mathbb{Q}(\varsigma_n)}$, where the matrix representation is relative to $\mathbb{Q}(\varsigma_n)(\sqrt[n]{a})$, which may be extended to a Galois extension $\mathbb{Q}(\varsigma_n)(\sqrt[n]{a}, \sqrt[n]{c})$ of degree n^2 over $\mathbb{Q}(\varsigma_n)$. This will show that there is a central extension $(a, \varsigma_n)_{\varsigma_n \mathbb{Q}(\varsigma_n)(\sqrt[n]{c})}$ of $(a, \varsigma_n)_{\varsigma_n \mathbb{Q}(\varsigma_n)}$. To be assured of the extension of the cyclic division ring, we need to know that such Galois field extensions exist by using Kummer theory (See, e.g. Dummit and Foote [49] section 14.7, p 626).

PROPOSITION 10. *(Dummit and Foote* [49] *section 14.7, p 626, propositions 36-37) Let F be any field that contains a primitive n-th root of unity ς_n and whose characteristic does not divide n. If K is a Galois cyclic extension F of dimension n over F, then $K = F(\sqrt[n]{c})$ for some element c of F.*

PROOF. Here is a sketch of the proof. The roots in a splitting field of $x^n - c \neq 0$ for $c \in F$ are $\varsigma_n^i \sqrt[n]{c}$, for $i = 1, 2, ..., n$ and since the automorphism group maps roots to roots, it follows that a generator for the cyclic group of order n is $\sigma(\sqrt[n]{c}) = \varsigma_n \sqrt[n]{c}$ and $\sigma^i \sqrt[n]{c}) = \varsigma_n^i \sqrt[n]{c})$. So it follows that the automorphism group is cyclic of order n and must be Galois of dimension n over F, since $F(\sqrt[n]{c})$ is the splitting field for a separable polynomial. The proof that any cyclic Galois extension K of F of dimension n is of the form indicated uses the 'Lagrange resolvent', defined as follows: For $a \in K$ and let σ be a generator of the automorphism group of K over F. Define the Lagrange resolvent of $(a, \varsigma_n) = \sum_{i=0}^{n-1} \varsigma_n^i \sigma^i(a)$. Then it follows after a short calculation that $\sigma^i(a, \varsigma_n) = \varsigma_n^{-i}(a, \varsigma_n)$, and $\sigma(a, \varsigma_n)^n = (\sigma(a, \varsigma_n))^n =$

$\varsigma_n^{-n}(a,\varsigma_n)^n = (a,\varsigma_n)^n$, which implies that $(a,\varsigma_n)^n \in F$, and is in no proper subfield. We claim that there is an element $a \in K$ such that (a,ς_n). There is a separable polynomial $f(x)$ in $F[x]$ of degree, which splits in K, so let θ be a root of $f(x)$, so that the elements of K may be represented uniquely in the form $\sum_{i=1}^{n} \alpha_i \theta^{i-1}$, for $\alpha_i \in F$, as $\{1, \theta, \theta^2, .., \theta^{n-1}\}$ is a basis for K/F. Considering (a,ς_n) and letting $a = \sum_{i=1}^{n} \alpha_i \theta^{i-1}$, compute the Lagrange resolvent (a,ς_n) as

$$\alpha_1 \left(\sum_{i=0}^{n-1} \varsigma_n^i \right) + \left(\sum_{j=1}^{n-1} \left(\sum_{i=2}^{n-1} \alpha_i \theta^i \right)^{\sigma^j} \right).$$

Using characters of groups, it can be shown that $\{1, \sigma, \sigma^2, ..., \sigma^n\}$ is linearly independent over K, which means the action on every r of K is linearly independent in the following sense:

$$\sum_{i=0}^{n-1} k_i r^{\sigma^i} = 0$$

for $k_i \in K$, is valid if and only if $k_i = 0$ for all i. Since $\sum_{i=0}^{n-1} \varsigma_n^i) = 0$, we have that $\sum_{i=2}^{n-1} \alpha_i \theta^i = 0$, which then implies that $\alpha_i = 0$, for all $i = 2, ..., n-1$. That being the case, we have that the Lagrange resolvent (a,ς_n) is zero if and only if $a \in F$. Since $(a,\varsigma_n)^n = c \in F$, then it follows that $\sqrt[n]{c} = \varsigma_n^i(a,\varsigma_n)$, for some element i. Then

$$F(\sqrt[n]{c}) = F(\varsigma_n^i(a,\varsigma_n)) = F((a,\varsigma_n)) = K,$$

since $(a,\varsigma_n)^{\sigma^i} = \varsigma_n^i(a,\varsigma_n)$, (a,ς_n) is not in any proper subfield of K. This completes the proof. □

There is a construction method that is relevant to Kummer theory that we shall use in our construction of central Galois extensions of skewfields of dimension $m = p^a$.

THEOREM 140. (*Theorem 1 of Lahtonen, Markin, McGuire [108]*). *Let $m = p^a$ be a prime power and let $K = \mathbb{Q}(\varsigma_m)$. Then there exist infinitely many cyclic Galois extensions L/K of degree m such that ς_m^i not a norm of L/K, for $0 < i < m$.*

Here is the method of construction: Let $m = p^a$ divide $t_i - 1$, where t_i is prime and all primes $t_i \in \lambda$ are distinct, where p^{a+1} does not divide $t_i - 1$. Take $\mathbb{Q}(\varsigma_m)\mathbb{Q}(\varsigma_{t_i}) = \mathbb{Q}(\varsigma_{mt_i})$, which is a Galois extension of $\mathbb{Q}(\varsigma_m)$ of dimension $t_i - 1 = \varphi(t_i)$. Let σ_i denote a generator of the cyclic group of order $t_i - 1$. Note that $\mathbb{Q}(\varsigma_{mt_i}) \cap \mathbb{Q}(\varsigma_{mt_j}) = \mathbb{Q}(\varsigma_m)$, for $t_i \neq t_j$. Let M_i denote the subfield of $\mathbb{Q}(\varsigma_{mt_i})$ fixed pointwise by σ_i^n. Then M_i is a cyclic Galois extension of $\mathbb{Q}(\varsigma_m)$ of degree m. Choose M_i/K, as $K(\sqrt[m]{a_i})$, for $a_i \in K$ and $M_j = K(\sqrt[m]{a_j})$, both cyclic Galois extensions of degree m.

Consider an example of the construction. Let $m = 2^2$ and consider the primes 5 and 29, noting that $m = 2^2$ divides $5 - 1$ and $29 - 1$ but 2^3 does not. Form $\mathbb{Q}(\varsigma_4)(\varsigma_5) = \mathbb{Q}(\varsigma_{4\cdot5})$ noting that this field is a cyclic Galois extension of $\mathbb{Q}(\varsigma_4)$ of degree $\varphi(5) = 4$. Similarly, form $\mathbb{Q}(\varsigma_4)(\varsigma_{29}) = \mathbb{Q}(\varsigma_{4\cdot5})$ noting that this field is a cyclic Galois extension of $\mathbb{Q}(\varsigma_4)$ of degree $\varphi(29) = 28$. Let σ denote a generator of the cyclic group of order 28. Choose the subfield M of $\mathbb{Q}(\varsigma_{4\cdot29})$ that σ^4 fixes pointwise, thereby inducing a cyclic Galois group of M of order 4, so that M is a cyclic Galois extension of $\mathbb{Q}(\varsigma_4)$ of degree 4. Note that $\mathbb{Q}(\varsigma_{4\cdot5}) \cap \mathbb{Q}(\varsigma_{4\cdot29}) = \mathbb{Q}(\varsigma_4)$ and $M \cap \mathbb{Q}(\varsigma_{4\cdot5}) \subset$

$\mathbb{Q}(\varsigma_{4\cdot29}) \cap \mathbb{Q}(\varsigma_{4\cdot5})$. It follows that $M\mathbb{Q}(\varsigma_{4\cdot5})$ is a Galois extension of $\mathbb{Q}(\varsigma_4)$ of dimension 4^2 with Abelian Galois group isomorphic to $z_4 \times z_4$. Finally, Kummer theory shows that $M = \mathbb{Q}(\varsigma_4)(\sqrt[4]{a_1})$ and $\mathbb{Q}(\varsigma_{4\cdot5}) = \mathbb{Q}(\varsigma_4)(\sqrt[4]{a_2})$ and $M\mathbb{Q}(\varsigma_{4\cdot5}) = \mathbb{Q}(\varsigma_4)(\sqrt[4]{a_1}, \sqrt[4]{a_2})$.

Generalizing this argument, let $m = p^a$ divide $t_i - 1$, where t_i is prime and all primes $t_i \in \lambda$ are distinct where p^{a+1} does not divide $t_i - 1$. Take $\mathbb{Q}(\varsigma_m)\mathbb{Q}(\varsigma_{t_i}) = \mathbb{Q}(\varsigma_{mt_i})$ which is a Galois extension of $\mathbb{Q}(\varsigma_m)$ of dimension $t_i - 1 = \varphi(t_i)$. Let σ_i denote a generator of the cyclic group of order $t_i - 1$. Note that $\mathbb{Q}(\varsigma_{mt_i}) \cap \mathbb{Q}(\varsigma_{mt_j}) = \mathbb{Q}(\varsigma_m)$, for $t_i \neq t_j$. Let M_i denote the subfield of $\mathbb{Q}(\varsigma_{mt_i})$ fixed pointwise by σ_i^n. Then M_i is a cyclic Galois extension of $\mathbb{Q}(\varsigma_m)$ of degree m. Choose M_i/K, as $K(\sqrt[m]{a_i})$, for $a_1 \in K$ and $M_j = K(\sqrt[m]{a_j})$, both cyclic Galois extensions of degree m. By the intersection property mentioned above, consider $K(\sqrt[m]{a_i}, \sqrt[m]{a_j})$ and then the polynomial as a product of the irreducible polynomials $(x^m - a_i)(x^m - a_j)$. $(x^m - a_i)$ has roots $\varsigma_m^j \sqrt[m]{a_i}$, for $j = 1, 2, ..., m$, so then polynomial is separable. The field is Galois, since it is the splitting field of this separable polynomial. For that reason, the field $K(\sqrt[m]{a_i}, \sqrt[m]{a_j})$ is Galois of degree m^2. Let τ_i denote a generator for the Galois group of order m in $K(\sqrt[m]{a_i})$. Let ρ be a set of primes with the above properties relative to p^a, Lahtonen, Markin, McGuire [108] show that ρ could be an infinite set using Dirichlet's theorem on arithmetic progressions (See (11, 13, 1&2). In the following corollary, the purpose of an ordered set will be made clear in the section on cyclic division algebras.

COROLLARY 40. *Let $m = p^a$ and let λ be an infinite set of primes $\{t_i \in \rho; i = 1, 2, ..., \infty\}$ such that p^a divides $t_i - 1$ but p^{a+1} does not divide $t_i - 1$. Then there exists an infinite set of elements a_i of $\mathbb{Q}(\varsigma_m)$ such that for any integer k, choose any k elements as an ordered set $\{b_1, ..., b_{k+1}\} \subset \{a_i; i = 1, ..., \infty\}$. It follows that there is a central Galois extension of $\mathbb{Q}(\varsigma_m)$, $\mathbb{Q}(\varsigma_m)(\sqrt[m]{b_1}, \sqrt[m]{b_2}, ..., \sqrt[m]{b_{k+1}})$ of degree m^k admitting an Abelian Galois group isomorphic to the direct product of cyclic groups isomorphic to z_m, $\Pi^{k+1} z_m$.*

3. Galois theory for division rings

In this section, we show how to use the extension theory for skewfields and effectively discuss a sort of Galois theory for cyclic division rings of finite degree. The idea here is to connect Galois theory of fields in Galois towers to Galois theory of corresponding division rings constructed using the fields in the field tower, for use in a geometric vein, when considering the theory of derivable nets.

We begin with a quadratic Galois tower of fields with base K. Assume that there are $k+1$ mutually non-isomorphic quadratic Galois field extensions $K(\theta_i)$, for $i = 1, 2, ...k + 1$ that are mutually non-isomorphic. Consider the tower as follows: $K \subset K(\theta_1) \subset K(\theta_1, \theta_2) \subset ... \subset K(\theta_1, \theta_2, ..., \theta_{k+1})$. The order is actually arbitrary, but ultimately, we wish to establish a related tower of division rings. The last field mentioned is a Galois extension of K and admits a Galois group isomorphic to an elementary Abelian 2-group $\mathcal{E}_{k+1}$ of order 2^{k+1}. Recalling André's theorem, we wish to find a set of mutually disjoint derivable nets that partition the spread together with two components $x = 0$ and $y = 0$. We consider the multiple replacement theorem and the set up, in this case for a Pappian plane Σ coordinatized by $K(\theta_1, \theta_2, ..., \theta_{k+1})$. To apply the theorem, an automorphism τ is required to form

Σ_τ. So, let τ be any non-identity element of $\mathcal{E}_{k+1}$. Since τ has order 2, there is a sub-field E_τ such that $Fix\langle\tau\rangle = E_\tau$ and E_τ has index 2 in the field $K(\theta_1, \theta_2, ..., \theta_{k+1})$ by results of Galois theory. We recall the results of the multiple replacement theorem. There is an index set λ such that

$$\Sigma = \{x = 0, y = 0\} \cup_{i \in \lambda} \{\{y = xbm_ib^{-\tau}; b \in S^*\} = \Gamma_i\},$$

and

$$\Sigma_\tau = \{x = 0, y = 0\} \cup_{i \in \lambda} \{\{y = bm_ib^{-\tau}x^\tau; b \in S^*\} = \Gamma_i^\sigma\},$$

where the unions are disjoint unions. In this setting, as τ has order 2, we see that the 2-dimensional vector space over $K(\theta_1, \theta_2, ..., \theta_{k+1})$ has dimension 4 over E_τ and that components are 2-dimensional E_τ-subspaces and the components of Σ_τ, $y = mx^\tau$ are also 2-dimensional E_τ-subspaces. For that reason, we have a set of mutually disjoint derivable nets which are regulus nets with respect to E_τ. The interesting thing is that we have $2^{k+1} - 1$ mutually disjoint sets of derivable nets which are regulus nets with respect to E_τ, for all $\tau \in \mathcal{E}_{k+1}$, $\tau \neq 1$, which partition the spread together with $x = 0$ and $y = 0$. We call such a set of $2^{k+1} - 1$ sets of Andrè nets an 'Andrè scheme'.

DEFINITION 112. *Given a Pappian spread coordinatized by a field F, a set of $2^{k+1} - 1$ sets of Andrè nets, each of which partition the spread together with $x = 0$ and $y = 0$, is said to be an 'Andrè scheme' of index $2^{k+1} - 1$.*

Since all Andrè nets are regulus nets, this might appear to be a contradiction. The question seems to be 'when is a regulus not a regulus?' When a derivable regulus net lies in $PG(3, E_\tau)$, it may be coordinatized by E_τ, that is, it is isomorphic to the classical regulus net with respect to E_τ. Note that the spread Σ lies in $PG(3, E_\tau)$, for all $\tau \in \mathcal{E}_{k+1}$, $\tau \neq 1$. This seems to contradict the concept of dimension 2 in translation planes. Note that E_τ cannot be isomorphic to E_ρ, for $\tau \neq \rho$. For example, consider $K(\theta_1, \theta_2)$, which has a Galois group of order 4 and three quadratic subfields. We know that $K(\theta_1)$ and $K(\theta_2)$ are not isomorphic. Can $K(\theta_1\theta_2)$ be isomorphic to $K(\theta_1)$? If all $\theta_i = \sqrt{a_i}$ an isomorphism would imply that $\sqrt{a_1}\sqrt{a_2} \to \sqrt{a_1}\delta$ and the square would show that $a_2 \to \delta^2$, a contradiction.

PROBLEM 73. *If $\theta_i = \sqrt{a_i}$, and all $K(\theta_i)$ are mutually non-isomorphic. Show that all quadratic extensions K within $K(\theta_1, ..., K_{k+1})$ are mutually non-isomorphic.*

PROBLEM 74. *Determine the translation planes that have spreads that can lie in more than one 3-dimensional projective space. For example, the extension theorem for skewfields shows that $(a, b)_{K(\theta_2)}$ coordinatizes a Desarguesian plane whose spread is in both $PG(3, (a, b)_K)$ and $PG(3, K(\theta_1, \theta_2))$, where $K(\theta_1)$ is in $(a, b)_K$.*

In this section, we develop some ideas on how to consider a Galois group of a central skewfield Galois extension of a skewfield F. All of the extensions will be central extensions of F using $Z(F)$. So, we consider a restricted situation and begin with central Galois quadratic extensions of a quaternion division ring $(a, b)_K$. If $x^2 - \alpha x - \beta$ is the associated irreducible polynomial, let σ denote the associated automorphism relative to the Galois quadratic extension $K(\theta)$, where $\theta^2 = \theta\alpha + \beta$. For example, if the characteristic is not 2, let $\alpha = 0$, then $\theta = \sqrt{a}$ for $a = \beta$ and the automorphism σ with fixed field K is defined by $\sigma : \sqrt{a} \to -\sqrt{a}$.

All of our field and skewfield extensions shall be by taking a set of mutually non-isomorphic quadratic Galois extensions of K and compose these to form a Galois extension. For example, consider $K(\theta_1, \theta_2, ..., \theta_{k+1})$, where $K(\theta_i)$ are Galois extensions for $i = 1, 2, ..., k + 1$, mutually non-isomorphic. Since $K(\theta_1, \theta_2, ..., \theta_{k+1})$ is the splitting field of a product of separable polynomials, it is necessarily Galois and we have, as a Galois group, an elementary Abelian 2-group of order 2^{k+1}. And, we note that we have that $K(\theta_1)(\theta_2, ..., \theta_{k+1})$ is a Galois extension of $K(\theta_1)$ whose Galois group fixing $K(\theta_1)$ is an elementary Abelian 2-group of order 2^k. We shall call such extensions 'central Galois extensions of K'.

DEFINITION 113. *If a non-commutative skewfield extension S of a skewfield F of dimension n^k/F admits an automorphism group of order n^k and has exactly $n^k - 1$ subskewfields that are non-commutative, and corresponds to the automorphism group exactly as in the field case, we shall say that the 'extension S is Galois over F'. If the extension is a central extension, in the sense that the associated irreducible polynomial makes the center $Z(S)$ of S an extension of center $Z(F)$ of F, we shall call S a 'central extension of F'. Note that the non-commutative skewfield extension could have a subfield, which does not change the definition of Galois extension in the non-commutative skewfield case.*

We shall develop quaternion division rings which are central Galois extensions. First an example: consider when $k = 3$ above, so that $K(\theta_1, \theta_2, \theta_3)$ is a central Galois extension of K and assume that $(a, b)_K$ is relative to $K(\theta_1)$. We have seen the geometric version of the extension of $(a, b)_K$, using the extension of skewfields theorem in part 2. And, we have seen the algebraic version of this in part 8 on lifting skewfields, where the emphasis is on central quadratic Galois extension $J(\tau)$ of skewfields J, where there is an associated automorphism of order 2 that fixes J pointwise, and a related Galois extension of $Z(J)$ to the center $Z(J(\tau)) = Z(J)(\tau)$. Recall from the multiple replacement theorem, we know that the Desarguesian plane coordinatized by $J(\tau)$ has a spread that together with $x = 0, y = 0$ is covered by a set of mutually disjoint pseudo-regulus derivable nets isomorphic to J. The requirement for that construction is the existence of an automorphism g of order 2 such that $Fix g = J$.

We continue the notation introduced previously.

NOTATION 9. *For a quaternion division ring $(a, b)_K^\alpha$, we shall also use the notation $(a, b)_K^{K(\theta_1)}$, where the α indicates the polynomial used to construct the field extension $K(\theta_1)$.*

DEFINITION 114. *Consider a central quadratic Galois tower of a field $K(\theta_1, \theta_2, \theta_3, ...\theta_{k+1})$, of length $k + 1$ as a composition of quadratic Galois fields $K(\theta_i)$, $i = 1, 2, ..., k + 1$. Using the extension theorem for quaternion division rings, there is a related quaternion division ring $(a, b)_{K(\theta_2,...,\theta_{k+1})}^{K(\theta_2,...,\theta_{k+1})(\theta_1)}$, which is $(a, b)_K^{K(\theta_1)} \otimes_K$ $K(\theta_2, \theta_3, ...\theta_{k+1})$, where $(a, b)_K^{K(\theta_1)}$ is the original quaternion division ring. For each link j in the tower, we may associate a quaternion division ring $(a, b)_{K(\theta_2,...,\theta_j)}^{K(\theta_2,...,\theta_j)(\theta_1)}$.*

Although, this procedure works in characteristic 2, for purposes of illustration, assume the characteristic is not 2. Then $\Pi_{i=1}^{k+1}(x^2 - c_i)$, for $c_i \in K$, for $i = 1, 2, ..., k+$

1 *is a separable polynomial with splitting field* $K(\theta_1, \theta_2, \theta_3, ...\theta_{k+1})$. *It follows that all of the extension quaternion division rings are central extensions.*

Also note that $(a, b)_K^{K(\theta_1)} \otimes_K K(\theta_2)$ *is* $(a, b)_{K(\theta_2)}^{K(\theta_2)(\theta_1)}$, *and therefore the Galois group of* $K(\theta_2)(\theta_1)$ *that fixes* $K(\theta_1)$ *(elementwise) is an automorphism group of* $(a, b)_{K(\theta_2)}^{K(\theta_2)(\theta_1)}$ *that fixes* $(a, b)_K^{K(\theta_1)}$ *pointwise.*

Similarly, $(a, b)_{K(\theta_2,...,\theta_{k+1})}^{K(\theta_2,...,\theta_{k+1})(\theta_1)}$ *admits an automorphism group of order* $\Pi^k z_2 = \mathcal{E}_k$, *fixing* $K(\theta_1)$ *and fixes* $(a, b)_K^{K(\theta_1)}$ *pointwise.*

When considering cyclic division rings of degree n, *assume that* n *is invertible, where* ς_n *is a primitive* n *-th root of unity contained in* K *and the characteristic of* K *does not divide* n. *Then the field extensions are cyclic Galois extensions of dimension* n, *and we use* $(a, b)_{\varsigma_n K(\theta_2,...,\theta_{k+1})}^{K(\theta_2,...,\theta_{k+1})(\theta_1)}$.

THEOREM 141. (*Galois division ring theorem*) *Let a division ring* D_K *of finite degree* n *over its center* K, *where* n *is invertible and the characteristic does not divide* n. *Assume that* D_0 *contains a cyclic field extension* $K(\theta_1)/K$ *of dimension* n *and there are* $k+1$ *cyclic field extensions* $K(\theta_i)$ *of dimension* n, *all of which are mutually non-isomorphic, for* $i = 1, 2, ..., k+1$.

(1) *Then the composition* $K(\theta_1, \theta_2, ..., \theta_{k+1})$ *is Galois with Galois group* $\Pi^{k+1} z_n$.
(2) $D_K \otimes_K K(\theta_2, ..., \theta_{k+1})$ *is a Galois division ring of degree* n *with Galois group* $\Pi^k z_n$; *If* E *is a subfield of* $K(\theta_2, ..., \theta_{k+1})$ *and* H *is a subgroup of* $\mathrm{Gal} K(\theta_2, ..., \theta_{k+1})$ *that fixes* $K(\theta_1)$ *pointwise, then* $D_K \otimes_K \mathrm{Fix} H = D_{\mathrm{Fix} H}$.

PROOF. We recall the extension of skewfields theorem, for division rings of finite degree n relies only on two things: The existence on a division ring D_0 of finite degree n containing a field $K(\theta_1)$ that is a cyclic Galois extension of dimension n over K and that there exists a second cyclic field extension $K(\theta_2)$ of dimension n such that $K(\theta_1)$ and $K(\theta_2)$ are not isomorphic. Then $K(\theta_1, \theta_2)$ is Galois with automorphism group $z_n \times z_n$, where z_n is the cyclic group of order n. Embed D_0 as a left matrix partial spread as a subplane covered net in $PG(2n - 1, K(\theta_1, \theta_2))$. Then using this structure, there is an central extension of D_0 to a division ring D_1 of degree n, and using Cohn and Dicks [30], $D_1 = D_0 \otimes_K K(\theta_2)$ so that D_2 has degree n and center $Z(D_1) = Z(D_0)(\theta_2) = K(\theta_2)$. It is now clear that the automorphism group z_n of $K(\theta_2)$ fixes D_0 elementwise and acts as an automorphism group of D_1. Considering z_n as the subgroup that fixes $K(\theta_1)$ pointwise, we see that the division ring D_1 is Galois. This idea may be employed more generally, if $K(\theta_i)$ is a cyclic Galois extension of dimension n over K for $i = 1, 2, ..., k+1$ where the extension fields are all mutually non-isomorphic. For that reason, we obtain a Galois extension field $K(\theta_1, ..., \theta_{k+1})$ with Galois group $\Pi^{k+1} z_n$, the Abelian group of the direct product of $k+1$ cyclic groups z_n of order n. Then $D_0 \otimes_K K(\theta_2, ..., \theta_{k+1}) = D_k$ is a division ring of degree n with automorphism group $\Pi^k z_n$, the subgroup that fixes $K(\theta_1)$ elementwise. Additionally, it follows that D_k is Galois; for each subfield E of $K(\theta_2, ...\theta_{i+1})$, there exists a subgroup H of $\Pi^k z_n$, $\mathrm{Fix} H = E$ that fixes $H(\theta_1)$ pointwise and a corresponding subdivision ring $D_E = D_0 \otimes_K E$ of index $|H|$ of D_k. And, therefore, all extensions are central. Hence, we have completed the proof of the theorem. $\square$

COROLLARY 41. *(Galois theorem for quaternion division rings)*. *We note that for quaternion rings of characteristic 2, the results are also valid mutatis mutandis. We now see that the automorphism group of $D_k = (a, b)^{K(\theta_2,...,\theta_{k+1})(\theta_1)}_{\varsigma_n K(\theta_2,...,\theta_{k+1})}$ is $\Pi^k z_2$ the Abelian group which is the direct product of k cyclic groups z_2 of order 2, fixing $K(\theta_1)$, where the automorphism group of $K(\theta_1, \theta_2, \theta_3, ...\theta_{k+1})$ is $\Pi^{k+1} z_2$, where the automorphism group of $K(\theta_1, \theta_2, ...\theta_{k+1})$ is $\Pi^{k+1} z_2$. Additionally, it now follows that D_k is a Galois division ring of degree 2 with center $K(\theta_2, \theta_3, ., \theta_{k+1})$. Given any non-trivial automorphism τ of D_k, there is a quaternion sub-ring D_τ of D_k of degree 2 with subfield E_τ such that $Fix\tau$ in $\Pi^k z_n$, has order equal to the index of D_τ in D_k.*

Therefore, there are $2^k - 1$ sub-quaternion division rings of index 2.

COROLLARY 42. *(Galois theory for cyclic division rings of degree n)* *For a cyclic division ring $(a, b)^{K(\theta_1)}_{\varsigma_n K}$, of finite degree n over a field K, where n is invertible and the characteristic does not divide n, let $K(\theta_1, \theta_2, \theta_3, ...\theta_{k+1})$ be a field with automorphism group $\Pi^{k+1} z_n$, where all fields $K(\theta_i)$, for $i = 1, 2, ..., k+1$ are cyclic Galois extensions of dimension n and mutually non-isomorphic. Using $(a, b)^{K(\theta_1)}_{\varsigma_n K}$, construct $D_k = (a, b)^{K(\theta_2,...,\theta_{k+1})(\theta_1)}_{\varsigma_n K(\theta_2,...,\theta_{k+1})}$ by repeated application of the extension of skewfields theorem for cyclic division rings. Then the automorphism group of*

$$(a, b)^{K(\theta_2,...,\theta_{k+1})(\theta_1)}_{\varsigma_n K(\theta_2,...,\theta_{k+1})}$$

is the subgroup $\Pi^{k+1} z_n$ of $\Pi^{k+1} z_n$ that fixes $K(\theta_1)$ elementwise.

Then $(a, b)^{K(\theta_2,...,\theta_{k+1})(\theta_1)}_{\varsigma_n K(\theta_2,...,\theta_{k+1})}$ is a Galois skewfield with Galois group $\Pi^k z_n$.

4. Galois division rings & applications of multiple replacement

Assuming the notation of the previous section, we let D_k denote a division ring of degree n with Galois group $\Pi^k z_n$ and consider any cyclic group $\langle \tau \rangle$ of $\Pi^k z_n$, letting Σ be the Desarguesian plane coordinatized by D_k with left components $x = 0, y = xm$ and let Σ_τ denote the Desarguesian plane with the right components $x = 0, y = mx^\tau$. Let E_τ be the subfield of $L = K(\theta_2, \theta_3, ..., \theta_{k+1})(\theta_1)$ fixing $K(\theta_1)$ pointwise; that is, $E_\tau = Fix_{E_\tau(\theta_1)} \langle \tau \rangle$ and let D_τ denote the associated sub-division ring $\ni D_\tau = Fix_{D_k} \langle \tau \rangle$. So, $y = x \cap y = x^\tau$ is a D_τ-space. The order of τ is $\frac{n}{d}$, where d divides n. If $\tau \neq 1$, then $d \neq n$. Then a set of disjoint replacement sets may be constructed using the multiple replacement theorem. Each component $y = xm$ and $y = nx^\tau$ are 1-dimensional D_k-subspaces and therefore are $\frac{n}{d}$-dimensional over D_τ. Hence, the vector space is of dimension $2\frac{n}{d}$ over D_τ. We see that $y = x^\tau$ is a subspace of a replacement net of dimension $\frac{n}{d}$ over a division ring D_τ. For that reason, the replacement nets are isomorphic to hyper pseudo-regulus nets of dimension $\frac{n}{d}$ over and with respect to D_τ (these would be pseudo-regulus derivable nets of type 2 in the quaternion cases).

Now consider the multiply replaced translation planes. The kernel of these planes is $Z(D_k) \cap Fix_{D_\tau} \tau = Z(D_\tau)$.

Since the index of $Z(D_\tau)$ is n^2 in D_τ, we see that the translation planes have their spreads in $PG(2\frac{n^3}{d} - 1, Z(D_\tau))$.

THEOREM 142. *Applying the multiple replacement theorem to the Desarguesian planes coordinatized left and right by D_k,*

(1) the replacements nets are isomorphic to hyper pseudo-regulus subplane covered nets of degree d, for $d \neq n$

(2) the non-trivial multiply replaced translation planes have spreads in $PG(2\frac{n^3}{d} - 1, Z(D_\tau))$.

COROLLARY 43. *For quaternion division rings, when $n = 2$ and $d = 1$, there are $2^k - 1$ non-identity automorphisms τ and associated sub-quaternion division rings of index 2.*

(1) For that reason, we have mutually disjoint pseudo-regulus derivable nets of type 2 with reference to a quaternion subring $(a, b)_{E_\tau}^{E_\tau(\theta_1)}$, where $E_\tau = Fix_{E_\tau(\theta_1)} \langle \tau \rangle$. And, there are $2^k - 1$ such sets of mutually disjoint pseudo-regulus nets.

(2) The multiple derived translation planes have spreads in

$$PG(15, Z(D_\tau)).$$

COROLLARY 44. *For cyclic division rings of degree $n \neq 2$, assume that 2 divides n and $d = \frac{n}{2}$.*

(1) There are $2^k - 1$ sets of mutually disjoint pseudo-regulus derivable nets of type 2 with reference to a cyclic division subring $D_\tau = (a, b)_{\varsigma_\tau E_\tau}^{E_\tau(\theta_1)}$ of index 2 in D_k. The group element τ has order 2, so $d = \frac{n}{2}$.

(2) The multiple derived translation planes have spreads in $PG(4n^2 - 1, Z(D_\tau))$.

COROLLARY 45. *When there are derivable nets constructed, there are central Galois quadratic extensions D_k of skewfields D_τ, where τ has order 2. Thus, sets of semifield planes in $PG(3, D_k)$ are obtained using skewfield lifting. There are infinitely many such semifield planes.*

REMARK 42. *In general from cyclic division rings D_τ of degree n, the associated multiply replaced planes have spreads in (1) for outer automorphism types $PG(2\frac{n^3}{d} - 1, Z(D_\tau))$ and (2) for the special inner automorphism types using cyclic division ring E_ρ, $PG(2n^2 - 1, Z(D_\rho))$.*

5. p-Adic numbers and Hensel's lemma

The proof that a cyclic algebra with center K that contains a cyclic field extension $K(\theta)$ of dimension n over K is actually a cyclic division algebra invariably reduces to show that $K \neq K(\theta)^{(\sigma^n - 1)/(\sigma - 1)}$, where σ is a generating automorphism of K, relative to the norm $N(K(\theta)/K)$. That is, there must be some element $c \in K$ such that c^i is not in $K(\theta)^{(\sigma^n - 1)/(\sigma - 1)}$, for $i = 1, ..., n - 1$.

For example, consider a quaternion division algebra $\mathcal{A}$ with center that $\mathcal{A}$ contains $K(\sqrt{5})$, an algebraic number field. The unique involution in the Galois group σ over maps $\sqrt{5} \to -\sqrt{5}$ and the norm over K is the set of elements of the form $a^2 - 5b^2$, for all $a, b \in K$. Note that $(a + b\sqrt{5})^\sigma(a + b\sqrt{5}) = a^2 - 5b^2$. A division algebra is obtained, provided $c \neq a^2 - 5b^2$ for $c \in K$. Note that using the matrix form of the quaternion algebra, we would consider the matrices, for some $c \in K$

$$\begin{bmatrix} u^\sigma & ct^\sigma \\ t & u \end{bmatrix} : \forall t, u \in K(\sqrt{5}).$$

This set will form a quaternion division ring if and only if the determinants of non-zero matrices are non-zero. We see that the determinant is $u^{\sigma+1} - ct^{\sigma+1} = 0$ if and only if $c = k^{\sigma+1} = a^2 - 5b^2$, for some $k \in K$.

In the article of F. Oggier, J-C. Belfiore, E. Viterbo [114], there are several cyclic division rings constructed. Here is a quaternion division ring $(a, c)_K^{K(\theta)}$:

$$(5, i)_{\mathbb{Q}(i)}^{\mathbb{Q}(i)(\sqrt{5})}, \ i = \sqrt{-1}.$$

The proof that i is in $\mathbb{Q}(i) - \mathbb{Q}(i)(\sqrt{5})^{1+\sigma}$, where σ is the automorphism mapping $\sqrt{5} \to -\sqrt{5}$, then considers the equation $a^2 - 5b^2 = i$ in $\mathbb{Q}(i)$, which is exactly what we are discussing above.

There is something of a standard procedure to lift problems over number fields and related fields. In this section, we provide the basics of this method and then show how Oggier et al. solve this non-norm problem.

The idea is to 'pass to a corresponding p-adic field $\mathbb{Q}_p$', for p a prime.

In about 1897, Kurt Hensel created the p-adic fields, which are completions of the field $\mathbb{Q}$ of rational numbers in an analogous manner as the construction of the field of real numbers $\mathbb{R}$ by using the norm of absolute value and the concept of Cauchy sequences of elements of $\mathbb{Q}$. p-adic fields and the generalizations of fields with valuation are necessary tools to create division algebras. We provide here a short introduction.

First consider the completion of the rational numbers that constructs the real number field. There are several ways to do this, but we consider the method of Cauchy sequences of rational numbers, using the norm of absolute value of a rational, $|x| = x$ if $x \geq 0$ and $-x$ if $x < 0$, for $x \in \mathbb{Q}$. We recall that a sequence $\{x_1, x_2, ..., x_n, ...\}$ is 'Cauchy' if and only if for each $\epsilon > 0$ in $\mathbb{Q}$, there exists an integer N such that $|x_n - x_m| < \epsilon$ for all $m, n > N$. Two Cauchy sequences are equivalent if and only if their difference tends to 0. Ultimately, as we recall, defining the set of 'real numbers $\mathbb{R}$' as equivalence classes of Cauchy sequences, with appropriate definitions of addition and multiplication, we obtain a 'complete field' i.e. satisfies the least upper bound condition relative to $\leq$.

Then it is easy to embed $\mathbb{Q}$ into $\mathbb{R}$ by mapping r to $\{r, r, ...r, \}$. Using the idea of sequences of rational numbers, we recall that all positive reals may be written in the form $\Sigma_{i=-N}^{\infty} a_i 10^{-i}$, where $0 \leq a_i \leq 9$, for all i such that $-N \leq i < \infty$, which we identify as $(a_{-N} a_{-N+1} ... a_0.a_1 a_2 ... a_n ....)_{10}$, called 'decimal form'.

Hensel shows that there are other complete fields extending the field of rational numbers obtained by using a different norm.

DEFINITION 115. *Let a be in the ring of integers $\boldsymbol{Z}$, and let p be a prime. Define $o_p(a) = c$ if and only if $p^c \parallel c$, the p-order at a. For $\frac{a}{b}$ for $b \neq 0$, $a, b \in \boldsymbol{Z}$, define $o_p(\frac{a}{b}) = o_p(a) - o_p(b)$. Now define the '$p$-adic norm' of any rational number x, $|x|_p = 0$ if $x = 0$ and $\frac{1}{p^{o_p(x)}}$ otherwise.*

The particular norm is 'non-Archimedean' in the sense that $|x + y|_p \leq \max(|x|_p, |y|_p)$, whereas the ordinary norm of absolute value clearly does not have this property.

DEFINITION 116. *A p-adic norm is a norm and with it a complete field may be constructed, called the 'p-adic numbers $\boldsymbol{Q}_p$' extending the field of rational numbers.*

The elements of the p-adic numbers may be written as polynomials in $\frac{1}{p}$, $\Sigma_{i=-k}^{0} b_i p^i$ plus a series $\Sigma_1^\infty b_j p^j$, so

$$\Sigma_{i=-k}^{0} b_i p^i + \Sigma_1^\infty b_j p^j, \ 0 \le b_k \le p - 1.$$

Now using the notation of $(b_{-k} b_{-k+1} \cdots b_0 . b_1 \cdots b_m \cdots)_p$ is the 'p-adic form/expansion'.

For pure elements of $\Sigma_1^\infty b_j p^j$ the point of separation is not used.

Just as in the real number case, the rational numbers are characterized by expansions of repeated blocks.

DEFINITION 117. Let $\mathbf{Z}_p = \{x \in \mathbf{Q}_p : |x|_p \le 1\}$ be called the 'p-adic integers'. These are elements of the form $b_0 + \Sigma_1^\infty b_j p^j$. The set $\mathbf{Z}_p^* = \{x \in \mathbf{Q}_p : |x|_p < 1\}$ is called the set of 'units' characterized by elements of the form $p\mathbf{Z}_p$.

The reader is directed to K. Conrad [33] for additional reading on the p-adic expansion.

What drives the construction is a lemma that, in particular, seems to be useful in finding non-norm elements of cyclic algebras.

THEOREM 143. (K. Conrad [31] Theorem 2.1) **(Hensel's Lemma). Elementary form:**

If $f(X) \in \mathbf{Z}_p[X]$ and $a \in \mathbf{Z}_p$ such that $f(a) \equiv 0 \bmod p$ and using the formal derivative $f'(X)$, $f'(a) \not\equiv 0 \bmod p$ then there exists a unique element $\alpha \in \mathbf{Z}_p$ such that $f(\alpha) = 0$ in $\mathbf{Z}_p$ and $\alpha \equiv a \bmod p$.

The unique element α is approximated by a sequence $\{a_n \in \mathbf{Z}_p, \text{ for } n = 1, 2, ...\}$ using the following:

$$f(a_n) \ \equiv \ 0 \bmod p^n$$
$$a_n \ \equiv \ a \bmod p.$$

Then the uniqueness is in the limit. This is proved by induction which is a constructive version of Hensel's Lemma.

PROBLEM 75. Homework: Show that $o_p(ac) = o_p(a) + o_p(c)$, and that $o_p(ad + bc) \ge \min(o_p(ad), o_p(bc))$.

Now the proof that

$$(5, i)_{\mathbb{Q}(i)}^{\mathbb{Q}(i)(\sqrt{5})}, \ i = \sqrt{-1},$$

is a quaternion division ring (as shown in F. Oggier, J-C. Belfiore, E. Viterbo [114]).

Consider the equation $x^2 + 1$ in $\mathbf{Z}_5$. Using Hensel's lemma, $2^2 + 1 \equiv 0 \bmod 5$ and $2 \cdot 2 \not\equiv 0 \bmod 5$.

Therefore, there is a unique solution α in $\mathbf{Z}_5$. The 2 is an approximation so $\alpha = 2 + \Sigma_{i=1}^\infty a_i 5^i$. What this means is that i is in $\mathbf{Z}_5$ and $\mathbb{Q}(i)$ may be considered in $\mathbf{Q}_5$. So consider $a^2 - 5b^2 = i$ in $\mathbf{Q}_5$. We don't know that a, b are in $\mathbf{Z}_5$, this is what we shall prove. Take the norm of both sides and apply the homework listed above. We do know that $i \in \mathbf{Z}_5$ so that $|i|_5 \le 1$. Therefore we have the following $\left|a^2 - 5b^2\right|_5 = \min(2 |a|_5, 2 |b|_5 + 1) \le 1$. If $2 |a|_5 > 1$ then $b = 0$ then $a^2 = i$ which implies that $a \in \mathbf{Z}_5$. If $b \ne 0$ then again so $|a|_5 \le 1$, and we have $a \in \mathbf{Z}_5$. Therefore, we have $5b^2 \in \mathbf{Z}_5$, which shows that $|10b|_5 \le 1$ so that $b \in \mathbf{Z}_5$. Therefore, $a^2 \equiv 2 \bmod 5$, a contradiction, as $a^2 = \pm 1 \bmod 5$, which completes the proof.

PROBLEM 76. *Consider $x^2 + 3x + 1$ in $\mathbf{Q}_{11}$. Show Hensel's lemma applies and find three approximations of the unique solution.*

There is a more extensive lemma, which is:

THEOREM 144. *(K. Conrad [31] Theorem 4.1- **Hensel's Lemma**). Let $f(X) \in \mathbf{Q}_p$ and $a \in \mathbf{Z}_p$ such that*

$$|f(a)|_p < |f'(a)|_p^2.$$

Then there is a unique $\alpha \in \mathbf{Z}_p$ such that $f(\alpha) = 0$ in $\mathbf{Z}_p$ and $|\alpha - a|_p < |f'(a)|_p$. Moreover,

$$(1) \quad |\alpha - a|_p = |f(a)/f'(a)|_p < |f'(a)|_p$$

$$(2) \quad |f'(\alpha)|_p = |f'(a)|_p.$$

Again, the reader can find many other interesting uses of p-adic integers in the work of Conrad [31]. In particular, the n-th roots of unity that lie in $\mathbf{Q}_p$ may be determined by Hensel's lemma:

THEOREM 145. *(K. Conrad [31] theorem 3.1) The roots of unity in $\mathbf{Q}_p$ are $(p-1)$-th roots of unity for p odd and ± 1 for $p = 2$.*

In another use of Hensel's lemma, Morandi, Sethuraman and Tignol [113] consider finding division algebras that have anti-automorphisms but no involution (order-reversing order 2). In this setting, the use of Hensel's lemma embeds a field $F = \mathbb{Q}(\omega_n)$ into $\mathbf{Q}_p$ (noting the previous lemma), ω_n a primitive n-th root of unity. Ultimately, $(a, \omega_n)_{\omega_n F} \otimes_F \mathbb{Q}_p = (a, \omega_n)_{\omega_n \mathbb{Q}_p}$ and $(a, \omega_n)_{\omega_n F}$ are division rings. It is noted that $\rho \in Gal(F/\mathbb{Q})$ extends to an automorphism of $D = (a, \omega_n)_{\omega_n F}$, where $\rho(a) = a$, but does not induce an involution on D (order inversing element of order 2), using aspects of Brauer theory.

REMARK 43. *We note that there is a multivariable Hensel's lemma, see e.g. [13], [131] (Chapter II, Theorem 1).*

We shall revisit the examples of cyclic division rings given here in the next section.

6. Field extensions of $\mathbf{Q}_p$

It should be mentioned that the idea of a non-Archimedean norm can be generalized and the theory of valuations on fields and division rings is a rich and important area. Everything that is mentioned in this section may be generalized (See, e.g. P. Ribenboim [126]). This section summarizes some of the work of E. Turner [139] in describing extensions of $\mathbf{Q}_p$. We shall discuss the particulars, but the main thrust of finding finite cyclic Galois field extensions K of $\mathbf{Q}_p$ of dimension n, is that the p-adic norm must be extended to K. The theory provides that there is a unique norm extension to K, which shall be denoted by $\|\|$, and we define a related norm as $\|\|'$, by $\|x\|' = \|\sigma(x)\|$, where $\sigma \in Gal(K/\mathbf{Q}_p)$, $x \in K$. Let α be a root of an associated monic irreducible polynomial. Since σ fixes α, we see that the '-norm of α is the extended norm of α. And, this shows that the extended norm of all/any root/s are all equal.

Accomplishing that, define the following sets

$$A = \left\{ x \in K : |x|_p \leq 1 \right\}.$$

A is called the 'integral closure of $\mathbf{Z}_p$', which is a ring with unique maximal ideal M, defined as follows:

$$M = \left\{ x \in K : |x|_p < 1 \right\}.$$

Then A/M is a finite field extension of $\mathbf{Z}_p/p\mathbf{Z}_p$ such that

$$[A/M : \mathbf{Z}_p/p\mathbf{Z}_p] \leq [K : \mathbf{Q}_p].$$

The field A/M is called the 'residue field of K'. Let the algebraic closure of $\mathbf{Q}_p$ be denoted by $\overline{\mathbf{Q}_p}$. The p-adic norm may also be extended to $\overline{\mathbf{Q}_p}$. If $K = \mathbf{Q}_p(\alpha)$, then α satisfies a monic irreducible polynomial $F(X)$ of degree n over $\mathbf{Q}_p$, and K is a cyclic Galois extension over $\mathbf{Q}_p$ of degree n.

Now, in general, if $K = F(\alpha)$, where F is a field, is a finite extension of F of degree n, with monic irreducible polynomial $f(X) \in F[X]$, then α satisfies $f(X) = \Sigma_{i=0}^n a_i x^i$, $a_n = 1$.

Now define a linear mapping σ from $K \to K$, by $\sigma(x) = \alpha x$ for all $x \in K$ and let A_α denote the corresponding n by n matrix with entries in F. And we have defined the 'norm of K to F' previously as $\|x\|' = \|\sigma(x)\|$.

Now $N_{K/F}(\alpha)$, is as follows:

$$N_{K/F}(\alpha) = \det A_\alpha.$$

Write A_α as .

$$\begin{bmatrix} 0 & 0 & 0 & \cdots & 0 & 0 & -\alpha_0 \\ 1 & 0 & 0 & \cdots & 0 & 0 & -a_1 \\ 0 & 1 & 0 & \cdots & 0 & 0 & -a_2 \\ \cdot & \cdot & \cdot & \cdots & 0 & 0 & -a_1 \\ \cdot & \cdot & \cdot & \cdots & 0 & 0 & \cdot \\ 0 & \cdot & \cdot & \cdots & 1 & 0 & \cdot \\ 0 & 0 & 0 & \cdots & 0 & 1 & -a_{n-1} \end{bmatrix}.$$

This will show that

$$N_{K/F}(\alpha) = \det A_\alpha = (-1)^n a_0.$$

Writing the splitting field over $\overline{\mathbf{Q}_p}$, then

$$f(\alpha) = \Pi_{i=1}^n (x - \alpha_i),$$

where $\alpha = \alpha_1$ and $\alpha^i = \alpha_i$ for $i = 1, ..., n$.

We note that since it follows that the norm in question is the standard norm of K to F, then $N_{K/F}(k) = k^{(\sigma^n - 1)/(\sigma - 1)}$, for all $k \in K$, writing K as a matrix field over F, is

$$N_{K/F}(\alpha) = \alpha^{1+\sigma+...+\sigma^{n-1}} = \Pi_{i=0}^{n-1} \alpha^{\sigma^i} = \Pi_{j=1}^n \alpha_i = (-1)^n a_0.$$

Now define the ord_p map

$$ord_p(k) = -\log_p |k|_p.$$

$$|\alpha|_p = \left| N_{K/F}(\alpha) \right|_p^{\frac{1}{n}},$$

for all $k \in K$. Therefore, $ord_p(\alpha) = -\log_p |\alpha|_p = -\log_p |N_{K/F}(\alpha)|^{\frac{1}{n}} = -\frac{1}{n}\log_p$ $|N_{K/F}(\alpha)| = -\frac{1}{n}\log_p |(-1)^n a_0|$.

So, for $\alpha \in K$, the image of K under the ord_p map is contained in the set $\frac{1}{n}\mathbf{Z}$. This image set is an additive subgroup and hence is of the form $\frac{1}{e}$, where e in an integer dividing n.

DEFINITION 118. *The integer e is called the 'index of ramification'. If $e = 1$, K is said to be an 'unramified extension', if e divides n and is not 1 or n, K is a 'ramified extension' and if $e = n$, K is a 'totally ramified extension'.*

The main theorem concerning totally ramified extensions is:

THEOREM 146. *([139] Theorem (4.14)) If K is a totally ramified extension of $\mathbf{Q}_p$ with $K = \mathbf{Q}_p(\alpha)$ then α satisfies an Eisenstein equation over $\mathbf{Z}_p$; that is, the monic irreducible polynomial for α, $f(X) = \Sigma_i^n a_i X^i$ of degree $e = n$, has coefficients $a_i \equiv 0 \mod p$ for $i \neq 0$ and $a_0 \not\equiv \mod p^2$.*

Conversely, if α is a root of an Eisenstein equation over $\mathbf{Q}_p$ of degree $e = n$ then $\mathbf{Q}_p(\alpha)$ is totally ramified over $\mathbf{Q}_p$ of degree e.

7. The quaternion division rings $(a, b)_{\mathbf{Q}_p}$

So, to find a quaternion division ring $(a, b)_{\mathbf{Q}_p}$, we first need to find an Eisenstein equation over $\mathbf{Q}_p$ of degree 2. Notice for any extension field K of $\mathbf{Q}_p$ of degree 2, either the index of ramification is either 1, and the extension is unramified, or the extension is totally ramified. Previously, for $\mathbf{Q}_5$ it was seen that $x^2 + 1$ satisfied Hensel's lemma for $x = 2$ and thus $Q(i)$ may be embedded in $\mathbf{Q}_5$, and we saw that $(5, i)_{\mathbb{Q}(i)}^{\mathbb{Q}(i)(\sqrt{5})}$ is a quaternion division ring with center $\mathbb{Q}(i)$. Consider $(5, i)_{\mathbb{Q}(i)}^{\mathbb{Q}(i)(\sqrt{5})} \otimes_{\mathbb{Q}(i)} \mathbf{Q}_5$. Since $x^2 + 5$ is an Eisenstein polynomial in $\mathbf{Q}_5$, there is a solution and $\mathbf{Q}_5(\sqrt{5})$ is totally ramified. Then question is whether $i = 2 + 5\mathbf{Z}_p$ is a non-norm element, and this has been proved in the section on Hensel's lemma. Therefore, $(5, i)_{\mathbb{Q}(i)}^{\mathbb{Q}(i)(\sqrt{5})} \otimes_{\mathbb{Q}(i)} \mathbf{Q}_5 = (5, i)_{\mathbf{Q}_5}^{\mathbf{Q}_5(\sqrt{5})}$ is a quaternion division ring with center $\mathbf{Q}_5$. Note that since this is a quaternion division ring, then $(5, i)_{\mathbb{Q}(i)}^{\mathbb{Q}(i)(\sqrt{5})}$ also is a quaternion division ring.

Similarly, take a polynomial $x^2 + a$, where $a \not\equiv 0 \mod p^2$ then there is a root in $\overline{\mathbf{Q}_p}$ and a field extension.

Much more generally, using aspects of quadratic forms, all quaternion division rings over $\mathbf{Q}_p$ may be completely determined.

The norm of absolute value used in the completion of the rational numbers $\mathbb{Q}$ to the field of real numbers $\mathbb{R}$ may be included as $p = \infty$ and the associated quaternion division rings classified together.

THEOREM 147. *(See Theorem (3.14) of Bayo [13]) For any prime p or $p = \infty$ there is a unique quaternion division ring over $\mathbf{Q}_p$, up to isomorphy (a rational quadratic form is isotropic if it evaluates a non-zero vector(null vector) to 0).*

    (i) For an odd prime p, the quaternion division ring is $(e, p)_{\mathbf{Q}_p}$, where $e \in \mathbf{Z}_p^*$ is   *an arbitrary element that reduces to a nonsquare modulo p.*

    (ii) For $p = 2$, the quaternion division ring is $(5, 2)_{\mathbf{Q}_2}$.

(iii) For $p = \infty$, it is $(-1, -1)_{\mathbb{R}}$, the real Hamiltonians.

The quaternion division rings over the field $\mathbb{Q}$ of rational numbers are connected as follows: If $(a, b)_{\mathbb{Q}}$ is a quaternion division ring with center $\mathbb{Q}$, form $(a, b)_{\mathbb{Q}} \otimes \mathbf{Q}_p \simeq (a, b)_{\mathbf{Q}_p}$ is either a quaternion division ring extension or isomorphic to the set of 2 by 2 matrices over $\mathbf{Q}_p$. How to tell when there is a division ring extends beyond the scope of this text, into the theory of quadratic forms and requires two or three deep results, which are extremely useful in the study of central division algebras and shall be stated here. The quaternion division rings with center $\mathbb{Q}$ are equivalent to certain nondegenerate ternary quadratic forms, which may be extended to quadratic forms over $\mathbf{Q}_p$.

THEOREM 148. *(Ostrowski's Theorem) Any non-trivial norm on $\mathbb{Q}$ is equivalent to precisely one of the p-adic norms, including $p = \infty$.*

THEOREM 149. *(Dirichlet's theorem on arithmetic progressions) Given a, n integers, which are coprime and $n \neq 0$, there are infinitely many primes p with $p \equiv a$ mod n.*

This is the theorem that is used by various of the mathematicians whose work is surveyed in the next section to generate appropriate Galois extensions, usually with $p \equiv -1 \bmod q^{\alpha}$, for p and q coprime.

THEOREM 150. *(Hasse-Minkowski Theorem). Let $\mathcal{L}$ be a rational quadratic form. Then $\mathcal{L}$ is isotropic (evaluates a non-zero vector(null vector) to 0) if and only if the extensions to $\mathbf{Q}_p$ are isotropic for every $p \in P$ (the set of positive primes union ∞).*

The interested reader is directed to J.P. Serre [131] for additional information and the proofs of these results.

Part 12

Extensions of skewfields

In this part, we try to provide some of the ways that skewfields may be extended. The quaternion division ring extensions are of interest with their connection to derivable nets. There are also extensions of cyclic division rings and general division rings that are of finite dimension over a field.

CHAPTER 14

Quaternion division ring extensions

Let $(a,b)_K^\alpha$ be any quaternion division ring L. An important part of the extension of skewfields theorem of (2,4,3) considers the fields H such that L can define a derivable net sharing the affine space of a right classical H-regulus net, where the two derivable nets share three components. Let $K(\theta)$ be the Galois quadratic extension of K, required for the construction of L, where σ is the associated involution in the Galois group over K. So to be able to represent L as a derivable net (necessarily of type 0), it is required that $H = K(\theta, \tau)$, containing $K(\theta)$ and $K(\tau)$ as distinct quadratic subfields of H such that σ extends to an automorphism group of H fixing $K(\tau)$ pointwise. It has been shown in (2,4,3) that there is a central extension $(a,b)_{K(\tau)}^\alpha$ of $(a,b)_K$, which is a central Galois extension in characteristic $\neq 2$. Because of the inherent problems with characteristic 2, these cases will be considered separately. Ultimately, we shall be interested in twisted T-copying $(a,b)_K^\alpha$ to $(a,b)_{K(z)}^\alpha$, where $K(z)$ is the rational function field in an independent variable z. The Galois quadratic field extension $K(\theta)$ of K defining the necessary automorphism σ that enables the matrix representation will then be extended to $K(\theta)(z)(\sqrt{z}) = K(z)(\theta, \sqrt{z})$, which is a Galois extension of $K(z)$ when the characteristic is $\neq 2$. For example, if $\theta = \sqrt{a}$ then we obtain a separable splitting field $(x - \sqrt{a})(x + \sqrt{a})(x - \sqrt{z})(x + \sqrt{z}) = (x^2 - a)(x^2 - z)$. When the characteristic is 2, it is necessary to make an additional field extension to construct a central Galois quadratic extension.

In this part, we consider what is called the rational function field extensions of skewfield planes. These extensions are also called twisted T-copies of the skewfield spreads and grew out of the idea of T-copies of finite spreads. If only concerned with skewfields, twisted T-copies are covered in Jacobson [64] (See below in the remark). We also are concerned with the very general theory of the construction of translation planes from skewfield planes, where the skewfield admits an automorphism. In particular, we are most interested in what happens to the quaternion division ring planes.

In parts 7 and 8, it was shown that the ideas of 'lifting' a finite translation plane of dimension 2 (Hiramine, Matsumoto and Oyama [62]) are not only valid for infinite translation plane over fields, the extension carries over to noncommutative skewfields. Here is the abbreviated version of the main result.

THEOREM 151. *The Main Theorem on Lifting Skewfields. Let $F(\theta)$ be a central Galois quadratic extension of a skewfield F, generated by an irreducible quadratic polynomial $x^2 \pm \alpha x - \beta$, where $\alpha, \beta \in Z(F)$, the center of F. The $\pm$ indicates that the form shows that both the plus and minus polynomials are irreducible if only one of them is irreducible. It follows that $Z(F(\theta)) = Z(F)(\theta)$, $\theta^2 = \theta\alpha + \beta$, there is*

*an induced involutory automorphism σ fixing F pointwise, where $\theta t + u$ represents
the elements of $F(\theta)$ and $(\theta t + u)^\sigma = (-\theta + \alpha)t + u$. The spread associated with
$F(\theta)$, denoted by $F_{\alpha,\beta}/F$, may be determined with the following set of matrices:*

$$x = 0, y = x \begin{bmatrix} u + \alpha t & \beta t \\ t & u \end{bmatrix}; \forall t, u \in F.$$

*F may be either a field or a non-commutative skewfield. Then there is a semifield
spread in $PG(3, F(\theta))$ of the following form:*

$$x = 0, y = x \begin{bmatrix} w^\sigma & \theta^\sigma s^\sigma \\ s & w \end{bmatrix}; \forall s, w \in F(\theta),$$

*denoted by $F(\theta)_{\theta^\sigma}/F(\theta)$ lifted from $F_{\alpha,\beta}/F$. The existence of $F(\theta)_{\theta^\sigma}/F(\theta)$ implies
that there is a set of semifield spreads*

$$x = 0, y = \begin{bmatrix} w^\sigma & (\theta\delta + \gamma)s^\sigma \\ s & w \end{bmatrix}; \forall s, w \in F(\theta),$$

*denoted by $F(\theta)_{\delta,\gamma}/F(\theta)$, for each pair of elements (δ, γ), for $\delta \neq 0$, $\gamma \in Z(F)$. These
semifield planes are induced by the following set of pre-skewfield spreads $PF_{\delta,\gamma}/F$
of $PG(3, F)$ defined as follows:*

$$x = 0, y = \begin{bmatrix} -\delta u + \gamma t & -\beta\delta t + (\alpha\delta + \gamma)u \\ t & u \end{bmatrix}; \forall t, u \in F.$$

The quaternion division rings are objects of special interest in the theory of non-commutative algebra but actually, translation planes coordinatized by quaternion division rings have not been extensively analyzed.

In particular, these Desarguesian affine planes admit a variety of new and interesting derivable nets. In part 10, the multiple replacement theorem is proved, which says that any Desarguesian plane coordinatized by a central Galois quadratic extension of a skewfield admits a set of mutually disjoint derivable nets that cover the spread except for two components, which are represented in the form $x = 0$ and $y = 0$. We shall see that the construction that we wish to consider in this part involves quasifibrations, which we discussed extensively in part 7. Of particular interest are the quasifibrations that are 'rationally extended' or T-copied by a transcendental extension of a finite spread. This means the finite matrix spread set of dimension 2 over $GF(q)$ is mimicked in form over $GF(q)(z)$, the rational function field over $GF(q)$. If the new differences of the matrices over $GF(q)(z)$ have non-zero determinants, the rational extension is a quasifibration or a spread. These ideas are important in the study of skewfield translation planes. Here we are also interested in twisted T-copies.

DEFINITION 119. *For arbitrary matrix spreads over fields K that admit automorphisms acting on the elements, such elements, such as u^σ, might not be recognized within a rational function field extension. In this setting, we allow $z^\sigma = z$ and extend the automorphism to $K(z)$, in this manner; $(\sum \alpha_i z)^\sigma = \sum \alpha_i^\sigma z)$. We shall call this method of possibly constructing a spread in $K(z)$ a 'twisted T-copy'. When the automorphism may be recognized, such as u^q where q is finite, there are possibly T-copies and twisted T-copies available. We note that T-copies are proper quasifibrations, whereas it is possible that twisted T-copies are spreads.*

REMARK 44. *Let L be a non-commutative skewfield and z an independent variable, then it is possible to form the ring $L[z]$, which then has a skewfield of fractions, provided, in general, that the ring is a principal ideal domain Cohn [29] (2.1.5). This construction might be considered analogous to T-copying, or twisted T-copying when considering skewfields. In the case that we are considering, that of skewfields that are finite-dimensional over their centers $Z(L)$, both of these constructions produce skewfields with centers $Z(L(z)) = (Z(L))(z)$.*

The ideas of T-copying and twisted T-copying were originated using geometric concepts with the goal of constructing quasifibrations. For Hughes-Kleinfeld look-alikes, when $\sigma\rho = 1$, twisted T-copying produces spreads. Further, when considering skewfields, one may see these ideas in Jacobson ([64] 1.9, 1.9.1 p. 33), where the twisting or allowing the automorphism of K to fix z in the rational function field is part of the construction, and the idea also occurs in Cohn [29].

We shall see these extensions occur, more generally, over division rings finite-dimensional over their centers, after a short discussion of Ore's method in the next section. The reader interested in ideas of Cohn and Ore is also directed to Jacobson chapter I, on skew polynomial rings which are PID and provide interesting generalizations of cyclic algebras. These also may be studied when considering derivable nets of type 1, which involve σ-derivations, where σ is an automorphism.

These ideas basically drive the present material, as we have proved the following fundamental result as part of the theorem on Hughes-Kleinfeld look-alikes in part 7.

THEOREM 152. *(**Rational Extension of** $(a,b)_F^\alpha$) Given any quaternion division ring Q, $(a,b)_F^\alpha$, where F is the center of Q, then there is an extension of $(a,b)_F^\alpha$ to $(a,b)_{F(z)}^\alpha$, where $F(z)$ is the rational function field extension of F. This extension is the twisted T-copy of $(a,b)_F$, or also the 'rational function field extension'.*

The matrix version is as follows: If we have a quaternion division ring $Q = (a,b)_K^\alpha$ in matrix form, the representation is as follows: Assume that there is a quadratic extension of $Z(Q)$, given by the irreducible polynomial $x^2 \pm \alpha x - a$, for $\alpha, a \in Z(Q)$, and there is a matrix spread set of $PG(3, Z(Q)(\theta))$, such that $\theta^2 = \alpha\theta + a$, that induces an involutory automorphism $\sigma : \theta \to -\theta + \alpha$. The matrix form of $(a,b)_F^\alpha$ is

$$x = 0, y = x \begin{bmatrix} u^\sigma & bt^\sigma \\ t & u \end{bmatrix}; t, u \in F(\theta).$$

Then the matrix spread set for the twisted T-copy extends

$$(a,b)_F^\alpha \text{ to } (a,b)_{F(z)}^\alpha,$$

$$x = 0, y = x \begin{bmatrix} u^\sigma & bt^\sigma \\ t & u \end{bmatrix}; t, u \in F(z)(\theta); b \in F(z)(\theta) - (F(z)(\theta))^{\sigma+1}.$$

The idea of the theorem on rational extensions, listed above, was conceived in Johnson and Jha [82], where it was proved that a quasifibration that is also a skewfield also becomes a spread, and we have seen this result in part 8. This reduces the problem of extending skewfields of finite dimension over their centers to proving that the certain matrix putative quasifibration sets have non-zero determinants.

That is to say, the material we are considering conceivably has extensions to cyclic division rings. In fact, the extension process of cyclic division rings using subplane covered nets extends the theory of derivable nets and is covered in (11,13, 1&2).

Consider again the construction of quasifibrations using T-copies. For almost all of the known quasifibrations that may be so constructed, we have noted in part 7 that every Galois tower of quadratic extensions using these quasifibrations as base geometries, has non-spread quasifibrations at each link in the tower. Hence, there are infinitely many mutually non-isomorphic quasifibrations. There is a set of quasifibrations related to the Hughes-Kleinfeld semifield planes that always have a proper quasifibration T-copy and, also furthermore, the twisted T-copies are always spreads.

In the following section, we show that we may rationally extend any quaternion division ring simply by using the 'form' of the matrix sets, as opposed to a proof using the collineation groups inherent in a skewfield translation plane.

Since we may extend any quaternion division ring $(a, b)_K$ to $(a, b)_{K(z)}$, the quaternion division ring may be further extended using algebraic extensions as well as transcendental extensions (multiple commuting independent variables). When we obtain central quadratic Galois extensions, the lifting theory may be employed to construct new semifield planes in 3-dimensional projective spaces over non-commutative skewfields.

We are also able to contribute to the set of quaternion division rings of characteristic 2 by extending the ideas of Conrad [34], to construct a variety of additional quaternion division rings $(\frac{\beta}{\alpha^2}, f(z))_{GF(q)(z)}^{GF(q^2)(z)}$, where K is a field of characteristic 2 and $x^2 + \alpha x + \beta$ is irreducible over K, and $f(z)$ is a polynomial of odd degree.

It may be noted that when quaternion division rings are considered as derivable nets, there are additional new interesting translation planes constructed by the derivations; as quadratic extensions $(a, b)_{K(\tau)}^{\alpha}$ of $(a, b)_K^{\alpha}$ determine derivable nets in the Desarguesian translation plane coordinatized by $(a, b)_{K(\tau)}^{\alpha}$.

COROLLARY 46. *Any quaternion division ring over $F(\theta)(x_1)$ may be extended to a quaternion division ring over $F(\theta)(x_1, x_2, ..., x_n)$, where x_i, $i = 1, 2, ..., n$ are commuting independent variables.*

COROLLARY 47. *If K is an extension field of $F(\theta)$ that admits σ as an automorphism, by finitely many algebraic or transcendental extensions, then any quaternion division ring over $F(\theta)$ (the center is Z), extends to a division ring over K. That is, $(a, b)_F$ extends to $(a, b)_{K^-}$, where $K^-(\theta) = K$, and $K^- = Fix \langle \sigma \rangle / K$.*

PROPOSITION 11. *Let F be a field, $F(\theta)$ a Galois extension, and $F(\tau)$ a quadratic extension, where $F(\theta)$ and $F(\tau)$ are not isomorphic. Form $F(\theta)F(\tau) = F(\theta, \tau)$. Then $F(\theta, \tau)$ admits the automorphism σ of order 2 obtained from $F(\theta)$.*

PROOF. If both $F(\theta)$ and $F(\tau)$ are Galois, then $F(\theta, \tau)$ is Galois. Therefore, assume that the characteristic is 2 and $F(\tau)$ is not Galois. It follows that τ is a non-square c in F and $F(\tau) = F(\sqrt{c})$. Therefore, $\{1, \sqrt{c}\}$ is a basis for $F(\theta, \tau)$ over $F(\theta)$. Let $\sqrt{c}r + m$, for all $r, m \in F(\theta)$ define all elements of $F(\theta, \tau)$. Define a putative automorphism ρ to fix $F(\sqrt{c})$ pointwise; that is, ρ as follows: $(\sqrt{c}r + cm)^\rho = \sqrt{c}r^\sigma + m^\sigma$. Clearly, this mapping is bijective and additive.

$$((\sqrt{c}r + m)(\sqrt{c}s + t))^\rho = \sqrt{c}(ms + rt)^\sigma + (crs + mt)^\sigma,$$

which is just

$$\sqrt{c}(m^\sigma s^\sigma + r^\sigma t^\sigma) + (cr^\sigma s^\sigma + m^\sigma t^\sigma) = (\sqrt{c}r + m)^\rho(\sqrt{c}s + t)^\rho,$$ as ρ also fixes c.

Finally, the endomorphism ρ is clearly injective by a basis argument and by our definition, this completes the proof of this result   □

So, we see that every quaternion division ring $(a,b)_K^\alpha$ has an extension quaternion division ring $(a,b)_{K(z)}^\alpha$, where $K(z)$ is the rational function field over K. And, if there is a Galois tower of length 2 over K of quadratic extensions $K(\theta)(\tau)$, then $K(\tau)$ and $K(\theta)$ and $K(\theta\tau)$ are quadratic subfields and thus there exist an extension of $(a,b)_K^\alpha$ to $(a,b)_{K(\tau)}^\alpha$, by assuming the fixed field of the non-trivial automorphism of $K(\theta)$ fixes $K(\tau)$ in $K(\theta,\tau)$. For example, let characteristic $\neq 2$ and let θ be a non-square in K but all non-squares in K have formal square roots in $K(\theta)$. All irreducible quadratic polynomials in K split in $K(\theta)$. For example, this would be the case when $K(\theta)$ is algebraically closed. The same would be true whenever K has a unique quadratic extension.

As promised in (2,3,4), we now are able to complete the discussion on transversals to derivable nets.

REMARK 45. *We have shown that the structure of derivable nets can be represented as either type 0, 1, 2, 3 derivable nets (2,4,1). In (3,5,11 & 12), there are comments of what happens when the derivable nets are of type 1; or 2. Also, the interested reader could now ask what dual translation planes containing type 0 derivable nets would look like (3,5,1&2).*

THEOREM 153. *In (8,10,2), it is shown that given any quaternion division ring* $(a,b)_K^\alpha$ *that admits a central Galois quadratic extension* $(a,b)_{K(\rho)}^{K(\rho)(\theta)}$, *there is an associated semifield spread in* $PG(3, (a,b)_{K(\rho)}^{K(\rho)(\theta)})$.

Consequently, the dual translation plane of the semifield field spread contains the classical pseudo-regulus net relative to $(a,b)_{K(\rho)}^{K(\rho)(\theta)}$.

1. Ore's method of localization of a ring

In this section, we shall give two generalizations of the geometric method of twisted T-copies, first using Ore's method of localization of a ring. We formalize the definition of an S-inverting homomorphism adapted from Cohn's text.

DEFINITION 120. *Let R and R' be rings and S a subset of R. A homomorphism $f : R \to R'$ is said to be 'S-inverting' if for each $s \in S$, sf is invertible, in the sense that it has a two-sided inverse. The elements sf are 'units'.*

PROPOSITION 12. *(Cohn [29] 1.3.1 p. 14) Given a ring R and a subset S of R, there exists a ring R_S with an S-inverting homomorphism $\lambda : R \to R_S$, which is universal S-inverting, in the sense that, for each S-inverting homomorphism $f : R \to R'$, there is a unique homomorphism $g : R_S \to R'$ such that $f = \lambda g$.*

We shall give the background for Ore's method directly from Cohn [29]. Note that the definition of 'localization of a ring R at a multiplicative subset S' is defined within the statement the following theorem.

THEOREM 154. *(Cohn* [**29**] *(1.3.2) p.15) Let R be a ring and S a set such that*
(D.1) S is multiplicative.
(D.2) For any $a \in R$ and $s \in S$, $aR \cap sR \neq \phi$.
(D.3) For any $a \in R$ and $s \in S$, $sa = 0$ implies $at = 0$ for some $t \in S$.
Then the universal S-inverting ring/'the localization of R at S, R_S' may be
constructed as follows:
On $R \times S$ define the relation $(a, s) \sim (a', s')$ whenever $au = a'u'$, $su = s'u' \in S$
for some $u, u' \in R$.
This is an equivalence relation on $R \times S$ and the quotient set $R \times S/ \sim$ is R_S. In
particular, the elements of R_S may be written as fractions $a/s = as^{-1}$. The kernel
$\lambda = \{a \in R; at = 0 \text{ for some } t \in S\}$.

COROLLARY 48. *(*[**29**] *(1.3.3) Let R be an integral domain such that $aR \cap bR \neq 0$*
for $a, b \in R^*$.
Then the localization of R at R^* is a skewfield (Cohn uses the word 'field' where
this text uses 'skewfield') F and the universal homomorphism $\lambda:R \to F$ is an em-
bedding.

We have mentioned type 1 derivable nets in part 2 and 3. We continue with
the discussion of skew polynomial rings over division rings, for which Ore's method
becomes very useful.

2. Skew polynomial rings

This section depends on Cohn [**29**], chapter 2 and Jacobson [**64**] chapter 1. Here,
we shall briefly mention and consider skew polynomial rings over division rings D
that admit an automorphism σ and a σ-derivation δ (twisted derivation/derivative
function when $\sigma = 1$, see (3,5, 10&11)).

DEFINITION 121. *Define a 'skew polynomial ring $D[t; \alpha, \delta]$' where t is an inde-*
terminant, the elements of the ring are left polynomials $\Sigma_{i=0}^{n} \alpha_i t^i$ in t with coefficients
α_i in D, $i = 0, 1, .., n$ as follows:

$$ta = a^\sigma t + \delta a, \forall a \in D.$$

By induction, the expressions for $t^i a$ may be given. This ring is shown to have
a left division algorithm in Jacobson, which shows that the ring is a left PID
(principal ideal domain). Hence, the skewfield of fractions exists by Ore's method,
therefore creating a division ring into which $D[t; \alpha, \delta]$ may be embedded. These
skewfields are important for the determination of natural locations of derivable nets
of type 1. The following is a possible semifield spread/Desarguesian spread over a
field admitting a derivation, if and only if the covering property may be proved.

Let c be a non-square in F, F of characteristic 2 and $A(v) = (\delta/\delta\sqrt{c}) \, v$, (the
partial derivative with respect to $\sqrt{c}$), then $A(A(v)) = 0$. Furthermore, the following
is an additive quasifibration.

$$x = 0, \ y = \begin{bmatrix} w + A(v) & A(w) + \sqrt{c}v \\ v & w \end{bmatrix} ; \forall w, v \in F(\sqrt{c}).$$

The theory of skew polynomial rings due to Ore (Cohn [**29**] chapter 2, Jacobson
[**64**], chapter 1) concerns the following rings. Let D be a division ring, σ an automorphism of D and δ a σ-derivation. Then $R = D[t; \sigma, \delta]$ becomes a left and right

principle ideal domain. Since PIDs satisfy the Ore condition and therefore admit a skewfield S of fractions, any such skewfield then defines a derivable net by considering $PG(3, S)$, and applying the contraction procedure, there is an associated derivable net. Recall over fields, it was shown in the section on type 1 derivable nets that, for fields, if a σ-derivation has the form $\delta(u) = u^\sigma c - cu$, then δ is identically zero, by basis changing.

DEFINITION 122. *A σ-derivation over a skewfield D is an 'inner σ-derivation' if and and only if there is an element c so that $\delta(u) = u^\sigma c - cu$, for all $u \in D$.*

Consider a basis change $t' = ut + v$, where $u \neq 0$, v in $R = D[t; \sigma, \delta]$, $t = u^{-1}t - u^{-1}v$, then notice first that defining δ' as

$$\delta' a = (ua^\sigma u^{-1}v - va + u\delta a),$$

and if σ is an inner automorphism kak^{-1}, choosing $u = k^{-1}$ provides that $\sigma = 1$ by changing basis $\mathfrak{B}$ to $u^{-1}\mathfrak{B}u$. Then $\delta' a = -av + va + u\delta a$, and if $v = 0$ then we have a differential polynomial.

Furthermore, considering the general situation, if $\delta a = a^\sigma c - ca$

$$-ua^\sigma u^{-1}v + va + u\delta a = -ua^\sigma u^{-1}v + va + a^\sigma c - ca$$

and choosing $u = 1$ and $v = c$, shows that we may choose $\delta = 0$. Hence, we have:

THEOREM 155. *If in $R = D[t; \sigma, \delta]$, σ is an inner automorphism, then with a basis change, $\sigma = 1$. And, if δ is an inner derivation, then with a basis change, $\delta = 0$.*

DEFINITION 123. *We note that if there is a recoordination of t as $t' = tu + v$ for u not zero and v in D, allowing that σ becomes 1, this skew polynomial ring is said to be a 'differential polynomial ring'. Similarly, when δ becomes 0 by a change of variables, the ring is called a 'twisted polynomial ring'.*

The argument that $\sigma = 1$ over fields in type 1 derivable nets rested on a contradiction that the σ-derivation over a field must have a non-zero vector v_0 so that $v_0 \neq v_0^\sigma$, obtain the equation $v_0 k - kv_0^\sigma = 0$ for $k \neq 0$. Essentially the same argument shows the following:

THEOREM 156. *If in $R = D[t; \sigma, \delta]$ the restriction of σ to the center of D, $Z(D)$, does not fix $Z(D)$ elementwise then δ may be taken to be 0 and we have a twisted polynomial ring.*

THEOREM 157. *If D has finite dimension over the center $Z(D)$ then $R = D[t; \sigma, \delta]$ is either a differential polynomial ring or a twisted polynomial ring.*

PROOF. If σ fixes $Z(D)$ pointwise, then using the Skolem-Noether theorem, σ is an inner automorphism and therefore $\sigma = 1$ by a basis change and we have a differential polynomial ring. If σ does not fix $Z(D)$ pointwise, we have a twisted polynomial ring by the previous result. □

Take any skew polynomial ring and form the associated skewfield of fractions. Considering this as a right classical pseudo-regulus net, we may determine a left type 1 derivable net.

The material of type 1 derivable nets and Chapter 2 of Cohn [**29**] have considerable overlap. The theorems on Ore domains, in particular theorem (2.3.1) and the exercises on pages 65, 71 of Cohn are relevant to the ideas presented in this book. In Cohn, we note that equation (9) on page 50, is identical to our 'twisted derivative function' definition.

- Note that above it was mentioned that over a field, a twisted derivative function, since corresponding to a derivable net of type 1, forces the automorphism $\sigma = 1$. Theorem 2.16 of Cohn also speaks to this situation. Also, the automorphisms σ and σ-derivations δ of a division ring place considerable restrictions on σ and δ. There are examples of onto inner derivations in Lazerson [**109**]. Also, in Cohn [**29**] page 71, exercises 4,5,6.

PROBLEM 77. *It is an open problem to find a derivable net with twisted derivative function. There would need to be a division ring D of infinite dimension over the center $Z(D)$ with a non-trivial outer automorphism σ and an outer σ-derivation δ that maps $Z(D)$ to 0 (See Cohn [**29**] , (2.13), and exercise #4 p. 56).*

*There is a deep theory on various types of derivations in rings, but the analysis of infinite dimensional algebras admitting both outer automorphisms σ and outer σ-derivations might not lie within this theory. For the interested reader, here are some articles on derivations in rings, the last of which is a historical account, V.K. Kharchenko [**105**], Chen-Lian Chuang [**27**], and M. Ashraf, S.Ali, C. Haetinger [**3**].*

3. Generalized cyclic algebras

In this section, we define generalized cyclic algebras and division rings, which are related to skew polynomial rings. Generalized cyclic division rings are defined using quotients of two-sided maximal elements producing associated maximal ideals within principal ideal domains R, so that R/I is a simple ring. Assume that $\sigma \neq 1$ and δ is not inner, then $R = D[t; \sigma, \delta]$ is a *PID*, which is of infinite dimension over $Z(D)$. Considering the finite-dimensional cases lead to considering the twisted polynomial rings and the differential polynomials. In this section, the twisted polynomial rings are considered and the centers of these rings are determined. The reader interested in analogous results for differential polynomials is directed to Jacobson [**64**].

Following p. 19 of Jacobson, consider then $R = D[t; \sigma]$, a skew polynomial ring such that D is finite-dimensional n over the center C, and such that the automorphism σ restricted to C is of finite order r. By the Skolem-Noether theorem, as σ^r is an algebra automorphism, σ^r is an inner automorphism $I_u : a \to uau^{-1}$. Normalizing u so that $\sigma u = u$, and letting $F = C \cap Fix\sigma$, the center $Z(R)$ of R is the set of polynomials

$$\gamma_0 + \gamma_1 u^{-1}t^r + \gamma_2 u^{-2}t^{r-1} + ..., \text{ for } \gamma_i \in F.$$

Then $Z(R) = F[z]$ where $z = u^{-1}t^r$. We may replace u by γu for $\gamma \neq 0$ in F. It follows that $t^r - u$ is a two-sided maximal element $(R(t^r - u) = (t^r - u)R)$, so that $R(t^r - u)$ is a maximal ideal and $R/R(t^r - u)$ is, by definition, a '**generalized cyclic algebra**', which is denoted by $(D, \sigma, \gamma u)$. If $D = C$, $u = 1$, a cyclic algebra is obtained. In general, where $(D, \sigma, \gamma u)$ becomes a division algebra of degree nr; $[(D, \sigma, \gamma u) : Z(R)] = (nr)^2$.

The following theorem is then of fundamental interest.

THEOREM 158. *(Jacobson* [**64**] *(1.4.6.) Let $R = D[t; \sigma]$, where D is a division ring of finite dimension n over its center C. Then $(D, \sigma, \gamma u)$, the localization R_S for S the monoid of non-zero elements of $Z(R) = F[z]$, $z = u^{-1}t^r$, is a division ring whose center is the field of fractions $F(z)$ of $F[z]$. Moreover, the map of R into R_S is canonical.*

4. Extending division rings using rational function fields

We have mentioned that our twisted T-copies include the quaternion division ring extensions. In Jacobson [**64**], it is shown that any division ring D that is finite dimension over a field F, may be extended to a division ring over a rational function field, as follows:

PROPOSITION 13. *Jacobson* [**64**] *(1.9.1), (the version presented basically reinterprets material on pp33). If D is a finite-dimensional division ring of degree n over a field F, let $E = F(s_1, s_2, ..., s_{k+1})$, where the s_i are commuting indeterminants, then D may be extended to a division ring $D_E = D \otimes_F E$ of degree n over E.*

(1) Furthermore, if D is a central division algebra over F then D_E/E is a central division algebra.

(2) If D admits an automorphism σ, define $\sigma s_i = s_i$ to extend σ to an automorphism of D_E/E.

(3) Let τ be any permutation of $\{s_1, s_2, ...s_{k+1}\}$, then τ defines an automorphism of $F(s_1, ..., s_{k+1})$ and is Galois over $Fix_E\tau$. D_E admits τ as a automorphism, where τ fixes D elementwise, so $D \subset Fix_{D_E}\tau$.

Using (2) and the multiple replacement theorem for the Desarguesian plane Σ coordinatized by D, with the Desarguesian plane Σ_σ, we have a set of mutually disjoint replacement partial spreads. The non-trivial sets of replacements construct translation planes that may be written over a vector space over $Fix\tau \cap Z(D)$, where $Z(D)$ is the center of D.

Using (3), there is a wide variety of possible translation planes that may be constructed using the multiple replacement theorem. Since we are interested in derivable nets, assume that τ has order 2 (for example, $\tau : s_1 \longleftrightarrow s_2$, and $s_i \to s_i$, for $i > 2$). Then $Fix_E\tau = F(s_1s_2, s_3, ..., s_{k+1})$. If F is the center of D then $E = F(s_1, ..., s_{k+1})$ is the center of D_E. Then $Fix_{D_E}\tau = D \otimes_F F(s_1s_2, s_3, ..., s_{k+1})$ is a sub-division algebra of index 2 in D_E, since we have a Galois extension E/F. Therefore, considering Σ and Σ_τ, using the notation of the multiple replacement theorem, we have a 2-dimensional D-vector space (left and right) and $(y = x) \cap (y = x^\tau)$ are intersections of 1-dimensional D_E-subspaces. Then the ambient vector space is now 4-dimensional over $Fix_{D_E}\tau$. So, $y = x$ and $y = x^\tau$ (See again the notation of the multiple replacement theorem) are 2-dimensional subspaces over $Fix_{D_E}\tau$ and $y = x \cap (y = x^\tau)$ is 1-dimensional over $Fix_{D_E}\tau$. This means that $y = xm$, for $m \in D_E$ becomes a left 2-dimensional $Fix_{D_E}\tau$-subspace and $y = mx^\tau$ becomes a right 2-dimensional $Fix_{D_E}\tau$-subspace. Therefore, we have a set of mutually disjoint pseudo-regulus derivable nets relative to $D \otimes_F F(s_1s_2, s_3, ..., s_{k+1})$ in the Desarguesian plane that becomes a spread in $PG(3, Fix_{D_E}\tau)$. Then the constructed translation planes are of dimension 8 over $F(s_1 s_2, s_3, ..., s_{k+1})$, which is the intersection $Fix_{D_E} \tau \cap (Z(D_E) = E)$, with spreads in $PG(7, F(s_1 s_2, s_3, ..., s_{k+1}))$. We state this formally and more generally. In the following, we continue with the same assumptions.

THEOREM 159. *If there is an automorphism ρ of $F(s_1, ..., s_{k+1})$, which cyclically permutes the set $\{s_1, ..., s_{k+1}\}$ and induces a Galois group that is cyclic and of order $k+1$ over $Fix_E \rho$, then $D \otimes_F E$ admits ρ as an automorphism fixing D pointwise and $Fix_{D_E} \tau$ is a sub-division algebra of index $k+1$ in D_E. This implies that we have a $2(k+1)$-dimensional vector space over $Fix_{D_E} \tau$ and the associated Desarguesian plane has spread in $PG(2(k+1) - 1, Fix_{D_E} \tau)$. In this case, the associated proper multiply replaced spreads are in $PG(2(k+1)^2 - 1, Fix_{D_E} \tau \cap Z(D_E))$.*

PROBLEM 78. *Let D be a division ring with center F and form $E = F(z_1, z_2, z_3, z_4)$. Let $\sigma : z_1 \leftrightarrow z_2, z_3 \leftrightarrow z_4$. Use the multiple replacement theorem to determine a set of multiply replacement sets in D_E. What are the replacement sets? In what projective space are the proper multiply replaced spreads contained?*

We shall revisit this technique in another section on quaternion division ring extensions.

Therefore, we have:

COROLLARY 49. *Consider also any cyclic division ring $(a, b)_{\varsigma_n K}$, containing $K(\theta_1)$, a cyclic Galois extension of K of degree n, where n is invertible, K is any field that contains a primitive n-th root of unity ς_n, and where the characteristic of K does not divide n. Let $K(\theta_1)(t)$ denote the rational field of fractions over $K(\theta_1)$, with independent variable t.*

THEOREM 160.

(1) Then $(a, b)_{\varsigma_n K}$ extends to $(a, b)_{\varsigma_n K(t)}$.

(2) If the characteristic is 2 (or more generally any characteristic) and $(a, b)_K^\alpha$ is a quaternion division ring, then there is an extension to $(a, b)_{K(t)}^\alpha$.

(3) Let D be any division ring of finite degree n, over a field $K(\theta_1)$, which is a Galois cyclic division ring of degree n over K, where $Z(D) = K$, then D/K extends to $D/K(t)$.

Using these ideas, there are many interesting division rings that may be constructed. Here are a few such constructions from Jacobson.

PROPOSITION 14. *Jacobson (1.9.2) [64]. Let E be a cyclic Galois field extension of F with Galois group $\langle \sigma \rangle$. Extend E to $E(z)$, where z is an indeterminant and extend σ to $E(z)$ by defining $\sigma z = z$. Then the cyclic algebra $(E(z), \sigma, z)$ is a central division algebra.*

We may provide some constructions of quaternion division rings over $K(z)$, the rational function field of K. The examples are not extensions of other quaternion division rings and are not twisted T-copies or T-copies, but rather a hybrid of both construction types. The idea of this construction is similar to that of a homework problem for even characteristic given in the summer course of K. Conrad [34]. The above example 3, p. 34, of Jacobson [64] is also similar. This idea will also produce central division algebras of degree n, by taking finite fields $GF(q)$ and cyclic field extensions $GF(q^n)$, which will produce cyclic division algebras $(GF(q^n)(z), \sigma, f(z))$, for appropriate elements $f(z) \in GF(q)[z]$. Each of these central cyclic division algebras may be extended by taking rational extension fields of two variables, defining an automorphism on the associated fields.

REMARK 46. *If a polynomial $f(y) = y^2 \pm y\alpha - \beta$ is irreducible over a field F. Assume that $\theta^2 = \theta\alpha + \beta$, and if the characteristic is 2 then $\alpha \neq 0$. Form the rational function field $F(\theta)(z)$ over $F(\theta)$. Then the quadratic extension $F(z)(\theta)$ is Galois and the automorphism σ_θ that maps $\theta \to -\theta + \alpha$ extends and fixes $F(z)$ pointwise, where the Galois quadratic extension of $F(z)$ is $F(z))(\theta)$.*

PROOF. This is immediate, by defining the automorphism of $F(\theta)$ to fix z. □

THEOREM 161. *Given any finite field $F = GF(q)$, and let $g(x) = x^2 \pm \alpha x - \beta$, for $\alpha, \beta \in GF(q)$ denote the irreducible quadratic polynomial that extends F to $GF(q^2) = GF(q)(\theta)$. Let σ denote the associated automorphism $x \to x^q$. For uniformity, assume that $\alpha \neq 0$. Form the rational function field $F(z)(\theta)$. Extend σ to $F(\theta)(z)$ by allowing that $\sigma z = z$. Let $f(z) \in F[z]$, where the degree of $f(z)$ is an odd integer. Then*

$$\begin{bmatrix} u^\sigma & f(z)t^\sigma \\ t & u \end{bmatrix}; \forall t, u \in F(z)(\theta),$$

is a quaternion division ring $(\frac{\beta}{\alpha^2}, f(z))_{GF(q)(z)}^{GF(q^2)(z)}$.

PROOF. We note that the matrix is additive and multiplicative, since σ fixes $F(z)$ elementwise and σ has order 2, and has a 1 and a 0. If we have a quasifibration, it follows that we have a skewfield of degree 2 and hence we have a quaternion division ring. Therefore, it remains to show that the determinants are all non-zero. So, assume that there exists an element $k \in F(z)(\theta) = F(\theta)(z)$ such that $f(z) = k^{\sigma+1}$. Then, there is a polynomial equation of the form

$$l^{\sigma+1} = f(z)m^{\sigma+1}.$$

Since σ fixes z, the left side is a polynomial of degree 2 and the right side is a polynomial of odd degree. Therefore, we have the proof of the theorem. □

COROLLARY 50. *We note that the result above is not stated in the most general form, as any Galois quadratic extension E of F will provide a quaternion division ring$(\frac{\beta}{\alpha^2}, f(z))_{F(z)}^{E(z)}$, for $f \in E[z]$ of odd dimension and σ the induced automorphism of E, with exactly the same proof.*

We note that although this above example appears to be a twisted T-extension, it is not, as z needs to be in subfield $GF(q)(z)$. Then since we have pointed out that we now can form a twisted T-extension, to extend any quaternion division ring plane, we see that we now have the following corollary:

COROLLARY 51. *$(\frac{\beta}{\alpha^2}, f(z))_{GF(q)(z)(t_1,t_2,...,t_k)}$ is a quaternion division ring, for any set of commuting independent variables t_i, for $i = 1, 2, ..., t_k$.*

Consider $(a, b)_K^{K(\theta)}$ and extend to $(a, b)_{K(t_1,t_2)}^{K((t_1,t_2)(\theta)}$, and let τ denote the automorphism of $K(t_1, t_2)$ that interchanges t_1 and t_2. Then $K(t_1, t_2)/Fix\tau$ is a Galois quadratic extension. Now extend $Fix\tau$ and $K(t_1, t_2)$ to $(Fix\tau)(\theta)$ and $K(t_1, t_2)(\theta)$, respectively. Therefore, it follows that $D = (a, b)_{K(t_1, t_2)}^{K((t_1, t_2)(\theta)}$ admits an automorphism ρ of order 2 that fixes $(a, b)_{Fix\rho}^{Fix\rho(\theta)}$ pointwise. So, $D = (a, b)_K^{K(\theta)} \otimes K(t_1, t_2)$ and $Fix\rho = (a, b)_K^{K(\theta)} \otimes Fix\tau$, so $Fix\rho$ is a 2-dimensional right and left D-subspace.

Let Σ denote the Desarguesian plane coordinatized by D, and let Σ_ρ be the corresponding Desarguesian plane as modeled in the multiple replacement theorem. Then $(y = x)$ and $(y = x^\rho)$ are 1-dimensional D-subspaces (left and right), which are 2-dimensional $Fix\rho$ subspaces, so it follows that $(y = x) \cap (y = x^\rho)$ is a 1-dimensional $Fix\rho$ subspace. Therefore, we have a set of mutually disjoint derivable nets which are isomorphic to a pseudo-regulus nets isomorphic to $(a, b)_{Fix\tau}^{Fix\tau(\theta)} = Fix\rho$. We note that we now have a 4-dimensional vector space $V_4/Fix\rho$, so the rest of the assertion follows from the classification of subplane covered nets.

This analysis also works for any division ring, or cyclic division ring to produce Desarguesian translation planes coordinatized by non-commutative skewfields that admit automorphisms ρ of order 2 that define a set of mutually disjoint derivable nets. The derivable nets are all isomorphic to a classical pseudo-regulus net with respect to D. The extension of $(a, b)_{Fix\tau}^{Fix\tau(\theta)}$ to $(a, b)_{K(t_1, t_2)}^{K((t_1, t_2)(\theta)}$ depends only on the extension of $Fix\tau$ to $K(t_1, t_2)$, and hence is a central Galois quadratic extension, the same is true of $(a, b)_{\varsigma_n K}$ or $(K(\theta), \sigma, b)$.

THEOREM 162. *If D is a finite-dimensional cyclic division ring of dimension n over the center F, a quaternion division ring, or a symbol division ring with center F, extend D by $D(t_1, t_2)$. Then there is an automorphism ρ of $D(t_1, t_2)$ of order 2 such that $Fix\rho$ has index 2 in $D(t_1, t_2)$.*

(1) The associated Desarguesian translation plane admits a set of mutually disjoint derivable nets isomorphic to $Fix\rho$, the classical pseudo-regulus derivable net relative $Fix\rho$; the components are 2-dimensional left $Fix\rho$ subspaces and the Baer subspaces are 2-dimensional right $Fix\rho$-subspaces.

   The non-trivial multiply derived translation planes have kernels $Fix\rho \cap Z(D)$.

(2) In the quaternion division ring case, the symbol division ring or cyclic division rings case, the extensions from $Fix\rho$ to D are central Galois quadratic extensions and hence produce infinitely many semifield planes within $PG(3, D)$ by lifting.

REMARK 47. *There are other ways to obtain automorphisms of order 2 of division rings. We illustrate the idea over a quaternion division ring $(a, b)_K$. Extend to $(a, b)_{K(t)}$. Then extend $K(t)$ to $K(t)(\sqrt{t})$. If the characteristic is $\neq 2$, we have an automorphism of $K(t)(\sqrt{t})$, that maps $\sqrt{t} \to -\sqrt{t}$. We then obtain an extension to $(a, b)_{K(t)(\sqrt{t})}$. And, this extension admits an automorphism and therefore satisfies the conditions of the multiple replacement theorem.*

As a corollary to the previous work, Jacobson constructs a symbol division ring, as follows.

COROLLARY 52. *(Jacobson [64] (1.9.3)) Let F be a field containing a primitive n-th root of unity ς_n and assume that the characteristic of F does not divide n. Construct $F(z, w)$, where z and w are commuting independent indeterminants.*

Form the symbol algebra $D = F(z, w) \langle x, y \rangle$, such that

$$x^n = z, y^n = w \text{ and } xy = \varsigma_n yx.$$

Then D is a central division ring over $F(z, w)$.

5. Derivable nets over twisted formal Laurent series

There are extensions of the ideas given here over fields to over non-commutative skewfields admitting automorphisms and derivations. For example, the reader is directed to Jacobson [**64**] (1.10), (1.11) pp. 37-38.

DEFINITION 124. *Let K be a field that admits an automorphism σ. Consider the formal Laurent series $\Sigma_{i=N}^{\infty} a_i x^i$, $\forall a_i \in K$, N an integer, notated by $K((x))$. Define $xt = t^\sigma x$, and use induction to show that $x^j t = t^{\sigma^j} x^j$. This is the non-commutative division ring '$K((x, \sigma))$-formal twisted Laurent series'. Let the fixed point subfield of σ be F. Then, if σ has infinite order, $Z(((K, \sigma))) = F$. If the order of σ is a finite integer e, then $Z(K((x, \sigma))) = F((x))$.*

As an example of a field with automorphism, take the p–adic field $\mathbb{Q}_p$, and consider the Frobenius automorphism σ, of infinite order with fixed field $GF(p)$.

REMARK 48. *For formal twisted Laurent series, we would be interested in the case where the automorphism is of finite order. Certainly if the automorphism has order 2, for $K((\sigma, x))$ then $Fix\sigma$ is a subfield of K so that K is a Galois quadratic extension, then the center is $F((x))$. This would be a quaternion division ring.*

PROBLEM 79. *Similarly, let K be a cyclic Galois extension of degree of a field F, so that $K = F(\sqrt[n]{a})$, by Kummer theory. If σ is an automorphism of order n, then show $F(\sqrt[n]{a})((\sigma, x))$ has center $F((\sigma, x))$, a division ring of degree n.*

If σ has finite order n, there are associated inner automorphisms of finite order dividing n. In particular, if n is even there are very interesting sets of mutually disjoint derivable nets in the associated Desarguesian translation planes. This can be of two types, both type 2, variations of the classical pseudo-regulus net derivable nets and of classical regulus variety. This work will be discussed usually using cyclic division rings and quaternion division rings, but it is interesting that the formal twisted Laurent division rings provide such immediate and rich examples that construct translation planes admitting multiple derivation.

So, using any skewfield $((K, \sigma))$, as a right classical pseudo-regulus net, there are at least type 2 and type 3 derivable nets relative to this skewfield. There may now be associated quaternion division rings masquerading as derivable nets. What about type 1 examples? Are there generalized quaternion division rings in quadratic skewfield extensions?

Morandi, Sethuraman and Tignol [**113**] and Hanke [**61**] have constructed some interesting twisted formal Laurent series division rings. This former work uses the idea of a valuation of a division ring and applications provide a variety of new examples.

Coming back to the question of fields K that admit a unique quadratic extension, consider the field $\mathcal{C}$ of complex numbers and let σ denote the automorphism that maps $i \rightarrow -i$. Then the formal Laurent series $\mathcal{C}((x))$ has a quadratic extension $\mathcal{C}(x))(\sqrt{x})$, where $((\mathcal{C}))$ has the form $\Sigma_{i=n}^{\infty} a_i t^i$, $\forall a_i \in K$.

- More generally, let K^- denote the fixed field of σ in K. Then K is a quadratic extension of K^-. Let $\{x_1, x_2, ..., x_n, ...\}$ be a set of commuting independent variables. Then the multiple rational field extension $K(x_1, x_2, ..., x_n, ...)$ admits σ as an automorphism that fixes $K^-(x_1, x_2, ..., x_n, ...)$ pointwise.

  Choose any x_j and take the quadratic extension

  $$K(x_1, x_2, ..., x_n, ...)(\sqrt{x_j}).$$

  This field also admits σ as an automorphism fixing

  $$K^-(x_1, x_2, ..., x_n, ...)$$

  pointwise. Then $(a, b)_F$ may be extended to the quaternion division ring

  $$(a, b)_{K^-(x_1, x_2, ..., x_n, ...)(\sqrt{x_j})}.$$

  As $(a, b)_{K^-(x_1, x_2, ..., x_n, ...)}$ is a non-commutative skewfield, we may now lift this skewfield since $\sqrt{x_j}$ commutes with

  $$(a, b)_{K^-(x_1, x_2, ..., x_n, ...)}$$

  to construct a semifield spread in $PG(3, (a, b)_{K^-(x_1, x_2, ..., x_n, ...)})$.

  This process can be extended by taking $\sqrt[2^k]{x_i}$, sequentially, each of which constructs a semifield spread in a projective 3-space over a non-commutative skewfield, and all of these are completely original. For characteristic $\neq 2$, all quadratic extensions are Galois and are central extensions of the previous quaternion division ring. Other ways to construct semifield spreads in 3-dimensional projective spaces over non-commutative skewfields, would be to take any quadratic algebraic extension of $F(\theta)$ which then admits σ as an automorphism.

  For characteristic not equal to 2, any Galois extension of dimension 4 of a Galois chain of length 2 will work. Characteristic 2 is not much different and does not require a Galois chain of length 2. Then any series of algebraic or transcendental extensions will work. Additionally, there are infinitely many mutually non-isomorphic new semifield spreads.

PROOF. The only question that remains is whether a series of finite sets of quaternion division rings over multiple rational field extensions converges to a quaternion division ring when $n \to \infty$. But, since all of the various rational expressions use finite sums of powers of the x_i, the number of variables is never the question. The method of construction forces the original quaternion division ring to be a sub-division ring in every succeeding field extension. We give a more complete proof in the characteristic 2 case in the next section. The idea is to form $F(z)(\sqrt{z})$, then find a Galois extension of $F(z, \sqrt{z})$, $F(z, \sqrt{z})(\tau)$ in the characteristic 2 case, and then note that $(a, b)_F$ extends to $(a, b)_{F(z)}$ then to $(a, b)_{F(z, \sqrt{z})(\tau)}$. □

6. Lifted semifields in $PG(3, L(\rho))$

By the lifting theorem, there are semifield translation planes of the following form in $PG(3, L(\rho))$, where $L(\rho)$ is a central quadratic Galois extension of a skewfield L.

$$x \;=\; 0, y = x \begin{bmatrix} w^{\sigma_\rho} & (\rho\delta + \gamma)s^{\sigma_\rho} \\ s & w \end{bmatrix} ; \forall s, w \in L(\rho)$$

for each element $\rho\delta + \gamma \;\in\; Z(L(\rho)) = Z(L)(\rho)$,

$$\delta \;\neq\; 0, \gamma \in Z(L).$$

Recall that these semifields are lifted from the Desarguesian plane coordinatized by $L(\rho)$ and associated pre-skewfields variations. These associated pre-skewfields, when extended by a unit, also are coordinatized by skewfields that represent the same skewfield written by a different generator of $L(\theta)$.

We have ways to construct central Galois quadratic extensions of quaternion division rings, as there are ways to obtain automorphisms of order 2 of division rings. We illustrate the idea over a quaternion division ring $(a, b)_K$. Extend to $(a, b)_{K(t)}$. Then extend $K(t)$ to $K(t)(\sqrt{t})$. If the characteristic is $\neq 2$, we have an automorphism of $K(t)(\sqrt{t})$, that maps $\sqrt{t} \to -\sqrt{t}$. We then obtain an extension to $(a, b)_{K(t)(\sqrt{t})}$. The characteristic 2 case will be completed in the proof of the following result.

COROLLARY 53. *Given any quaternion division ring, there is an extension that has a central Galois quadratic extension and thus may be lifted to a family of semifield translation planes over non-commutative 3-dimensional projective spaces.*

PROOF. Use essentially the same proof as above. Given $(a, b)_K$ then extend to $(a, b)_{K(z)}$. Then, for characteristic $\neq 2$, note that $(a, b)_{K(z)(\sqrt{z})}$ is a central Galois extension of $(a, b)_{K(z)}$. For characteristic 2, it is still a quaternion division ring extension, but not one which is a corresponding Galois extension necessary and used in the construction of the family of semifield spreads. If K has an irreducible polynomial $x^2 + x + \beta$, producing $K(\theta)$ so that $\theta^2 = \theta + \beta$, so $K(z)$ has the same irreducible polynomial producing $K(z)(\theta)$, we claim that

$$(**) : x^2 + x + z + \sqrt{z} + \beta$$

is irreducible over $K(z)(\sqrt{z})$, which is a field, but not a Galois field over $K(z)$. Hence, $\sqrt{z}d + e = 0$ for $d, e \in K(z)$ if and only if $d = e = 0$. So let $x = \sqrt{z}r + t$, where $r, t \in K(z)$. Then assume that

$$x^2 + x + z + \sqrt{z} + \beta = z(r + 1)^2 + \sqrt{z}(r + 1) + t^2 + t + \beta = 0.$$

Then $r = 1$ (from the $\sqrt{z}$-coefficient) and $t^2 + t + \beta = 0$, a contradiction. Let τ be a formal root of $(**)$.

Then a Galois extension of $K(z)(\sqrt{z})$ by τ becomes a central extension $(a, b)_{K(z)(\sqrt{z})(\tau)}$ of $(a, b)_{K(z)(\sqrt{z})}$. This completes the proof. $\square$

SUMMARY 1. *Take any quaternion division ring $(a, b)_K$ of any characteristic. We have shown that there is a quaternion division ring $(a, b)_{K+}$ containing $(a, b)_K$ that admits a central Galois quadratic extension of $(a, b)_{K+}$, $(a, b)_{K+(\tau)}$. Using $(a, b)_{K+(\tau)}$ there is a lifting construction from the quaternion spread in $PG(3, (a, b)_{K+})$ that produces an infinite set of semifield translation planes with spreads in $PG(3, (a, b)_{K+(\tau)})$. Each of these semifield planes are derivable by pseudo-regulus nets that are not regulus nets. The associated derived translation planes are said to have kernels that are blended 'left–right'-kernels in the sense that there is a proper* *decomposition of the spread into left and right $K^+(\tau)$-vector spaces. Thus, the derived spreads are not in $PG(3, (a,b)_{K^+(\tau)})$.*

We also note that this same result may be obtained using symbol division rings or cyclic division rings.

7. A garden of division rings & multiple replacements

We have determined infinite classes of new translation planes. In addition, here we review various other new examples of cyclic division rings, all of which relate to the multiple replacement theorem.

- Certainly for any proper division rings, and a number of cyclic division rings to which our results on the construction of translation planes will apply, at least with respect to inner automorphisms, there are new translation planes. For cyclic division rings of finite degree n, there are always inner automorphisms of order n.

- In (12,14,5) on twisted T-extensions, it is shown that every quaternion division ring $(a,b)_K$ has an extension $(a,b)_{K(z)(w)}$, where $K(z)$ is the rational function field of K such that $(a,b)_{K(z)(w)}$ has a central Galois quadratic extension $(a,b)_{K(z)(w,\tau)}$. Thus, using the central extension to coordinatize a Desarguesian translation plane, there is a set of net semifield spreads in $PG(3, (a,b)_{K(z)(w,\tau)})$.

In the following, we relate the types of division rings created by five groups of researchers, the union of which seems to be a reasonable representation of current research both on spacetime coding theory, aspects of first order logic, and central algebras. We recall the spreads for the multiply replaced translation planes. In general, from cyclic division ring D_τ of degree n, the associated multiply replaced planes have spreads in (1) for outer automorphism types $PG(2\frac{n^3}{d} - 1, Z(D_\tau))$ and (2) for the special inner automorphism types using cyclic division ring E_ρ, $PG(2n^2 - 1, Z(D_\rho))$.

There are a number of coding theory articles relating division rings to spacetime coding. The work of S. Pumplün and T. Unger [129], considers quaternion division rings for their study, and is recommended as a bridge to this area.

- In the article of F. Oggier, J-C. Belfiore, E. Viterbo [114], there are several cyclic division rings constructed. Basically, the degree $n = 2, 3, 4, 5, 6$. The cyclic Galois extension fields that are used over which the matrix sets are considered are all over $\mathbb{Q}(i$ or $j = \varsigma_3$,or θ, where $\theta = 2\cos\left(\frac{2\pi}{n}\right)$), for $\mathbb{Q}$ the field of rational numbers). Recalling that the n th roots of unity are $\varsigma_n^k = e^{2\pi ki/n} = \cos(\frac{2\pi k}{n}) + i\sin(\frac{2\pi k}{n})$, for $k = 0, 1, ...n-1$, then $\varsigma_n + \varsigma_n^{-1} = 2\cos\left(\frac{2\pi}{n}\right)$.

  Here are the specific division rings:

$$n = 2: (5, i)_{\mathbb{Q}(i)}^{\mathbb{Q}(i)(\sqrt{5})}, \ i = \sqrt{-1}.$$

In this setting, letting $\theta_1 = \sqrt{5}$, choose a set of positive integers m_i, $i = 1, 2, .., k+1$ such that $m_1 = 5$. Take the product $\Pi_{i=1}^{k+1}(x^2 - m_i)$ together with $x^2 + 1$, where all individual field extensions are mutually non-isomorphic to construct the Galois field extension $\mathbb{Q}(i, \theta_1, \theta_2, ..., \theta_{k+1}) =$

$\mathbb{Q}(i)(\sqrt{5})(\theta_2, ..., \theta_{k+1})$. This will construct the quaternion division ring $D_k = (5, i)_{\mathbb{Q}(i)(\theta_2,...,\theta_{k+1})}^{\mathbb{Q}(i)(\theta_2,...,\theta_{k+1})(\sqrt{5})}$. This quaternion division ring is Galois with group $\mathcal{E}_k$, the elementary Abelian 2-group of order 2^k, which is the automorphism group of the associated Galois field extension that fixes $\mathbb{Q}(i)(\sqrt{5})$ pointwise. Using the multiple replacement theorem, we obtain a Desarguesian translation plane admitting $2^k - 1$ sets of mutually disjoint derivable nets, each derivable net isomorphic to a quaternion division ring pseudo-regulus net. There are $2^k - 1$ automorphisms of order 2, τ and the same number of quaternion division subrings $(5, i)_{Fix\tau}^{Fix\tau(\sqrt{5})}$ of index 2 in D_k. These automorphisms produce multiply derived translation planes with spreads in $PG(15, Z(D_k))$.

Furthermore, there are inner automorphisms of order 2 obtained by $\rho(\theta_1) = \theta_1 r \theta_1^{-1}$, on each of these division rings, where $\theta_1 = \sqrt{5}$. In this setting, the multiply derived translation planes have spreads in $PG(7, Z(D_k))$.

$$n = 3 : (\mathbb{Q}(\varsigma_3)(\theta)/\mathbb{Q}(\varsigma_3), \sigma, \varsigma_3),$$

$$\text{where } \theta = 2\cos(\frac{2\pi}{7}) = \varsigma_7 + \varsigma_7^{-1},$$

$$\sigma : \varsigma_7 + \varsigma_7^{-1} \rightarrow \varsigma_7^2 + \varsigma_7^{-2}.$$

Note that $\varsigma_7 + \varsigma_7^{-1}$ generates the unique

cyclic group of order 3

in the cyclotomic extension $\mathbb{Q}(\varsigma_7)/\mathbb{Q}$.

Since $K = \mathbb{Q}(\varsigma_3)$ contains a primitive 3 root of unity, then Kummer theory applies and there are cyclic towers of degree 3, forming a Galois Abelian extension $E_k = K(\theta_1, \theta_2, .., \theta_{k+1})$, thus creating a Galois cyclic division ring admitting an automorphism group $\Pi^k z_3$, the subgroup of the Galois group of the field fixing $K(\theta_1)$ pointwise, where $\theta = \theta_1$. Furthermore, there are $3^k - 1$ automorphisms τ of order 3. We have a set of $3^k - 1$ cyclic division subrings of index 3. In this setting, there are translation planes with spreads in $PG(2n^3 - 1, Z(E_k))$, for $n = 3$, so in $PG(53, Z(E_k))$, using multiple replacement theorem. Similarly, using analogous inner automorphisms, there are translation planes with spreads in $PG(2n^2 - 1, Z(E_k))$, so in $PG(17, Z(E_k))$.

$$n = 4 : (\mathbb{Q}(i)(\theta)/\mathbb{Q}(i), \sigma, i),$$

$$\text{where } \theta = 2\cos(\frac{2\pi}{15}) = \varsigma_{15} + \varsigma_{15}^{-1},$$

$$\sigma : \varsigma_{15} + \varsigma_{15}^{-1} \rightarrow \varsigma_{15}^2 + \varsigma_{15}^{-2}.$$

$$\text{Recall that } i = \varsigma_4.$$

Since $i = \varsigma_4$, our previous remarks apply to show that the multiple replacement theorem applies in this setting. And, since 2 divides n, we obtain sets of mutually disjoint derivable nets isomorphic to pseudo-regulus nets that are isomorphic to certain cyclic division ring nets. Applying the arguments of the previous, there will be sets of translation planes with spreads in $PG(2\frac{n^3}{d} - 1, Z(D_\tau))$ for $n = 4$, and $d = 1$ or 2 and in $PG(2n^2 - 1, Z(E\tau))$, for certain cyclic division rings of degree 4 and their Galois extensions. Therefore, spreads in $PG(127, Z(D_\tau))$, $PG(63, Z(D_\rho))$, and $PG(31, Z(D_\upsilon))$.

$$n = 5 : (\mathbb{Q}(i)(\theta)/\mathbb{Q}(i), \sigma, \frac{3+2i}{2+3i}),$$

where $\theta = 2\cos(\frac{2\pi}{11}) = \varsigma_{11} + \varsigma_{11}^{-1}$,

$$\sigma : \varsigma_{11} + \varsigma_{11}^{-1} \to \varsigma_{11}^2 + \varsigma_{11}^{-2}.$$

Note $\mathbb{Q}(\varsigma_{11} + \varsigma_{11}^{-1})/\mathbb{Q}$ is of degree 5.

PROBLEM 80. *Determine if the case* $n = 5$ *and/or certain Galois extensions can be used in the multiple replacement theorem for outer automorphisms.*

Let θ define an inner automorphism $\rho(r) = \theta r \theta^{-1}$. The order is 5. And $Z((\mathbb{Q}(i)(\theta)/\mathbb{Q}(i), \sigma, \frac{3+2i}{2+3i}) \cap Fix\rho$ is $\mathbb{Q}(i)(\theta)$. Use the multiple replacement theorem to determine the multiple replacement nets and the projective spaces containing the spreads for the proper multiply replacement translation planes. The associated multiply replaced planes have spreads in $PG(2 \cdot 5^2 - 1, Z(E_\tau))$, using the Galois extensions E_τ.

•

$$n = 6 : (\mathbb{Q}(\varsigma_3)(\theta)/\mathbb{Q}(\varsigma_3), \sigma, -\varsigma_3),$$

where $\theta = 2\cos(\frac{\pi}{14}) = \varsigma_{28} + \varsigma_{28}^{-1}$,

$$\sigma : \varsigma_{28} + \varsigma_{28}^{-1} \to \varsigma_{28}^2 + \varsigma_{28}^{-2}.$$

The Euler ϕ-function $\phi(28) = 12$, so that

the extension $\mathbb{Q}(\varsigma_{28} + \varsigma_{28}^{-1})/\mathbb{Q}$

is of degree 6.

Similarly, since $\mathbb{Q}(\varsigma_3)(\varsigma_2) = \mathbb{Q}(\varsigma_6)$, it follows that Kummer theory would also work on this situation, allowing application to the multiple replacement theorem and since 2 divides 6, there would be Desarguesian planes admitting sets of mutually disjoint derivable pseudo-regulus nets isomorphic to cyclic division rings of degree 6.

Also, here there are inner automorphisms of order 6. There are multiply replaced translation planes in $PG(\frac{2n^3}{d} - 1, Z(D_\tau))$ and $PG(2n^2 - 1, Z(E_\rho))$ where $n = 6$ and $d = 1, 2, 3$.

- It is also pointed out that Lahtonen, Markin, McGuire [108] have constructed infinitely many cyclic division rings and therefore the inner automorphism theorem produces infinitely many classes of translation planes.
- In the following, we show that each of the previous cyclic division rings of dimension n admit an extension of dimension n, thereby permitting both inner and outer automorphism constructions to provide infinitely many new translation planes.

From the section on Kummer Theory, using Lahtonen, Markin, McGuire [108], and the extension theorem for skewfields, we are able to draw the following conclusion:

COROLLARY 54. *Let $n = p^a$ and let ρ be an infinite set of distinct primes $\{t_i; i = 1, 2, ..., \infty\}$ such that p^a divides $t_i - 1$, but p^{a+1} does not divide $t_i - 1$. Then there exists an infinite set of elements a_i of $\mathbb{Q}(\varsigma_n)$ such that for any integer k, choose any $k + 1$ elements as an ordered set $\{b_1, ..., b_{k+1}\} \subset \{a_i; i = 1, ..., \infty\}$. Then there is a central Galois extension of $\mathbb{Q}(\varsigma_n)$, $\mathbb{Q}(\varsigma_n)(\sqrt[n]{b_1}, \sqrt[n]{b_2}, ..., \sqrt[n]{b_{k+1}})$ of degree n^{k+1} admitting an Abelian Galois group isomorphic to the direct product of cyclic groups isomorphic to z_n, $\Pi^{k+1} z_n$. Additionally, from Lahtonen, Markin, McGuire [108], it is shown, for $K = \mathbb{Q}(\varsigma_m)$, that the elements ς_n^j are in $K - K(\sqrt[n]{a_i})^{\frac{\tau_i^n - 1}{\tau_i - 1}}$, for all $i \in \rho$ and for all $j = 1, 2, ..., n - 1$.*

Rather than give a formal proof, we make some comments that will be useful in the proofs of the results to follow.

In particular, the use of localization is available to prove this last part regarding non-norm elements. By the extension of skewfields theory, the more general remark follows, showing that the extension

$$\mathbb{Q}(\varsigma_n)(\sqrt[n]{b_1}, \sqrt[n]{b_2}, ..., \sqrt[n]{b_{k+1}})$$

and

$$(b_1, \varsigma_n)_{\varsigma_n \mathbb{Q}(\varsigma_n)(\sqrt[n]{b_2}, ..., \sqrt[n]{b_{k+1}})},$$

relative to

$$\mathbb{Q}(\varsigma_n)(\sqrt[n]{b_1})(\sqrt[n]{b_2}, ..., \sqrt[n]{b_{k+1}})$$

is a central Galois extension of

$$(b_1, \varsigma_n)_{\varsigma_n \mathbb{Q}(\varsigma_n)}.$$

This Galois extension admits the automorphism group $\Pi^k z_n$ that corresponds to the Galois group of

$$\mathbb{Q}(\varsigma_n)(\sqrt[n]{b_1})(\sqrt[n]{b_2}, ..., \sqrt[n]{b_{k+1}})$$

that fixes $\mathbb{Q}(\varsigma_n)(\sqrt[n]{b_1})$ pointwise. Using the same idea we see that

$$(b_i, \varsigma_n)_{\varsigma_n K}(\sqrt[n]{b_2}, .. \widehat{\sqrt[n]{b_i}}, \sqrt[n]{b_{k+1}})$$

is a central Galois extension of $(b_i, \varsigma_n)_{\varsigma_n K}$ and shows that all sub-cyclic algebras of

$$(b_i, \varsigma_n)_{\varsigma_n K}(\sqrt[n]{b_2}, .. \widehat{\sqrt[n]{b_i}}, \sqrt[n]{b_{k+1}})$$

are also cyclic division rings. The decorated term indicates a missing element. This then provides an alternative method to establish that ς_n^j are in $K - K(\sqrt[n]{a_i})^{\frac{\tau_i^n - 1}{\tau_i - 1}}$, since this is true for all cyclic subalgebras (which we know from above are actually cyclic division algebras). Note further that Lahtonen, Markin, McGuire [108], assert that there are infinitely many such primes, for any $p^a = n$, which is valid by Dirichlet's theorem on arithmetic progressions. Also, note that there are infinitely many central Galois extensions of dimension n.

REMARK 49. *The construction of the cyclic Galois extensions of $\mathbb{Q}(\varsigma_n)$ is attributed to C. Grover, C. Ling, R. Vehkalahti [58] and is as follows, as mentioned previously. Let p^a divide $t - 1$, where t is a prime and p^{a+1} does not divide $t - 1$. Form $\mathbb{Q}(\varsigma_n)(\varsigma_t)$, which is a cyclotomic extension with cyclic group of order $t - 1$, since t is prime. It is noted that this field is also $\mathbb{Q}(\varsigma_{nt})$, since n and t are relatively* *prime. Since this is a Galois extension and n divides t − 1, there is a subfield containing* $\mathbb{Q}(\varsigma_n)$ *which is* $Fix\sigma^n$, *where* σ *is a generating automorphism of order* $t-1$, *this then must have degree n over* $\mathbb{Q}(\varsigma_n)$.

Putting these two results together, and using the extension of cyclic division rings theorem, using the assumptions and notation of the previous corollary, we obtain the following:

THEOREM 163. *For any prime power* $n = p^a$, *there exist infinitely many Galois chains of fields containing* $K(\sqrt[n]{a_i})$ *each of dimension n over the previous. There are also infinitely many central Galois extensions of cyclic division rings in a Galois chain, each of dimension n over the previous. At the* $k + 1$ *link, there is a set of elements* $\{b_i; i = 1, 2, ..., k\}$ *such that there is a cyclic division ring*

$$D_i = (b_i, \varsigma_n)_{\varsigma_n K(\sqrt[n]{b_2}, .. \widehat{\sqrt[n]{b_i}}, \sqrt[n]{b_{k+1}})},$$

where the notation $\widehat{\sqrt[n]{b_i}}$ *indicates that* $\sqrt[n]{b_i}$ *is missing, the automorphism group is still* $\Pi^k z_n$ *but corresponds to the Galois subgroup of the field extension that fixes* $\mathbb{Q}(\varsigma_n)(\sqrt[n]{b_i})$ *pointwise. Now we have infinitely many Galois cyclic division rings* D_i.

Consequently, we have:

COROLLARY 55. *For any* $n = p^a$, *for any prime p, there exist Desarguesian translation planes* Σ *that admit Abelian automorphism groups of order* n^k *so by the multiple replacement theorem, choose any cyclic group of order d dividing n with generator* τ. *Using* Σ_τ, *these Desarguesian planes admit a set* Λ *of mutually disjoint replaceable nets that are hyper pseudo-regulus nets defined over subplane covered nets. The set* Λ *covers the translation plane components except for two carrier components* $x = 0, y = 0$. *Choose any proper subset* δ *of* Λ *and multiply replace the corresponding hyper pseudo-regulus nets. Then there is a constructed translation plane that has a blended left/right kernel, assuming the Desarguesian plane is coordinatized by a non-commutative skewfield.*

Again, for a cyclic division subring $D_\tau = Fix\tau$ *of degree d, the associated multiply replaced planes have spreads in (1) for outer automorphism types* $PG(2\frac{n^3}{d} - 1, Z(D_\tau))$ *and (2) for the special inner automorphism types using cyclic division ring* E_ρ, $PG(2n^2 - 1, Z(D_\rho))$, *so, in this setting* $n = p^a$, *and* $d = p^b$, *for* $0 \leq b < a$.

When $n = 2^a$, *there is an automorphism* τ_i *of order 2, and, for the right Desarguesian plane, the replacement nets are derivable nets of type 2 isomorphic to the left classical quaternion division ring* D_τ.

THEOREM 164. *Each of the k central Galois quadric extensions of a skewfield in the previous theorem produces a semifield plane with spread in*

$$PG(3, (a, b)_{\varsigma_n K(\theta_2, ... \theta_k)}).$$

This occurs exactly when n is an even integer.

Consequently, there are k mutually non-isomorphic sets of semifield planes with spreads in $PG(3, (a, b)_{\varsigma_n K(\theta_2, ... \theta_k)})$.

Let τ_i *denote the automorphisms of order 2 and let* $K(n_i)$ *denote the subfields on index 2. Let* $\{1, \partial_i\}$ *denote a basis for* $(a, b)_{\varsigma_n K(\theta_2, ... \theta_k)}$ *over* $(a, b)_{\varsigma_n K(n_i)}$ *as well as a basis for* $K(\theta_1, ..., \theta_{k+1})$ *over* $K(n_i)$, *for* $i = 2, ..., k + 1$.

COROLLARY 56.

(1) *The lifted semifield spreads* $(semi(\delta_i, \gamma_i), \tau_i)$ *for* $\delta_i \neq 0$, $\gamma_i \in K(n_i)$ *have the following form:*

$$x = 0, y = x \begin{bmatrix} u^{\tau_i} & (\delta_i \partial_i + \gamma_i) t^{\tau_i} \\ t & u \end{bmatrix} : \forall t, u \in (a, b)_{\varsigma_n K(\theta_2, \dots \theta_k)}.$$

(2) *The spread is in* $PG(3, (a, b)_{\varsigma_n K(\theta_2, \dots \theta_k)})$. *Each of these semifield planes has a derivable net consisting of the spread components when* $t = 0$. *This net is a type 2 pseudo-regulus net corresponding to the left classical pseudo-regulus net corresponding to* $(a, b)_{\varsigma_n K(\theta_2, \dots \theta_k)}$.

    There is also a carrier derivable net of the same type of class 2 obtained when $u = 0$.

(3) *The derived planes obtained from the derivable nets have spreads with mixed left/right kernel* $((a, b)_{\varsigma_n K(\theta_2, \dots \theta_k)_{Left}}, (a, b)_{\varsigma_n K(\theta_2, \dots \theta_k)_{right}})$, *which makes the kernel of these translation planes* $K(\theta_2, \dots \theta_k)$. *Consequently, the spreads for* $(semi(\delta_i, \gamma_i), \tau_i)$ *are in* $PG(2^2 n^2 - 1, K(\theta_2, \dots \theta_k))$.

PROOF. This corollary now follows directly from part 8 on lifting skewfields, by looking carefully at the matrices constructed there. □

There are three other sources of cyclic division rings available.

- In the lifting results, we ask the question whether the quaternion division rings of Milliet [112] have quadratic central extensions. By the results on rational function fields, part 12, we know that there is a quaternion extension of the quaternion division ring of Milliet that admits a quadratic central extension. Milliet's work is also connected to mathematical logic and the terms NIP (not the independence property), NIP_2, and dp-rank have complex definitions.

- Additionally, Milliet [112] constructs cyclic division rings of dimension $n = 3^2$ over the centers K. If there is a Galois extension of dimension n^2, where the Galois group is a product of $z_n \times z_n$, we have the situation considered in the extension. Thus, the two results on construction of translation planes using inner automorphism and outer automorphism both might apply, depending on the existence of Galois extensions of K. Of course, extensions to rational function fields of many commuting independent variables are also available that produce automorphisms and therefore satisfy the conditions of the multiple replacement theorem.

- Both Milliet's work and that of C. D'Elbée [38] consider cyclic division rings of dp-rank.

- C. D'Elbée [38] constructs cyclic division rings of degree q, where q is a prime. Moreover, the center K contains a primitive q-th. root of unity and the method of construction uses arguments on using primes p such that q divides $p - 1$ (Dirichlet's theorem) to obtain appropriate field extensions, similar to that which have discussed previously when considering the cyclic division rings of Lahtonen, Markin, McGuire [108]. D'Elbée, mentions W. Johnson's [100] classification of dp-minimal fields F, where F is either algebraically closed, real-closed or admits a definable Henselian valuation. This result is used in the study. The division rings associated in this work are then realized using Hahn series, which are generalizations of formal power series.

- D'Elbée also reinvents Albert's non-cyclic division rings of degree 16 to show there exist such division rings of finite dp-rank.

- The work of Morandi, Sethuraman and Tignol [113] considers finding division algebras that have anti-automorphisms but no involution. The type of division algebras that they list as examples is also similar to those that have been described previously, but over different fields, and also permit application of the ideas of this text. Here is the corollary to their main result, and here note some similarity of ideas on extensions of skewfields, and on Galois division rings given in this text.

COROLLARY 57. (Morandi, Sethuraman and Tignol [113] Corollary 3.6) Let F contain an $(n > 2)$-th root of unity ω_n and F a Galois extension of $\mathbb{Q}$ and $a \in F^*$ is such that

(1) there is an automorphism ρ, such that $\rho(a) = a$ in $Gal(F/\mathbb{Q})$ of order 2,
(2) $\tau(a) = a^{-1}$ for some $\tau \in Gal(F/\mathbb{Q})$,
(3) the symbol algebra $D = (a, \omega_n)_{\omega_n F}$ is a division algebra.

Then D has an anti-automorphism but no involution.

Of the examples that are constructed, there are examples where the n divides $p - 1$ and F is embedded into $\mathbb{Q}_p$ by Hensel's lemma, and division algebras are constructed $(a, \omega_n)_{\omega,F} \otimes_F \mathbb{Q}_p = (a, \omega_n)_{\omega_n \mathbb{Q}_p}$. It is noted that ρ extends to an automorphism of D, but does not induce an involution on D. Moreover, a result in this article shows (Proposition 3.5) that, for a symbol division ring $(a, b)_{\varsigma_n F}$, every automorphism in $Gal(F/\mathbb{Q})$ that fixes a induces an automorphism of the division ring. Therefore, since there is an automorphism ρ of D, the multiple replacement requirements are satisfied for D coordinatizing a Desarguesian spread producing sets of mutually disjoint replacement sets. And, of course, the inner automorphisms of finite order also produce interesting mutually disjoint replacement sets.

Furthermore, the use of Hensel's lemma is used in various ways embedding number fields in $\mathbb{Q}_p$ for various primes p. Using other non-Archimedean valuations, there are also constructions of division rings using 'local invariants' and by use of the description of Brauer groups employing this method. Finally, there are many interesting division rings using valuation theory applied to twisted Laurent series.

- Furthermore, the extension theory of Jacobson [64] and the variety of constructed division rings with automorphisms all provide new examples of translation planes, some of which have been discussed in the previous sections. There are various results on skew polynomial rings of the form $D[t; \delta]$ where δ is a derivation that will provide many new derivable nets of type 1, but which we have not been able to include for space reasons. In particular, results on purely inseparable extension fields of F of exponent 1 and characteristic p are not covered but which also provide many varieties of derivations. It turns out that division rings of finite dimension over F (that is, over the purely inseparable fields) with center F, then also inherit the derivations. Both the fields and the division rings provide examples of derivable nets of type 1 (See [64] 1.8).

CHAPTER 15

General ideas on Klein extensions

For the last part of this text, we consider the general idea of flocks over skewfields, where previously there has not been any theory at all. For this, non-commutative algebraic structures are defined over skewfields, which could be possibly commutative (fields) or non-commutative. Are there flocks over non-commutative quadratic cones or are there flocks of non-commutative hyperbolic quadrics? Here the term 'quadric' is used in a very general manner, as they need not be quadrics at all.

We shall see that, from the standpoint of translation planes, moving to a non-commutative skewfield situation seems quite natural. Whether such geometries have meaning projectively remains to be seen.

1. Klein varieties

DEFINITION 125. *Let F be a skewfield, commutative or non-commutative and let V_4 be an F-bimodule. Consider $PG(3,F)_S$. Note that we normally consider F to be a left vector space, so that the projective space has left 1-dimensional S-subspaces as points. But, F could be considered as a right S-module, so that $PG(3,F)_S$ could have right 1-dimensional F-subspaces as points. To be clear, we shall use the term 'left points' and right points'. This came up with the study of derivable nets. The embedding/contraction method has a left/right sort of theory, and the contraction vector space V_4 is considered a left F-space, but this is purely arbitrary and we could have considered V_4 as a right F-space. The difference between what happens with the derivable net and the derived net, is simply that the components and Baer subplanes are right and left, or left and right subspaces, respectively, depending on convention. The associated projective spaces could be considered changed as well, by taking the dual projective space.*

In the following, the mental image should be that we are considering a projective space, where both left points and right points occur within the space, sort of as a union of the projective space and the dual projective space.

DEFINITION 126. *In the above setting, when F is a field, we considered an α-hyperbolic quadric, with projective points (x_1, x_2, x_3, x_4) such that $x_1 x_4^\alpha = x_2^\alpha x_3$.*

Using the field/skewfield F^α, but, when F is a non-commutative skewfield, we shall generalize all of this in two ways: First assume that all $x_3 x_4 \neq 0$. In the commutative case we would have $x_3^{-1} x_1 = x_4^{-\alpha} x_2$. Now under the skewfield F^α, we would have $\delta(x_1, x_2, x_3, x_4) = (\delta^\alpha x_1, \delta x_2, \delta^\alpha x_3, \delta x_4)$. But now $(\delta^\alpha x_3)^{-1}(\delta^\alpha x_1) = x_3^{-1} x_1 = (\delta x_4)^1 (\delta x_2) = x_4^{-1} x_2$. If $x_3 x_4 = 0$, then use the standard definition.

Hence, for skewfields, we define a 'left α-hyperbolic quadric' with the left points satisfying the union of the following sets of points:

$$\text{Use } x_1 x_4^\alpha = x_2^\alpha x_3 \text{ when } x_3 x_4 = 0,$$
$$\text{and use } x_3^{-1} x_1 = x_4^{-\alpha} x_2^\alpha, \text{ for } x_3 x_4 \neq 0.$$

Then for the 'right α-hyperbolic quadric', we would have the right points satisfying the following condition::

$$\text{Use } x_1 x_4^\alpha = x_2^\alpha x_3 \text{ when } x_1 x_2 = 0$$
$$\text{and use } x_3 x_1^{-1} = x_4^\alpha x_2^{-a} \text{ for } x_1 x_2 \neq 0.$$

Now form something called the 'left α-cones' ('right α-cone') by using a left α-hyperbolic quadric form (right α-hyperbolic quadric form) and let say $x_2 = x_3$. So, when the skewfield is a field, right and left α-quadrics are simply α-quadrics. Use LC to indicate a left cone and RC for a right cone.

DEFINITION 127. *A flock of an LC α-cone is a set of left planes of $PG(3, F)_S$, as a left F-space (or rather, as the corresponding 4-dimensional vector space V_4 as a left K-vector space), that forms a partition of the LC α-quadric. Similarly a flock of an RC α-cone uses a set of right planes of $PG(3, F)_S$, as a right F-space with a corresponding partition. We shall agree to write the left planes with the x_i on the left and the right planes with the x_i on the right.*

The terms that we are using have no intrinsic meaning other than we are trying to find a flock type geometry using a certain variety such that there is a connection with a translation plane admitting a central collineation group, one of whose orbits together with the axis or axis/coaxis is a derivable net. We have the required groups, relative to the types of derivable nets. We are then trying to determine translation planes and geometric flocks corresponding to type 2 pseudo-regulus nets for a non-commutative skewfield or type 2 regulus nets for a field. Of course, there may be connections with type 0 and type 1 derivable nets, but these may not be connected to generalizations of quadrics.

The above generalizations will allow connections with central collineation groups whose orbits under the group F^α, when F is non-commutative can be visualized as 'points' under the generalization of homogeneous coordinates, thus allowing a right and left flock system using left and right points as orbits that are the Baer subspaces of the derivable nets.

2. Translation planes admitting twisted pseudo-regulus nets

In this section, we generalize the material on twisted hyperbolic flocks to non-commutative twisted hyperbolic flocks. Throughout this part, F is a non-commutative skewfield.

Assume that we have a twisted hyperbolic quadric given by points

$$(x_1, x_2, x_3, x_4) \text{ such that } x_1 x_4^\alpha = x_2 x_3^\alpha.$$

Then this represents projective points with homogeneous coordinates provided the coordinate system is commutative. For left skewfield systems, we need to change this to

$$\text{Use } x_1 x_4^\alpha = x_2^\alpha x_3 \text{ when } x_3 x_4 = 0,$$
$$\text{and use } x_3^{-1} x_1 = x_4^{-\alpha} x_2^\alpha, \text{ for } x_3 x_4 \neq 0, \text{ defining } 0^{-1} = 0.$$

Then for 'right α-hyperbolic quadrics', we would have the right points satisfying the following condition:

$$\text{Use } x_1 x_4^\alpha = x_2^\alpha x_3 \text{ when } x_1 x_2 = 0$$
$$\text{and use } x_3 x_1^{-1} = x_4^\alpha x_2^{-a} \text{ for } x_1 x_2 \neq 0, \text{ defining } 0^{-1} = 0.$$

DEFINITION 128. *A non-commutative left flock NCF_L is a partition of the left points the left non-commutative twisted hyperbolic quadric THQ_L of $PG(3,K)_S$, where F is a skewfield by a set of mutually disjoint left planes. A right NCF_R of the corresponding THQ_R is defined accordingly.*

REMARK 50. *Note that α could be 1, in this setting, thus also generalizing the idea of a flock of a quadratic cone over a non-commutative quadric as well as a non-commutative hyperbolic flock of a non-commutative hyperbolic quadric.*

We have previously mentioned that the form for the spreads of the translation planes over non-commutative skewfields are essentially identical in form to the field situation. This theory could be developed when F is a skewfield initially, except that when F is non-commutative, the connection with a generalization of an α-hyperbolic flock is not clear.

THEOREM 165. *Let Σ be a translation plane with spread in $PG(3,F)$, for F an arbitrary skewfield. Let α denote an outer automorphism of F, possibly trivial. Assume that Σ admits an affine homology group one orbit of which, together with the axis and coaxis, is a twisted pseudo-regulus net. Then all orbits are twisted pseudo-regulus nets and the spread may be coordinatized in the following form: Let V_4 be the associated 4-dimensional vector space over K. Letting x and y denote 2-vectors, then the spread is:*

$$x = 0, y = x \begin{bmatrix} u^\alpha & 0 \\ 0 & u \end{bmatrix}, \text{ and}$$

$$y = x \begin{bmatrix} f(t) & g(t) \\ 1 & t \end{bmatrix} \begin{bmatrix} v^\alpha & 0 \\ 0 & v \end{bmatrix};$$

$$\forall u, t, v, uv \neq 0, \text{ of } F,$$

and functions f, g on F. Moreover, f is bijective. Furthermore,

$$x = 0, y = x \begin{bmatrix} u^\alpha & 0 \\ 0 & u \end{bmatrix}$$

is a left non-commutative hyperbolic quadric.

PROOF. To obtain a spread, the differences of the matrices involved in the components must be non-singular, which implies that f is injective. To see that f is bijective, consider the vector $(1, -a, 0, 1)$ for $a \in K$. This point is on

$$y = x \begin{bmatrix} f(t) & g(t) \\ 1 & t \end{bmatrix} \begin{bmatrix} v^\alpha & 0 \\ 0 & v \end{bmatrix},$$

for some $v \neq 0$, so that $(f(t) - a)v^\alpha = 0$, and $(g(t) - at)v = 1$, so that $f(t) = a$, for some t, proving that f is bijective, just as in the field case. Since the twisted regulus net may be assumed to be generic and then is known to have the given form by part 3. Finally, the points of $x = 0, y = x \begin{bmatrix} u^\alpha & 0 \\ 0 & u \end{bmatrix}$, have the form $(0, 0, x_3, x_4)$ and $(x_1, x_2, x_1 u^\alpha, x_2 u)$.

$$
\begin{aligned}
\text{Use } x_1 x_4^\alpha &= x_2^\alpha x_3 \text{ when } x_3 x_4 = 0, \\
\text{and use } x_3^{-1} x_1 &= x_4^{-\alpha} x_2^\alpha, \text{ for } x_3 x_4 \neq 0,\ 0^{-1} = 0,
\end{aligned}
$$

to see that we have a twisted hyperbolic quadric, commutative or non-commutative. We have completed the proof. □

DEFINITION 129. *A translation plane satisfying the conclusions of the previous theorem shall be called a plane admitting a 'twisted pseudo-regulus inducing homology group' when the homology group is not commutative and a plane admitting a 'twisted regulus inducing homology group' when the homology group is commutative.*

DEFINITION 130. *A translation plane with spread in $PG(3, F)$ is called a 'cyclic $q+1$-homology group plane' or more simply a 'cyclic $q+1$-plane' when the homology group is cyclic of order $q + 1$.*

Recall in part 8, there is a construction of classes of semifield planes in $PG(3, F(\theta))$, where F is a non-commutative skewfield and $F(\theta)$ is a central Galois quadratic extension, with automorphism τ of order 2. Every quaternion division ring has an extension, which has a central Galois quadratic extension (12,14,5). Furthermore, we have constructed central division ring planes of degree n over the center F, which have this property provided n is even (12,14,6). The spreads have the following form: For each generator of the center $Z(F)$ of F, $(\theta\delta + \gamma)$ for $\delta \neq 0$, $\gamma \in Z(F)$, we have the spread:

$$
x = 0, y = 0, y = x \begin{bmatrix} u^\tau & (\theta\delta + \gamma)t^\tau \\ t & u \end{bmatrix}; \forall t, u \in F(\theta),\ (t, u) \neq (0, 0).
$$

Let $\begin{bmatrix} f(t) & g(t) \\ 1 & t \end{bmatrix} = \begin{bmatrix} t^\tau & (\theta\delta + \gamma) \\ 1 & t \end{bmatrix}$, then $\begin{bmatrix} t^\tau & (\theta\delta + \gamma) \\ 1 & t \end{bmatrix} \begin{bmatrix} v^\tau & 0 \\ 0 & v \end{bmatrix} = \begin{bmatrix} (tv)^\tau & (\theta\delta + \gamma)v \\ v^\tau & tv \end{bmatrix}$.

Let $v^\tau = s$, and $tv = w$ to obtain $\begin{bmatrix} w^\tau & (\theta\delta + \gamma)s^\tau \\ s & w \end{bmatrix}$, since $\tau^{-1} = \tau$. Therefore, we have the following result.

THEOREM 166. *Let K be a non-commutative skewfield and let $F(\theta)$ be a central Galois quadratic extension of K. Then there is a class of corresponding semifield planes lifted from $F(\theta)$ that are translation planes admitting twisted pseudo-regulus-inducing homology groups.*

PROBLEM 81. *Discuss whether the translation planes above are connected to right flocks of a right twisted hyperbolic quadric.*

3. Skewfield α-flocks

The types of translation planes admitting twisted pseudo-regulus homology groups also admit twisted pseudo-regulus elation groups as the spreads

$$x = 0, y = 0, y = x \begin{bmatrix} u^\tau & (\theta\delta + \gamma)t^\tau \\ t & u \end{bmatrix}; \forall t, u \in F(\theta), (t, u) \neq (0, 0),$$

admit the τ-pseudo-regulus-inducing elation group

$$\left\langle \begin{bmatrix} I_2 & 0_2 \\ 0_2 & \begin{bmatrix} u^\tau & 0_2 \\ 0_2 & u \end{bmatrix} \end{bmatrix}; u \in F \right\rangle,$$

where K is a skewfield. This may be seen, since the spread is a semifield spread.

DEFINITION 131. *A translation plane with spread in $PG(3, F)$, for F a non-commutative skewfield admitting an automorphism α, will be said to admit a 'twisted pseudo-regulus-inducing elation group' if and only if the spread has the following form:*

$$x = 0, y = x \begin{bmatrix} u^\alpha + g(t) & f(t) \\ t & u \end{bmatrix}; t, u \in F$$

where f and g are functions on F such that $g(0) = f(0) = 0$. Note that all orbits of the twisted pseudo-regulus-inducing elation group together with the axis $x = 0$ are twisted pseudo-reguli.

The semifield spreads in $PG(3, F(\theta) = K)$,

$$x = 0, y = 0, y = x \begin{bmatrix} u^\tau & (\theta\delta + \gamma)t^\tau \\ t & u \end{bmatrix}; \forall t, u \in F(\theta), (t, u) \neq (0, 0),$$

provide infinitely many examples of such spreads.

PROBLEM 82. *Are the semifield planes above connected to flocks of left α-quadrics? Discuss the possibilities.*

REMARK 51. *We note that the Baer group spreads over non-commutative skew-fields are also interesting objects as they relate directly to the translation planes considered in the previous two sections.*

4. Skewfield 'cyclic' translation planes

In this last section, we consider what the analogues of cyclic $q+1$-planes would be like over non-commutative skewfields K. For this generalization, we consider the representation introduced in part 10. Let K be a non-commutative skewfield and let $K(\theta)$ be a central Galois quadratic extension of K. We now use parts 8 and 10, first to represent $K(\theta) = u + t\theta$, where $u, t \in K$ and $\theta^2 = \theta f + g$, where $f, g \in Z(K)$, the center of K. We assume that K is irreducible over $x^2 \pm xg-$, so that $Z(K(\theta))$ is $Z(K)(\theta)$. Therefore, the matrix representation shall be taken as

$$\begin{bmatrix} u + tg & ft \\ t & u \end{bmatrix}; u, t \in K, f, g \in Z(K),$$

the involution σ : $\theta \to -\theta + g$ and fixes K elementwise.

Recall that the skewfield determinant for $t \neq 0$ is $((u + tg)t^{-1}u - ft)t$ and is u^2 when $t = 0$ (8,10,2). Thinking of how to consider the non-commutative analogue of a cyclic group of order $q + 1$ (in a translation plane of order q^2), we recall the foundation of the multiple replacement theorem of part 10 where Σ and Σ_σ are the left and right Desarguesian spreads. We consider the group:

$$G = \langle (x,y) \rightarrow (bxe, bye^\sigma); b, e \in K^* \rangle.$$

This is the non-commutative analogue of the cyclic group of order $q+1$ or the non-commutative version of the group with elements $T = \begin{bmatrix} u+tg & ft \\ t & u \end{bmatrix}$, in a field L so that $\delta_{t,u} = 1$, or equivalently $T^{\sigma+1} = 1$. We now represent $x = \begin{bmatrix} x_1 + tx_2 & fx_2 \\ x_2 & x_1 \end{bmatrix} = (x_1, x_2)$, and letting $b = (b_1, b_2)$ and $c = (c_1, c_2)$, represented in matrix form. In the following theorem, we assume the previous.

THEOREM 167. *Let π be a translation plane with spread in $PG(3,K)$, where K is a non-commutative skewfield and $K(\theta)$ is a central Galois quadric extension. Assume that π admits the group G as a collineation group, where the spread for π has the form:*

$$x = 0, \ y = x \begin{bmatrix} F(t,u) & G(t,u) \\ t & u \end{bmatrix}; t, u \in K$$

where we have the Desarguesian partial spread in Σ:

$$x = 0, y = 0, y = x$$

such that the group G maps $y = x$ to $\{y = xcc^{-\sigma}; c \in K^*\}$ and maps $y = x^\sigma$ in Σ_σ $\{y = b^{-1}b^\sigma x^\sigma\}$. Since $\sigma^2 = 1$, these sets define the components set and Baer subspace set, respectively, of a derivable net which is a pseudo-regulus net relative to K.

Assume that we consider an orbit set under G as $\{n_i; i \in \lambda\}$, where n_i are elements of $GL(2,K)_{left}$. Represent all terms as matrices with entries in K, so the translation plane has the following form:

$$x = 0, y = 0, \cup_{i\in\lambda}\{y = xcn_ic^{-\sigma}; c \in K^*(\theta)\}.$$

Then we see that the translation plane is a union of pseudo-regulus nets with carrying lines $x = 0$, and $y = 0$, where the corresponding translation plane of derived nets has the following form:

$$x = 0, y = 0, \cup_{i\in\lambda}\{y = cn_ic^{-\sigma}x^\sigma; c \in K^*(\theta)\}.$$

PROOF. We need to show that

$$x = 0, y = 0, \{y = xcn_ic^{-\sigma}; c \in K^*(\theta)\}$$

is a derivable net with derived net

$$x = 0, y = 0, \{y = cn_ic^{-\sigma}x^\sigma; c \in K^*(\theta)\}.$$

We note that n_i is in $GL(2,K)_{left}$, which means that $(x, xn_i) \rightarrow (bx, bxn_i)$, so that all components n_i are fixed by the kernel group of Σ and similarly all components of $y = n_ix^\sigma$ are fixed by the kernel group of Σ_σ: $(x,y) \rightarrow (xc, yc^\sigma)$ as $(x, n_ix^\sigma) \rightarrow (xc, (n_ix^\sigma)c^\sigma)$ and all elements are matrices so we see that $y = n_ix^\sigma$ is fixed by this kernel group of Σ_σ. Now we may follow the proof of the multiple replacement theorem, as the components are 2-dimensional left subspaces over K and the Baer subspaces are 2-dimensional right subspaces over K. The intersection $(y = x) \cap (y =$

x^σ) is a 1-dimensional vector space over $Fix\sigma = K$. Hence, we have a derivable net which is a pseudo-regulus net with respect to K by part 1.   □

PROBLEM 83. *So, we do have many examples of Desarguesian spreads of the above type. The question is whether these translation planes are related to generalized hyperbolic fibrations over skewfields (whatever these may be). Discuss how to determine any connections.*

5. Spreads of large dimension

Here we list some of the various new infinite spreads that are obtained as applications of the ideas of derivation and subplane covered nets. These spreads were obtained using various instances of the multiple replacement theorem, with inner and outer automorphisms. These results may be coupled with the Galois theory of skewfields. The central Galois quadratic extensions of skewfields include quaternion division rings and cyclic division rings, each of which give a variety of translation planes with large dimension spreads and the dimension 2 extensions provide classes of derivable semifield spreads over non-commutative $PG(3, K)$, K a non-commutative skewfield. The main results of the section on Galois theory and applications are as follows:

THEOREM 168. *Applying the multiple replacement theorem to the Desarguesian planes Σ and Σ_τ coordinatized left and right by a division ring D, where τ is an automorphism of D. If the division ring is of finite degree n then:*

(1) The replacement nets are isomorphic to hyper pseudo-regulus subplane covered nets of degree d, for $d \neq n$ and d divides n.

(2) The multiply replaced translation planes have spreads in $PG(2dn^2 - 1, Z(D))$.

COROLLARY 58. *For quaternion division rings $(a, b)^{K(\theta_2,...,\theta_{k+1})(\theta_1)}_{K(\theta_2,...,\theta_{k+1})}$ of characteristic $\neq 2$, there are $2^k - 1$ non-identity automorphisms τ and associated sub-quaternion division rings of index 2.*

Consequently, we have mutually disjoint pseudo-regulus derivable nets of type 2 with reference to a quaternion subring $(a, b)^{E_\tau(\theta_1)}_{E_\tau}$, where $Fix E_\tau = \langle \tau \rangle$. And, there are $2^k - 1$ such sets of mutually disjoint pseudo-regulus nets.

COROLLARY 59. *The multiple derived translation planes have spreads in $PG(15, Z(D))$.*

COROLLARY 60. *For cyclic division rings D of degree $n \neq 2$, assume that 2 divides n.*

(1) There are $2^k - 1$ sets of mutually disjoint pseudo-regulus derivable nets of type 2 with reference to a pseudo-regulus net with reference to a cyclic division subring $(a, b)^{E_\tau(\theta_1)}_{\varsigma_\tau E_\tau}$ of index 2 in D. The group element τ has order 2, so $d = 2$.

(2) The multiply derived translation planes have spreads in $PG(4n^2 - 1, Z(D))$.

Apart from many examples of quaternion division ring planes, there are examples of cyclic division rings of degree $n > 2$ in (11,12,4), for which the spreads in $PG(2dn^2 - 1, D_\tau)$ can be determined. For example:

THEOREM 169. *Let D be a finite-dimensional cyclic division ring of dimension n over the center F, a quaternion division ring, or a symbol division ring with center F. Extend D by $D(t_1, t_2)$, t_1, t_2 commuting independent variables. Then there is an automorphism of $D(t_1, t_2)$ of order 2 such that $Fix\tau$ has index 2 in $D(t_1, t_2)$.*

(1) The associated Desarguesian translation plane admits a set of mutually disjoint derivable nets isomorphic to $Fix\tau$, the classical pseudo-regulus derivable net relative $Fix\tau$; the components are 2-dimensional left $Fix\tau$ subspaces and the Baer subspaces are 2-dimensional right $Fix\tau$-subspaces.

The non-trivial multiply derived translation planes have kernels $Fix\tau \cap Z(D)$.

(2) In the quaternion division ring, the symbol division ring, or cyclic division rings cases, the extensions from $Fix\tau$ to D are central Galois quadratic extensions and hence produce infinitely many semifield planes within $PG(3, D)$.

For Desarguesian translation planes coordinatized by non-commutative skew-fields S admitting infinite automorphisms σ, there are still replaceable nets. And the multiply replaced translation planes have kernels over $Fix\sigma \cap Z(S)$. Certainly, it would be possible to have multiple replacement translation planes with spreads in projective spaces of infinite dimensions.

There are infinite numbers of new translation planes about which essentially nothing is known, other than their existence. There are Desarguesian planes that admit arbitrary numbers of sets of mutually disjoint replaceable nets arising from quaternion division ring extensions and cyclic division ring extension. Using rational function field extensions, any finite-dimensional division ring admits extensions admitting automorphisms of order 2, ultimately leading to mutually disjoint derivable nets which must be pseudo-regulus nets.

6. Finite and infinite theories of flocks

As we have seen, the theory of flocks of quadratic cones and hyperbolic flocks are much different in the finite and infinite cases. The continuing development of the theory of finite flocks of quadratic cones is truly amazing. The generalization to infinite flocks of quadratic cones and to flocks of α-cones has very interesting developments. In this part, we discuss some of the differences between the finite and infinite theories. The interested reader is further directed to the general article on flocks by N.L. Johnson and S.E. Payne [96], and the recent work of Johnson and Jha [93], and Johnson [89], [90], [91].

7. Division ring connections

The infinite case gives us the quaternion division ring planes, which are derivable and also can coordinatize derivable nets, if there is an algebraic extension of the associated Galois quadratic extension fine. In general, the infinite setting is not only complementary but admits completely different theory in many ways.

If we notice that a quaternion division ring $(a, b)_K$ contained in a central extension $(a, b)_{K(\theta)}$, this means for the corresponding partial spread in $PG(3, (a, b)_K)$ that $(a, b)_K$ corresponds to a derivable net that lies in a Desarguesian translation plane coordinatized by $(a, b)_{K(\theta)}$. Additionally, a quaternion division ring translation plane (Desarguesian plane) is derivable where the derivable net is a twisted

σ-derivable net, where σ is the involutory automorphism of K. So, there is a corresponding flock of an σ-cone. The form of the derivable net also shows that there is a corresponding flock of a twisted hyperbolic quadric.

When there is an extension $(a, b)_{K(\theta)}$, the multiple replacement theorem in part 10 shows there is a set of mutually disjoint twisted derivable nets that together with two components define the associated spread.

There are also interesting results connecting cyclic division algebras and subplane covered nets, which form an affine space containing the algebras, so to speak. There are skewfield extensions that appear from this embedding that could be of benefit to other areas of combinatorial geometry. The multiple replacement theorem shows that for every Desarguesian spread admitting an automorphism, the associated spread may be partitioned into a set of mutually disjoint replaceable nets whose union together with two components constitute the spread. For finite Desarguesian spreads of dimension 2, this is the theorem of Andrè, where the replaceable nets are regulus nets. When the dimension $n > 2$, this is the theorem of Ostrom, where the replaceable nets are hyper-regulus nets. It is shown that whenever a skewfield has a central Galois quadratic extension, there is an analogous set of mutually disjoint derivable nets, which are now pseudo-regulus nets. And, when the dimension n is finite and larger than 2, the replaceable nets are pseudo-hyper regulus nets. There are also many new types of replacement nets using inner automorphisms of non-commutative skewfields coordinatizing Desarguesian affine planes. We have new semifield planes with spreads in non-commutative $PG(3, K)$, where K is a non-commutative skewfields. Similarly, there are new translation planes with spreads in $PG(2^2n^2 - 1, K)$, and new translation planes with what are called blended/kernels and left/right kernels.

8. Group theoretic differences

Let F be a flock of a quadric cone in $PG(3, q)$. So, there is an associated translation plane with spread in $PG(3, q)$ that is the union of a set of regulus nets, called the 'base reguli' sharing one component. This component is the axis of an elation group E, called a regulus-inducing elation group. In the finite case, Gevaert, Johnson and Thas [57] have shown that either the flock is linear and the associated translation plane is Desarguesian, or the spread admits a regulus net exactly when the regulus net is a base regulus, which forms the basis of the theory that shows the elation group E is normal in the full collineation group (that leaves the zero vector invariant) G_π and the full collineation group G_F is $G_\pi/EGF(q)^*$, when the flock is not linear. The more general result is on flocks of α-cones, as we have seen in the material on α-flokki in part 4.

THEOREM 170. *Let π be an α-flokki plane of order q^2, where $q > 5$. Then one of the following occurs:*

(1) the full collineation group permutes the q α-derivable nets,
(2) $\alpha = 1$ and the plane is Desarguesian or
(3) $\alpha \neq 1$ and the α-flokki plane is a Hughes-Kleinfeld, or transposed Hughes-Kleinfeld semifield plane that corresponds to a linear flokki.

In the infinite case, when F is a flock over an infinite field K, there is still an associated translation plane with spread in $PG(3, K)$ with regulus-inducing elation group E. However, there are examples when E is not normal in G_π, as has been discussed in the sections on bilinear flocks of quadratic cones. All of these examples have associated translation planes admitting $SL(2, K)$, generated by regulus-inducing elation groups that leave invariant one of the base reguli $\mathcal{B}$ but admit a set of regulus-inducing elation groups E_M with axis M for each component $M \in \mathcal{B}$. Since $\mathcal{B}$ is a regulus net, by deriving $\mathcal{B}$, there is an infinite class of translation planes with spreads in $PG(3, K)$, admitting regulus-inducing Baer-elation groups also generating $SL(2, K)$.

In part 9, we consider all of these ideas for infinite fields, particularly ordered fields K, and show that there exists a wide class of flocks of quadratic cones in $PG(3, K)$ that have extremely interesting and different properties that finite flocks cannot have. In particular, these flocks are simultaneously (1) bilinear, (2) admit $SL(2, K)$, generated by elations, (3) do not have a normal regulus-inducing elation group in the collineation group and (4) there are ordered fields where generalized quadrangles of the flock type do not exist. In general, we consider whether there are flocks of any type that are bilinear; flocks of quadratic cones, α-flocks, flock of hyperbolic quadrics or their twisted varieties, or elliptic flocks. We consider this as one of the main parts of this text.

The following theorem has been proved in the section on double covers (9,11,5).

THEOREM 171. *Let F be a flock of a quadratic cone over K, where K is an ordered field such that positive elements are squares with full collineation group G_F. Let π denote the associated translation plane and let G_π denote the full collineation group. Let K^* denote the kernel homology group and E a regulus-inducing elation group.*

Then one of the following holds:

(1) π is Pappian and F is linear.

(2) π is a bilinear flock spread, there is a regular double cover, and $SL(2, K)$ is generated by the elation groups.

(3) $G_\pi/EK^* \simeq G_F$; the groups of the flocks G_F and of the group of the flock translation plane, G_π correspond, as in the finite case.

The connections between the groups of flocks of hyperbolic quadrics and of the associated translation planes have not been pursued in the infinite case, due to the tremendous variation of examples.

Derivable nets in translation planes are often connected to flocks of certain types, as in the regulus type derivable nets, these are also connected with conics in Pappian projective planes.

Here is a list of certain differences between the finite and infinite cases. *i.F* and *i.I* stand for finite, and infinite situations, respectively.

9. Specific differences in the theory

- 1.*F*. J.A. Thas [137] has shown that apart from linear flocks that correspond to finite Desarguesian translation planes, there is exactly one class of non-linear flocks of quadratic cones whose planes share a point, namely the Kantor-Knuth flocks. In particular, this says that there cannot be a bilinear flock of a quadratic cone.

- 1.*I*. The $F(a_1, a_2, b_1, b_2)$ spreads of Biliotti and Johnson [15] over ordered fields K are bilinear flocks of a quadratic cone over K.

- 2.F. For any non-linear finite flock of a quadratic cone, the associated translation planes admit exactly the regulus nets determined by a regulus-inducing elation group E, Gevaert, Johnson [56]. Hence, the group E is normal in the collineation group G and if K is the kernel group, the corresponding collineation group of the flock is isomorphic to G/EK.

- 2.I. The flocks of part 1.I. admit $SL(2, K)$ as a collineation group, generated by regulus-inducing elation groups, where the group leaves invariant a regulus net N. Each component of N has an associated regulus-inducing elation group and therefore corresponds to a flock of a quadratic cone. Consequently, there are infinitely many flocks of a quadratic cone corresponding to these $SL(2, K)$-planes.

- 3.F.(a) Similarly, the finite translation planes of order q^2 (of flock type or not) admitting $SL(2, q)$ are completely determined by the work of W. Walker [141] and H. Schaeffer [130], when the plane is in $PG(3, q)$.

- 3.F.(b) The more general case has been resolved for arbitrary translation planes by D. Foulser and N.L. Johnson [54], [53]. In particular, the case when $SL(2, q)$ is generated by elations implies that the plane is Desarguesian by Foulser, Johnson, Ostrom [55]). Analogous research on the group $S_z(q)$ by W. Büttner [20] is recommended to the interested reader.

- 4.F. For each flock of a quadratic cone, there is an associated generalized quadrangle of flock type, and each is an elation generalized quadrangle. A recoordinatization that preserves the center of the elation group will create a set of q flocks, related using BLT-sets of odd order. The corresponding translation planes are constructed using the construction techniques of s-inversion for odd order and s-square for even order.

- 4.I. There are infinite flocks that do not have corresponding flock generalized quadrangles; there are associated translation planes whose s-inversions are not all flocks. In this setting, the s-inversions are proper quasifibrations but not translation planes.

- 5.F. Thas [134] for even order and Orr [116] for odd order, have shown that any finite elliptic flock is linear.

- 5.I. There is a variety of non-linear flocks of elliptic quadrics over the field of real numbers, due to Biliotti and Johnson [15].

- 6.F. and 6. I. In trying to generalize theories of finite geometry, when falling somewhat short of this, often the associated geometries become maximal partial spreads, called quasifibrations. This is true when considering infinite flocks of quadratic cones, infinite α-flocks. The constructions in the infinite case involve T-copying the finite spread to the associated rational function field. There is also a construction, twisted T-copying that also sometimes produces spreads, such as in the quaternion division ring case, and infinitely non-surjective quasifibrations on the other extreme, which means that any quadratic Galois chain based on such a quasifibration can never admit a spread in the chain.

- 7.F. The theory of finite hyperbolic fibrations with constant back half of Baker, Ebert and Pentilla [11], producing flocks of quadratic cones have an intersection connection with translation planes. That is, this theory is equivalent to the theory of translation planes of order q^2 admitting a cyclic affine homology group of order $q + 1$. These translation planes admit a set of $q - 1$ mutually disjoint regulus nets, and all multiple derivations produce the same hyperbolic fibration.

- 7.I. There is something of a corresponding theory of infinite hyperbolic fibrations with constant back half using associated quasifibrations.

- 8.F and 8.I. There are results on general oval cone flocks and infinite generalizations in the infinite cases in Johnson and Lin [**95**] and Johnson and Payne [**96**].

10. Hyperbolic flocks, finite and infinite

The connections between the groups of flocks of hyperbolic quadrics and of the associated translation planes have not been pursued in the infinite case, again due to the tremendous variation of examples.

- $H.F.$ Thas [**136**], Bader, Lunardon [**8**] have classified the finite flocks of hyperbolic quadrics. This general theorem used collaborating but independent work of Bader [**5**], Baker-Ebert [**9**], Bonisoli [**26**], Johnson [**76**]). The set of such corresponding translation planes is the set of Dickson nearfield planes and three of the irregular nearfield planes.

We shall use the notation $H(i)$ I to denote the differences in the infinite case.

However, in the infinite case, there are hyperbolic flock planes of the following types.

- $H(1).I$. There are Bol translation planes corresponding to Andrè planes that are not nearfield planes.

- $H(2).I$. There are infinite hyperbolic flock planes that are Bol but not nearfield nor Andrè planes.

- $H(3).I$. there are infinite non-linear hyperbolic flock planes of characteristic 2.

- $H(4).I$. There are infinite hyperbolic flock planes are not Bol.

- $H(5).I$. There are infinite nearfield spreads in $PG(3, GF(p)(z))$, where $GF(p)(z)$ is the rational function field over $GF(p)$ (See T. Grundhoefer and M. Gruinger [**59**]). These spreads are also hyperbolic flock spreads.

11. Remarks on corrections to earlier work

In Jha and Johnson [**70**], there was a first attempt at a classification of derivable nets for fields only. However, there were some errors, that have been corrected, regarding type 0 planes and assertions on the type of type 1 planes, available only in characteristic 2. In particular, it was stated that there can be no type 0 derivable nets, in the field case. This is not correct and the proof stating that has been turned into the theorem on 'extending skewfields', (2, 4,10). Also, there was a sign error in the algebra on type 1 derivable nets that led to the conclusion that only characteristic 2 field planes could be involved. This, of course, also is false, and has been corrected, in the section on type 1 derivable nets in (3, 5, 10). We have also given examples and constructions on all of the various classes.

It might be important to note the places wherein the errors might have migrated into other works.

There are five previous articles/books that should be corrected in view of the current results. The main classification theorems should replace the relevant theorems in the following (See references for explicit information):

(a) The Handbook [**94**], 49.16 and 49.18.
(b) Combinatorics of Spreads and Parallelisms [**88**], pp. 342-343, asserts type 1 only for characteristic 2.
(c) Transversal Spreads, [**87**] Remark 17 and Theorem 8.
(d) Lifting Quasifibrations II Corollary 1, and Theorem 4 [**70**].
(e) Infinite Baer Nets,[**71**] examples are correct but not complete.

Finally, it should be noted that in the author's work on infinite nests of reguli [**85**], there were results (5.2) and (11.1) that had the assumption that when K is a full field and $K(\sqrt{\gamma})$ a full field quadratic extension with automorphism σ that $\sigma : n \to n^{\sigma+1}$ should be surjective onto K. The proof does not require this assumption and should be eliminated. As these results are used in this text, this correction has been noted in the pertinent sections.

For examples, the reader is directed to Jha and Johnson [**72**] p. 664:

EXAMPLE 5. *Let P be isomorphic to GF(p) where p is an odd prime. Let F be any algebraic field extension of P which is not algebraically closed and which is not a series of quadratic extensions of extensions of P. Then F is a full field.*

EXAMPLE 6. *Let F be an ordered field which admits an ordered quadratic extension K such that the positive elements of each field have square roots in the field. Then both F and K are full fields.*

Bibliography

[1] A.A. Albert, A Construction of Non-Cyclic Normal Division Algebras, Bull. American Math. Soc. vol. **38** (1932), 449-456.

[2] Amitsur, S.A., On Central Division Algebras, Israel Journal of Math. vol. **12** (1972), 408-420.

[3] M. Ashraf, S. Ali, C. Haetinger, On derivations in rings, The Aligarth Bull. of Math. Vol **25, no. 2**, 2006.

[4] J. Andrè, Uber nicht-Desarguessche Ebenen mit transitiver Translationsgruppe, Math. Zeitschrift, **60** (1954), 156-186.

[5] L. Bader, Some new examples of flocks of $\mathbb{Q}^+(3,q)$, Geom. Ded. **27** (1988), 213-218.

[6] L. Bader, G. Lunardon, and J.A. Thas, Derivation of flocks of quadratic cones, Forum Math. **2** (1990), no. 2 163-174.

[7] L. Bader, N. Durante, M. Law, G. Lunardon, T. Penttila, Flocks and Partial Flocks of Hyperbolic Quadrics, Journal of Algebraic Combinatorics, **16** (2002), 21-30.

[8] L. Bader and G. Lunardon, On the flocks of $\mathbb{Q}^+(3,q)$, Geom. Ded. **29** (1989), 177-183.

[9] R. D. Baker, G.L. Ebert, A nonlinear flock in the Minkowski plane of order 11, Congre. Numer. **58** (1987), 75-81.

[10] R.D. Baker, G.L. Ebert, Nests of size $(q-1)$ and another family of translation planes, J. London Math. Soc. 2 **38** (1988), 341-355.

[11] R.D. Baker, G.L. Ebert, and T. Penttila, Hyperbolic fibrations and q-clans, Des. Codes, Cryptogr. 34 (2005), no. 2-3, 295-305.

[12] A. Barlotti, On the definition of Baer subplanes of infinite planes, Journal Geometry **3** (1979) 87-92.

[13] Bayo, Nadir, Classification of quaternion algebras over the field of rational numbers, Bachelor Thesis, ETH Zürich, 2017.

[14] M. Biliotti and N.L. Johnson, Variations on a Theme of Dembowski, Mostly Finite Geometries, Marcel Dekker **190** (1997), 139-168.

[15] M. Biliotti and N.L. Johnson, Bilinear flocks of quadratic cones, J. Geom. **64** (1999) 16-50.

[16] M. Biliotti, V. Jha and N.L. Johnson, Foundations of Translation Planes. Monographs and Textbooks in Pure and Applied Mathematics, Vol. **243**, Marcel Dekker, New York, Basel, 2001, xvi +542 pp.

[17] A. Bonisoli. The regular subgroups of the sharply 3-transitive finite permutation groups, Annals Discrete Math. **37** (1988), 75-86.

[18] R.P. Burn. Bol quasifields and Pappus Theorem Math. Z. **105** (1968), 351-364.

[19] A.A. Bruen, Inversive geometry and some new translation planes, I. Geom. Dedicata **7** (1978), no. 1, 81-98.

[20] W. Büttner, On Translation Planes that contain $S_z(q)$ in their translation complement, Geom. Dedicata **11** (1981), no. 3.

[21] I. Cardinali, O. Polverino, R. Trombetti, Semifield planes of order q^4 with kernel F_{q^2} and center F_q, European Journal of Combinatorics, Vol. **27**, Issue 6, 2006, 940-961.

[22] G. Cherlin, T. Grundhoefer, A. Nesin, H. Volklein, Sharply Transitive Linear Groups over algebraically closed fields, Proceedings of the American Math. Soc. **111**(2) (1991), 541-549.

[23] W.E. Cherowitzo and L. Holder, Bilinear Flocks, J. Geometry. **105** (2014), no. 3 625-634.

[24] W.E. Cherowitzo, N.L. Johnson, Net Replacements in the Hughes-Kleinfeld Semifield Planes, J. Geometry **97**, 45-57 (2010).

[25] W. E. Cherowitzo and N.L. Johnson, Parallelisms of Quadric Sets, Innov. Incidence Geom. Vol. **12** (2011), 21-34.

[26] W.E. Cherowitzo, N.L. Johnson, O. Vega, α-Flokki and partial α-Flokki, Innov. Incidence Geom. Vol. **15** (2017), 5-29.

[27] C-L. Chuang, Identities with skew derivations, Journal of Algebra **224** 292-335 (2000).

[28] J. Cofman, Baer subplanes and Baer collineations of derivable projective planes, Abh. Math. Sem. Hamburg **44** (1975), 187-192.

[29] P.M. Cohn, Skew Fields, Theory of general division rings. Cambridge University Press, 1995

[30] P.M. Cohn and W. Dicks, On Central Extensions of Skew Fields, Journal of Algebra 63 143-151 (1980).

[31] K. Conrad, Hensel's Lemma, Math. Dept. UConn, Expository article.

[32] K. Conrad, Quaternion Algebras, Math. Dept. UConn, Expository article.

[33] K. Conrad, The p-Adic expansion of rational numbers, Math. Dept. UConn, Expository article.

[34] K. Conrad, Ross Program Lecture (2004), homework set #4. Quaternion Algebras, Math. Dept. UConn.

[35] K. Conrad, Totally Ramified Primes and Eisenstein Polynomials, Math. Dept. UConn, Expository article.

[36] M. Cordero and R.F. Figueroa, On some new classes of semifield planes, Osaka J. Math. **30** (1993), no. 2 171-178.

[37] T. Czerwinski and D. Oakden, The translation planes of order twenty-five, Journal of Combinatorial Theory Series A, **59** (1992) 193-217.

[38] C. D'Elbée, Cyclic and Non-cyclic division algebras of finite dp-rank, arXiv:2106.09767v.1 (2021).

[39] F. De Clerck and N.L. Johnson, Subplane Covered Nets and semipartial geometries, Discrete Math. 106/107 (1992), 127-134.

[40] F. De Clerck and H. Van Maldeghem, On flocks of infinite quadratic cones. Bull. Belgian Math. Soc. -Simon Stevin, **1 (3)**, (1994), 399–415.

[41] F. De Clerck and J.A. Thas, Partial geometries in finite projective spaces, Arch. Math. **30** (1978), 537-540.

[42] P. Dembowski, Finite Geometries, Springer-Verlag, New York, Inc. 1968.

[43] L.E. Dickson, Linear algebras in which division is always uniquely possible, Bull. Amer. Math. Soc. **12** (1905-6), 441-442..

[44] J. Dieudonné, Les determinants sur un corp non-commutatif, Bull. Soc. Math. France **71** (1943), 27-45.

[45] J. Dieudonné, La géométrie des group classiques, Springer, Berlin-Göttingen-Heildelberg, 1955.

[46] G. Donati, N. Durante, Absolute points of correlations of $PG(3, q^n)$, J. of Algebraic Comb. Online, (2021), **54** (1-3).

[47] N. Durante, Geometry of sesquilinear forms. Notes of a course given at the workshop Finite Geometry and friends, Bruxelles, June 2019.

[48] P.K. Draxl, Skew Fields, Cambridge Univ. Press, Cambridge, 1983.

[49] D.S. Dummit, R.M. Foote, Abstract Algebra, **3rd** Edition 2004, 932 pp.

[50] G.L. Ebert, Nests, covers, and translation planes, Ars. Combin. **25** (1988), 213-233.

[51] J.C. Fisher and J.A. Thas, Flocks in $PG(3, q)$, Math. Z. **169** (1979), 1-11.

[52] D.A. Foulser, Baer p-elements in translation planes, Journal of Algebra. **31** (1974) 354-366.

[53] D.A Foulser and N.L. Johnson, The translation planes of order q^2 that admit $SL(2, q)$, *I*. Odd order, J. Geometry, **18** (1982), no. 2, 122-139.

[54] D.A Foulser and N.L. Johnson, The translation planes of order q^2 that admit $SL(2, q)$, *II*. Even order, J. Algebra **86** (1984), no. 2 385-406

[55] D.A. Foulser, N.L. Johnson, T.G. Ostrom, A characterization of the Desarguesian planes of order q^2 by $SL(2, q)$, Int. J. Math. Math Sci. **6** (1983), no. 3 605-608.

[56] H. Gevaert and N.L. Johnson, Flocks of quadratic cones, generalized quadrangles and translation planes, Geom. Dedicata **27** (1988), 301-317.

[57] H. Gevaert, N.L. Johnson and J.A. Thas, Spreads covered by reguli, Simon Stevin **62** (1988), 51-62.

[58] C. Grover, C. Ling, R. Vehkalahti, Non-Commutative Ring Learning With Errors From Cyclic Algebras, arXiv: 2008.0834.

[59] T. Grundhoefer and M. Gruinger, Wild Nearfields of dimension 2 over rational function fields, Journal of Geometry **107** (2) (2016), 317-328.

[60] P. Gvozdevsky, Commutator lengths in general linear group over a skewfield, Journal of Mathematical Sciences (to appear 2022), 1-12.

[61] T. Hanke, A twisted Laurent Series Ring that is a noncrossed product, Israel Journal of Math. (to appear).

[62] Y. Hiramine, M. Matsumoto, and T. Oyama, On some extension of 1-spread sets, Osaka J. Math. **24** (1987), no. 1 123-137.

[63] D.R. Hughes, A class of non-Desarguesian projective planes, Can. J. Math. **9** (1957), 378-388.

[64] N. Jacobson, finite-dimensional Division Algebras over Fields, Springer Books, Heidelberg, 1996.

[65] V. Jha and N.L. Johnson, Notes on the Derived Walker Planes, Journal Combin. Theory, Ser. A **42**, 320-323 (1986).

[66] V. Jha and N.L. Johnson, Automorphism groups of flocks of oval cones, Geometriae Dedicata volume **61**, 71–85 (1996).

[67] V. Jha and N.L. Johnson, Infinite flocks of a quadratic cone, Journal of Geometry, **57** (1996), 123-150.

[68] V. Jha and N.L. Johnson, Rigidity in Conical Flocks, Journal of Combinatorial Theory, Series A **73**, 60–76 (1996).

[69] V. Jha and N.L. Johnson, Conical, Ruled and deficiency-one flocks, Bull. Belgian Math. Soc., Simon Stevin, **6** (1999), 187-218.

[70] V. Jha and N.L. Johnson, Lifting Quasifibrations II – Non-Normalizing Baer involutions, Note Mat. **20** (2000/01), no. 2, 51–68.

[71] V. Jha and N.L. Johnson, Infinite Baer Nets, J. Geom. **68** (2000), no. 1-2, 114-141.

[72] V. Jha and N.L. Johnson, Infinite flocks of quadratic cones-II generalized Fisher flocks, July 2002, Journal of the Korean Mathematical Society **39(4)**, 653-664.

[73] N.L. Johnson, Lezioni sui Piani di Traslazione, Quaderni dei Departimento di Matematica, dell'Universita di Lecce, Q. **3** 1986 (Italian).

[74] N.L. Johnson, The maximal special linear groups which act on translation planes, Boll. Unione Mat. Ital. A **6** (1986), no. 3, 349-352.

[75] N.L. Johnson, Derivable nets and 3-dimensional projective spaces, Abh. Math. Sem. Hamburg, **58** (1988), 245-253.

[76] N.L. Johnson, Flocks of Hyperbolic Quadrics and Translation Planes Admitting Affine Homologies, J. Geom. **34**(1/2) (1989), 50-73.

[77] N.L. Johnson, Translation Planes admitting Baer Groups and partial flocks of quadric sets, Simon Stevin **63** (1989), 167-188.

[78] N.L. Johnson, Derivation is a Polarity, Journal of Geometry volume 35, 97–102 (1989).

[79] N.L. Johnson, Flocks and partial Flocks of Quadric sets, Finite Geometries and Combinatorial Designs, Amer. Math. Soc. Providence, RI, 1990, pp. 111-116.

[80] N.L. Johnson, Derivable nets and 3-dimensional projective spaces. II. The structure, Archiv d. Math. **55** (1990), 94–104.

[81] N.L. Johnson, Extending partial flocks containing linear subflocks, J. Geom. **55,** (1996), 99-106.

[82] N.L. Johnson, Lifting quasifibrations, Note di Mat. **16** (1996), 25–41.

[83] N.L. Johnson, Flocks of infinite hyperbolic quadrics, Journal of Algebraic Combinatorics **6** (1997), 27-51.

[84] N.L. Johnson, Derivable Nets can be embedded in nonderivable planes, Trends in Mathematics (1998), Birkhauser Verlag, Basel Switzerland, 123-144.

[85] N.L. Johnson, Infinite Nests of Reguli, Geom. Dedicata **70**, 221-267, 1998.

[86] N.L. Johnson, Subplane Covered Nets, Pure and Applied Mathematics, Vol. 222 Marcel Dekker, 2000.

[87] N.L. Johnson, Transversal Spreads, Bull. Belg. Math. Soc. **9** (2002), 109–142.

[88] N.L. Johnson, Combinatorics of Spreads and Parallelisms, Pure and Applied Mathematics, Vol. 295, Marcel Dekker, 2010.

[89] N.L. Johnson, Twisted Hyperbolic Flocks, Innovations in Incidence Geom. Vol. **19** (2021), No. 1, 1-23.

[90] N.L. Johnson, Twisted Quadrics and α-Flocks, Algebraic Combinatorics, Volume **5** (2022) no. 5, pp. 803-826.

[91] N.L. Johnson, Classifying Derivable Nets, Innovations in Incidence Geometries, **19-2** (2022) 59-94.

[92] N.L. Johnson, M. Cordero, Transitive Partial Hyperbolic Flocks of deficiency-one, Note di Mat. **29** (2009), n.1 89-98.

[93] N.L. Johnson, V. Jha, Lifting Skewfields, Journal of Geometry, **113: 5** (2022).

[94] N.L. Johnson, V. Jha and M. Biliotti, The Handbook of Finite Translation Planes, Pure and Applied Mathematics, Vol. 289, Marcel Dekker, 2007.

[95] N.L Johnson and X. Lin, Flocks of quadratic and Semi-elliptic Cones, Mostly Finite Geometries, Marcel Dekker **190** (1997), 275-304.

[96] N.L. Johnson and S.E. Payne, Flocks of Laguerre and Associated Geometries, Mostly Finite Geometries, Marcel Dekker **190** (1997), 51-122.

[97] N.L. Johnson, R. Pomareda, Andrè planes and nests of reguli, Geom. Dedicata **31** (1989) no. 3, 245-260.

[98] N.L. Johnson, R. Pomareda, A maximal partial flock of deficiency-one of the hyperbolic quadric in $PG(3,9)$, Simon Stevin, no. **2 64** (1990), 169-177.

[99] N.L. Johnson, R. Pomareda, and F.W. Wilke, j-planes, Journal of Combinational Theory, Series A **56** 271-284 (1991).

[100] W. Johnson, The canonical topology on dp-minimal fields, J. Math. Log. **82(1):** 1850007, **23**, 2018.

[101] D. Jungnickel, Maximal partial spreads and transversal-free translation nets, Journal Combin. Theory Series A, **62, 1** (1993), 66-92.

[102] W. M. Kantor and T. Penttila, Flokki planes and cubic polynomials, Note di Mat. **29** (2009), suppl. n. 1, 211-222.

[103] W.M. Kantor, Ovoids and translation planes, Canad. J. Math. **34**, No. 5 (1982), 1192-1207.

[104] W.M. Kantor, Some generalized quadrangles with parameters q^2, q, Math. Zeit **192**, (1986), no. 1, 45-50.

[105] V.K. Kharchenko, Automorphisms and derivations of associative rings, Kluwer Academic, Dordrecht, 1991.

[106] N. Knarr, A geometric construction of generalized quadrangles from polar spaces of rank three, Resultate d. Math. **21** (1992), n. 3-4, 332-344.

[107] N. Knarr, Derivable Affine planes and Translation Planes, Bull. Belg. Math. Soc. **7** (2000), 61-71.

[108] J. Lahtonen, N. Markin, and G. McGuire, "Construction of multiblock space–time codes from division algebras with roots of unity as nonnorm elements," IEEE Transactions on Information Theory, vol. **54**, no. 11, pp. 5231–5235, Nov. 2008.

[109] E.E. Lazerson, Onto inner derivations in division rings, Bull. Amer. Soc. **67** (1961), 356-358.

[110] H. Lüneburg, Translation Planes, Springer-Verlag, Berlin, 1980.

[111] R. Mathon and G.F. Royle, The translation planes of order 49, Designs Codes and Cryptography, **5** (1995) 57-72.

[112] C. Milliet, NIP and NTP_2 Division Rings of Prime Characteristic, (preprint) 2019.

[113] P.J. Morandi, B.A. Sethuraman, and J.P. Tignol, Division Algebras with no Anti-automorphism but with Involution. Advances in Geom. **5(3)** (20050, 485-495.

[114] F. Oggier, J-C. Belfiore, E. Viterbo, Cyclic Division Algebras: A Tool for Space-Time Coding, Foundations and Trends in Communications and Information Theory, Vol. **4**, No. 1 (2007) 1–95.

[115] O. Ore, Linear equations in noncommutative fields, Ann. Math. **32** (1931), 463-77.

[116] W.F. Orr, The Miquelian inversive plane IP(q) and the associated projective planes, PhD Thesis, Univ. Wisconsin, 1973.

[117] T.G. Ostrom, Semi-translation planes, Trans. Amer. Math. Soc. **111** (1964), 1-18.

[118] T.G. Ostrom, Derivable Nets, Canad. Bull. Math. **8** (1965), 601-613.

[119] T.G. Ostrom, Vector spaces and construction of finite projective planes, Arch. Math. **19** (1968), 1-25.

[120] T. G. Ostrom, Finite Translation Planes, Lectures Notes in Mathematics, vol. **158**, Springer-Verlag, Berlin, 1970.

[121] S.E. Payne, Generalized Quadrangles as Group Coset Geometries, Congress. Numer. **XXIX** Proc. 12th S.E. Conf. Comb. Graph Theory and Combinatorics (1980), pp. 717-734.

[122] S.E. Payne and L.A. Rogers, Local group actions on generalized quadrangles, Simon Stevin **64** (1990), no. 3-4, 249-284.

[123] S.E. Payne and J.A. Thas, Conical flocks, partial flocks, derivations and generalized quadrangles. Geom. Dedicate 38 (1991), 229-243.

[124] T. Pentilla and L. Storme, Monomial flocks and herds containing a monomial oval, J. Combinatorial Theory Series A **83**, 21-41 (1998).

[125] G. Pickert. Projektive Ebeben. Math. Wissenshaften, Springer-Verlag. Berlin-Gottingen-Heidelberg, 1955.

[126] P. Ribenboim, The Theory of Classical Valuations, Springer-Verlag, New York Heidelberg, Berling, 1999.

[127] R. Riesinger, Faserungen, die aus Reguli mit einem gemeinsamen Geradenparr zusammengesetzt sind, J. Geom. 45 (1992), 137-157.

[128] G.F. Royle, An orderly algorithm and some applications to finite geometry, Discrete Math. **185** (1998), 105-115.

[129] S. Pumplün and T. Unger. Space-time block codes from nonassociative division algebras. Adv. Math. Commun., **5(3)**:449–471, 2011.

[130] H.J. Schaeffer, Translationsebenen, auf denen die Gruppe $SL(2, p^n)$ Operiert, Diplomarbeit, Univ. Tübingen, 1975.

[131] J.P. Serre, A Course in Arithemetic, Springer, 1985.

[132] F.A. Sherk and G. Pabst, Indicator sets, reguli, and a new class of spreads, Canad. J. Math. **29** (1977), no. 1 132-154.

[133] P. Sziklai, Partial flocks of the quadratic cone, J. Comb. Theory Ser. A 113 (2006), no. 4, 698-702.

[134] J.A. Thas, Flocks of finite egglike inversive planes, in: Finite geometric structures and their applications (ed. A. Barlotti), Ed. Cremonese Roma (1973), 189-191.

[135] J.A. Thas, Flocks of nonsingular Ruled Quadrics in $PG(3, q)$, Atti. Accad. Lincei Rend. 59 (1975), 83-85.

[136] J.A. Thas, Flocks, Maximal Exterior Sets and Inversive Planes, Finite Geom. and Combinatorial Designs, Amer. Math. Soc. Providence, RI, 1990, pp. 187-218.

[137] J.A. Thas, Generalized quadrangles and flocks of cones, European J. Combin. **8** (1987), 441-452.

[138] J.A. Thas and F. De Clerck, Partial geometries satisfying the axiom of Pasch, Simon Stevin **51** (1990), 123-137.

[139] E. Turner, The p-Adic numbers and Finite Field Extensions of Q_p, Expository article, REU program, 2011.

[140] J. Voight, Quaternion algebras, Dartmouth, 2020.

[141] M. Walker, On translation planes and their collineation groups, Ph.D thesis, Univ. London 1973.

[142] M. Walker, A class of translation planes, Geom. Dedicata **5** (1976), 135-166.

[143] Jill C.D.S. Yaqub. Translation ovoids over skewfields. Geom. Ded. **36** (1990), 261-271.

Index

(F,K)-bimodule, 69
1-Class retraction., 16
1-parallel class retraction, 15
1-test, 184
1-test failure for quasifibrations, 190
2nd cone for flokki , 99
2nd-cone triple, 100

Inner automorphism net replacement
    theorem, 275

A-nest; q+1 -nest, 233
Abelian torsion group, 57
Additive quasifibration, 274
Affine dual translation plane, 8
Affine plane, 3
Affine restriction, 4
Affine space AG(V,K)., 13
Albert's non-cyclic division rings, 274
alpha cone, 94
Alpha conic, 141
alpha conic, 94
Alpha quadric, 141
Alpha-conic, 88
Alpha-flock, 88
Alpha-flokk plane, 93
Alpha-flokki, 91
Alpha-flokki planes, 91
Alpha-regulus-inducing elation group, 88
Andre' planes, 169, 276
Andre' scheme, 295
Apha-Klein quadric, 88
Artin-Wedderburn theorem , 57

Baer groups for flokki planes, 100
Baer subplane, 4
Base reguli, 233, 235
Bilinear flock, 91, 222
Bilinear flock of a quadratic cone, 252
Binary system., 9
Blended kernel, 68
Blended kernel of a translation plane, 261
Blended left/right kernel., 261

Blocking set, 92
BLT-set, 116
Bol axiom, 9
Bol translation plane., 9
Brauer equivalency relation, 57
Brauer group of a field, 57

Cardano's equations, 94
Carrier derivable net of class i, i=0,1,2,3, 82
Cayley-Dickson process, 197
Central extension of dimension n, 58
Central extension of skewfield, 58
Central simple algebra, 56
Chains of reguli, 232
Circle geometry, 83, 236
Classical Derivable net of type 3, 72
Classical pseudo-regulus net, 18, 47
Classical semifield spread with derivable net
    admitting a derivation, 150
Classification of derivable nets, part II, 65
Classification of finte hyperbolic flocks, 139
Classification of subplane covered nets, 17
Classification theorem for derivable nets,
    part I., 51
Classsical derivative type spread, 140
Co-dimension, 13
Co-dimension 2 construction, 15
Co-dimension 2 construction., 15
Collineation group of a derivable net, 16
Commutator replaceable net, 276
Conical flock plane, 91
Conjecture on Galois chains of
    quasifibrations, 175
Contraction of embedded nets, 16
Contraction theorem of subplane covered
    nets, 18
Coordinate system for a derivable net, 5
Critical linear subflock, 235
Critical partial flock extension, 236
Critical partial flock extension theorem, 236
Critical partial subflock, 235
Cyclic Division Ring, 274

351

Printed in the United States
by Baker & Taylor Publisher Services